W9-AYM-150

FILTRATION

Part I

CHEMICAL PROCESSING AND ENGINEERING

An International Series of Monographs and Textbooks

Volume 1: Chemical Process Economics, Second Edition, Revised and Expanded
by John Happel and Donald G. Jordan

Volume 2: Gas Cleaning for Air Quality Control: Industrial and Environmental Health and Safety Requirements
edited by J. M. Marchello and John J. Kelly

Volume 3: Models for Flow Systems and Chemical Reactors
by C. Y. Wen and L. T. Fan

Volume 4: Thermodynamics of Fluids: An Introduction to Equilibrium Theory
by K. C. Chao and R. A. Greenkorn

Volume 5: Petroleum Refining: Technology and Economics
by James H. Gary and Glenn E. Handwerk

Volume 6: Petroleum Economics and Engineering: An Introduction
by Hussein K. Abdel-Aal and Robert Schmelzlee

Volume 7: Control of Air Pollution Sources
by J. M. Marchello

Volume 8: Gas-Solids Handling in the Process Industry
edited by J. M. Marchello and Albert Gomezplata

Volume 9: Viscoelastic Fluids: An Introduction to Their Properties and Behavior
by Ronald Darby

Volume 10: Filtration: Principles and Practices (in two parts)
edited by Clyde Orr

IN PREPARATION

Chemical Reactions as a Means of Separation
—Sulfur Removal, *edited by Billy L. Crynes*

FILTRATION
Principles and Practices

(in two parts)

Part I

edited by
Clyde Orr
School of Chemical Engineering
Georgia Institute of Technology
Atlanta, Georgia

MARCEL DEKKER, INC. New York and Basel

Library of Congress Cataloging in Publication Data
Main entry under title:

Filtration.

Includes bibliographies and indexes.
1. Filters and filtration. I. Orr, Clyde.
TP156.F5F54 660.2'8424 75-18059
ISBN 0-8247-6283-5

MARCEL DEKKER, INC.

270 Madison Avenue, New York, New York 10016

Current printing (last digit):
10 9 8 7 6 5 4 3 2 1

PRINTED IN THE UNITED STATES OF AMERICA

PREFACE

Filtration is both a process of major contemporary importance and one with its beginnings rooted in antiquity. Hardly a modern industry exists without some dependence on a filtering operation. Large-scale filters are employed in diverse processes annually treating millions of tons of minerals, chemicals, and liquid wastes. Others protect the environment from disagreeable or deleterious gas-borne emissions. Small filters are found in every automobile, and exceedingly small ones ensure the reliable performance of sophisticated navigational and space oriented equipment. Still others capture bacteria or remove traces of solid matter from gas and liquid streams so that delicate instruments can be assembled without defect and beverages, medicinals, and foods can be had free of contamination.

The earliest Chinese writings describe a crude form of filtration as do ancient Hebrew scrolls. Egyptian records by the third century A. D. refer to filters for dyes consisting of an earthen vat with perforated bottom which supported a grass matt and upon it a layer of lime. Clay was added to the solution being filtered to serve as a filter aid. By the Middle Ages gravity filtration through cloth was in common usage, although straining was the term commonly applied to the operation. Our word "filter," or fylter or filtre, probably derives from the Latin filtrum which is closely related to feltrum, meaning felt or compressed wool, and both possibly derived from the Greek word signifying hair.

The first patent on a filter may be that issued by the French Government in 1789 to one Joseph Amy. Sponge was apparently the basis for the patent claim. A British patent of 1791 describes an operation identified as filtration by ascent, the invention here being a vessel containing coarse gravel at the bottom followed by graded sand above. A central pipe conducted water from the top into the gravel; discharge of filtrate occurred from the side of the vessel near the top of the sand. The discharge could be collected and run back through the filter, thus anticipating modern backwashing operations. Continuous filter patents date from the 1870s when many forms were devised involving drums with cloth on the outside and on the inside, drums covered with hot-air hoods for drying, and drums provided with various mechanisms for cleaning the cloth.

Today filters can be had that function from cryogenic to quite elevated temperatures, that remove micrometer size particles in plant-scale operations, that separate low-molecular-weight materials from liquids containing polymeric and other high-molecular-weight species, and that withstand most corrosive or solvent fluids. So extensive have been the developments in the area of filtration and so specialized have become the applications that one versed in gas filtration usage rarely is more than vaguely familiar with liquid filtration processes, for example. To understand many of the present uses of ultra-fine filtration, one needs to know something of microbiology. The emergence of low-pressure membrane ultrafiltration as a unit operation is such a recent development that little of its potential has yet been realized.

Each chapter of Filtration--Principles and Practices has been written by an eminent authority in his specialty, one who is actively engaged in research, development, or the application of his subject to current requirements. The book attempts, as its title suggests, to cover theory as well as the practical considerations that enter into actual applications, making it useful, hopefully, to the research worker, the developer seeking improvements in a process, and to the plant engineer looking for tips that may make an operation more satisfactory. Hopefully, it also points the way to future developments.

I am very grateful indeed to the chapter authors who have so generously contributed of their time, knowledge, and experience.

Clyde Orr

CONTENTS

	Preface	iii
	List of Contributors	vii
1	GAS FILTRATION THEORY	1
	Joseph Pich	
I.	Introduction	2
II.	General Approach	6
III.	Structure of Fibrous Filters	12
IV.	Stationary Filtration	18
V.	Nonstationary Filtration	113
VI.	Membrane Filters	128
VII.	Nuclepore Filters	146
	References	149
2	LIQUID FILTRATION THEORY AND FILTRATION PRETREATMENT	169
	Richard J. Akers and Anthony S. Ward	
I.	Introduction	170
II.	Filtration Process Theory	179
III.	Modification of Cake Properties	197
IV.	Filter Aids	236
V.	Miscellaneous Pretreatment Techniques	242
	Notation	244
	References	247
3	FILTER MEDIA	251
	A. Rushton and P. V. R. Griffiths	
I.	Introduction	252
II.	Media Classification and Requirements	254
III.	Criteria of Choice	256
IV.	Measurement of Pore Size and Particle Retention	257
V.	Permeability of Clean Media	260
VI.	Particle Deposition Mechanisms	288
VII.	Flow Resistance of Used Media	291
VIII.	Effect of Media Structure on Cake Resistance	296

IX. Media Bridging 299
Notation 302
References 306

4 INDUSTRIAL GAS FILTRATION 309

Koichi Iinoya

I. Introduction 310
II. Bag Filters and Baghouses 310
III. Fibrous Mat Filters 324
IV. Mist Collectors 335
V. Aggregate and Porous Filters 337
VI. Performance Criteria 337
VII. Gas Cooling 347
VIII. Typical Applications and Cost Data 351
Notation 354
References 356

5 FILTRATION IN THE CHEMICAL PROCESS INDUSTRY 361

Frank M. Tiller, Antoine Alciatore, and Mompei Shirato

I. Introduction 362
II. Analysis of Filtration Operations 365
III. Product Specification 366
IV. Slurry and Cake Properties 368
V. Initial Choice of Equipment Class 369
VI. Modification of Slurry Properties 370
VII. Washing and Deliquoring 375
VIII. Description of the Equipment 377
IX. Flow Equations for Porous Media 394
X. Batch Cake Filtration 406
XI. Cake Porosity 416
XII. Cake Washing 437
XIII. Experimental Testing Procedures 442
XIV. Cycle Analysis 452
XV. Continuous Filtration 459
Notation 469
References 472

6 ULTRAFILTRATION 475

Richard P. de Filippi

I. Introduction 476
II. Distinguishing Characteristics 476
III. Membranes 480
IV. Mass Transfer 486
V. Engineering and Design of Systems 504
VI. Applications 509
Notation 516
References 517

Author Index 519
Subject Index 529

CONTRIBUTORS

RICHARD J. AKERS Department of Chemical Engineering, Loughborough University of Technology, Loughborough, Leicestershire, England

ANTOINE ALCIATORE Dicalite Division, GREFCO, Inc., New Orleans, Louisiana

RICHARD P. DE FILIPPI Abcor, Inc., Cambridge, Massachusetts*

P. V. R. GRIFFITHS Chemical Engineering Department, The University of Manchester Institute of Science and Technology, Manchester, England

KOICHI IINOYA Department of Chemical Engineering, Kyoto University, Kyoto, Japan

JOSEF PICH The J. Heyrovsky Institute of Physical Chemistry and Electrochemistry, Czechoslovak Academy of Sciences, Prague, Czechoslovakia

A. RUSHTON Chemical Engineering Department, The University of Manchester Institute of Science and Technology, Manchester, England

MOMPEI SHIRATO Department of Chemical Engineering, Nagoya University, Chikusa, Nagoya, Japan

FRANK M. TILLER Department of Chemical Engineering, University of Houston, Houston, Texas

ANTHONY S. WARD Department of Chemical Engineering, Loughborough Institute of Technology, Loughborough, Leicestershire, England

*Current affiliation: Arthur D. Little, Inc., Cambridge, Massachusetts.

Chapter 1

GAS FILTRATION THEORY

Josef Pich

The J. Heyrovsky Institute of
Physical Chemistry and Electrochemistry
Czechoslovak Academy of Sciences
Prague, Czechoslovakia

I. Introduction 2

II. General Approach 6
A. Diffusion Deposition 8
B. Direct Interception 8
C. Inertial Deposition 8
D. Gravitational Deposition 9
E. Electrostatic Deposition 9
F. London-van der Waals Deposition 9
G. Capture Coefficients 9

III. Structure of Fibrous Filters 12
A. Fiber Diameter and Fiber Diameter Distribution 12
B. Shape of Fibers 13
C. Pore Size and Pore Size Distribution 13
D. Spatial Arrangement of Fibers 14
E. Specific Surface 15
F. Fiber Concentration 16
G. Filter Porosity 16

IV. Stationary Filtration 18
A. Filtration in the Continuum Region 18
B. Flow Through Fibrous Filters 20
C. Velocity Field Around an Isolated Cylinder 22
D. Velocity Field in a System of Cylinders 23
E. Theory of Drag Forces 27
F. Pressure Drop of Fibrous Filters 29

G. Diffusional Mechanism 34
H. Direct Interception 39
I. Inertial Mechanism 41
J. Gravitational Mechanism 49
K. Electrostatic Mechanisms 50
L. Deposition Due to London-van der Waals Forces 59
M. Filtration in Noncontinuum Regions 61
N. Combined Filtration Mechanisms 75
O. Interference Effect 82
P. Fiber Capture Coefficient and Filter Efficiency 84
Q. Characteristics of Filters 85
R. Penetration of Particles into Filters 101
S. Other Filtration Theories 103

V. Nonstationary Filtration 113
A. Adhesion and Reentrainment of Particles 114
B. Clogging of Filters 118
C. Capillary Phenomena in Filters 127
D. Loss of Electric Charge and Filter Destruction 128

VI. Membrane Filters 128
A. Preparation and Properties 129
B. Structure 132
C. Flow 136
D. Deposition 138
E. Characteristics 144

VII. Nuclepore Filters 146

References 149

I. INTRODUCTION

Filtration can be defined as the process of separating dispersed particles from a dispersing fluid by means of porous media. The dispersing medium can be a gas (or gas mixture, most frequently air) or a liquid. With regard to the type of the medium, this process can be divided into the filtration of aerosols and lyosols. In this chapter attention is focused on aerosol filtration although there are several common features in filtration of both types of disperse systems. From a phenomenological point of view, the filtration process can be characterized by several parameters.

The pressure drop of a filter Δp is defined by

$$\Delta p = p_1 - p_2 \tag{1}$$

where p_1 is the gas pressure before the filter and p_2 that behind the filter. This quantity is dependent only on the properties of the fluid and on the properties of a porous substance used as the filter in the case of a clean filter. As filtration proceeds the pressure drop becomes dependent also on the properties of particles deposited in or on the filter.

If G_1 is the flux of particles into the filter, G_2 the flux of particles from the filter, and G_3 the quantity of particles retained by the filter in unit time, $G_1 = G_2 + G_3$ from the law of conservation. For a monodisperse system of particles, the filter efficiency E is then defined by

$$E = \frac{G_3}{G_1} = \frac{G_1 - G_2}{G_1} = \frac{G_3}{G_3 + G_2} \qquad (2)$$

The first equality in Eq. (2) defines E in terms of captured and incoming particles, the second in terms of incoming and outgoing particles, and the third in terms of captured and outgoing particles. The quantities G_1, G_2, and G_3 can be expressed in terms of number, weight, activity, etc. The related quantity is the penetration of the filter P defined by

$$E + P = 1 \qquad (3)$$

Sometimes the coefficient $P^* = P^{-1} = (1 - E)^{-1}$ is used describing the lowering of the particle concentration after passage through the filter. For example, $E = 0.99999$ may be expressed as $P = 10^{-5}$ and $P^* = 10^5$, indicating that the particle concentration after a passage through this filter will decrease 10^5 times.

Filter capacity is defined as a quantity of deposited particles (usually expressed in grams or kilograms) which the filter is capable of accumulating before reaching a certain pressure drop. The filter capacity is approximately equal to a quantity of particles accumulated on the filter between two subsequent operations of filter regeneration. The capacity of the same filter for small particles is always smaller than for large particles. Hence, the filter capacity should be specified for particles of given size.

Economic indexes include the cost of the filter device, the

consumption of energy and material (consumption of energy for overcoming the filter resistance, filter cleaning, and regeneration), and the cost of the gas cleaning (usually expressed as a cost of cleaning to a given degree of 1000 m^3 of gas per hour).

There are, of course, further important factors such as the chemical composition of the filter, its physical and chemical properties, etc. For a comparison of different filters a quantity Q* called the filter quality is used which is defined by

$$Q^* = \frac{-\ln P}{\Delta p} \tag{4}$$

so that the better filter is characterized by a higher value of Q*. A certain disadvantage of the definition of filter quality given by Eq. (4) is that Q* is not dimensionless. This circumstance has been recently discussed by Juda and Chrosciel [1].

There are three objects that take part in the filtration process: the dispersed particles, the dispersing medium, and the porous substance. The individual dispersed particles and their assembly are characterized by the following factors: the diameter of the particles d_p or their size distribution; the shape of the particles; the mass m and the density $\mathcal{S}_p$ of the particles; the electric charge, dielectric constant, and chemical composition of the particles; and their concentration (number, weight, volume, and active concentration). The fluid flow is characterized by the velocity U_o, density $\mathcal{S}_g$, absolute temperature T, pressure p, dynamic viscosity η, kinematic viscosity $\nu = \eta/\mathcal{S}_g$, and humidity. The porous substrate is characterized by its geometrical dimensions—the filter surface area A and filter thickness L—by the size of structural units and their distribution and arrangement in the filter, and by the porosity of the filter ε and its specific surface, electric charge, dielectric constant, and chemical composition. The basic parameters of the filter, the pressure drop Δp and filter efficiency E, generally depend on nearly all the above mentioned factors. The determination of this dependence of Δp and E on quantities characterizing the dispersed particles, medium, and porous substance can be considered as the basic problem of filtration.

Theoretically, two phases are distinguished in the process of filtration. In the first phase the deposition of particles takes place in a pure filter of a certain structure, which is always approximated by a certain model. The structural changes caused by the deposited particles are assumed to be too small to influence both the basic parameters, Δp and E. In this phase both Δp and E do not depend on time, and, consequently, this phase of the filtration process is called stationary filtration. From a practical point of view the filtration process can be treated approximately as a stationary process in the initial stage of filtration for low concentrations of the inlet aerosol. In stationary filtration it is usually assumed that the collision efficiency of particles with the structural elements of the filter is unity, so that a particle that touches the collecting surface remains in contact with it and is not separated in the further course of the filtration process.

In reality, however, the filtration process is much more complicated, especially in the later stages. Structural changes take place in the filter as a result of the deposition of individual particles; both basic parameters Δp and E change with time (both parameters may decrease or increase); and the filter becomes clogged. This phase of the filtration process is called nonstationary filtration. Investigation of nonstationary filtration includes studies of the collision efficiency of particles with the collecting surface and, in this context, studies of adhesion of particles, processes on the collecting surface, and, in general all so-called secondary processes which lead to the change of pressure drop and filter efficiency with time.

Porous substances used as filters differ as to their structure, which may be fibrous, porous (vacuol), or granulated or a combination of these basic structural types. Because of their practical importance and widespread use, only the first two structural types will be discussed here, i.e., fibrous and membrane filters. The theory and the use of membrane filters will be discussed in a subsequent part of this chapter. The first part is devoted to the theory of fibrous filters which is the most fully developed to the present. The theory of fibrous filters has been reviewed by a large number of scientists. These reviews are arranged chronologically and listed under Refs. 2 through 18. In this review the original papers relating to individual problems are summarized and referred to—whenever possible—in chronological sequence to give a clear picture of individual problems and the state of their solution.

II. GENERAL APPROACH

In order to solve the basic problem of filtration theory it is necessary to express both the pressure drop Δp and the filter efficiency E as a function of quantities describing the properties of particles, dispersing medium, and the filter. Concerning the pressure drop, the methods of fluid flow physics through porous media are applied. In fact, the theory of pressure drop of fibrous aerosol filters is a part of a more general theory of permeability of porous media.

The situation is more complicated for the second basic parameter, the filter efficiency E. This quantity describes the retention of particles in a porous substance which is in reality a system having a complex geometry. The basic structural unit of fibrous filters is a fiber which can be geometrically approximated by a cylinder. Hence, it is natural to begin any discussion of particle deposition in a system of complex geometry with an analysis of particle deposition on a simple cylindrical body.

Essentially, the deposition of particles from a flowing fluid on bodies of simple geometric shape and the deposition in filters of an arbitrary structure involves an interfacial mass transfer of small particles in a nonuniform fluid in a system with both a simple and a complex geometry. In systems with a simple geometry the interfacial mass transfer between the gas suspension and a solid body (collector—in this case the cylinder) is usually described by two quantities: capture coefficient E_f (the index f denoting fiber) and a local capture coefficient. These quantities can be defined in the following way: Let J be the rate of deposition, defined as the number of deposited particles on a unit surface in a unit time. The integral of J over the surface of the body (cylinder) S_c gives the number of particles Φ_f captured by the body in unit time so that $\Phi_f = \int_{S_c} J\, dS_c$. The flux of particles which would pass through a projection S_M of the collector into the plane perpendicular to the fluid flow is $n_0 U_0 S_M$, where n_0 is the number concentration of particles and U_0 the velocity of fluid flow, both quantities taken at an infinite distance from the collector. Then the capture coefficient of the body (fiber) is defined by

$$\Phi_f = E_f n_0 U_0 S_M = E_f n_0 U_0 d_f \tag{5}$$

where d_f is the diameter of the fiber placed perpendicular to the fluid flow. The last equality in Eq. (5) is related to the unit length of the fiber. An equivalent definition of E_f is the ratio of the area of the stream σ from which all incoming particles are captured (capture cross-section) to the projected area of the body S_M. Hence the capture cross-section is given by $\sigma = E_f S_M$. The coefficient E_f is usually called the collection efficiency or the deposition efficiency of the fiber. This terminology has been criticized by Pich [19] since in many cases $E_f > 1$. Several investigators, e.g., Fortier [20], Stechkina [21], Mazin [22], and Gupalo [23], use the term capture coefficient, which seems appropriate. The local capture coefficient is defined in a similar way (see, e.g., Levin [24]), which describes the distribution of deposited particles on the collecting surface. In order to describe the deposition of polydisperse systems these two coefficients must be generalized to obtain an integral capture coefficient and an integral local capture coefficient as discussed by Mazin [22].

Similarly, the interface mass transfer by small particles can be described by a local rate of filtration T and a total rate of filtration in systems with complex geometry (filters of arbitrary structure). The local rate of filtration T—called the local rate of removal of particles per unit volume of filter by Spielman and Goren [25]—gives the number of particles captured by a unit volume of the filter in unit time. Hence, the total rate of filtration is $\Phi = \int_{V_f} T \, dV_f$, where V_f is the volume of the filter. The rate Φ is related to the flux of particles entering a filter of surface A in unit time, N_1U_0A, by

$$\Phi = EN_1U_0A \tag{6}$$

where the coefficient of proportionality E is the efficiency of the filter and Eq. (6) is a special case of Eq. (2). N_1 is the average particle concentration at the filter inlet and, if N_2 is that at the filter exit, the filter penetration P is $P = N_2/N_1$, which is again a special case of Eq. (3).

It has been established that in the deposition of particles from a flowing fluid several mechanisms are acting, the most important of which are described below.

A. Diffusion Deposition

The trajectories of individual small particles do not coincide with the streamlines of the fluid because of Brownian motion. With decreasing particle size the intensity of Brownian motion increases and so does, as a consequence, the intensity of diffusion deposition.

B. Direct Interception

This mechanism involves the finite size of particles. A particle is intercepted as it approaches the collecting surface to a distance equal to its radius. A special case of this mechanism is the so-called sieve effect, or sieve mechanism, which occurs if the distance between fibers is smaller than the particle diameter d_p.

C. Inertial Deposition

The presence of a body in the flowing fluid results in a curvature of the streamlines in the neighborhood of the body. The individual particles do not, due to their inertia, follow the curved streamlines but are projected against the body and may deposit there.

It is obvious that the intensity of this mechanism increases with increasing particle size and velocity of flow.

D. Gravitational Deposition

Individual particles have a certain sedimentation velocity due to gravity. As a consequence, the trajectories of particles deviate from the streamlines of the fluid and, owing to this deviation, the particles may touch a fiber.

E. Electrostatic Deposition

Both the particles and the fibers in the filter may carry electric charges. Deposition of particles on the fibers may take place because of the forces acting between charges or induced forces.

F. London-van der Waals Deposition

When the distance between a particle and a fiber is very small, deposition may be influenced by London-van der Waals intermolecular forces.

G. Capture Coefficients

Each of the above mechanisms is described by one or several dimensionless parameters, the numerical value of which determines the intensity of the individual mechanisms. To each dimensionless parameter corresponds a capture coefficient that quantitatively describes the rate of particle deposition due to the given mechanism. These partial capture coefficients are denoted here by E_D, E_R, E_I, etc., so that E_D is a capture coefficient of the fiber due to diffusion deposition, E_R that due to direct interception, and so on. The symbol E_f is reserved for a fiber capture coefficient without distinguishing the mechanism of particle deposition. If several mechanisms are acting simultaneously, the resulting (total) capture coefficient is an unknown function of partial capture coefficients.

A knowledge of particle trajectories in the vicinity of the

collector is usually required in order to calculate these quantities. The trajectory of a particle traveling through a fluid is governed by Newton's law of motion

$$m \frac{dV}{dt} = F_e - F_M \tag{7}$$

where F_e is the force externally applied to the particle and F_M the force of the medium resistance. Let $F_e = F_d + F_s$, where F_d is a deterministic and F_s a stochastic part of the force F_e. In the quasistationary state $V - U = BF_M$, where B is the mobility of the particle with velocity V, and Eq. (7) becomes

$$m \frac{dV}{dt} = F_d + F_s + \frac{1}{B} (U - V) \tag{8}$$

Using the dimensionless variables $t' = tU_0/a$, $V' = V/U_0$, and $U' = U/U_0$, Eq. (8) becomes

$$2Stk \frac{dV'}{dt'} + V' - U' = \frac{B}{U_0} (F_d + F_s) \tag{9}$$

where $Stk = mU_0B/d_f$ is a Stokes number (inertial parameter). In a certain sense this equation is the basis of all filtration theories. Several important cases can be distinguished.

If the stochastic force F_s due to the random molecular collisions resulting in the Brownian motion of particles is significant, then Eq. (9) is a Langevin equation. There are two possible ways to calculate the rate of particle deposition in this case. The first approach starts with the solution of the Langevin equation as discussed by Chandrasekhar [26]. Then according to Dawson [27] a Monte-Carlo technique can be used to obtain the trajectories of particles and, consequently, the rate of deposition. The second approach usually employed is based on Einstein's conclusion that the flux of particles follows Fick's law of diffusion. The flux of particles can be obtained by solving the equation of convective diffusion

$$\frac{\partial n}{\partial t} + V_o \operatorname{grad} n = D \nabla^2 n \tag{10}$$

where V_0 is the velocity of the ordered motion of the particle, D the diffusion coefficient of the particle, and ∇^2 the Laplace operator. In dimensionless form, Eq. (10) becomes $dn'/dt' = 2\nabla^2 n'/Pe$, where $Pe = U_0 d_f/D$ is a Péclet number. Sedunov [28] has shown that solutions of this equation under ordinary circumstances give the equivalent collection by working directly with the Langevin equation.

The concept of limiting trajectory can be applied for the determination of the fiber capture coefficient in the case of purely deterministic force applied to the particle ($F_s = 0$). The force F_d can be of gravitational or electrostatic character, as will be discussed later.

If all external forces are absent or can be neglected ($F_d = F_s = 0$), Eq. (9) provides a basis for a theory of purely inertial deposition of particles.

Finally, if inertial forces are also negligible ($Stk = F_d = F_s = 0$), Eq. (9) reduces to $V' - U' = 0$, which is the basis of the theory of direct interception.

There is no need to calculate the limiting (critical) trajectories of particles in some cases. According to Duchin and Deryaguin [29], if the inertia of particles may be neglected and the field of external forces is selenoidal, then the number concentration along the particle trajectories is constant, i.e., $n = n_0$. Then the total flux of particles Φ_f to the surface of the collector can be expressed by

$$\Phi_f = n_0 \int_{S_c} (U_n + BF_n)\, dS_c \tag{11}$$

where U_n is the radial component of the gas velocity and F_n the radial component of the electrostatic or some other central force. This method is especially suitable for the calculation of fiber capture coefficients due to the electrostatic deposition of particles since the field of charged bodies is selenoidal. The capture coefficient of the fiber can be calculated, at least in principle, in this way, knowing the velicity field U around the fiber.

The scheme for calculating the efficiency of the entire filter is as follows:

1. The velocity field around a cylinder, or around the model by which the filter is approximated, is calculated.
2. The capture coefficient of the fiber due to the different deposition mechanisms is calculated.
3. If several deposition mechanisms are acting simultaneously, the resulting (total) capture coefficient is estimated.
4. If the calculation is based on a model of an isolated cylinder the influence of the neighboring fibers on the deposition process is estimated (interference effect).
5. The calculated capture coefficient of the fiber in the filter is related to the filter efficiency.

The theory of fibrous filters will be discussed essentially in this sequence.

III. STRUCTURE OF FIBROUS FILTERS

The theory of the structure of fibrous filters is a part of the general theory of the structure of porous media discussed, e.g., by Matheron [30]. Here briefly summarized and discussed are the quantities usually employed for the description of filter structures.

A. Fiber Diameter and Fiber Diameter Distribution

Fiber diameter $d_f = 2a$ (a = radius of the fiber) is a basic structural quantity. The fibers are approximated by cylinders, the system of cylinders of the same d_f being monodisperse and that of different diameters polydisperse. The log-normal diameter distribution of fibers is applicable according to Werner and Clarenburg [31] for Johns-Manville glass fiber filters, for glass fiber filters according to Farrow [32], and for organic (PVC) fiber filters with geometric standard deviations $\sigma_g = 1.88$ and 1.66 according to Simon et al. [33]. The width of the diameter distribution characterized by σ_g is important for fundamental experimental studies when the goal, for example, is to check the theoretical predictions. Dawson [27], for comparison between theory and experiment, excludes all filters where $\sigma_g > 1.2$.

B. Shape of Fibers

The shape of fibers may be different, but usually a circular cylinder is assumed. Studies of the properties of a system consisting of noncircular cylinders are rather limited. Kuwabara [34] studied flow in a system of elliptic cylinders; Glushkov [35] calculated the pressure drop of a fiber filter consisting of noncircular cylinders in free molecule flow; and an extensive study of the effects of fiber cross-sectional shape on resistance to the flow of fluids through fiber mats has been reported by Labrecque [36].

C. Pore Size and Pore Size Distribution

The term "pore size" has a conditional character and care must be exercised in using it. Employing the concept of hydraulic radius, the mean pore diameter λ_s is defined by $\lambda_s = 4V_p/S$, where V_p is the volume of voids and S the total surface of the porous material. As the specific surface $S_0 = S/V_f$, where V_f is the filter volume, this equation becomes $\lambda_s = 4\varepsilon/S_0$. The ratio ε/S_o is called the hydraulic radius.

Benarie [37,38] defined the effective diameter of a pore as follows: If there are n capillaries of a circular cross-section having radii $r_1, r_2, \cdots, r_n$ and length $L_1, L_2, \cdots, L_n$, then the effective radius r is defined as a radius of a capillary having the length $L = L_1 + L_2 + \cdots + L_n$ and giving the same volume flow rate at the same pressure drop. In such a definition the effective radius is dependent not only on the system geometry but also on the type of flow.

Piekaar and Clarenburg [39] used the concept of the hydraulic radius for the description of pore size. The pore size distribution has been investigated experimentally using different methods—optical microscopy, electron microscopy, permeability methods, and mercury porosimetry. The log-normal pore size distribution was found experimentally by Skau, et al. [40], Corte [41], and Benarie [37]; calculated by Piekaar and Clarenburg [39]; and assumed by Clarenburg and van der Wal [42]. The geometric standard deviation of the interfiber pore distribution was employed by Benarie [37,38] for calculating the effect of filter nonuniformity on filtration efficiency.

D. Spatial Arrangement of Fibers

In simple cases the arrangement of fibers can be described by the following quantities:

1. *Orientation*

The orientation of fibers is usually related to the direction of gas flow. Sullivan [43] and Fowler and Hertel [44] suggested for the description of fiber orientation the orientation factor θ^* defined by $\theta^* = \overline{(\sin^2 \phi)}$, where ϕ is the angle between the direction of gas flow and the normal to the element of the cylinder surface. Usually two extreme cases are distinguished: transverse flow, in which the direction of the gas flow is perpendicular to the axes of the cylinders, and longitudinal (axial) flow with the flow parallel to the cylinder axes. Corresponding values of the orientation factor are $\theta^* = 1/2$ and $\theta^* = 1$, respectively. Another factor is the angle between the axes of individual cylinders, the simplest case being a system of parallel cylinders. In addition to parallel cylinders, Spielman and Goren [25] distinguished three cases. The first is fiber axes all lying in planes perpendicular to the flow direction but having completely random angles in these planes. This case is of great practical importance since many fibrous media are prepared by depositing fibers onto a flat surface with their axes almost completely parallel to the surface. The second case is fiber axes lying in planes parallel to the direction of flow but having completely random angles in those planes. Finally, in the third case the fiber axes are oriented completely randomly in all directions.

2. *Interfiber Distance*

Let 2b be the distance between the cylinder axes in a system of parallel cylinders. The ratio $Fu = 2a/2b = d_f/2b$ has a basic significance for flow through these systems. Fuchs [13] concluded that at small Reynolds numbers, Re, the flow is independent of Re but essentially dependent on the parameter Fu. Obviously $0 < \mathrm{Fu} \leq 1$. The parameter Fu is related to the porosity of the filter ε, being $\mathrm{Fu} \ll 1$ for high-porosity systems and $1 - \mathrm{Fu} \ll 1$ ($\mathrm{Fu} \sim 1$) for low-porosity systems.

3. *Kinds of Arrangement*

An infinite array of parallel cylinders represents a "structure of the first order," and the porosity of the system is related to Fu by

$$\varepsilon = 1 - \frac{\pi}{4}\,\mathrm{Fu} \tag{12}$$

A system of infinite arrays of parallel cylinders represents a "structure of the second order." The most common arrangements of this kind are: 1. Square structure (arrangement), where the axes of parallel cylinders lie at the corners of a square. In this case

$$\varepsilon = 1 - \frac{\pi}{4}\,\mathrm{Fu}^2 \tag{13}$$

2. Triangular structure (staggered arrangement), which is described by

$$\varepsilon = 1 - \frac{\pi}{2\sqrt{3}}\,\mathrm{Fu}^2 \tag{14}$$

and 3. Nonarranged structure (2b is not constant), described by Fuchs and Stechkina [45] as

$$\varepsilon = 1 - \mathrm{Fu}^2 \tag{15}$$

E. Specific Surface

Specific surface is defined by $S_0 = S/V_f$. The relation of this quantity to filter porosity and fiber radius given by Sullivan [43] is

$$S_0 = \frac{2(1 - \varepsilon)}{a} \tag{16}$$

F. Fiber Concentration

Different fiber concentrations are employed (the number of cylinders in a volume, surface, and length unit, the number of cylinder arrays in a unit length, etc.). If N is the number of cylinders per unit length in one array and M is the number of cylinder arrays in a unit length, then $N = M = 1/2b = Fu/2a$ for a square structure according to Keller [46]. The product NM gives the number of cylinders per surface unit and is numerically equal to the length of cylinders in a unit volume, $\beta/\pi a^2$. For a nonarranged structure, the surface concentration of cylinders is $1/\pi b^2$ according to Kuwabara [47].

G. Filter Porosity

If V_f is the volume of the filter, V_s the volume of fibers, and V_p the voids volume, then $V_f = V_s + V_p$, and the porosity of the filter is defined by

$$\varepsilon = \frac{V_p}{V_f} = 1 - \frac{V_s}{V_f} = 1 - \beta = \frac{S_f - S^*}{S_f} \tag{17}$$

The quantity β is the packing density of the filter (sometimes called filter solidity), S_f the density of the fiber material, and S^* the density of the filter. The porosity of high-efficiency filters is usually large [14] so that β is small ($\beta < 10\%$). The porosity of voids noninterconnected is sometimes called the inert porosity. The introduction of mean porosity is necessary for nonuniform filters, according to Benarie [37]. Then, using the quantities defined above, it is possible to derive useful relationships.

The total length of a fiber which it is possible to obtain from the volume V_s is $L_f = V_s/\pi a^2$. As $V_s = \beta V_f$, there can be obtained $L_f = V_f\beta/\pi a^2$. Hence, the length of fiber in a unit volume of the filter is

$$\ell_f = \frac{\beta}{\pi a^2} \tag{18}$$

If the mean length of the fiber is $\tilde{\ell}$, the total number of fibers in the filter N then is $N\tilde{\ell} = L_f$. Using Eq. (18) there is obtained

$$N = \frac{\beta V_f}{\pi a^2 \tilde{\ell}} \tag{19}$$

Hence, the number of fibers in a unit volume n is given by

$$n = \frac{N}{V_f} = \frac{\beta}{\pi a^2 \tilde{\ell}} \tag{20}$$

in agreement with Piekaar and Clarenburg [39] for monodisperse systems. These authors also derived a similar expression for a polydisperse system, viz.

$$n = \frac{4\beta}{\pi \tilde{\ell} \overline{d_f^2}} \tag{21}$$

The number of pores per surface unit n_p according to Piekaar and Clarenburg for a polydisperse system of pores is given by

$$n_p = \frac{8\beta \overline{d_f}}{\pi^2 \tilde{\ell} \overline{d_f^2}} \left(\frac{8\beta \tilde{\ell} \overline{d_f}}{\pi \overline{d_f^2}} - 1 \right) - \frac{4\beta \overline{d_f}}{\pi \tilde{\ell} \overline{d_f^2}} \tag{22}$$

expressing n_p as a function of ε, d_f, $\overline{d_f^2}$, and $\tilde{\ell}$. For a monodisperse system of pores, $d_f = \overline{d_f} = \overline{d_f^2}$, and Eq. (22) reduces to

$$n_p = \frac{8\beta}{\pi^2 \tilde{\ell} d_f} \left(\frac{8\beta \tilde{\ell}}{\pi d_f} - 1 \right) - \frac{\beta}{\pi \tilde{\ell} d_f} \tag{23}$$

The geometric mean hydraulic radius of pores r_{hg} according to these authors is given by

$$r_{hg} = \left(\frac{0.0295}{n_p} \right)^{1/2} \tag{24}$$

where n_p is given by Eqs. (22) or (23). The mean surface of the pore $\overline{A}_p$ is given by $\overline{A}_p = 1/n_p$, and also the spectrum of pore surface is log-normal according to Piekaar and Clarenburg.

An attempt to describe a fiber filter structure statistically was made by Radushkevich [48] who investigated the distribution of the free lengths of fibers (the length between fiber intersections). This investigator concluded that the distribution is of the Clausius type (valid for the distribution of the free path of the gas molecules). The cumulative (integral) distribution function of free lengths is given by

$$G(x) = e^{-x/\delta} \tag{25}$$

where x is the free length of the fiber and the parameter δ is a mean free length given by $\delta = (\pi/4)(d_f/\beta)$. The number of intersections (contacts) of fibers in a unit volume n_c is given by Radushkevich [48] as

$$n_c = \left(\frac{4\beta}{\pi}\right)^2 \frac{1}{d_f^3} \tag{26}$$

IV. STATIONARY FILTRATION

The first step to a solution of the basic problem of filtration is a calculation of the velocity field around an individual fiber (cylinder) or in a system of fibers. This calculation is based on the equations of motion for a gas, viz., the Navier-Stokes equation in various approximations, most frequently that of Oseen (low Reynolds numbers) or of Euler (high Reynolds numbers, usually potential flow). As is well known, these equations can be applied only when the gas can be considered a continuous medium (continuum mechanics). This assumption is fulfilled if the value of the Knudsen number $Kn = \lambda/a$, where λ is the mean free path of the gas molecules and $a = d_f/2$ is a fiber radius, is negligible.

A. Filtration in the Continuum Region

According to the classification of different regimes of gas flow as suggested by Devienne [49], the flowing gas can be treated as a continuous medium if the value of the Knudsen number is in the range $0 < Kn < 10^{-3}$. Of course, this limit ($Kn = 10^{-3}$) is a matter of convention. Accepting this classification, it follows that within this range the relations, equations, and the whole approach of classical hydrodynamics can be applied. Since, under normal conditions, the value of λ for air [13] is $\lambda = 0.653 \times 10^{-5}$ cm this condition is fulfilled in filters having relatively thick fibers ($a > 65.3$ μm). The technology of fiber preparation has advanced so much that fibers have been prepared having a thickness near the value of λ or even smaller. The classical (continuum) mechanics can no longer be applied in these cases for the description of gas flow, and the methods of mechanics of rarefied gases must be used. Since the character of the gas flow is of basic importance for the description of particle deposition, the filtration process can be divided into filtration in the continuum region and filtration in noncontinuum regions. Hence, the theory of filtration in the continuum region is based on continuum mechanics, and the theory of filtration in noncontinuum regions on the mechanics of rarefied gases. Corresponding to the state of development of continuum mechanics and mechanics of rarefied gases, the theory of filtration in the continuum region is much more fully developed than is the description of filtration processes in noncontinuum regions.

The first papers dealing with the theory of aerosol filtration in the continuum region appeared in 1931 [50,51]. A systematic investigation of the theory of fibrous filters was started by Langmuir [52,53] during the Second World War. His abilities were remarkable, as after many years other investigators using modern mathematical approaches have come to essentially the same results as did Langmuir using intuition. The historical aspects of the development of filtration theory have been discussed by Davies [2].

For the calculation of filter efficiency Langmuir developed a method of isolated fibers, the principle of which is as follows: The velocity field around an isolated fiber placed perpendicular to the direction of flow is used for the calculation of the rate of particle deposition on the fiber due to the various deposition mechanisms. The influence of neighboring fibers—called by Chen [4] an interference effect—is expressed by the introduction of empirical or semi-empirical corrections, and the final phase of the calculation passes from a determination of the rate of deposition on the fiber to an evaluation of filter efficiency. The drawback of this approach is that the interference effect is described in an empirical manner, and that this approach can be used only for high-porosity filters where the influence of neighboring fibers is small. On the other hand, there is a basic advantage to this approach, as the results obtained for an isolated fiber have a significance beyond that of the theory of filters. Beyond filtration, the possible application of these results and the methods involved are, for example, (1) investigation of icing on aircraft and electric transmission lines caused by the deposition of supercooled water drops; (2) description of the growth of rain drops due to gravitational and kinematic coagulation, which is one of the basic mechanisms in rain precipitation, and other problems of the physics of clouds; and (3) analytical purposes where a knowledge of the deposition of particles from a flowing gas can be used for dispersity and concentration determination of substances in the stream. A further advantage of the method of isolated fibers is that computation of the rate of deposition on a isolated fiber, which can be exactly defined geometrically, can be verified experimentally with a high degree of accuracy. Well-defined systems in which the deposition of particles occur is practically realized in experiments with single fibers and in experiments with model filters. On the other hand, there is danger of oversimplification as far as fibrous filters are concerned. Various models simulating the structure of real filters have been suggested for this reason. These models will be described subsequently in connection with the discussion of individual original papers.

B. Flow Through Fibrous Filters

Theory of gas flow through fibrous aerosol filters is a part of the physics of fluid flow through porous media as summarized by Muskat [54], Carman [55], Scheidegger [56], and Collins [57]. Usually, filtration takes place at low velocities so that the concept of low-speed, incompressible, Newtonian flow governed by the D'Arcy law can be applied. The D'Arcy law in integral form can be expressed as

$$\Delta p = \frac{1}{k^*} \eta U_0 L \tag{27}$$

where k* is the permeability of the given porous substance. Permeability is a basic quantity which in the continuum region is a descriptive constant for each porous substance. Its dimension in the cgs system of units is square centimeters. Sometimes the darcy is used as a measuring unit, 1 darcy being equal to 9.87×10^{-9} cm^2. With regard to the analogy between D'Arcy's and Ohm's laws as pointed out by Langmuir [53] and discussed, e.g., by Pich and Spurny [58], it can be shown that when porous substances are arranged in parallel the resulting permeability is the sum of the individual permeabilities, and that when arranged in series the reciprocal values of permeabilities are additive. The possibility of application of this quantity for comparison of different filters has been discussed by Pich [17].

In the continuum region (negligibly small Knudsen numbers), permeability is dependent only on the structural properties of the porous substance, mainly on such quantities as porosity, size of the structural units (fibers), and specific surface. Several theories have been developed to determine this dependency, from which essentially two have been applied in the theory of aerosol filters.

The first is the channel theory of permeability, which assumes that the porous substance can be modeled as a system of capillaries (channels) and that the flow can be described, starting with Poiseuille's law for the flow through a circular capillary. This approach is usually based on a concept of a hydraulic radius. It has been developed by Kozeny [59] and Carman [60]; modified by Wiggins et al. [61], Fowler and Hertel [44], Sullivan and Hertel [62], and Sullivan [43]; and summarized in a review by Sullivan and Hertel [63]. This approach has been followed by Clarenburg and coworkers [31, 42, 64-67] in connection with the theory of aerosol filters.

The second approach involves the drag theory of permeability. In this case a certain geometrical model of the porous substance is assumed as realistically as possible but simultaneously sufficiently simple to enable the calculation of the velocity field. Knowing the velocity distribution, the drag forces acting on individual structural units (fibers) can be determined sufficiently for a determination of the permeability. Since, according to Langmuir [53],

Brinkman [68], and others, the application of the channel theory to high-porosity filters is rather problematical, attention is focused here on the drag theory of permeability. Extension of this approach to the noncontinuum regions is discussed in a later section.

C. Velocity Field Around an Isolated Cylinder

The simplest model of a high-porosity fibrous aerosol filter in which the influence of neighboring fibers is neglected is an isolated, infinitely long cylinder placed perpendicular to the fluid flow. There is considerable literature dealing with flow around isolated cylinders. The texts of Illingworth [69], Van Dyke [70], and Happel and Brenner [71] should be consulted for modern mathematical approaches to this problem.

The starting point for calculation of the velocity field is Oseen's approximation (small Reynolds numbers) of the Navier-Stokes equations, since for Stokes' approximation of the Navier-Stokes equations there is no solution for an isolated cylinder in an unbounded fluid (Stokes paradox). The exact solution of Oseen's equations for a cylinder was found by Bairstow, et al. [72] and by Faxen [73]. However, since these solutions are too complex for further use, usually approximations of the velocity field around a cylinder are employed. These approximations have been derived, in chronological sequence, by Lamb [74], Thom [75], Davies [76], Tomotika and Aoi [77], Imai [78], Tomotika [79], and Kaplun [80]. Usually the approximation derived by Lamb [74] is employed, which is expressed by

$$\underline{\psi}(S,\theta) = \frac{aU_o \sin\theta}{2(2 - \ln \mathrm{Re})} \left(\frac{a}{S} - \frac{S}{a} + 2\,\frac{S}{a} \ln \frac{S}{a} \right) \tag{28}$$

where S and θ are the polar coordinates and $\underline{\psi}(S,\theta)$ is the stream function related to velocity components in polar coordinates, $u_S(S,\theta)$ and $u_\theta(S,\theta)$, by $u_S = (1/S)(\partial\underline{\psi}/\partial\theta)$ and $u_\theta = -\,\partial\underline{\psi}/\partial S$. Furthermore, U_o is an undisturbed flow velocity (velocity at infinite distance from the cylinder); $a = d_f/2$ the radius of cylinder; Re = $d_f U_o/\nu$ the Reynolds number, $\nu = \eta/S_g$, where ν is the kinematic viscosity; η the dynamic viscosity; and S_g the density of the fluid. The validity of Eq. (28) is limited to conditions near the cylinder, i.e., for $(S - a)/a < 1$ and small Re. The approximations for the velocity field of other workers [75-80] are usually more accurate but also more complicated.

The other extreme case—large Reynolds numbers—corresponds to ideal flow. In this case the flow of an ideal, incompressible, and irrotational fluid (potential flow) around an isolated cylinder is given by

$$\underline{\psi}(S,\theta) = aU_o\left(\frac{S}{a} - \frac{a}{S}\right) \sin\theta \qquad (29)$$

D. Velocity Field in a System of Cylinders

1. *Array of Cylinders*

Other than an isolated cylinder, the simplest system is a single array of cylinders (structure of the first order). Usually an infinite row of circular, equidistant, parallel cylinders perpendicular to the gas flow is investigated. Two extreme cases of structures of the first order should be distinguished: high-porosity systems (Fu << 1 or $\beta << 1$) and low-porosity systems (1 - Fu << 1 or $\varepsilon << 1$). The first case was investigated by Tamada and Fujikawa [81,82] using Oseen's approximation. These results are complicated, and even simpler expressions describing the velocity distribution in the limit of negligible inertial forces, due to Miyagi [83], are too lengthy to quote here. Flow in a row of elliptic cylinders was investigated by Kuwabara [34].

The second extreme case (low-porosity structure of the first order) was studied by Keller [46].

2. *Arrays of Parallel Cylinders (Structures of Second Order)*

a. High-Porosity Systems ($\beta << 1$)

(1) *Transverse Flow* The cell models discussed in detail by Happel and Brenner [71] are usually employed in this case. Each cylinder is assumed to be surrounded by a concentric circular

cylinder, and some of the boundary conditions are specified on the surface of this concentric cylinder (cell). The radius of the cell b is fixed by requiring that the packing density (solidity) of the cell be equal to the packing density of the filter β in that region. As $(a/b)^2 = Fu^2 = \beta$, the radius of the cell b can be expressed by $b = a/Fu = a/\beta^{1/2}$. Kuwabara [47] assumed that vorticity vanishes at the cell surface, and Happel [84] assumed that the shear stress vanishes at the cell surface, i.e., for $S = b$. Their result can be expressed by

$$\underline{\psi}(S,\theta) = \frac{aU_o \sin\theta}{2(-C - \frac{1}{2}\ln\beta)}\left(\frac{a}{S} - \frac{S}{a} + 2\frac{S}{a}\ln\frac{S}{a}\right) \tag{30}$$

where C = 0.75 according to Kuwabara and C = 0.5 according to Happel. A comparison of Eq. (30) with Eq. (28) shows three essential properties of the Kuwabara-Happel (K-H) velocity field. With regard to the dependence $\underline{\psi} = \underline{\psi}(S,\theta)$, the K-H field is the same as that of Lamb (Eq. 28). This is advantageous as identical methods with those that use Lamb's field can be applied to the integration of the diffusion equation and the equations of motion. The K-H field (Eq. 30) does not include Re, which is consistent with experiments [13]. Since Eq. (30) describes the velocity field in a system of fibers, the influence of the neighboring fibers is simultaneously taken into account and the correction for interference effect (see p. 78) becomes superfluous.

Experimental verification of the K-H velocity field, Eq. (30), has been carried out by Kirsch and Fuchs [85] by taking photographs of granules (20 to 30 μm spherical particles of polymethyl methacrylate) in a viscous fluid (95% glycerol). Much better agreement was found with Kuwabara's field than with that of Happel. The value of the constant C appearing in Eq. (30), being 0.75 according to Kuwabara, was found experimentally to lie between the limits 0.73 and 0.77.

The velocity field in high-porosity systems of parallel cylinders for a square arrangement of cylinders was investigated by

Hasimoto [86].

A different approach to calculation of the velocity distribution in a system of cylinders was suggested by Spielman and Goren [25]. For transverse flow the result can be expressed by

$$\underline{\psi}(S,\theta) = aU_0 \sin\theta \left\{ \frac{S}{a} - \frac{a}{S}\left[1 + \frac{2K_1(z)}{zK_0(z)}\right] + \frac{2K_1[(S/a)z]}{zK_0(z)} \right\} \tag{31}$$

where $z = X^{1/2}$ and X is the resistivity of the filter, as will be defined later. The quantity z, and therefore filter resistivity, is given implicitly as a function of the filter packing density β by

$$\beta = \frac{z^2}{2z^2 + 4z\, K_1(z)/K_0(z)} \tag{32}$$

where $K_0(z)$ and $K_1(z)$ are modified Bessel functions of zero and first order, respectively. According to Spielman and Goren [25], the field, Eq. (31), is expected to apply rigorously to the case where the axes of all cylinders are parallel to one another, and approximately to the case where cylinders have a random orientation in planes perpendicular to the direction of flow. For small values of $z = X^{1/2}$, expanding the Bessel functions for small values of their arguments, the S-G field, Eq. (31), becomes functionally the same as that of Lamb, Eq. (28), and the K-H field, Eq. (30). Extending the analysis leading to the K-H field, Eq. (30), to more typical porosities, all these velocity fields can be summarized by

$$\underline{\psi}(S,\theta) = \frac{aU_0 \sin\theta}{2H^*}\left(\frac{a}{S} - \frac{S}{a} + 2\frac{S}{a}\ln\frac{S}{a}\right) \tag{33}$$

where H* is a hydrodynamic factor given by

$$H^* = 2 - \ln Re \quad \text{(Lamb)} \tag{33a}$$

$$H^* = \frac{-3}{4} - \frac{1}{2}\ln\beta + \beta - \frac{1}{4}\beta^2 \quad \text{(Kuwabara)} \tag{33b}$$

$$H^* = \frac{-1}{2} - \frac{1}{2}\ln\beta + \beta^2/1 + \beta^2 \quad \text{(Happel)} \tag{33c}$$

$$H^* = \frac{K_0(z)}{z\, K_1(z)} \quad \text{(Spielman and Goren)} \tag{33d}$$

For systems of very high porosity the hydrodynamic factor H* is reduced to the factor H given by:

$$H = 2 - \ell n\ \mathrm{Re} \quad \text{(Lamb)} \tag{33e}$$

$$H = \frac{-3}{4} - \frac{1}{2}\ell n\ \beta \quad \text{(Kuwabara)} \tag{33f}$$

$$H = \frac{-1}{2} - \frac{1}{2}\ell n\ \beta \quad \text{(Happel)} \tag{33g}$$

$$H = -0.5772 - \ell n\left(\frac{1}{2} z\right) \quad \text{(Spielman and Goren)} \tag{33h}$$

It is adequate to approximate the velocity field of Eq. (33) by using only the first term in a power series expansion by powers $(S - a)/a$ in some cases. The result is

$$\underline{\psi}(S,\theta) = \frac{aU_o \sin\theta}{H^*}\left(\frac{S}{a} - 1\right)^2 \tag{34}$$

(2) *Longitudinal (Axial Flow)* The velocity distribution for flow parallel with the axes of cylinders in a cylinder system has been investigated as follows: for a nonarranged structure by Happel [84]; for a square structure by Emersleben [87,88] and Sparrow and Loefler [89]; and for a triangular arrangement by Sparrow and Loefler [89]. These results are partially summarized and discussed by Happel and Brenner [71]. Parallel flow has been investigated by Leonard and Lemlich [90].

b. Low-Porosity Systems

Approximations of the velocity field for transverse flow in low-porosity systems of parallel cylinders have been described by Fuchs [13] and later by Keller [46] and Gupalo [23]. These velocity distributions provide the basis for the theory of low-porosity filters as discussed later.

E. Theory of Drag Forces

The drag force acting upon a body in a flowing fluid is an important characteristic. Its evaluation is usually one of the primary objectives of any investigation. A knowledge of this force is important from at least two points of view: it is a part of the general theory relating to the resistance to motion of nonspherical particles discussed, e.g., by Fuchs [13] and especially by Happel and Brenner [71]; and it can be used for the calculation of the pressure drop of filters on the basis of a drag theory of permeability of porous media discussed, e.g., by Scheidegger [56].

Let F be the drag force acting on a unit length of cylinder placed perpendicular to fluid flow. It is useful to define a dimensionless drag F* by

$$F^* = \frac{F}{\eta U_o} \tag{35}$$

As the familiar drag coefficient C_D is defined by $F = C_D d_f (S_g U_o^2/2)$, the relation between F* and C_D is given by

$$F^* = \frac{1}{2} C_D Re \tag{36}$$

where Re is the Reynolds number.

The drag of an isolated cylinder in transverse flow was investigated by Lamb [74] and Davies [76]. Starting from different approximations of the velocity field around a cylinder, these authors came to the same result, expressed as

$$F^* = \frac{4\pi}{2 - \ln Re} \tag{37}$$

This equation has been verified experimentally by White [91,92], Finn [93], and Tritton [94], good agreement with Eq. (37) being found.

The drag of a single cylinder in a high-porosity row of parallel, equidistant cylinders was investigated by Tamada and Fujikawa [81] and Miyagi [83]. Their result can be expressed by

$$F^* = \frac{4\pi}{-0.64 - \ell n\ Fu + 0.82\ Fu^2} \tag{38}$$

This equation was checked experimentally by Kirsch and Fuchs [95] and excellent agreement of experimental data with Eq. (38) was found at $Fu \leq 0.7$.

For a low porosity row of parallel equidistant cylinders, according to Keller [46], F^* is given by

$$F^* = \frac{9\pi/2\sqrt{2}}{(1 - Fu)^{5/2}} \tag{39}$$

Fair agreement of this equation with experiments for large values of Fu (low-porosity system) was found also by Kirsch and Fuchs [95]. For example, for $Fu = 0.888$ the experimental value of F^* is $(2.3 \pm 0.1) \times 10^3$ and the value calculated from Eq. (39) is 2.36×10^3. These investigators emphasized that good agreement between theory and experiment is observed only if the face velocity of flow is assumed to be U_o.

In the case of structures of the second order, Kuwabara [47], using the cell model with a vorticity-free boundary, derived the equation

$$F^* = \frac{4\pi}{-0.75 - \frac{1}{2}\ell n\ \beta + \beta - \frac{1}{4}\beta^2} \tag{40}$$

Happel [84], using the cell model with shear-free boundary, concluded that F^* is given by

$$F^* = \frac{4\pi}{-0.5 - \frac{1}{2}\ell n\ \beta + \beta^2/(1 + \beta^2)} \tag{41}$$

Concerning this equation, there is a slight error in the papers by Kirsch and Fuchs [95], but it was detected by Dawson [27]. The same problem for the square arrangement of parallel cylinders was investigated by Hasimoto [86]. His result can be expressed by

$$F^* = \frac{4\pi}{-0.6174 - \ell n\ Fu + \frac{\pi}{4}Fu^2} \tag{42}$$

For a square arrangement $\beta = \pi F u^2/4$, and this equation becomes

$$F^* = \frac{4\pi}{-0.7384 - \frac{1}{2}\ln\beta + \beta} \tag{43}$$

giving approximately the same values of F* as Kuwabara's equation, Eq. (40), for high porosity systems.

Spielman and Goren [25] using their velocity field, Eq. (31), concluded that F* is given by

$$F^* = 4\pi\left[\frac{1}{2}z^2 + \frac{zK_1(z)}{K_0(z)}\right] \tag{44}$$

where again z is used implicitly as a function of filter packing density as given by Eq. (32).

F. Pressure Drop of Fibrous Filters

The pressure drop Δp of a filter is of the same importance as efficiency and sometimes even more so. It is not difficult to prepare a filter having practically absolute efficiency, but the high value of Δp would make any practical use of such a filter nearly impossible. The pressure drop is related to permeability by Eq. (27), so, knowing filter permeability, the pressure drop can be easily calculated for arbitrary conditions in the range of validity of the D'Arcy law.

Another suitable quantity for describing the flow properties of fibrous filters was proposed by Davies [96], basing his treatment on a dimensional analysis of the D'Arcy law. The quantity is given by

$$X = \frac{a^2\Delta p}{\eta U_o L} \tag{45}$$

and is called the coefficient of resistance by Billings [97] and the resistivity by Dawson [27]. The great advantage of this quantity is that X is dimensionless. Combining Eqs. (27) and (45) gives for the relation between resistivity and permeability

$$X = \frac{a^2}{k^*} \tag{46}$$

Modifications of these quantities are sometimes employed. For example, Happel and Brenner [71] used the D'Arcy resistance coefficient as being equal to k^{*-1}, while Spielman and Goren [25] used dimensionless pressure drop or dimensionless pressure gradient as being equal to $X/4\beta$.

According to Iberall [98], Happel and Brenner [71], and Dawson [27], the pressure drop Δp is related to the drag force acting on a unit length of the fiber F by

$$\frac{\Delta p}{L} = F \frac{\beta}{\pi a^2} = F^* \frac{\eta U_o \beta}{\pi a^2} \tag{47}$$

The theoretical equation for Δp corresponding to a certain model of the filter can be derived using this equation and any expression for F^* given in the previous section. Furthermore, expressing different equations for pressure drop using the concept of resistivity X as defined by Eq. (45) is advantageous. Then the theoretical equations for filter resistivity can be summarized, in chronological sequence, as below.

Langmuir [52] derived for a system of parallel cylinders the equation

$$X = \frac{2B^*\beta}{-\frac{3}{4} - \frac{1}{2}\ln \beta + \beta - \frac{1}{4}\beta^2} \tag{48}$$

where B^* is a numerical coefficient, to account for the nonparallel orientation of fibers with respect to the gas flow. It was concluded that the effect of the transverse position of the fibers with respect to the direction of the flow was to increase the resistivity by a factor not greater than 2. Hence, $1 \le B^* \le 2$, and for parallel (axial) flow $B^* = 1$.

Iberall [98] assumed that cylindrical fibers are equipartitioned in three perpendicular directions, one of which is the direction of flow. This result can be expressed by

$$X = \frac{4\beta}{3(1 - \beta)} \frac{4 - \ln Re}{2 - \ln Re} \tag{49}$$

Reynolds number dependency results from the application of Lamb's equation, Eq. (37) for the drag of isolated cylinders.

Fuchs and Stechkina [45] used Kuwabara's and Happel's expressions, Eqs. (40) and (41), respectively, for F^* and Eq. (47), obtaining for high-porosity systems ($\beta << 1$) the equation

$$X = \frac{4\beta}{-C - \frac{1}{2} \ln \beta} \tag{50}$$

where $C = 0.5$ (Happel) and $C = 0.75$ (Kuwabara). Fuchs and Stechkina checked this equation by comparing the theoretical values of Δp with the experimental data of Chen [4], Davies [96], Sullivan [43], and Stern et al. [99]. They concluded that experimental values are 20 to 50% smaller than theoretical ones. Fuchs and Stechkina tried to explain this difference by three factors: parallel flow (not all fibers are perpendicular to the flow), inhomogeneity of the filters, and errors in the experimental determination of the packing density (solidity) β. Finally, they pointed out that a value of $C = 0.5$ is in somewhat better agreement with experiments on real filters than the value $C = 0.75$. Another check of Eq. (50) was reported by Billings [97]. Experimental values always smaller than the theoretical ones were found, good agreement with experiment being obtained with $C = 0.25$.

Happel [84] derived for parallel (axial) flow an equation identical with Eq. (48) with $B^* = 1$ using a free surface model. The same equation follows also from Sparrow and Loefler's [89] results for high-porosity systems ($\beta << 1$). For the case of transverse flow, Happel's results can be expressed by

$$X = \frac{4\beta}{-\frac{1}{2} - \frac{1}{2} \ln \beta + \beta^2/(1 + \beta^2)} \tag{51}$$

Using Eq. (40) (vorticity-free boundary) the following equation is obtained, also for transverse flow:

$$X = \frac{4\beta}{-\frac{3}{4} - \frac{1}{2} \ln \beta + \beta - \frac{1}{4} \beta^2} \tag{52}$$

It is interesting that this latter equation is identical with Eq. (48), where $B^* = 2$. According to the Langmuir analysis both the Kuwabara and Happel solutions for transverse flow should slightly over-estimate the resistivity. For small solidity values ($\beta \ll 1$), Eqs. (51) and (52) reduce to the Fuchs-Stechkina equation, Eq. (50).

Spielman and Goren [25], using the velocity field as given by Eq. (31), concluded that in the case of transverse flow the resistivity is given implicitly by

$$X = 4\beta\left[\frac{1}{2}X + \frac{X^{1/2}K_1(X^{1/2})}{K_0(X^{1/2})}\right] \tag{53}$$

this equation being identical with Eq. (32). The above equation gives for $\beta < 0.1$ smaller values of X than either the Kuwabara or the Happel solution. Spielman and Goren claimed their theory to be better than the unit-cell model, as the majority of experimental data from the literature fall lower than all theoretical predictions and Eq. (53) yields the lowest values of X in the range $\beta < 0.1$. In the case where fiber axes are completely randomly oriented in all directions, the resistivity is, according to Spielman and Goren, given by

$$X = 4\beta\left[\frac{1}{3}X + \frac{5}{6}X^{1/2}\frac{K_1(X^{1/2})}{K_0(X^{1/2})}\right] \tag{54}$$

Kirsch and Fuchs [95] conducted experiments with a filter model consisting of regular arrays of parallel cylinders with diameters from 0.015 to 0.07 cm. Their measurements of resistivity, correlated with F^* by $X = F^*\beta/\pi$, essentially agreed with the equations of Kuwabara, Eqs. (40) and (52), Hasimoto, Eq. (42), and Tamada and Fujikawa, Eq. (38), and were, therefore, significantly above the predictions of the equations of Happel, Eqs. (41) and (51), and Spielman and Goren, Eqs. (44) and (53). When the wires were moved out of parallel, the resistivity decreased somewhat below the prediction of Happel, Eq. (51). The resistivity decreased even more when the wires remained parallel but the spacing between them was made nonuniform.

By inspecting the equations for F* relating to a system of cylinders, it can be concluded that in the continuum region and within the range of validity of the D'Arcy law the resistivity is dependent only on the filter solidity, i.e., $X = X(\beta)$. This circumstance has been pointed out by Davies [96]. The dependence according to different theoretical equations has been constructed by Dawson [27]. It follows that the dependence $X = X(\beta)$ is a series of increasing curves, the highest being that according to Tamada and Fujikawa, Eq. (38), Kuwabara, Eq. (40), and Hasimoto, Eq. (42). These curves coincide for $\beta < 0.05$. The curve according to Happel, Eq. (41), is lower, and the curve according to Spielman and Goren, Eq. (44), is the lowest.

There are a large number of empirical or semi-empirical equations for resistivity of fibrous aerosol filters. The well-known Kozeny-Carman equation for fibrous filters was expressed by Sullivan and Hertel [63] by

$$X = \frac{22\beta^2}{(1 - \beta)^3} \qquad (55)$$

However, as pointed out by Brinkman [68], this equation can be applied only for dense filters with solidity $\beta > 0.12$. Davies [96] suggested, on the basis of numerous experimental data, the empirical but useful equation

$$X = 16\beta^{3/2}(1 + 56\beta^3) \qquad (56)$$

Other empirical equations have been suggested by Sullivan [43], Silverman and First [100], Blasewitz and Judson [101], Kimura and Iinoya [102], Whitby et al. [103], and others as summarized by Chen [4] and Billings [97].

G. Diffusional Mechanism

Particle deposition on a single cylinder or in a system of cylinders due to particle diffusion is described by the capture coefficient E_D defined by Eq. (5) as being a function of the Péclet number $Pe = d_f U_o / D$. The Péclet number, or, more exactly, its reciprocal value $N_D = Pe^{-1}$, is a dimensionless parameter characterizing the intensity of diffusion deposition. The Péclet number can be expressed as $Pe = ReSc$, where $Sc = \nu/D$ is a Schmidt number and $\nu = \eta/S_g$ is the kinematic viscosity of the gas. As shown by Pich [19], under normal conditions ($t = 20°C$, $p = 760$ mmHg) the magnitude of Sc for aerosols ranges from $10 < Sc < 10^7$. In the case of laminar flow, $Re < Re_{cr}$; and because Sc is finite, then for deposition from laminar flow, $0 < Pe < Pe_{cr}$. The rate of deposition is finite for $U_o \to 0$ as diffusion deposition may take place in a medium at rest. Hence, from the definition of the capture coefficient, Eq. (5), it follows that for $U_o \to 0$, i.e., for $Pe \to 0$, then $E_D \to \infty$.

The case of the isolated cylinder will be discussed first. For the function $E_D = E_D(Pe)$ over the range $0 < Pe < Pe_{cr}$ three regions which have so far been investigated theoretically may be distinguished.

1. *Pe << 1, Re < 1*

For this case of deposition at small Pe values and viscous flow, Stechkina [104] derived the equation

$$E_D = \frac{2\pi}{Pe(1.502 - \ln Pe)} \tag{57}$$

from which it follows that, first, for $Pe \to 0$, $E_D \to \infty$ in agreement with the definition of E_D; second, to a first approximation the nondimensional flux of particles on a cylinder is independent of Re; and, third, the function $E_D = E_D(Pe)$ given by Eq. (57) is a decreasing one. In the range of validity of Eq. (57) and even for $Pe < 1$, $E_D > 1$.

2. *Pe >> 1, Re < 1*

This region of large Pe values, which usually occurs in fibrous filters, has been investigated a number of times. Langmuir [52] was perhaps the first to study this dependence in connection with filtration theory. His result can be expressed by

$$E_D = \frac{1}{2(2 - \ln Re)} \left[2(1 + x) \ln(1 + x) - (1 + x) + \frac{1}{1 + x}\right] \tag{58}$$

where $x = 1.308[(2 - \ln Re)^{1/3}/Pe^{1/3}]$. For large Pe, i.e., for $x << 1$, this equation reduces to

$$E_D = 1.71 \frac{1}{(2 - \ln Re)^{1/3}} Pe^{-2/3} \tag{59}$$

Johnstone and Roberts [105] and Ranz [106] applied the analogy between heat and mass transfer and suggested a relation given by

$$E_D = \frac{1}{Pe} + 1.727 \frac{Re^{1/6}}{Pc^{2/3}} \tag{60}$$

Calculations based on this equation are in good agreement with Eq. (58). Friedlander [107] and Natanson [108], using a modern concept of the diffusion boundary layer, derived functionally the same relation as Langmuir's reduced equation, Eq. (59), so that the results of all three authors can be summarized by

$$E_D = C \frac{1}{(2 - \ln Re)^{1/3}} Pe^{-2/3} \tag{61}$$

where C = 1.71 (Langmuir), C = 2.22 (Friedlander), and C = 2.92 (Natanson). A similar equation was derived by Torgeson [109], which is

$$E_D = 0.75 \left(\frac{4\pi}{2 - \ln Re}\right)^{0.4} Pe^{-0.6} \tag{62}$$

From a mathematical point of view the most precise equation in this region was derived by Stechkina [21], the equation being

$$E_D = \frac{2.9}{(2 - \ln Re)^{1/3}} Pe^{-2/3} + 0.624 Pe^{-1} \tag{63}$$

According to Stechkina this equation is valid for $Pe > 80(2 - \ln Re)$.

3. *Pe >> 1, Re >> 1*

This region of diffusive deposition from potential flow was investigated by Boussinesque [110], Lewis and Smith [111], Stairmand [112], and Natanson [108]. The results of all these studies can be summarized by

$$E_D = C \frac{1}{Pe^{1/2}} \tag{64}$$

where C = 2.83 (Stairmand), C = 3.19 (Boussinesque, Natanson), and $C = \pi/2$ (Lewis and Smith, if the sticking factor is taken as unity).

The first calculation of diffusion deposition in a system of parallel cylinders was reported by Fuchs and Stechkina [45]. Using the method by which Natanson's equation, Eq. (61), was derived and employing the Kuwabara-Happel velocity field, Eq. (30), these authors deduced the equation

$$E_D = \frac{2.9}{(-C - \frac{1}{2} \ln \beta)^{1/3}} Pe^{-2/3} \tag{65}$$

where again C = 0.75 (Kuwabara) and C = 0.5 (Happel). A refinement of this calculation as given by Stechkina [21] is

$$E_D = \frac{2.9}{(-C - \frac{1}{2} \ln \beta)^{1/3}} Pe^{-2/3} + 0.624\ Pe^{-1} \tag{66}$$

For Pe > 10 this equation reduces with sufficient accuracy to Eq. (65).

Spielman and Goren [25] adapted Natanson's solution, Eq. (61), to their velocity field, Eq. (31), and calculated the numerical coefficient in the resulting equation for different models of filter structure. Their result is

$$E_D = \mu \frac{2.9\ Pe^{-2/3}}{\xi^{1/3}} \tag{67}$$

where $\xi = K_0(X^{1/2})/X^{1/2}K_1(X^{1/2})$, X is the resistivity related to porosity by Eq. (53), $\mu = 1$ for the case where all fibers are normal to flow, $\mu = 0$ (all fibers parallel to flow), $\mu = 0.824$ (fiber axes oriented randomly in two-dimensional planes parallel to flow), and $\mu = 0.911$ (fiber axes oriented randomly in three dimensions. In the same manner Eq. (66) has been adapted, giving

$$E_D = \mu\left(\frac{2.9\ Pe^{-2/3}}{H^{*1/3}} + 0.62\ Pe^{-1}\right) \tag{68}$$

where the hydrodynamic factor H* is given by Eq. (33d).

Certain criticisms have been raised by Wilson and Cavanagh [113,114] that the theory of Brownian motion has been wrongly introduced into the theories of filtration and coagulation. This has been challenged by Binek [115], Thalhammer [116], Davies [117,118, 119], and Megaw [120] and replied to by Wilson and Cavanagh [121, 122,123,124,125].

Studies of "pure" diffusion deposition where the influence of other deposition mechanisms is excluded as far as possible are very limited, despite numerous experimental studies of particle deposition in fibrous aerosol filters (details will be discussed later). Concerning the diffusion deposition of particles on isolated cylinders, Dobry and Finn [126] concluded that their experimental data are in good agreement with Eq. (59) for $Pe \geq 5$.

Experiments on diffusion deposition in model filters have been reported by Kirsch and Fuchs [127]. The first model filter employed by these investigators consisted of a system of parallel, staggered cylinders placed normal to the flow, with cylinder diameters from $d_f = 0.043$ mm to 0.5 mm. For neutral, highly dispersed particles of NaCl (standard geometric deviation smaller than 1.25) and monodisperse dioctylsebacate particles in the size range $0.003 < \bar{d}_p < 0.018$ μm, the parameter of direct interception, $N_R = d_p/d_f$, becomes very small and the deposition can be treated as "pure" diffusion deposition (no electrical effects). Good agreement with Eq. (65) was found for $Pe \geq 2$. The second filter was a "fan" model, consisting of a system of parallel, equidistant rows of circular cylinders, the cylinders being parallel and equidistant in each row but oriented at random in different rows. The capture coefficient E_D^f (index f for "fan" model) was found to be independent of β in the range $0.01 < \beta < 0.15$ and described by the empirical relation $E_D^f = 2.7\ Pe^{-2/3}$.

Radushkevich and Kolganov [128] claimed to have found good agreement with Natanson's (C = 2.92), Eq. (61), in experiments with model filters having an orderly arrangement of fibers of $d_f = 4.3$ µm and 15 µm and using aerosols with d_p from 0.1 to 0.6 µm. However, the value of the numerical coefficient C used in Eq. (61) was C = 1.84, instead of Natanson's value C = 2.92.

Diffusion deposition in real filters, as tested by Sadoff and Almlof [129] using biological aerosols, leads to the conclusion that the capture coefficient E_D is related to filter solidity by $E_D \sim \beta^{-0.4}$. Hence compressing the filter (increasing β) leads to a decrease of E_D. On the other hand, Kirsch and Fuchs [127], measuring E_D for different commercial filters, concluded that compressing the filter does not change E_D. They attempted to solve the problem of diffusion deposition in real filters in the following manner, using the empirical equation

$$\frac{E_D^f}{E_D^r} = \frac{F^{*f}}{F^{*r}} = x \tag{69}$$

where the index r signifies "real." F^* is a dimensionless force acting on unit fiber length, which for the "fan" model is given by

$$F^{*f} = \frac{F^f}{\eta U_o} = 4\pi\left[-\frac{1}{2}\,\ln\frac{2\beta}{\pi} + \frac{2\beta}{\pi} - \frac{(\beta)^2}{\pi^2} - \frac{3}{4}\right]^{-1} \tag{70}$$

and F^{*r} is related to Δp by

$$F^{*r} = \frac{\pi \Delta p \overline{a^2}}{\eta U_o \beta L} \tag{71}$$

(see also Eq. 47). As the absence of uniformity of any kind in the filter structure leads to decrease of F^* from F^{*f}, these investigators interpret x as a measure of the degree of nonuniformity of the filter. Combining Eqs. (69), (70), and (71) with Eq. (176) (see subsequent section) and using their conclusion that $E_D^f = 2.7\ Pe^{-2/3}$, they obtained for the filter penetration P the equation

$$P = \exp\left[-1.37\left(\pi\eta U_o Pe^{2/3}\right)^{-1}\Delta P\bar{a}\left(-\frac{1}{2}\ln\frac{2\beta}{\pi}-\frac{3}{4}\right)\right] \tag{72}$$

H. Direct Interception

This mechanism represents the size of the depositing particles. A particle is intercepted as soon as the distance to the surface of the fiber is equal to its radius. This fact—deposition of particles of finite size—is included by a change of boundary conditions when the equation of convective diffusion or the equations of particle motion are being solved. If this is not possible (because of mathematical difficulties) the mechanism has to be studied independently.

The mechanism of direct interception is characterized by a dimensionless parameter $N_R = d_p/d_f$, and the capture coefficient due to direct interception E_R is a function of this parameter, i.e., $E_R = E_R(N_R)$ at otherwise constant conditions. The determination of this function is relatively easy in two extreme cases according to Rodebush [130] and Fuchs [13]. In the first case, Stk $\rightarrow \infty$, the inertia of particles is such as to make them travel directly toward the cylinder. Then $E_R = N_R$. In the second case, Stk $\rightarrow 0$, the particles have essentially no inertia and travel along the streamlines of the flowing fluid. As the stream function $\underline{\psi}(S,\theta)$ gives the area (volume per unit length of the cylinder) rate of flow between the critical streamline, $\underline{\psi} = \underline{\psi}_{cr}$, and the stagnation streamline, $\underline{\psi} = 0$, there is obtained

$$E_R = \frac{1}{aU_o}\underline{\psi}_{cr} = \frac{1}{aU_o}\underline{\psi}\left(a + r, \frac{\pi}{2}\right) \tag{73}$$

Apparently Eq. (73) was derived first by Natanson [131]. Using this equation and Eq. (29) for the velocity distribution in the potential flow gives

$$E_R = 1 + N_R - \frac{1}{1 + N_R} \tag{74}$$

in agreement with Fuchs [13] and Gillespie [132]. The latter equation can be approximated by $E_R \sim 2N_R$ for small values of N_R.

In the case of viscous flow, using Eq. (73) and Eq. (33) leads to

$$E_R = \frac{1}{2H^*}\left[(1 + N_R)^{-1} - (1 + N_R) + 2(1 + N_R)\,\ell n(1 + N_R)\right] \tag{75}$$

where H* is the hydrodynamic factor given by Eqs. (33a)-(33d). Equation (75) with $H^* = 2 - \ell n\,Re$ was derived by Langmuir [52] and rederived by Torgeson [109], but Torgeson's approximation of this equation for $N_R \ll 1$ is incorrect. Equation (75) with $H^* = H = 0.75 - \frac{1}{2}\,\ell n\,\beta$ was derived by Pich [133] and Stechkina et al. [134].

An approximation for the capture coefficient E_R can be obtained using Eqs. (73) and (34). The result is

$$E_R = \frac{1}{H^*} N_R^2 \tag{76}$$

in agreement with Spielman and Goren [25]. Equation (76) with $H^* = 2 - \ell n\,Re$ was derived originally by Natanson [131] and with H* given by Eq. (33b) by Pich [133]. The same result can also be obtained by expanding the function of Eq. (75) in an infinite series and retaining only the first term of the expansion (small N_R approximation).

From Eqs. (74), (75), and (76), it follows that for the potential field around an isolated cylinder $E_R = E_R(N_R)$, for viscous flow through a system of cylinders $E_R = E_R(N_R, \beta)$.

Experimental investigation of the direct interception mechanism has been reported by Radushkevich and Kolganov [128]. They indirectly measured the capture coefficient E_R in experiments with model filters having fiber diameters d_f of 4.3 and 15 μm using polystyrene particles with diameter d_p from 0.1 to 0.6 μm at a face velocity U_o = 15 cm/sec. They concluded that E_R increases linearly with the interception parameter N_R, the coefficient of proportionality between E_R and N_R being less than 2. This is in disagreement with Eq. (74) as well as with Eq. (76).

I. Inertial Mechanism

Particle deposition from a flowing fluid due to the inertia of the particles is of great importance and interest in science and technology. The mechanism is also the basis for many processes occurring in nature, as well as in man-made devices. Investigation of this process has therefore attracted much theoretical as well as experimental attention. As far as the author knows, there are five reviews dealing with the subject. Chronologically, the first is by Ranz [135]; it discusses the principles of inertial impaction, their engineering aspects, and applications related mainly to problems of gas cleaning. The second is by Mazin [22], treating the topic in connection with the physical principles of aircraft icing. The third is by Levin [24], discussing the meteorological aspects of particle deposition, including the determination of drop size distribution from inertial deposition on bodies of different geometrical shape. A further review was published by Golovin and Putnam [136], summarizing in part the open literature and various technical reports on the inertial impaction of small aerosol particles. The data presented here have possible practical application to the design of filters, heat exchangers, and combustion devices. The same applies to the fifth review, one published by Chamberlain [137].

An aerosol particle in a fluid stream tends to move in a straight line because of its inertia. Consequently, as the fluid flow curves around an obstacle (this discussion is limited to a cylinder or to cylinder systems), the particle begins moving relative to the fluid and may deposit on the obstacle. At constant conditions, there is a certain limiting trajectory separating the trajectories of particles that are captured from the trajectories of particles that miss the fiber and are not captured. The capture coefficient for particles of finite size due to their inertia E_{IR} can then be defined as the ratio of the number of captured particles to the number of particles that would be captured if the particles moved only in a linear direction. This quantity is sometimes called "target efficiency" and can be expressed by $E_{IR} = h/a$, where h is the distance of the limiting trajectory at infinity from the axis of the flow. The quantity h has sometimes been referred to as the "effective radius of the fiber." The bases for the calculation of E_{IR} are, of course, the equations of motion, Eq. (9), from which the limiting trajectories of the particles can be calculated, at least in principle.

An analysis of these equations of motion based on similarity theory leads to several parameters affecting the capture coefficient E_{IR}. The first is the Reynolds number of the cylinder Re, and the second is the parameter of interception N_R, which represents the finite size of particles. The third parameter is called the Stokes number or the inertia parameter (inertial impaction parameter) and is given by

$$\mathrm{Stk} = \frac{\ell_i}{d_f} = \frac{\tau U_o}{d_f} = \frac{1}{18}\,\frac{\mathcal{S}_p d_p^2 U_o}{\eta d_f} \tag{77}$$

where ℓ_i is the stopping distance of the particle and τ the relaxation time of the particle. The last parameter ϕ^* is given by

$$\phi^* = \frac{\mathrm{Re}_p^2}{\mathrm{Stk}} = 18\,\frac{\mathcal{S}_g}{\mathcal{S}_p}\,\mathrm{Re} \tag{78}$$

where $\mathrm{Re}_p = d_p U_o \mathcal{S}_g/\eta$ is the particle Reynolds number. The parameter ϕ^* accounts for conditions where velocities or particle sizes are so great that particle drag cannot be described by Stokes law. For particles obeying Stokes law (Stokes particles), $\phi^* = 0$. When rarification effects are important (e.g., particle impaction at reduced pressures) further parameters (the Knudsen number of the particle and Knudsen number of the cylinder) must be taken into account. Hence in the continuum region

$$E_{IR} = E_{IR}(\mathrm{Re},\ N_R,\ \mathrm{Stk},\ \phi^*) \tag{79}$$

In cases where particle diameter can be neglected in comparison with the cylinder diameter ($N_R \to 0$), a mathematical abstraction of the deposition of inertial point particles can be applied. In this case the coefficient E_{IR} reduces to the capture coefficient E_I describing "pure" inertial impaction, so that $E_I = E_I(Re, Stk, \phi^*)$. Furthermore, for Stokes particles, $E_I = E_I(Re, Stk)$; and finally, for potential flow, $E_I = E_I(Stk)$.

From a mathematical point of view, solution of the impaction problem consists of solving the particle motion equations in the given velocity field. These equations are so complex that, even in the case of Stokes particles, they are not easily integrable owing not only to their nonlinearity but also to the fact that some functions involved in them are solutions of the complicated Navier-Stokes equations describing the flow field. Two important extreme cases can be distinguished: the case of potential flow (high Re) and the case of creeping flow around a cylinder (low Re). Solution of the motion equations encounters serious difficulties even using these approximations, so analytical solutions have been obtained only in exceptional cases. Usually numerical step-by-step or computer calculations have been made. Assuming that the solution of the particle motion equations has been found, then knowledge of particle trajectories yields the following data: the theoretical capture coefficient E_{IR} or E_I, the local capture coefficient as a function of the angular coordinate describing the distribution of the impacted particles on the cylinder surface, angles of particle impingement, and the velocity at particle impact.

1. *Impaction on Isolated Cylinder in Potential Flow*

This is the first extreme case where the velocity field around a cylinder is approximated by Eq. (29). The problem has been treated theoretically by a large number of authors [50,51,138-149]. Chronologically, the development was initiated by Albrecht [50] and Sell [51], followed by Langmuir and Blodgett [140], and continued by a group at NACA [142,144] where inertial deposition from potential flow on bodies of different geometrical shape was extensively investigated (see also the references in Ref. 136) mainly in connection with the problem of aircraft icing. The same subject was pursued by Mazin [22,141]. Results are usually shown graphically by plotting E_I against Stk (see the comparison in Golovin and Putnam [136] and in Fuchs [13]). Ranz [135] concluded that, using $Stk^{1/2}$ as an independent variable, the function $E_I = E_I(Stk^{1/2})$ has in semi-logarithmic coordinates an "S" shape characteristic for different impaction systems.

In 1956 Robinson [146] suggested a concept of aerosol "liquid" that was employed by Yuryev [147] to derive the approximate but analytical expression

$$E_I = 1 - \frac{1}{Stk}\int_0^{\infty} \frac{e^{-x/Stk}}{1 + x^2}\, dx \qquad (80)$$

Values of E_I calculated from this equation agree very well with the numerical calculations of Mazin [141] in the range $0.5 < Stk < 100$.

Yoshioka et al. [148] calculated the capture coefficient E_{IR} of a single cylinder assuming potential flow, for $0.006 < N_R < 1$, and $Stk > 0.08$. They also calculated the local capture coefficient as a function of the angular coordinate θ. The theory was verified by experiments with copper, nichrome, and iron cylindrical collectors.

Subramanyam and Kuloor [149], assuming potential flow and point particles ($N_R = 0$), calculated the coefficient E_I for non-Stokes particles ($\phi^* \neq 0$) in the range $Stk < 20$. Their theoretical results are well represented by

$$E_I = \frac{100\ Stk}{Stk + 1.5} \qquad (81)$$

and agree closely with calculations of Langmuir and Blodgett [140] and Brun et al. [144]. Moreover, these authors calculated the local capture coefficient as a function of the angular coordinate and the angles of particle impingement.

2. *Impaction on Isolated Cylinder in Viscous Flow*

Calculation of the capture coefficient E_I due to "pure" inertial impaction or the coefficient E_{IR} in the case of viscous flow is more difficult as the corresponding velocity field is more complex than for potential flow. The problem has been treated by several authors [96,145,150-154].

Davies [96] established for Re = 0.2 the following equation:

$$E_{IR} = 0.16\left[N_R + (0.5 + 0.8\ N_R)Stk - 0.105\ N_R\ Stk^2\right] \tag{82}$$

Davies and Peetz [145] illuminated the effect of Re on the capture coefficient of a cylinder by comparing results for Re = 2000, 10, and 0.2. The capture coefficient is lowest in viscous flow for any value of Stk.

Torgeson [150], modifying Davies' approach [96], obtained

$$E_{IR} = E_R\left[1 + N_R^{-3/2}\ Stk(0.5 + 0.8\ N_R)\right] \tag{83}$$

where E_R is given by Torgeson's incorrect approximation of Eq. (75).

Yoshioka et al. [152] using Lamb's velocity field, Eq. (28), calculated numerically (Runge-Kutta method) the capture coefficient E_{IR} for Re = 0.2, 0.5, and 1.0 for N_R values $0.06 \le N_R \le 1$ and Stokes numbers $0 < Stk < \infty$. The calculated function $E_{IR} = E_{IR}(Stk, N_R, Re)$ for isolated cylinder was corrected by Chen's equation for interference effects (see p. 78).

In a second paper, Yoshioka et al. [153] extended these calculations to higher Reynolds numbers. Using the results obtained in a previous paper [152] for Re ≤ 1, the results of Davies and Peetz [145] for Re = 10, and the results of Yoshioka et al. [148] for the potential flow range (Re = 1000), the former investigators constructed interpolation charts enabling the determination of approximations of E_{IR} for the whole range of Reynolds numbers ($0.2 < Re < \infty$) as a function of Re, Stk, and N_R).

Householder and Goldschmidt [154] using Davies' approximation [76] of the velocity field around a cylinder calculated numerically (Runge-Kutta method) the function $E_{IR} = E_{IR}(Stk, N_R)$ for Re = 0.2

and N_R values of 1, 2, 3, 4, 5, 6, and 7. Using a potential flow approximation, Eq. (29), the function E_{IR} for the same values of N_R was also calculated. Furthermore, they attempted to extend the calculation of E_{IR} for all Reynolds numbers, constructing a relationship that would satisfy both extreme cases. They pointed out that, as the independent variables of the function E_{IR} are Re and N_R, the Stokes number is an undesirable parameter because it contains both Re and N_R (this was pointed out also by Yoshioka et al. [153]. For this reason, Householder and Goldschmidt expressed their data as $E_{IR} = E_{IR}(Re, N_R, S_p/S_g)$. It is, of course, questionable that the assumption that particles do not disturb the velocity field of the cylinder is still valid for $N_R = 7$, for example.

3. *Impaction in a Cylinder System*

Dawson [27] using the Spielman-Goren velocity field, Eq. (31), calculated numerically the function $E_{IR} = E_{IR}(Stk, N_R, \beta)$ for selected values of Stk in the range $0 < Stk < 20$, values of N_R being $0.01 < N_R < 0.2$, and packing density values $0.005 < \beta < 0.2$. E_{IR} was found to be an increasing function of all three parameters.

Harrop and Stenhouse [155] based their numerical calculations of $E_{IR} = E_{IR}(Stk, N_R, \beta)$ on Happel's velocity field, Eq. (30), and calculated E_{IR} as a function of Stk at $N_R = 0.05$ for β-values of 1%, 3%, 6%, and 11%. They concluded that E_{IR} is an increasing function of all three parameters and that, as the parameter N_R increases, the effect of β becomes less significant.

Stechkina et al. [134], assuming $Stk \ll 1$ and using Kuwabara's velocity field, Eq. (33b), derived for the function E_{IR} the equation

$$E_{IR} = \frac{1}{2H^{*2}}\left[(29.6 - 28\ \beta^{0.62})N_R - 27.5\ N_R^{2.8}\right]Stk \tag{84}$$

where H* is given by Eq. (33b). Equation (84) is valid for $0.01 \le N_R \le 0.4$ and packing density values of $0.0035 \le \beta \le 0.111$. From this equation it follows that at $N_R \to 0$, i.e., in the absence of interception, $E_{IR} \to 0$ as it should at small (subcritical) values of the Stokes number. However, these same investigators in a subsequent communication [156] rejected Eq. (84), stating it to be incorrect. They stated that at very small Stk values the corresponding capture coefficient is negative, i.e., particle inertia decreases the coefficient E_{IR} for particle deposition on a cylinder placed perpendicular to the flow. No arguments or calculations supporting this conclusion were given, however.

4. *Experiments and Empirical Correlations*

There are many reports [157-174] of experimental investigations of the inertial deposition of particles, the results usually being plotted graphically with measured E_{IR} or E_I (point particles) values as a function of Stk. Landahl and Hermann [157] expressed their results for Re = 10 by the empirical relationship

$$E_I = \frac{Stk^3}{Stk^3 + 0.77\ Stk^2 + 0.22} \tag{85}$$

Whitby et al. [171] and Whitby [175] concluded on the basis of experimental data that a plot of E_I against $Stk^{1/2}$ on log probability paper is linear. Thus experimental inertial impaction data can be characterized by the parameters $Stk_m^{1/2}$, which is the value of $Stk^{1/2}$ at $E_I = 50\%$, and the geometric standard deviation σ_g. For $Re < 0.2$ the median value of $Stk_m^{1/2}$ is given by $Stk_m^{1/2} = 1.4$, and for $0.2 < Re < 150$ by $Stk_m^{1/2} = 1.253\ Re^{-0.0685}$. In both cases $\sigma_g = 1.65$. However, there are inconsistencies in this paper.

May and Clifford [172] presented a calculation of the stopping distance ℓ_i for non-Stokes particles as required for the Stk calculation (see Eq. 77). Apart from experimental data, they presented interesting photographs of streamlined smoke-flow past cylinders, spheres, ribbons, and discs.

Starr [173] concluded, on the basis of experimental data, that E_I as a function of Stk lay between potential and viscous flow and that there were finite values of E_I for Stk less than the critical Stk. Friedlander and Pasceri [18], correlating experimental data of Wong and Johnstone [162], concluded that impaction does not become important unless the Stokes number is larger than about 0.5. They correlated these data by

$$E_I = 0.075\ Stk^{6/5} \tag{86}$$

for $0.8 < Stk < 2$, $Re < 1$, and $N_R < 0.2$.

Stenhouse et al. [174] using model filters verified experimentally the theory of inertial deposition in a cylinder system based on Happel's velocity field as developed by Harrop and Stenhouse [155]. They concluded that theory and experiment are compatible with respect to the packing density effect.

A special investigation of the distribution of deposited particles on cylinders was reported by Asset and Hutchins [176]. They defined the ratio N_L/N_W between the number of particles counted on the leeward (rear) side of the cylinder and the number counted over an equal area on the windward (front) side of the cylinder and studied how this quantity varied with the Stokes number. It was concluded that when $0.002 \leq Stk \leq 0.085$, the ratio N_L/N_W varied between 1.7 and 131%. Deposition mechanisms other than impaction were undoubtedly acting on the windward side. When $0.085 \leq Stk \leq 0.18$, the ratio N_L/N_W varied from 0.3 to 1.7%. Finally, in the range $0.18 \leq Stk \leq 85$, there was no deposition on the rear and $N_L/N_W = 0$.

Another special investigation was reported by Rosinski and Church [177], who studied the angle of approach of aerosol particles to solid surfaces during aerodynamic capture.

5. *Critical Values of the Stokes Number*

The critical value of the Stokes number Stk_{cr} is defined by

$$Stk < Stk_{cr} \qquad E_I = 0 \tag{87}$$

This means that particles must have a certain minimal inertia given by the value Stk_{cr} for deposition on an obstacle to occur. Consequently, the curve $E_I = E_I(Stk)$ does not intersect the origin but cuts the x-axis at the point Stk_{cr}.

Taylor [138] was one of the first who came to a conclusion

about the existence of Stk_{cr}. Taylor showed that if the velocity field around a critical (singular) point of a body can be expressed by $u_x = -k'x$ and $u_y = k'y$ (hyperbolic flow, then $Stk_{cr} = (1/8)k'$. Langmuir and Blodgett [140] concluded that for potential flow around a cylinder $Stk_{cr} = 1/16$. Levin [24,178,179] suggested a general analytical method for the determination of Stk_{cr} for potential flow around various bodies (cylinders, spheres, ellipsoids, etc.), and for a cylinder his value of Stk_{cr} is identical with that found by Langmuir and Blodgett [140].

In the case of viscous flow, by numerical integration of the motion equations for a particle (Lamb's velocity field was used), Natanson [180] derived for Re = 0.1 the value $Stk_{cr} = 2.15 \pm 0.05$.

Voloshchuk [181] using the concept of an aerosol "liquid" concluded that Stk_{cr} is located in the range

$$0.377(2 - \ell n \text{ Re}) \leq Stk_{cr} \leq 0.796(2 - \ell n \text{ Re}) \tag{88}$$

This estimate is in good agreement with Natanson's value mentioned above and with the location of a sharp rise of E_I in the curves of $E_I = E_I(Stk)$ as obtained by Yoshioka et al. [152].

In reality, usually there are other mechanisms of deposition involved, so that even for $Stk < Stk_{cr}$ the total capture coefficient $E_f \neq 0$.

J. Gravitational Mechanism

Particles can deposit from a flowing gas on collector surfaces under the finfluence of the gravity force. This deposition of particles due to gravitational forces is expected to be considerable for large particles and small flow velocities. The intensity of the mechanism is described by a dimensionless parameter N_G which, for Stokes particles, is given by

$$N_G = \frac{V_s}{U_o} = \frac{d_p^2 S_p g}{18\eta U_o} \tag{89}$$

where V_s is the terminal settling velocity of the particle.

The capture coefficient of a fiber E_G describing the rate of particle deposition due to gravitational forces is a function of this parameter, so, at constant conditions, $E_G = E_G(N_G)$. This function is dependent on the angle between the vector of the gravity force and the direction of gas flow.

Ranz and Wong [161] concluded that in the case of down-flow (horizontal cylinder transverse to the gas flow, the direction of flow and gravity coinciding) and point particles, E_G is given by

$$E_G = N_G \tag{90}$$

Chen [4], without giving specific arguments, stated that for randomly oriented cylinders the capture coefficient should be that calculated from Eq. (90) multiplied by the ratio of cross-sectional area projected in the vertical direction to that projected in the direction of flow.

Thomas and Yoder [182] found that the efficiencies of the same filter were difrerent in the case of down-flow and up-flow, and they tried to interpret these differences by the intervention of the gravitational mechanism.

Stechkina et al. [134], without giving details, stated that the capture coefficient E_G, to a first approximation, is equal to $(1 + N_R)N_G$ in the case of down-flow and to $-(1 + N_R)N_G$ in the case of up-flow. If the interception mechanism can be neglected ($N_R \to 0$), the first case reduces to Eq. (90). In the case of horizontal flow, E_G is positive and small (of the order of magnitude N_G^2). Hence, the theory of the gravitational mechanism is incomplete as yet.

K. Electrostatic Mechanisms

Aerosol particles and the fibers of a filter often carry electrostatic charges that may considerably influence particle deposition. The electrostatic charge on the fibers is, in the majority of cases, not stable, and it decreases with time mainly due to the conductivity of the fibers, passage of ionized gases, X or radioactive radiation, deposition of charged particles, and humidity. A charge on fibers and particles may influence the filtration process by altering particle trajectories and by altering particle adherence to fiber surfaces.

As deposition due to electrostatic forces is of importance not only in aerosol filtration but also in other processes (deposition of particles on high-voltage lines, electrostatic effects in thermal coagulation, etc.), a large number of papers have been addressed to this problem. A part of this general problem—electrostatis deposition from a gas stream on bodies of cylindrical shape—has been reviewed by Whitby and Liu [183] and Voorhoeve [184].

In order to describe this process, a knowledge of how electrostatic forces act between particles and fibers is indispensable. The order of magnitude of different forces acting on particles in electrical or magnetic fields has been discussed by Zebel [185]. The following cases can be distinguished in the electrostatic interaction between particle and fiber.

1. *Charged Particle, Charged Fiber*

The Coulombic attraction between a charged particle and an oppositely charged cylinder is described, according to Gillespie [132] and Natanson [186], by

$$F(S) = \frac{2Qq}{S} \tag{91}$$

where q is the particle charge, Q the charge per unit length of the fiber (sometimes the charge density μ_e is used, related to Q by $Q = \pi a^2 \mu_e$) and S the distance between particle and fiber.

2. *Charged Fiber, Neutral Particles*

The interaction between an electrically charged cylinder and its image in the aerosol particle is described [186] by

$$F(S) = 4Q^2 \frac{D_1 - 1}{D_1 + 2} \frac{r^3}{S^3} \tag{92}$$

where D_1 is the dielectric constant of the particle.

3. *Neutral Fiber, Charged Particles*

The interaction between a charge on an aerosol and its image in the fiber is given by Natanson [186] and Gillespie [132] as

$$F(S) \; \frac{q^2}{4(S - a)^2} \frac{D_2 - 1}{D_2 + 1} \tag{93}$$

where D_2 is the dielectric constant of the fiber. There are two other types of interaction described by Ranz and Wong [161]. The first consists of space-charge repulsion of the particle being collected by the unipolar-charged aerosol of which it is a part, and the second consists in the interaction between a charged particle and a collector having a charge induced by the unipolar-charged aerosol surrounding the collector. These two types of interaction are usually negligible, as stated by Whitby and Liu [183].

4. *Dimensionless Parameters*

As with mechanical filtration mechanisms (diffusion, interception, inertial deposition), it is advantageous to describe each electrostatic mechanism by a dimensionless parameter describing the intensity of the given mechanism. There are two approaches in deriving these parameters. The first, due to Ranz and Wong [161], interprets the parameters in terms of ratios of electrostatic forces to drag forces. The second, due to Pich [17], consists of making the motion equations of the particle dimensionless. Both approaches lead to identical results.

The dimensionless parameter describing deposition due to Coulombic forces N_{Qq} is

$$N_{Qq} = \frac{4QqB}{d_f U_o} = \frac{4Qq}{3\pi\eta d_p d_f U_o} \tag{94}$$

where B is the mechanical mobility of the particle, and U_o is the velocity of the flow far from the cylinder. The dimensionless parameter N_{Qo} describing deposition due to the image forces in a system consisting of charged fibers and neutral particles is given by

$$N_{Qo} = \frac{4}{3\pi} \frac{D_1 - 1}{D_1 + 2} \frac{d_p^2 Q^2}{d_f^3 U_o \eta} \tag{95}$$

Finally, the dimensionless parameter N_{0q} describing deposition due to the image forces in a system composed of neutral fibers and charged particles is

$$N_{0q} = \frac{q^2}{3\pi\eta d_p d_f^2 U_o} \frac{D_2 - 1}{D_2 + 1} \tag{96}$$

The right-hand expression of Eq. (94) and Eqs. (95) and (96) are valid only for Stokes particles.

5. *External Electric Field Absent*

The results obtained by different authors can be expressed by means of individual dimensionless parameters. Two cases should be distinguished, according to Natanson [186]. In the first, the limiting trajectories of particles are tangential to the cylinder surface, and deposition occurs chiefly on the windward side of the cylinder. In the second, limiting particle trajectories sweep past the cylinder and then curve back to the rear stagnation point; deposition occurs on the whole cylindrical surface. Only the second case, which is mathematically simpler, will be considered.

a. Coulombic Deposition

Denoting by E_{Qq} the capture coefficient due to Coulomb forces, there is obtained

$$E_{Qq} = \pi N_{Qq} = \frac{4Qq}{3\eta d_p d_f U_o} \tag{97}$$

Equation (97) was derived by Kraemer and Johnstone [187], Natanson [186], and Torgeson [188]. Levin [24,189], using the method of

Duchin-Deryaguin [29] (Eq. 11), derived a general equation describing the deposition of charged particles on charged bodies of arbitrary shape. This relationship is

$$E_{Qq} = \frac{4\pi qQB}{U_o S_M} \tag{98}$$

where S_M is the projection of the collector onto a plane perpendicular to the flow. Equation (98) reduces to Eq. (97) for Stokes particles and cylindrical collectors ($S_M = d_f$).

b. Image Forces, Charged Fiber and Neutral Particle System

Denoting by E_{Qo} the corresponding capture coefficient, Natanson's result [186] for $E_{Qo} >> 1$ is given by

$$E_{Qo} = \left(\frac{D_1 - 1}{D_1 + 2} \frac{2Q^2 d_p^2}{d_f^3 \eta U_o} \right)^{1/3} \tag{99}$$

Introducing the stationary sedimentation velocity of the particles V_s into Eq. (99) instead of U_o gives the Cochet's equation [190] which describes the deposition of sedimenting particles on a horizontal, charged cylinder. Using the parameter N_{Qo}, Eq. (99) becomes

$$E_{Qo} = \left(\frac{3\pi}{2}\right)^{1/3} N_{Qo}^{1/3} \tag{100}$$

and for $E_{Qo} << 1$

$$E_{Qo} = \pi N_{Qo} \tag{101}$$

In these cases the deposition is independent of the velocity field.

c. Image Forces, Charged Particles and Neutral Fiber System

Denoting the corresponding capture coefficient by E_{0q}, the equation derived by Natanson [186] for potential flow, assuming $E_{0q} << 1$, can be expressed by

$$E_{0q} = (6\pi)^{1/3} N_{0q}^{1/3} \tag{102}$$

and for viscous flow by

$$E_{0q} = \frac{2}{(2 - \ell n \ Re)^{1/2}} N_{0q}^{1/2} \tag{103}$$

It should be noted that there are some errors in the tabulation of these equations as presented by Whitby and Liu [183]. Approximate relations describing electrostatic deposition have also been derived by Gillespie [132]. Dawkins [191] reported a computer solution for a cylinder in a potential flow field.

Radushkevich [192], using a phenomenological approach, assumed that the cylindrical charge Q decreases exponentially with time, and he attempted to find how the efficiency of an entire filter decreased. However, some of his assumptions (the additivity of electrical and mechanical effects and, especially, the proportionality between Q and the deposition rate for any type of electrical interaction) have no justification and are in disagreement with Eqs. (99) and (101).

The electrostatic mechanism has been studied experimentally by a number of authors [193-199]. Goyer et al. [193] investigated the filtration of a charged (25 to 150 electronic charges) dioctylphthalate aerosol ($r = 0.25$ to 0.55 μm). An appreciable reduction in penetration was observed. The effect of charges on filter penetration was greater the smaller the particle. Rossano and Silverman [194] found an increase in efficiency due to particle and fiber charge, but the data do not show a dependence of filter efficiency on flow velocity. Silverman et al. [195] used the motion of a fabric over a fibrous filter to produce a continuously charged filter. Filter efficiency was found to increase inversely with velocity and directly with fiber charge. Gillespie [132] also found a considerable decrease in penetration due to electrostatic effects. Lundgren and Whitby [196,197] found qualitative agreement with Natanson's equation, Eq. (103), the theoretical values being low by about 40%. The empirical equation $E_{0q} = 1.5 \ N_{0q}^{1/2}$ for viscous flow around a cylinder was proposed to fit the experimental data. These investigators concluded that at certain experimental conditions particle charging caused a tenfold increase in the capture coefficient of a fiber leading to a one-hundred-fold decrease in filter penetration. Ogorodnikov and Basmanov [198] observed a high initial efficiency of FPP (perchlorvinyl fibers) fibrous filters for submicron aerosols caused by electric charges of the filters. For example, for particles with $d_p = 0.34$ μm at velocity $U_o = 1$ cm/sec the filter penetration was $P = 10^{-4}\%$, and after irradiating the filter, which reduced the charges, the penetration increased 10^4 times.

6. *Fibrous Filters with External Electric Field*

Thomas and Woodfin [200] and Havlicek [201] concluded that placing a fibrous filter between two charged electrodes resulted in markedly improved efficiency for conventional fiber filters. Rivers [202] compared experimental data on a commercial filter with theoretical calculations taking into account polarization and Coulombic forces. These data showed a moderate increase in filter efficiency when a strong electric field was imposed across the filter.

Recently, Makino and Iinoya [203] investigated the efficiency of a dielectric fibrous filter (glass fibers) for a stearic acid aerosol using an external electric field. The effect of the orientation of the fibers (fibers placed randomly and perpendicularly to the gas flow) was investigated, and a semi-empirical equation for the efficiency was established. The efficiency of the electric fibrous filter was found to be higher than that of the filter without an electric field. In another paper, Makino and Iinoya [204] investigated the capture coefficient of a single electrically conductive fiber. The capture coefficient was found to be about 30 times greater than that of the fiber without an external electric field.

Instead of placing a conventional fiber filter between electrodes, Walkenhorst and Zebel [205] employed a very open arrangement of 30 to 100 layers of regularly spaced fibers, having imposed on them a strong external electric field E_o. As the interfiber distance 2b was large, it was possible to apply the theory of particle deposition on a single fiber to a loosely packed filter of this type. An approximate theory for such filters was developed by Zebel [206,

207]. According to Zebel, the total electrical field E around a fiber, caused by a homogeneous external electrical field E_o and by a polarization of a dielectric fiber, is given by

$$E_S = \left(\frac{D_2 - 1}{D_2 + 1} \frac{a^2}{S^2} + 1\right) E_o \cos\theta \qquad (104)$$

$$E_\theta = \left(\frac{D_2 - 1}{D_2 + 1} \frac{a^2}{S^2} - 1\right) E_o \sin\theta$$

where E_S and E_θ are the field components in polar coordinates. If $D_2 \to \infty$ then $\mu = (D_2 - 1)/(D_2 + 1) = 1$ and the fiber is metallic. In this case the tangential component E_θ disappears on the cylinder surface, and the lines of a vector E_o meet the cylinder surface normally.

The components of force F_S and E_θ acting on a charged particle in the field, Eq. (104), are $F_S = qE_S$ and $F_\theta = qE_\theta$. Zebel solved the corresponding motion equation for point, charged particles both for potential and viscous flow, assuming that inertial and image forces could be neglected. In this way the limiting trajectories of particles were calculated and the deposition (capture) coefficient determined. Denoting the corresponding capture coefficient by $E_{E_o q}$, Zebel's results can be summarized in terms of a dimensionless parameter characterizing the particle deposition given by

$$N_{E_o q} = \frac{\mu G - 1}{G + 1} \qquad (105)$$

where $G = E_o qB/U_o$ and $E_{E_o q} = f(N_{E_o q})$.

If the vector E_o has the same direction as the vector U_o, this dependence for potential flow is given by

$$E_{E_o q} = 0 \qquad \text{for } N_{E_{E_o q}} < -1 \qquad (106a)$$

$$E_{E_o q} = 1 + N_{E_o q} \qquad \text{for } -1 \le N_{E_o q} \le 1 \qquad (106b)$$

and

$$E_{E_oq} = 2N_{E_oq}^{1/2} \qquad \text{for } N_{E_oq} \geq 1 \tag{106c}$$

Equation (106b) is for $G = E_oqB/U_o > 0$, valid also for viscous flow. For neutral, polarized particles depositing on a cylinder having an impressed external electric field, the solution of particle motion equations would require the application of numerical methods, as pointed out by Zebel.

Having calculated the capture coefficient E_{E_oq}, it is possible to calculate the penetration P of the whole fiber system by means of the equation

$$P = \left(1 - \frac{a}{b} E_{E_oq}\right)^m \tag{107}$$

where m is the number of fiber layers.

Walkenhorst and Zebel [205] tested filters having m = 20 to 100 layers, fiber diameters $2a = d_f = 50$ μm, field intensity E_o = 6,000 V/cm, and flow velocity U_o = 10 cm/sec. The results seem to confirm Eq. (107).

The theory has been further refined with the mechanism of diffusion deposition included by Zebel [208]. Experiments on the motion of charged particles around a cylinder in an electrical field have also been reported by Hochrainer et al. [209]. Situations were investigated where an uncharged and a charged cylinder were placed in an electrical field, the direction of which was parallel to the gas velocity U_o. The experimental results were found to be in satisfactory agreement with Zebel's theory except for particle velocities attained during approach to the forward stagnation point of the cylinder. Hochrainer [210], using microcinematography, also found satisfactory agreement between theoretical and experimental particle trajectories. Filtration by this type of filters has recently been reviewed from the practical point of view by Walkenhorst [211,212].

L. Deposition Due to London-van der Waals Forces

Molecular interaction between particles and fibers may play a role in the filtration process, according to Tunitskii and Petryanov [213]. Molecular forces may have double significance in filtration as do electrostatic forces. They may influence the process of particle deposition and may also be of importance in keeping (sticking) the particle on the fiber. The molecular component of the adhesion force acting between a deposited particle and a fiber will be discussed in a later section dealing with nonstationary filtration.

The force resulting from the London (London-van der Waals) attraction between a plane and a sphere has been calculated by Hamaker [214], neglecting the retardation effect. The result is

$$F = \frac{2}{3} \frac{Q_o r^3}{(S'^2 - r^2)^2} \tag{108}$$

where S' is the distance between the particle center and a plane, and Q_o the Hamaker's constant of interaction, being typically of the order 10^{-13} to 10^{-12} erg. Assuming that the cylinder can be treated as a plane wall (this is a good approximation if the particles are small compared to the cylinder, i.e., $N_R << 1$), Eq. (108) becomes

$$F = \frac{2}{3} \frac{Q_o r^3}{[(S - a)^2 - r^2]^2} \tag{109}$$

where S is the radial coordinate from the particle center. Far from the cylinder ($S - a >> 2r$), Eq. (109) becomes

$$F \approx \frac{2}{3} \frac{Q_o r^3}{(S - a)^4} \tag{110}$$

and near the cylinder ($S - a << 2r$) Eq. (109) can be approximated by

$$F \approx \frac{Q_o r}{6(S - a - r)^2} \tag{111}$$

Natanson [186] used Eq. (110) to describe the deposition due to the London forces. A dimensionless parameter N_M characterizing the intensity of particle deposition due to the London forces can be derived in the same way as the dimensionless electrostatic parameters. The result is

$$N_M = \frac{2}{3}\frac{Q_o r^3 B}{a^4 U_o} = \frac{Q_o r^2}{9\pi a^4 U_o \eta} \tag{112}$$

where B is again the mobility of the particle. The right-hand term in Eq. (112) is valid, of course, only for Stokes particles ($B = 1/6\pi\eta r$). The capture coefficient of the fiber due to the London force E_M is a function of this dimensionless complex N_M, e.g., at constant conditions $E_M = E_M(N_M)$. Using this concept based on the theory of similarity, Natanson's result can be expressed by

$$E_M = (30\pi)^{1/5} N_M^{1/5} \tag{113}$$

for potential flow, Eq. (29), and

$$E_M = \left(\frac{3\pi}{2}\right)^{1/3} \frac{1}{(2 - \ln Re)^{2/3}} N_M^{1/3} \tag{114}$$

for viscous flow (Lamb's velocity field, Eq. (28), was used by Natanson).

Spielman and Goren [215], using the velocity field given by Eq. (34) with the hydrodynamic factor H* given by Eq. (33d), concluded that E_M could be expressed by

$$E_M = \frac{1}{H^*} N_R^2 f(N_A) \tag{115}$$

where $f(N_A)$ is an undetermined function of the dimensionless complex N_A called by Spielman and Goren the adhesion number and given by

$$N_A = \frac{2H^* a^2 Q_o}{9\pi\eta U_o r^4} = \frac{2HN_M}{N_R^6} \tag{116}$$

Equation (116) suggests a possible method for correlating experimental data. For large values of N_A, Eq. (116) becomes

$$E_M = \left(\frac{3\pi}{4}\right)^{1/3} \frac{1}{H^*} N_R^2 N_A^{1/3} = \left(\frac{3\pi}{2}\right)^{1/3} \frac{1}{H^{*2/3}} N_M^{1/3} \tag{117}$$

For Lamb's velocity field ($H = 2 - \ln Re$) this equation exactly agrees with Natanson's expression, Eq. (114).

For small values of N_A it is expected that

$$E_M = c \frac{1}{H^*} N_R^2 N_A^{0.059} \tag{118}$$

where c is an undetermined numerical constant.

The function $f(N_A)$ appearing in Eq. (115) is expected to be an increasing function of its argument, coming to the form of Eq. (117) for very large values of N_A and to the form of Eq. (118) for very small N_A values.

M. Filtration in Noncontinuum Regions

The theory of filtration dealt with thus far is based on the assumption that the gas can be treated as a continuum. As already stated this assumption is fulfilled if, according to Devienne [49], the Knudsen number Kn is less than 10^{-3}. For greater values of $Kn = \lambda/a = 2\lambda/d_f$ there are other regimes of gas flow as classified by Devienne [49].

The region $10^{-3} < Kn < 0.25$ covers the flow of partly diluted gases. Two methods can be applied for calculation of the velocity field in this case. The first is based on application of the equations of motion of diluted gases (Burnett's equations) with unchanged boundary conditions (zero velocity at the surface of a body). The second uses the equations of continuum mechanics (Navier-Stokes equations) with changed boundary conditions. The change of boundary conditions corresponds to a slip of the gas at the surface of the obstacle, i.e., there is a velocity discontinuity. The theory of this discontinuity has been developed by Maxwell [216], Epstein [217], and Bakanov and Deryaguin [218] (see also the book by Chapman and Cowling [219]).

The region $0.25 < Kn < 10$ is denoted as a transition region. The case where Kn is of the order of unity is the most difficult, and as yet there is no general theory for this type of flow.

The region $Kn > 10$ is termed one of free molecule flow. In this case the number of collisions of the gas molecules with the surface of the body is large in comparison with the number of collisions between molecules. The properties of this type of flow have been described by Patterson [220] and partly by Devienne [49].

1. *Small Knudsen Numbers*

For calculating the pressure drop and filter efficiency in this case the equations of classical hydrodynamics are usually applied (no application of Burnett's equations to filter theory is known to the author), the boundary conditions being, however, more generally expressed so that the slip of the gas at the fiber surface may be taken into account. It is obvious that the relations obtained in this way are more general and that from these relations follow the equations of filter theory in the continuum region as a special case when Kn values become negligibly small.

The slip effect in aerosol filtration theory has been approximately accounted for by Langmuir [52] and by Dorman [221]. However, sound theoretical studies of filtration at small Kn based on the mechanics of rarefied gases have come later. As in the case of filtration theory in the continuum region, filtration theory in the range of small Kn numbers is based on the velocity distribution around the fiber or in filters.

Natanson [131] calculated rigorously the velocity around an isolated, infinitely long cylinder placed perpendicular to gas flow, taking into account the gas slipping at the cylinder surface. The result is

$$\underline{\psi}(S,\theta) = \frac{aU_0 \sin\theta \left[\frac{a}{S} - \frac{S}{a} + 2\left(1 + 2\frac{\xi}{a}\right)\frac{S}{a}\ \ln \frac{S}{a}\right]}{2(2 - \ln \mathrm{Re}) + 4\frac{\xi}{a}(2.5 - \ln \mathrm{Re})} \tag{119}$$

where ξ is the coefficient of slip.

Pich [17,133], generalizing Kuwabara's boundary conditions in order to include the slip effect, derived an equation describing the velocity distribution in a system of parallel cylinders at small Kn. The result is

$$\underline{\psi}(S,\theta) = \frac{aU_o \sin\theta}{2K}\left[\left(1 - \frac{\beta}{2} + \beta\frac{\xi}{a}\right)\frac{a}{S} - \left(1 - \beta\right)\frac{S}{a} + 2\left(1 + 2\frac{\xi}{a}\right)\frac{S}{a}\ln\frac{S}{a} - \frac{\beta}{2}\left(1 + 2\frac{\xi}{a}\right)\frac{S^3}{a^3}\right] \tag{120}$$

where

$$K = -\frac{3}{4} - \frac{1}{2}\ln\beta + \beta - \frac{1}{4}\beta^2 + \frac{\xi}{a}\left(-\frac{1}{2} - \ln\beta + \frac{1}{2}\beta^2\right) \tag{120a}$$

For a high porosity system ($\beta << 1$), near the vicinity of the fiber [$(S - a)/a << 1$] and at small Kn ($\xi/a << 1$), Eq. (120) can be simplified to

$$\underline{\psi}(S,\theta) = \frac{aU_o \sin\theta\left[\frac{a}{S} - \frac{S}{a} + 2\left(1 + 2\frac{\xi}{a}\right)\frac{S}{a}\ln\frac{S}{a}\right]}{2\left(-\frac{3}{4} - \frac{1}{2}\ln\beta\right) + 2\frac{\xi}{a}\left(-\frac{1}{2} - \ln\beta\right)} \tag{121}$$

Introducing again the hydrodynamic factor H, both fields, Eqs. (119) and (121), can be summarized by

$$\underline{\psi}(S,\theta) = \frac{aU_o \sin\theta\left[\frac{a}{S} - \frac{S}{a} + 2\left(1 + 2\frac{\xi}{a}\right)\frac{S}{a}\ln\frac{S}{a}\right]}{2H + 4\frac{\xi}{a}\left(H + \frac{1}{2}\right)} \tag{122}$$

where

$$H = 2 - \ln \mathrm{Re} \quad \text{(Natanson)} \tag{122a}$$

$$H = (-3/4) - (1/2)\ln\beta \quad \text{(Pich)} \tag{122b}$$

The kinetic theory of the slip coefficient ξ was developed by Maxwell [216] and Epstein [217]. Their result is

$$\xi = 0.998\,\frac{2 - f}{f}\,\lambda \tag{123}$$

where f denotes the fraction of gas molecules undergoing diffuse reflection and (1 - f) the fraction undergoing specular reflection. For f = 1 Eq. (123) becomes $\xi = 0.998\lambda$. Bakanov and Deryaguin [218] derived the same equation but with a different numerical coefficient. Their result is $\xi = 1.0878\lambda$. Assuming that diffuse reflection occurs (f = 1) and using Eq. (123), Eq. (122) becomes

$$\underline{\psi}(S,\theta) = \frac{aU_o \sin\theta \left[\frac{a}{S} - \frac{S}{a} + 2\left(1 + 1.9996\mathrm{Kn}\right)\frac{S}{a}\,\ell n\,\frac{S}{a}\right]}{2H + 3.992\mathrm{Kn}\left(H + \frac{1}{2}\right)} \tag{124}$$

Obviously, if the slip effect can be neglected (Kn → 0), the Natanson-Pich velocity field, Eq. (124), reduces to the Lamb-Kuwabara velocity field, Eqs. (33a) and (33b). A useful approximation of the field of Eq. (124) for very near the cylinder can be obtained by expanding the function $\underline{\psi}(S,\theta)$ into series in powers of $(S - a)/a$. The result is

$$\underline{\psi}(S,\theta) = \frac{aU_o \frac{S - a}{a}\left(\frac{S - a}{a} + 1.996\mathrm{Kn}\right)\sin\theta}{H + 1.996\mathrm{Kn}\left(H + \frac{1}{2}\right)} \tag{125}$$

Again, for Kn → 0 this expression reduces to Eq. (34) in the continuum region.

Other equations, describing the velocity distribution in fibrous filters of different geometries (including the case of transverse flow) in the range of small Kn, were derived by Spielman and Goren [25]. However, these expressions are mathematically more complex and the implicit occurrence of the filter resistivity X as a function of β and Kn makes the application of this theory rather difficult. Concerning the theory of drag forces in the small Kn range, Natanson [131] derived for the force F acting on unit length of the cylinder (transverse flow) the equation

$$F = \frac{4\pi\eta U_o\left(1 + 2\,\frac{\xi}{a}\right)}{2 - \ell n\,\mathrm{Re} + 2\,\frac{\xi}{a}(2.5 - \ell n\,\mathrm{Re})} \tag{126}$$

Pich and Spurny [222] using another approach also derived Eq. (126).

Using Eq. (123) and assuming f = 1, Eq. (126) becomes

$$F = \frac{4\pi\eta U_o(1 + 1.996Kn)}{2 - \ln Re + 1.996Kn(2.5 - \ln Re)} \quad (127)$$

Pich [17], treating the same problem for a system of parallel cylinders, based the calculations on the velocity field of Eq. (120). This resulted in

$$F = \frac{4\pi\eta U_o(1 + 1.996Kn)}{-\frac{3}{4} - \frac{1}{2}\ln\beta + \beta - \frac{1}{4}\beta^2 + 0.998Kn\left(-\frac{1}{2} - \ln\beta + \frac{1}{2}\beta^2\right)} \quad (128)$$

Again, for $Kn \rightarrow 0$, Eq. (127) reduces to the Lamb-Davies equation, Eq. (37), and Eq. (128) reduces to Kuwabara's equation, Eq. (40). Hence, the dimensionless drag $F^* = F/\eta U_o$ at small Kn is $F^* = F^*(Re, Kn)$ for an isolated cylinder and $F^* = F^*(\beta, Kn)$ or $F^* = F^*(Fu, Kn)$ for a cylinder system.

Reported investigations of the pressure drop of fibrous aerosol filters in the range of small Kn are limited. Pich [223], basing calculations on the velocity field described by Eq. (120) and using the drag theory of permeability, derived the equation

$$\Delta p = \frac{4\pi\beta U_o L(1 + 1.996Kn)}{a^2\left[-\frac{3}{4} - \frac{1}{2}\ln\beta + \beta - \frac{1}{4}\beta^2 + 0.998Kn\left(-\frac{1}{2} - \ln\beta + \frac{1}{2}\beta^2\right)\right]} \quad (129)$$

Since at constant temperature $\lambda = const/p$, it follows from this equation that, with decreasing gas pressure p, the pressure drop decreases. This conclusion is in agreement with experiments of Stern et al. [99], Wheat [224], and Werner and Clarenburg [31]. Moreover, the reduced form of Eq. (129) ($\beta << 1$, $Kn << 1$), is functionally the same as an empirical equation suggested by Stern et al. [99]. Spielman and Goren [25] came theoretically to the same conclusion (decreasing Δp with p), basing their theory on the Brinkman model. Quantitatively, they concluded that, e.g., for $Kn \sim 0.2$, the pressure drop is reduced to approximately 0.7 of the nonslip value. These authors also calculated the curves of filter resistivity X as a function of filter solidity β, with Kn as a parameter. However, the construction of these curves for Kn $\sim$ 1 and Kn $\to \infty$ is misleading, because no predictions can be made in the framework of small Kn theory for these Kn values. For example, there is no obvious reason for these curves being equidistant at different values of Kn.

The pressure drop Δp at small Kn has also been experimentally investigated by Wheat [224] and Werner and Clarenburg [31]. These latter investigators concluded that Δp decreases with decreasing p. In order to describe this dependence analytically, Wheat [224] and Werner and Clarenburg [31] suggested empirical formulas, the construction of which suffers by the application of the Knudsen-Weber (Millikan) correction. The application of this correction to fibers approximated geometrically by cylinders is theoretically unjustified. The quantitative aspect of the Δp reduction with decreasing gas pressure will be discussed later in connection with the pressure characteristics of fibrous filters. Theoretical studies of filtration mechanisms in the range of small Kn values have been limited so far to investigations of the diffusion and direct interception mechanisms.

a. Diffusion Mechanism

Pich [225], using the velocity field defined by Eq. (125) and the concept of a diffusion boundary layer (Pe >> 1) and supposing diffuse reflection of the gas molecules from the fiber surface (f = 1), derived an equation which can be expressed by

$$E_D = \frac{2.86}{Y^{1/3}\,Pe^{2/3}}\left(1 + \frac{0.388\text{Kn}\,Pe^{1/3}}{Y^{1/3}}\right) \tag{130}$$

where Y = H + 1.996Kn(H + 1/2). The upper limit of the slip region is given by Kn = 0.25 and λ = const/p, and at pressure p = 760 mmHg, $\lambda = 0.653 \times 10^{-5}$ cm; then, from the condition Kn < 0.25, it follows that Eq. (130) can be strictly applied for pressures p(mmHg) >

397.02/d_f(μm). This applies approximately for $d_f \geq 0.5$ μm at normal pressure. Analyzing Eq. (130), it can be shown that $dE_D/d\lambda > 0$. Hence, the capture coefficient due to diffusion deposition E_D increases with decreasing gas pressure. This theoretical conclusion is in agreement with experiments of Stern et al. [99]. These experimenters measured the efficiency of filters with $d_f = 17$ μm, $\beta = 13\%$, and L = 0.084 cm at pressures of 760, 340, 102, 38, and 17 mmHg using particles in the diameter range 0.026 to 1.7 μm at varying velocities. Calculation of the filter efficiency for these conditions ($d_p = 0.138$ μm, $U_o = 10.16$ cm/sec) using Eq. (130) (no correction for the interference effect is required) and comparing the theoretical results with the experimental data of Stern et al. leads to the following conclusions: (1) The theory, in agreement with experiment, shows that the efficiency of diffusion filtration increases with decreasing gas pressure. (2) At normal pressure for a filter with $d_f = 17$ μm, the influence of gas slip is negligible and Eq. (130) gives practically the same values as the Fuchs-Stechkina relationship, Eq. (65). In fact, for Kn → 0, Eq. (130) practically reduces to the Fuchs-Stechkina equation or that of Natanson, Eq. (61). (3) At lower pressures, where a more substantial influence of gas slip can be expected, Eq. (130) yields higher values of efficiency than the corresponding equation in the continuum region, Eq. (65). Hence, diffusion increases with the gas slip. (4) Theory, at the present stage of development, shows that the dependence of efficiency on the gas pressure is stronger than that found by Stern et al. Decreasing the pressure to 340.4 mmHg gave practically no efficiency increase according to Stern et al., but theory gives an increase of about 5%. There are some obscurities in the Stern et al. paper, but, on the other hand, the theory is also incomplete as the concentration jump at the fiber surface is not taken into account. Assuming the experiments to be correct, a possible theoretical explanation is that the concentration discontinuity on the fiber surface is partly compensated for by the velocity discontinuity (gas (slipping).

b. Direct Interception

The first investigation of the mechanism of direct interception at small Kn numbers was reported by Natanson [131]. The result for an isolated cylinder is given by

$$E_R = \frac{N_R\left(N_R + 2\,\frac{\xi}{a}\right)}{2 - \ln \text{Re} + 2\,\frac{\xi}{a}(2.5 - \ln \text{Re})} \tag{131}$$

Pich [133] developed a theory of direct interception at small Kn for a system of parallel cylinders. In principle, any form of the velocity field, Eqs. (119), (120), (121), (122), (124), or (125), can be used as a basis for a calculation. Using, e.g., Eq. (121) the result is

$$E_R = \frac{\left(1 + N_R\right)^{-1} - \left(1 + N_R\right) + 2\left(1 + 2\,\frac{\xi}{a}\right)\left(1 + N_R\right)\,\ln\left(1 + N_R\right)}{2\left(-\frac{3}{4} - \frac{1}{2}\ln \beta\right) + 2\,\frac{\xi}{a}\left(-\frac{1}{2} - \ln \beta\right)} \tag{132}$$

Stechkina et al. [134] obtained the same result as Eq. (132). Introducing the hydrodynamic factor H and assuming diffuse reflection (f = 1), these equations can be unified by

$$E_R = \frac{\left(1 + N_R\right)^{-1} - \left(1 + N_R\right) + 2\left(1 + 1.996\text{Kn}\right)\left(1 + N_R\right)\,\ln\left(1 + N_R\right)}{2H + 3.992\text{Kn}\left(H + \frac{1}{2}\right)} \tag{133}$$

where $H = 2 - \ln \text{Re}$ (Natanson), $H = (-3/4) - (1/2)\ln \beta$ (Pich; Stechkina et al.). A simpler expression for small values of N_R is obtained from Eq. (133) by expansion in a series of powers of N_R. The result is

$$E_R = \frac{N_R\left(N_R + 1.996\text{Kn}\right)}{H + 1.996\text{Kn}\left(H + \frac{1}{2}\right)} \tag{134}$$

This same result is, of course, obtained by basing the calculation on the velocity field of Eq. (125). For Kn → 0, these equations reduce to corresponding equations in the continuum region. Hence, for an isolated cylinder $E_R = E_R(N_R, \beta, Kn)$. It can be shown that $dE_R/dKn > 0$, so it may be concluded that the intensity of the direct interception mechanism increases with decreasing gas pressure.

2. *Transition Region*

The transition region covers the interval of Kn numbers $0.25 < Kn < 10$. A typical value of Kn in this region is $Kn \sim 1$. From the theoretical point of view this is the most difficult case, and as a consequence no rigorous results concerning the flow field, drag forces, pressure drop, or filtration efficiency are available. Gas flow in this case is of mixed type between the continuum and free molecule flow.

An approximation for the velocity field around an isolated cylinder (transverse flow) at $Kn \sim 1$ was suggested by Velichko and Radushkevich [226]. These investigators used the concept of a "limiting sphere" and derived for the stream function the equation

$$\underline{\psi}(S,\theta) = \frac{aU_o \sin\theta}{H' - 1 + 2(2 - \ln Re)}\left[(H' - 2)\frac{S}{a} + \omega(1 + K_1 Kn)^2 \frac{a}{S} + 2\frac{S}{a}\ln\frac{S/a}{1 + K_1 Kn}\right] \tag{135}$$

where

$$H' = \omega(8Kn + 1) \tag{135a}$$

$$\omega = \frac{2\,Kn + \delta/2}{2\delta\,Kn + Kn + \delta/2} \tag{135b}$$

K_1 is a numerical constant of order unity and $\delta = 1 + \pi/4$. If $Kn \to 0$, then $H' = 1$, $\omega = 1$, and Eq. (135) reduces to Lamb's velocity field, Eq. (28).

Concerning the drag force acting on a cylinder at $Kn \sim 1$ (transverse flow), Liu and Passamaneck [227] derived for the force acting on the unit length of an isolated cylinder the equation

$$F = \frac{1}{2}\left(3\pi + \frac{1}{2}\pi^2\right)\frac{\eta U_o}{1.166 + Kn} \tag{136}$$

This equation for large values of Kn reduces to the correct limit in the free molecule region (see the next section), for for $Kn \to 0$, it does not reduce to the Lamb-Davies equation, Eq. (37), in the continuum region.

Pich [228], using the method of a "limiting sphere" (molecular layer) and assuming diffuse reflection of the gas molecules, derived for the drag on an isolated cylinder at $Kn \sim 1$ the equation

$$F = \frac{4\pi\eta U_o}{2 - \ln Re + 1.747Kn - \ln(1 + 0.749Kn)} \tag{137}$$

This equation does reduce to the correct limits in the continuum as well as in the free molecule flow regime. Both Liu and Passamaneck's theory and Pich's theory are in good agreement with the experiments of Coudeville et al. [229] (see the comparison in Pich's [228] paper).

As far as is known to the author, there are no theoretical studies of the pressure drop of fibrous aerosol filters in the transient region of Kn. Judging from the experiments of Petryanov et al. [230], also in the transient region, pressure drop decreases with increasing Kn, so that with decreasing gas pressure the pressure drop decreases.

Studies of filtration mechanisms in the transition region of Kn have been limited so far to investigations of the diffusion and direct interception mechanisms. The diffusion mechanism of particle deposition under these conditions was investigated by Radushkevich and Velichko [231]. These authors introduced a modified Péclet number $\overline{Pe}$ given by $\overline{Pe} = U_o(r + a + 2\delta^*)/2D$, where δ^* is the "mean free path of the particles" as introduced by Smoluchowski and given by $\delta^* = V_T\tau$ in which τ is the relaxation time for a particle and V_T, the median thermal velocity of particles, is given by $V_T = (8kT/\pi m)^{1/2}$, where m is the particle mass and k Boltzman's constant. The capture coefficient of a fiber due to diffusion deposition of particles has been defined by

$$\bar{E}_D = \frac{\Phi_f}{2(r + a + \delta^*)n_o U_o} \tag{138}$$

where, again, Φ_f is the flux of particles on a unit cylinder length. Assuming $U_o/V_T << 1$, they derived for $\bar{E}_D$ the following equation:

$$\bar{E}_D = \frac{\pi}{2} \frac{zI_o(\overline{Pe}) - I_o(\overline{Pe})}{zK_o(\overline{Pe}) + K_1(\overline{Pe})} \left[I_1(\overline{Pe})K_o(\overline{Pe}) + I_o(\overline{Pe})K_1(\overline{Pe})\right]$$

where $z = (V_T/2U_o)[(r + a)/(r + a + 2\delta^*)]$ and I_o and I_1 are modified Bessel functions of the first kind of zero and first order, respectively, and K_o and K_1 are modified Bessel functions of the second kind, zero and first order, respectively. The capture coefficient $\bar{E}_D$ given by Eq. (139) is greater than unity and decreases with increasing particle size and with increasing velocity.

The mechanism of direct interception at Kn ~ 1 was investigated by Velichko and Radushkevich [226]. These investigators, using their velocity field, Eq. (135), derived for the capture coefficient due to direct interception E_R the equation

$$E_R = \frac{1 + K_1 Kn}{H' - 1 + 2(2 - \ln Re)} \left[(H' - 2)q' + \frac{\omega}{q'} + 2q' \ln q' - (H' + \omega - 2)\right] \tag{140}$$

where $q' = 1 + N_R/(1 + K_1 Kn)$, $\delta = 1 + \pi/4$, and $K_1 \sim 1$. For $Kn \to 0$, there is obtained $q' = 1 + N_R$ and $\omega = 1$, and Eq. (140) reduces to Eq. (75).

Another approach for explaining high values of the capture coefficients with ultrathin fibers having diameters of the same order as the mean free path of the gas molecules recently has been suggested by Radushkevich [232]. The hypothesis is based upon the intensive Brownian vibrations of ultrathin fibers, due to which aerosol particles should be additionally captured by these fibers from the gas flowing through the filter. A rough estimate suggests that for a single fiber with $d_f = 0.05$ μm and fixed at one point, the number of particles depositing at its free end within one second increases 10 to 20 times in comparison with the number on the same fiber when motionless. However, the conclusion that increased particle deposition due to Brownian vibrations is confirmed by the higher values of deposition observed experimentally is unjustified at this stage of development.

Experiments investigating particle deposition in the transition region of Kn numbers have been carried out by Radushkevich and Kolganov [233,234]. The deposition of polydisperse particles of WO_3 (later also polystyrene and methylene blue aerosols) with diameters d_p = 0.04 to 0.34 μm on single asbestos cylinders with d_f = 0.04 to 0.08 μm from gas flowing with velocities U_o = 0.5, 2, 5, and 25 cm/sec was investigated. Under these conditions the Knudsen number is Kn ~ 1 and the interception parameter $N_R > 1$. The capture coefficient, defined in these papers by $\Phi_f = \tilde{E}(2a + d_p)n_o U_o$, was found to decrease with increasing particle size and to be independent of cylinder diameter d_f over the entire range of investigated particle sizes and flow velocities. Independence or a weak dependence of the capture coefficient on fiber diameter is concluded to be a characteristic feature of particle deposition in the transition region of Kn.

3. *Free Molecule Region*

The free molecule region is characterized by Kn → 10, according to Devienne's classification. The drag of a cylinder in free molecule flow (transverse flow) was investigated by Stalder et al. [235]. Assuming diffuse reflection of the gas molecules from the cylinder surface, the results for a biatomic gas can be expressed as

$$F = aS_g c_m^2 f(S) \tag{141}$$

$$f(S) = \sqrt{\pi}\, Se^{-S^2/2}\left[\frac{3}{2} I_o + \frac{1}{2} I_1 + S^2(I_o + I_1)\right] + \frac{\pi^{3/2}}{4} SD^{1/2} \tag{141a}$$

and

$$D = \frac{2S^4 I_o + 9S^2 I_o + 6I_o + 2S^4 I_1 + 7S^2 I_1}{6S^2 I_o + 6I_o + 6S^2 I_1} \tag{141b}$$

In this equation c_m is the most probable velocity of the gas molecules related to the mean velocity of the molecules $\bar{c}$ by $c_m = \frac{1}{2}\sqrt{\pi}\,\bar{c}$. $I_o = I_o(1/2S^2)$ and $I_1 = I_1(1/2S^2)$ are modified Bessel functions of the first kind, zero and first order, respectively. The parameter S is called the molecular speed ratio and is given by

$$S = \frac{U_o}{c_m} = \frac{U_o}{\frac{1}{2}\sqrt{\pi}\,\bar{c}} = \frac{U_o}{\frac{1}{2}\sqrt{\pi}\,\sqrt{8kT/m_g\pi}} = \frac{U_o}{\sqrt{2kT/m_g}} \tag{142}$$

where m_g is the mass of the molecule.

Pich [236] concluded that since, at normal temperatures, c_m is of the order 10^2 m/sec, the case $S \ll 1$ is most important for the physics of disperse systems. Expanding the function f(S) into a series and reducing it to the first power of S, he concluded that for $S \ll 1$ Eq. (141) is approximated by

$$F = \frac{6\pi + \pi^2}{8}\, aS_g\, \bar{c}\, U_o \tag{143}$$

Although the general equation for cylinder drag in a monoatomic gas is slightly different from Eq. (141), the approximation of Eq. (143) is also valid for a monoatomic gas. Equation (143) is a limit to which both Eqs. (136) and (137) reduce for cylinder drag in the transition region under very large Knudsen numbers.

The pressure drop of fibrous aerosol filters in the free molecule region was investigated by Glushkov [35]. For filters consisting of circular cylinders these results can be expressed by

$$\Delta p = \frac{8.38}{\pi}\,\frac{\beta}{(1-\beta)^2}\,\frac{1}{a\bar{c}}\,LU_o p \tag{144}$$

where p is the gas pressure. Pich [236] treated the same problem using Eq. (143). This result is

$$\Delta p = 2.29\,\frac{\eta\beta U_o L}{a\lambda} \tag{145}$$

For high porosity filters ($\beta \ll 1$) and, using the kinetic expressions [219] $\eta = 0.499 S_g \bar{c}\lambda$ and $p = S_g \overline{c^2}/3$ where S_g is the gas density and $\overline{c^2}$ - $1.18\bar{c}^2$, both Eqs. (144) and (145) can be summarized by

$$\Delta p = C \frac{\beta L U_o}{a\bar{c}} p \tag{146}$$

where C is a numerical constant having the value 2.67 according to Glushkov and 2.91 according to Pich. From Eq. (146) the following conclusions can be drawn: (1) In the free molecule region the pressure drop is proportional to the gas viscosity, velocity, and filter thickness, i.e., $\Delta p \sim \eta U_o L$, just as it is in the continuum region. (2) $\Delta p \sim \beta$, so that the pressure drop decreases with increasing filter porosity. (3) Unlike the continuum region where $\Delta p \sim 1/a^2$, in the free molecule region $\Delta p \sim 1/a$. (4) The pressure drop in the free molecule region decreases linearly with decreasing gas pressure at constant temperature. The experimental data of Petryanov et al. [230], who measured the pressure drop of fibrous FPA-15 filters with $d_f = 1.5$ μm up to pressures p = 12 mb, qualitatively agree with conclusions (1), (2), and (3).

Studies of deposition mechanisms in the free molecule region are limited to the investigation of particle deposition due to thermal (Brownian) motion as reported by Friedlander [237]. Friedlander derived equations for the capture coefficient due to Brownian deposition of particles E_D on bodies of simple geometric shape (spheres, ribbons, and cylinders). In the case of cylinders, this resulted in the expression

$$E_D = \left(\frac{\pi}{8}\right)^{1/2} \left\{ \frac{1}{x^{1/2}} e^{-x} I_o(x) + 2x^{1/2} e^{-x} [I_o(x) + I_1(x)] \right\} \tag{147}$$

where $x = (2/\pi)(U_o/V_T)^2$, V_T is again the median thermal velocity of particles given by $V_T = (8kT/\pi m)^{1/2}$, and I_o and I_1 are modified Bessel functions of the first kind, zero and first order, respectively. Friedlander concluded that E_D as given by Eq. (147) was greater than unity but approached unity as the velocity of flow U_o approached infinity. The capture coefficient E_D was independent of the cylinder diameter.

N. Combined Filtration Mechanisms

In the process of deposition on fibers, particles may be subjected to the simultaneous effect of all deposition mechanisms, playing a different role under different filtration conditions. One of the basic but most difficult problems in filtration theory is to find the total capture coefficient of a fiber E_f while allowing for the simultaneous effect of these mechanisms. In general, this problem has not been exactly solved owing to the great mathematical difficulties. The problem has been attacked in the manner described below.

Perhaps the simplest and most practical procedure was suggested by Ranz and Wong [161]. According to them, usually only one or two of the deposition mechanisms play a dominant role. In order to determine the most important mechanism under given filtration conditions, it is advantageous to describe the individual mechanisms by dimensionless parameters. Then, the dominant mechanism is the one having the greatest parameter, the remaining mechanisms being of less importance and are neglected in the first approximation. The question of the main filtration mechanism is very important, and this criterion for the primary parameter is very simple and practical. On the other hand, the degree of accuracy is essentially unknown, and the approximation becomes more and more incorrect with increasing values of parameters corresponding to mechanisms other than the main one.

The second and a widespread approach is to assume that the individual capture coefficients E_i $(i = 1, 2, \cdots, n)$ corresponding to different mechanisms are additive, i.e.,

$$E_f = E_1 + E_2 + \cdots + E_n \qquad (148)$$

This assumption would be justified if all the mechanisms were acting independently. In reality, however, this assumption leads to results of unknown accuracy, and in general the whole approach has no justification. There are only a few special cases where this approach has theoretical support. For example, Robinson [146] has shown that for $N_G \ll 1$ the capture coefficients of inertial and gravitational deposition are additive, i.e., $E_f = E_I + E_G$.

Some authors assumed that the total penetration P_f, defined by $P_f = 1 - E_f$, can be expressed as

$$P_f = P_1 \cdot P_2 \cdot P_3 \cdot \ldots \cdot P_n \tag{149}$$

where penetrations P_i $(i = 1, 2, \cdots, n)$ are defined by $P_i = 1 - E_i$ $(i = 1, 2, \cdots, n)$. For example, for the simultaneous action of diffusion, interception, and inertial deposition, putting $E_1 = E_D$, $E_2 = E_R$, and $E_3 = E_I$, from Eq. (149) it follows that

$$E_f = 1 - (1 - E_D) \cdot (1 - E_R) \cdot (1 - E_I) \tag{150}$$

Equation (149) is valid if n filters or, more generally, n separators are arranged in series. Then the total penetration is given by Eq. (149). The application of Eq. (149) for calculation of the total capture coefficient of a fiber implicitly assumes that the different deposition mechanisms follow one after another, e.g., that diffusion is followed by interception and this in turn by inertial deposition. No arguments justifying this application are known to the author except special cases (see the section dealing with membrane filters).

Chen [4] assumed that for small values of N_D and Stk the total capture coefficient (three mechanisms) could be expressed by

$$E_f = E_{DR} + E_{IR} \tag{151}$$

where E_{DR} is the capture coefficient due to the combined action of diffusion and interception and E_{IR} that due to the combined action of inertia and interception.

Fuchs [13] stated without giving arguments that E_f is greater than any of the partial capture coefficients and smaller than their sum, i.e.,

$$E_j \leq E_f \leq E_1 + E_2 + \cdots + E_n \quad (j = 1, 2, \cdots, n) \tag{152}$$

Moreover, this statement is in disagreement with Stechkina and

Fuchs' [243] Eq. (155) as the interaction term is always positive.

1. *Simultaneous Action of Diffusion and Direct Interception*

Apparently Langmuir [52] first approximately solved the problem of particle deposition on fibers due to the simultaneous action of particle diffusion and direct interception. Using the method of isolated fibers he derived the equation

$$E_{DR} = \frac{1}{2 - \ln Re}\left[(1 + 2x)\ln(1 + 2x) - \frac{x(2 + 2x)}{1 + 2x}\right] \tag{153}$$

where x is the solution of the equation

$$x\left(x - \frac{1}{2}N_R\right)^2 = 0.28(2 - \ln Re)Pe^{-1} \tag{153a}$$

Gillespie [132] assumed that both mechanisms are additive, i.e., $E_{DR} = E_D + E_R$. Assuming potential flow, his calculation of E_R is identical with Eq. (74). However, his calculation of E_D is erroneous as his solution of the diffusion equation for Pe = 0 cannot be applied to particle deposition in fibrous filters where $Pe \neq 0$ and usually $Pe \gg 1$ (typical values of Pe in Gillespie's calculations were $Pe \sim 10^4$).

Friedlander [238], in semi-empirical fashion, derived an equation that can be written

$$E_{DR} = \frac{6Re^{1/6}}{Pe^{2/3}} + 3N_R^2\,Re^{1/2} \tag{154}$$

The theoretical basis of this equation has been questioned by Natanson and Ushakova [239].

Radushkevich [240] derived an explicit expression $E_{DR} = E_{DR}(Pe, N_R, Re)$ but, for the limiting case of point particles ($N_R \to 0$), this equation does not reduce to the correct expression, Eq. (61). For this and other reasons, Radushkevich's paper has been criticized by Natanson and Ushakova [239], although some of Natanson and Ushakova's arguments are incorrect (see also Radushkevich's reply [241].

Pasceri and Friedlander [242] checked Eq. (154), comparing theoretical values with the experimental data of Thomas and Yoder [182].

Torgeson's attempt [150] to evaluate the function E_{DR} suffers by the incorrect approximation of Eq. (75) used in his calculations.

Stechkina and Fuchs [243], after giving a detailed mathematical analysis of the convective diffusion equation and using electronic computers, derived the equation

$$E_{DR} = E_D + E_R + 1.24 \frac{N_R^{2/3}}{H^{1/2}Pe^{1/2}} \tag{155}$$

where E_D is given by Eq. (66) and E_R by Eq. (75).

Friedlander [244], on the basis of similarity theory, concluded that for $Re < 1$, $N_R \ll 1$, and $Pe \gg 1$ the capture coefficient is given by

$$E_{DR} = \frac{2}{PeN_R} f\left[\frac{N_R^3 \, Pe}{2(2 - \ln Re)}\right] \tag{156}$$

where f is an undetermined function. This expression amounts to the similitude law for the diffusion deposition of particles of finite size. For fixed Re, the group $y_F = E_{DR}Pe\, N_R$ as a function of $x_F = N_R Pe^{1/3}$ should be a single-valued function $y_F = y_F(x_F)$. This concept of generalized (Friedlander's) coordinates y_F and x_F is extremely useful in correlating experimental data, although it also has disadvantages, as pointed out by Natanson and Ushakova [239]. Two asymptotes of the function $y_F = y_F(x_F)$ are known: the first, for $N_R \to 0$, is Eq. (61) (pure diffusion); and the second, corresponding to $Pe \to \infty$, is Eq. (76) (pure interception). Friedlander has shown that, for the forward stagnation point, Eq. (156) becomes

$$E_{DR} = \frac{2}{Pe\, N_R} \frac{e^{-x}}{\int_1^{\infty} e^{-xz^3}\, dz} \tag{157}$$

where

$$x = \frac{N_R^3 \, Pe}{6(2 - \ln Re)} \tag{157a}$$

Friedlander pointed out that the denominator in Eq. (157) can be evaluated in terms of an incomplete gamma function. Radushkevich and Kolganov [128], using an expansion of the gamma function into a series at small values of x and taking the exponential in the numerator as unity for small x values, approximated Eq. (157) by

$$E_{DR} = \frac{4.35}{2^{1/3}} \frac{1}{(2 - \ln Re)^{1/3}} \frac{Pe^{-2/3}}{0.893 - x} \tag{158}$$

Concerning empirical correlations, Aiba and Yasuda [245], approximating Langmuir's equation for diffusion deposition, Eq. (58), by $E_D = Re^{1/18} \, Pe^{-2/3}$; using for direct interception (see Eq. 75) $E_R = N_R^2 \, Re^{1/6}$; and assuming that $E_{DR} = E_D + E_R$, obtained

$$E_{DR} = Re^{1/18} Pe^{-2/3} + N_R^2 \, Re^{1/6} \tag{159}$$

They found a good agreement of this equation with experimental data from the literature.

Friedlander and Pasceri [18], using the experimental data of Chen [4] and Wong and Johnstone [162], suggested the empirical equation

$$E_{DR} N_R Pe = 1.3 N_R Pe^{1/3} + 0.7 (N_R Pe^{1/3})^3 \tag{160}$$

This equation becomes $y_F = 1.3x_F + 0.7x_F^3$ using the Friedlander coordinates x_F and y_F or, in equivalent form,

$$E_{DR} = 1.3 Pe^{-2/3} + 0.7 N_R^2 \tag{161}$$

This latter correlation differs somewhat from Eq. (154) as originally suggested by Friedlander [238], which included a weak Reynolds number dependence similar to the Aiba and Yasuda correlation, Eq. (159).

2. *Simultaneous Action of Diffusion, Interception, and Inertia*

Davies [96] derived on semi-empirical grounds for the capture coefficient E_{DRI} due to the combined action of diffusion, direct interception, and inertial deposition the equation

$$E_{DRI} = 0.16[N_R + (0.5 + 0.8N_R)(Pe^{-1} + Stk) - 0.105N_R(Pe^{-1} + Stk)^2] \tag{162}$$

Torgeson [150], modifying Davies' theory for inertial deposition and direct interception and using his own equation for pure diffusion deposition, Eq. (62), suggested the equation

$$E_{DRI} = 0.75E_{IR} + FGE_D \tag{163}$$

where E_D is given by Eq. (62) and E_{IR} is a function of the parameters N_R and Stk, which suffers by Torgeson's incorrect approximation of Eq. (75). The quantities F and G are complicated functions of Re, Pe, Stk, N_R, and the drag coefficient of the fiber C_D which were calculated and plotted as a graph. The complex $C_D Re/2$ was determined from experimental pressure drop data.

Dawson [27] assumed that the capture coefficient E_{DRI} could be expressed as

$$E_{DRI} = E_{DR} + E_{IR} - E_R \tag{164}$$

where E_{DR} is given by the Stechkina-Fuchs equation, Eq. (155), so that Eq. (164) becomes

$$E_{DRI} = E_{IR} + E_D + 1.24 \frac{1}{H^{1/2}\, Pe^{1/2}} N_R^{3/2} \tag{165}$$

E_D is given by Stechkina's equation, Eq. (66), and the function E_{IR} was calculated numerically as $E_{IR} = E_{IR}(Stk, N_R, \beta)$ using the Spielman-Goren velocity field. On this basis Dawson calculated velocity characteristics $E_{DRI} = E_{DRI}(U_o)$ showing minima at fixed particle sizes. A shift of the minimum to lower velocities with increasing particle size was observed. Moreover, "isoefficiency curves" in the coordinate system of particle diameter d_p vs. flow velocity U_o were calculated.

Stechkina et al. [134] assumed that E_{DRI} could be expressed as

$$E_{DRI} = E_{DR} + E_{IR} \tag{166}$$

where the capture coefficient E_{DR} due to the simultaneous action of diffusion and interception was given by Eq. (155) and E_{IR} by Eq. (84). However, caution should be exercised in using this equation, as Eq. (84) was subsequently found by Stechkina et al. [156] to be incorrect.

Yoshioka et al. [153] assumed that E_{DRI} could be expressed by

$$E_{DRI} = E_D + E_{IR} \tag{167}$$

where E_D is the first portion of Friedlander's semi-empirical equation, Eq. (154), corresponding to diffusion deposition and given by $E_D = 6Re^{1/2}Pe^{-2/3}$. For calculation of E_{IR} Yoshioka et al. [153] used their own numerical results.

Dorman [221] also followed the semi-empirical approach. Realizing that in potential flow $E_D \sim U^{-1/2}$, that E_R is independent of the flow velocity U_o, and assuming that $E_I \sim U_o^2$, Dorman described the penetration of the filter P by

$$P = \exp\left(-k_D' U_o^{-1/2} - k_R' - k_I' U_o^2\right) L \tag{168}$$

where k_D', k_R', k_I' are quantities describing the diffusion, interception, and inertial mechanisms, respectively. Expressing this equation in terms of percentage penetration P(%), there is obtained

$$\log P(\%) = 2 - k_D L U_o^{-1/2} - k_R L - k_I L U_o^2 \tag{169}$$

Taking into account that at the point U_{om} the velocity characteristic has a minimum, i.e., $dP(\%)/dU_o = 0$, then from Eq. (169) it follows that $k_D = 4k_I U_{om}^{5/2}$. Hence, Eq. (169) can be written as

$$2 - \log P(\%) = k_I L \left(U_o^2 + 4U_{om}^{5/2} U_o^{-1/2}\right) + k_R L \tag{170}$$

From Eq. (170) it follows that a plot of 2 - log P(%) vs.

$U_o^2 + 4U_{om}^{5/2}\, U_o^{-1/2}$ should yield a straight line the slope of which gives k_I and, therefore, k_D, while the intercept gives k_R. In this manner Dorman was successful in the interpretation of the data of Ramskill and Anderson [159] and those of Thomas and Lapple [246].

Concerning the simultaneous action of several filtration mechanisms, Stechkina et al. [134] assumed that in the range of maximum penetration (see the discussion of filter selectivity) the capture coefficient E_{DRIG} due to the simultaneous action of diffusion, interception, inertial, and gravitational deposition can be expressed by

$$E_{DRIG} = E_{DR} + E_{IR} + E_{GR} \tag{171}$$

where E_{DR} is given by Eq. (155) and E_{IR} by Eq. (84), where E_{GR} is equal to $(1 + N_R)N_G$ in the case of down-flow and $-(1 + N_R)N_G$ for up-flow, and where N_G is the parameter of the gravitational deposition given by Eq. (89).

Experimentally, the problem of particle deposition where two, three, or more mechanisms are acting simultaneously has been studied by a large number of investigators using different filters, aerosols, and measuring techniques. These papers, arranged chronologically, are listed under Refs. 247, 161, 4, 163, 182, 246, and 248. Investigation of the filtration process using biological aerosols (sterilization) has been reported by several authors whose papers, again arranged chronologically, are listed under Refs. 249-253. Finally, investigation of filtration of radioactive aerosols has been reported by Blasewitz and Judson [101], Zumach [254], and Shleien et al. [255].

O. Interference Effect

The capture coefficient of a fiber embedded in a filter $E_{j\beta}$ differs from the capture coefficient of an isolated fiber E_j (j = D, R, I, G, and other subscripts corresponding to different mechanisms) for two main reasons: (1) The velocity field around individual fibers, and (2) the median gas velocity in the filter is higher than that corresponding to an isolated fiber. The resulting influence of neighboring fibers on the deposition process for any selected fiber is called an interference effect. Many investigators have tried to evaluate this effect, usually on empirical grounds.

Langmuir [52] allowed for the interference effect by means of a constant in the equation for efficiency. The value of this constant was calculated from Δp determined experimentally. Davies [96], using experimental data, concluded that the interference effect is the same for different filtration mechanisms and could be expressed by

$$E_{j\beta} = E_j(0.16 + 10.9\beta - 17\beta^2) \tag{172}$$

Chen [4] concluded that the interference effect was positive, i.e., that the presence of neighboring fibers led to an increase of efficiency, the increase being a function of the filter porosity and, probably, of Re. He expressed the interference effect by the empirical equation

$$E_{j\beta} = E_j(1 + 4.5\beta) \tag{173}$$

for $\beta < 0.09$. The value 4.5 is subject to small changes for different mechanisms.

Sadoff and Almlof [129], again on empirical grounds, came to the somewhat surprising result that the interference effect for the diffusion mechanism was negative, i.e., that diffusion deposition decreased with increasing β. To this unexpected result also came Dorman [14], who expressed the interference effect for the diffusion mechanism by the empirical equation

$$E_{D\beta} = E_D(1 - 4\beta) \tag{174a}$$

The influence of β on the interception effect was described by

$$E_{R\beta} = E_R(1 + 30\beta) \tag{174b}$$

and the interference effect for the inertial mechanism by

$$E_{I\beta} = E_I(1 + 110\beta) \tag{174c}$$

All these equations are of empirical character and, therefore, of limited value. Consequently, the interference effect is a weak point in any analysis involving isolated fibers.

A significant advance was made by Fuchs and Stechkina [45] who suggested basing the theory of aerosol filters on the velocity field in a system of cylinders, where the correction for the interference effect becomes superfluous. Since then, modern filtration theories have usually been based on the velocity field in a system of parallel cylinders, this field being approximated by Eq. (33).

P. Fiber Capture Coefficient and Filter Efficiency

The last step in the calculation of filter efficiency E or filter penetration P is the transition from the computed total capture coefficient of the individual fiber in the filter E_f to the overall filter. Sometimes another quantity α, called the coefficient of absorption of particles by the filter, is used; it is related to E by

$$E = 1 - P = 1 - e^{-\alpha} \tag{175}$$

Hence α is related to P by $\alpha = -\ln P$.

Assuming a homogeneous filter with monodisperse fibers placed perpendicular to the gas flow, the following relationship between α and E_f can be easily derived:

$$\alpha = \frac{4}{\pi} \frac{1 - \varepsilon}{\varepsilon} \frac{L}{d_f} E_{f\beta} \tag{176}$$

This equation was developed first by Langmuir [52] and later by Davies [96], Chen [4], and others. Equation (176) can be expressed by $\alpha = S'E_{f\beta}$, where $S' = (4/\pi)[\beta/(1 - \beta)](L/d_f)$ and called a solidarity factor by Whitby and coworkers [175].

Wong et al. [163], Landt [256], and Friedlander [18] derived a similar expression:

$$\alpha = \frac{4}{\pi} (1 - \varepsilon) \frac{L}{d_f} E_{f\beta} \tag{177}$$

This latter equation follows from Eq. (176) for small β, i.e., for $\varepsilon \to 1$. According to Friedlander [18], $E_{f\beta}$ should be interpreted as an effective capture coefficient with an unknown functional dependence on β and, therefore, Eq. (176) or (177) is acceptable provided that the treatment is consistent. For filters consisting of fibers of different radii, Chen [4] suggested that the fiber radius $a = d_f/2$ appearing in Eq. (176) should be replaced by a_2^2/a_1, where a_2 is the mean square radius and a_1 the mean arithmetic radius of the fibers. Leers [257] introduced a symbol for the filtration constant k_L defined by

$$\alpha = \frac{L}{k_L} \tag{178}$$

The physical meaning of this constant follows from Eq. (178). For $L = k_L$, $E = 63.2\%$.

For model filters consisting of series of fiber layers arranged geometrically (square structure), Zebel [206] and Stenhouse et al. [174] quote the equation

$$E = 1 - (1 - E_{f\beta}Fu)^n \tag{179}$$

where $Fu = a/b$ is the Fuchs' number, $2b$ is the distance between the cylinder axes, and n is the number of fiber layers in the filter. As $n = L/2b$ and for a square structure $\beta = (\pi/4)Fu^2$, Eq. (179) can be expressed as

$$E = 1 - \left(1 - 2E_{f\beta}\sqrt{\beta/\pi}\right)^{(2L/d_f)\sqrt{\beta/\pi}} \tag{180}$$

Q. Characteristics of Filters

Practical application of filtration theory is concerned with construction of new fibrous filters, selection of filter materials, exploitation of the filtration devices, etc. Such application requires an analysis of the derived equations for pressure drop and filter efficiency to determine how these basic parameters, or the filter quality defined by Eq. (4), depend on the properties of the particles, the gas flow, and the filter. As already noted, three factors (particles, gas flow, and filter) take part in the filtration process, and, consequently, all these dependencies (characteristics) can be divided into three groups. The first group includes the dependencies of filter efficiency on the properties of the particles (size, shape, density, charge, etc.). The pressure drop of a clean filter is independent of the properties of the particles; the changes in Δp of a clogged filter with size, shape, weight, and other properties of the particles will be discussed in the section dealing with nonstationary filtration. The second group contains the dependencies of Δp and E on the properties of the gas flow (velocity of flow, pressure, temperature, and viscosity). Finally, the third group includes the dependencies of Δp and E on the properties of the filter (filter thickness, fiber diameter, porosity, kind of fiber arrangement, etc.). Besides the theoretical analysis, all these dependencies can be measured experimentally, at least in principle. The most important dependence contained in the first group is that of filter efficiency on the size of the particles, called here a selective characteristic.

1. *Selective Characteristics*

The selective characteristics of a fiber, or that of the filter, is defined as a dependence of the capture coefficient or the filter efficiency on particle size. This dependence has been studied theoretically by several authors [52,258,238,259,175,134,156], arranged chronologically. Related experimental papers, again arranged chronologically, are [260-266].

Taking into account just the mechanisms of diffusion, direct interception, and inertial deposition, it follows from the definition of corresponding dimensionless parameters that the diffusion parameter $N_D = Pe^{-1}$ decreases with increasing particle size, while the parameters of interception N_R and inertial deposition Stk increase with particle size. These mechanisms act simultaneously; the corresponding capture coefficients are increasing functions of their parameters, and the filter efficiency is probably an increasing function of individual capture coefficients. Hence, it can be expected that selective characteristics will exhibit a minimum. Langmuir [52] was apparently the first to come to this conclusion on theoretical grounds.

Landt [258] tried to interpret analytically the selectivity curves having minimum dependence on the filtration velocity as obtained experimentally by Thomas and Yoder [182]. Landt, in a rough and in some points incorrect analysis, explained these curves as a result of the combined action of diffusion, direct interception, and gravitational deposition, assuming mechanisms to be additive.

Pich [259] reported a qualitative analysis of selective characteristics based on Eq. (154) as derived by Friedlander [238]. Expressing the diffusion coefficient D by Einstein's relation D = kTB, assuming Stokes particles (i.e., $B = (3\pi\eta d_p)^{-1}$), and substituting for Re, Pe, and N_R in Eq. (154), he obtained the fact that the dependence $E_{DR} = E_{DR}(d_p)$—i.e., the fiber selectivity—exhibited a minimum. For the position of the minimum there was obtained

$$d_{pm} = \frac{0.85(kT)^{1/4} d_f^{3/8} \nu^{1/8}}{(3\pi\eta)^{1/4} U_o^{3/8}} \tag{181}$$

where, again, $\nu = \eta/S_g$ is the kinematic viscosity of the gas. The position of the selectivity minimum d_{pm} depends, therefore, both on the fiber diameter and on the flow velocity. From Eq. (181) it follows that with increasing velocity d_{pm} is shifted toward smaller particle sizes. Furthermore, it has been shown that E_{DR} in the minimum decreases with increasing velocity and that the curve of selectivity is narrowed with increasing U_o. The dependence of d_{pm} on U_o is important for the experimental checking of theory. Analysis of the dependence of the minimum position on the fiber diameter leads to the following conclusions: (1) With increasing fiber diameter the selectivity minimum is shifted toward larger particle sizes; (2) with increasing fiber diameter the value of E_{DR} in the minimum decreases; (3) with increasing fiber diameter the selectivity curve is widened in the vicinity of the minimum; and (4) the necessary condition for avoiding the shift of the selectivity minimum is d_f/U_o = const.

Whitby [175] calculated selectivity curves according to various filtration theories. He described a grapho-analytical procedure for obtaining certain quantities (fiber diameter and drag coefficient of the fiber) needed in computing the selectivity curves. These quantities for any operating Reynolds numbers were computed from experimental pressure drop vs. flow velocity data.

Stechkina et al. [134,156], using Eq. (166) and the concept of a fan model, calculated selectivity curves corresponding to the combined action of diffusion, direct interception, and inertial deposition. These curves show a distinct minimum, the position of which is approximately $r_m \approx 10^{-5}$ to 3×10^{-5} cm. They also concluded that the position of the minimum $r_m = d_{pm}/2$ depends on fiber diameter and on flow velocity. With increasing velocity r_m shifts toward smaller sizes, and the minimum becomes wider, i.e., the selectivity of the filter decreases. Numerical calculations based on Eq. (166) have shown that in the vicinity of the minimum the inertial deposition can be neglected. Therefore, at moderate velocities the position of the minimum can be found by equating E_D^f and E_R^f.

From an experimental point of view, apparently the first to discover the existence of a minimum in the selectivity curve was Freundlich [260] in 1926. He found the maximum penetrating particle diameter to be between 0.2 and 0.4 μm. Chen [4], in experiments with monodisperse particles of 0.15 μm diameter and glass fiber filters of average fiber diameter d_f = 2.5 μm, found a minimum in the selectivity curve when the velocity was below 4 cm/sec. Thomas and Yoder [182], in experiments with dioctylphthalate droplets and glass fiber filters, also found a minimum in the selectivity curve lying at 0.5 μm diameter for a velocity of 0.94 cm/sec and 0.7 μm for a velocity of 0.094 cm/sec. Fitzgerald and Detwiller [261], using particles of potassium permanganate, found the minimum in the selectivity of filter papers (Whatman Nos. 40 and 41) to be in the range 0.01 to 0.04 μm at a velocity of 150 cm/sec.

Radushkevich and Kolganov [262,263] investigated selectivity in experiments with a model filter consisting of a large number (up to 372) of separate layers of polymer fibers having a mean diameter 1.5 μm. A polydisperse, polystyrene aerosol with $0.1 < d_p < 1$ μm was used at velocities U_o = 0.4, 5, and 25 cm/sec. The maximum penetration was found for particle diameters of 0.2 to 0.3 μm at U_o = 0.4 cm/sec, shifting to lower particle sizes with increasing velocity. Calculation of the capture coefficient of a fiber embedded in a model filter showed that it remained practically constant for filters with densities of 33 to 75 fibers/mm, i.e., no effect of neighboring fibers was noted.

Later Radushkevich and Kolganov [264] investigated in detail the influence of velocity on the selective characteristics of isolated cylinders. These investigators measured the selectivity of individual fibers with a mean diameter of 4.3 μm for spherical polystyrol particles with d_p = 0, 1 to 0.6 μm in the broad range of velocities from 0.05 to 15 cm/sec. A minimum in the selectivity curves was found for all velocities less than 15 cm/sec, the minimum being shifted to smaller particle sizes with increasing velocity. For U_o = 15 cm/sec there was no minimum, and the capture coefficient steadily increased with particle size. As, under these conditions, the inertial and electrostatical mechanisms are negligible (Stk numbers are from 0.0043 to 0.0387, no electrical effects), they interpreted these results as being the result of the combined action of diffusion and interception. They concluded that the two mechanisms can be experimentally separated into a range of "pure diffusion" (U_o < 15 cm/sec, $d_p \leq 0.1$ μm) and a range of "pure interception" (U_o = 5 to 15 cm/sec, d_p = 0.1 to 0.6 μm). For the range of "pure interception," they found the empirical equation $E_R = CN_R$, where C is a constant increasing slightly with velocity.

Concerning the relationship between selectivity of the fiber and selectivity of the filter, Kolganov et al. [265] concluded that the selectivity of isolated fibers is greater than the selectivity of fibers embedded in the filter, i.e., fibers in a filter partially lose their selectivity. They also concluded that filter selectivity increases with the homogeneity of the filter.

Further experiments investigating the behavior of fibrous aerosol filters in the range of maximum penetration have been reported by Kirsch et al. [266].

The capture coefficient of a fiber and the efficiency of a filter is, of course, dependent also on other properties of the particles, e.g., on the shape of the particles. In all filtration theories it is assumed (usually implicitly) that the particles are of spherical shape. The influence of particle shape on filter efficiency was investigated by Benarie [267]. Benarie concluded that, for particle Reynolds numbers $Re_p \leq 2$, acicular particles reach the fibers completely unoriented and behave like spheres of the same mass having a diameter equal to the particle length multiplied by 0.285. The value 0.285 was derived theoretically.

2. *Velocity*

Considering the second group of dependencies, the dependence of the capture coefficient on flow velocity or the filter efficiency dependence on gas velocity—called also the velocity characteristics of the fiber or the filter—will be discussed first.

From the definition of the parameter N_D, it follows that its value—and, consequently, the capture coefficient of diffusion deposition—decreases with increasing gas velocity. The parameter of interception is independent of the velocity, and the parameter Stk rises with velocity so that the capture coefficient of inertial deposition increases with velocity. Hence, the mechanisms of diffusion and inertial deposition are dependent on velocity in an adverse sense. It may therefore be expected that the velocity characteristics exhibit a minimum like those of selectivity. The decreasing branch of this curve corresponds to the diffusion mechanism, the increasing branch to the mechanism of inertial deposition or to direct interception. The parameter N_R is independent of velocity, but the capture coefficient E_R of itself is slightly dependent on velocity through the Reynolds number; this relates to an isolated fiber or to filters at high values of Re.

Ramskill and Anderson [159], realizing this, graphically constructed a "filtration diagram" (penetration vs. velocity dependence), describing schematically the velocity characteristics. They distinguished certain important velocities. One, U_{o1}, is the velocity where the inertial mechanism should be measurable, and another, U_{o2}, is the velocity above which the inertial mechanism exerts no additional effect on filter efficiency. It is difficult to obtain U_{o1} from penetration vs. velocity curves, as it is concealed by diffusion and interception effects. The best approximation to U_{o1} is obtained by taking it as the velocity U_{om} where the inertial mechanism is becoming the controlling factor, i.e., where the efficiency starts to increase with increasing velocity. Hence, U_{om} gives the position of the minimum (maximum penetration). On semiempirical grounds they concluded that

$$U_{om} = C \frac{d_f}{d_p^2 S_p} \tag{182}$$

where C is a constant for the given filter.

Pich [259] analyzing Friedlander's equation, Eq. (154), derived for U_{om} the expression

$$U_{om} = \frac{2(kT)^{2/3} d_f \nu^{1/3}}{(3\pi\eta)^{2/3} d_p^{8/3}} \tag{183}$$

Hence, the position of the minimum is dependent on particle size and fiber diameter. Analysis of Eqs. (182) and (183) shows that: (1) With increasing particle size the minimum is shifted toward lower velocities; (2) E_{DR} at the minimum rises with increasing particle size; (3) with increasing fiber diameter the position of the minimum shifts toward higher velocities; (4) E_{DR} in the minimum decreases with increasing fiber diameter; and (5) to avoid shifting of the minimum due to a change in particle size and fiber diameter, the condition $d_f/d_p^{8/3}$ = const must be fulfilled. Conclusions (1) and (3) are in agreement with Eq. (182) and conclusions (1) to (4) agree with the filtration diagram of Ramskill and Anderson [159].

Stechkina et al. [134], on theoretical grounds and using Eq. (166), concluded that the capture coefficient E_{DRI} due to the combined mechanisms of diffusion, interception, and inertial deposition as a function of flow velocity shows a minimum. The position of this minimum is determined by the condition $E_D^f = E_I^f$, as in this case E_R^f is constant.

Velocity characteristics were investigated experimentally by Ramskill and Anderson [159], measuring the penetration of different filters for aerosols of sulfuric acid or dioctylphthalate of d_p = 0.2 to 0.8 μm at velocities up to 285 cm/sec. In several instances they found a minimum in the velocity characteristics. Smith and Surprenant [268] using 0.3 μm dioctylphthalate particles found a distinct minimum in the velocity characteristics of the Whatman No. 41 filter.

Chen [4] reported experiments employing glass fiber filters having an average fiber diameter of 2.5 μm with monodisperse aerosols (four discrete sizes in the range d_p = 0.15 to 0.72 μm) with velocities from 0.87 to 47 cm/sec. He detected a so-called isoefficiency point, i.e., the velocity where all the investigated particle sizes (d_p = 0.15, 0.30, 0.52, and 0.72 μm) have the same capture coefficient. In Chen's experiment this isoefficiency point was found to be about 5 cm/sec.

Humphrey and Gaden [250], in experiments with resin bonded mats of glass fibers of average diameter 16 μm and using biological aerosols *(B. Subtilis)* of d_p = 1 μm, found a minimum in the velocity characteristics between 30 to 60 cm/sec.

Stern et al. [99] reported experiments with IPC fibrous filters (Institute of Paper Chemistry filters, impregnated with an organic adhesive) of average fiber diameter 17 μm and using nondisperse polystyrene particles of d_p from 0.026 to 1.71 μm. They found a distinct minimum in the velocity characteristics for all investigated filters. Moreover, for all investigated filters they found that a unique velocity existed where filter efficiency was the same for all particle sizes, i.e., the existence of an isoefficiency point similar to Chen [4]. This isoefficiency point was at about 17.8 cm/sec at ambient pressure and shifted to lower velocities with a reduction in gas pressure.

Engelhardt and Mohrman [269], using dioctylphthalate particles with d_p = 0.6 to 1.6 μm, found a distinct minimum in the velocity characteristics of asbestos and glass fiber filters located at about 0.6 cm/sec. Moreover, they observed that with further increases of velocity the filter efficiency again decreased.

Lindeken et al. [270] measured the velocity characteristics of Whatman No. 41 paper using monodisperse polystyrene latex particles with d_p = 0.088, 0.188, 0.264, 0.365, and 0.557 μm. A well-developed minimum in the velocity characteristics was found in the velocity range of 10.2 to 15.2 cm/sec.

In addition to the influences of particle size and fiber diameter on the position and shape of the velocity characteristics, there should also be an influence from the filter porosity. This has been well illustrated by Stenhouse et al. [174] in experiments with model filters (series of grids of parallel wires) and monodisperse NaCℓ particles. From their experimentally determined velocity characteristics of a fiber in the filter at packing densities β = 1%, 3%, and 11% with particles of d_p = 3.4 μm, it follows that the capture coefficient exhibits a minimum at about 30 cm/sec. The capture coefficient increases with increasing values of β, and, simultaneously, the minimum shifts to lower velocities.

3. *Pressure*

The dependence of both basic parameters Δp and E on the pressure of the filtered gas is denoted as the pressure characteristic of the filter.

Concerning the dependence $\Delta p = \Delta p(p)$, Eqs. (50), (129), and (146) will be taken as representative of the continuum, slip flow, and free molecule region, respectively. From these equations it follows that in the continuum region Δp is independent of the gas pressure; while in the slip flow and free molecule region, Δp decreases with increasing Knudsen numbers, i.e., with decreasing gas pressure. This conclusion is in agreement with the experiments of Wheat [224], Werner and Clarenburg [31], Stern et al. [99], and Petryanov et al. [230]. A detailed analysis of this dependence and comparison of the theory with available experimental data has been reported by Pich [271]. Some aspects of the flow of rarefied gases through porous media have also been discussed by Wilson et al. [272].

The dependence of individual mechanisms on gas pressure must be analyzed in order to determine the dependence of filter efficiency or fiber capture coefficient on gas pressure. Concerning the diffusion mechanism, the deciding quantity is the diffusion coefficient of the particles D. For Stokes particles in the continuum region, D is given by

$$D = \frac{kT}{3\pi\eta d_p} \tag{184}$$

It is necessary to apply Cunningham's correction for small particle Knudsen numbers, so

$$D = \frac{kT}{3\pi\eta d_p}\left(1 + 1.246\,\frac{\lambda}{r}\right) \tag{185}$$

In the transition region D is given by the Knudsen-Weber (Millikan) equation

$$D = \frac{kT}{3\pi\eta d_p}\left(1 + 1.246\,\frac{\lambda}{r} + 0.42\,\frac{\lambda}{r}\,e^{-0.87r/\lambda}\right) \tag{186}$$

The numerical constants in Eqs. (184) and (185) characterize the reflection of the gas molecules from the particle surface [13]. Since, at constant temperature, λ is inversely proportional to the gas pressure [219], it follows from these equations that D increases with decreasing pressure. Since E_D is proportional to N_D and $N_D = Pe^{-1}$ to D, it follows also that E_D increases with decreasing pressure. Similar analyses for the mechanisms of direct interception, inertial deposition, and electrostatic deposition have been reported by Petryanov et al. [230]. These investigators concluded that the intensity of all three mechanisms increases with decreasing gas pressure. Hence, filter efficiency increases with decreasing pressure and Δp decreases, so filter quality, defined by Eq. (4), rises considerably. These conclusions were checked experimentally by Petryanov et al. [230] using atmospheric aerosols labeled by radioactive daughter products of radon. The quality of FPA-15 filters (mean fiber diameter $d_f = 1.5\ \mu m$) defined by Eq. (4) was measured at a pressure of 13 mmHg and compared with the measured quality of the same filter at p = 760 mmHg. About a fivefold increase in filter quality was found. Also, Stern et al. [99] in experiments with IPC filters having $d_f = 17\ \mu m$ concluded that filter efficiency for a given particle size and flow velocity increases with a reduction in pressure.

The reverse case, the filtration of compressed gases was investigated by Brink et al. [273], where it was found that fiber mist eliminators can be successfully used to purify gases and solve difficult air pollution problems involving methanol synthesis gas, sulfonation and chlorination process gases, nitric acid process gases, chlorine, and compressed air.

4. *Temperature*

The temperature characteristics of a filter or that of a fiber can be defined in terms of the efficiency or capture coefficient dependence on the temperature of the filtered gas while other conditions remain constant. This dependence was studied theoretically by Pich and Binek [274] using the following assumptions: (1) Relatively dry air passes through the filter; (2) no phase transitions occur in the aerosol during the change of temperature, i.e., the particles are not subject to evaporation or condensation so that d_p = const; and (3) the filter is resistant to high temperatures, so that no changes occur in the structure of the filter. On the basis of these assumptions, the parameter of interception N_R is independent of temperature. For Stokes particles the parameters N_D and Stk are affected only by the viscosity η, which is a function of the temperature. Using Sutherland's equation [219] there may be written

$$\eta = \frac{B_s T^{3/2}}{T + C_s} \tag{187}$$

where B_s and C_s are constants. Using this relation the following expressions are obtained for the temperature dependence of parameters N_D and Stk:

$$N_D = \frac{k}{6\pi r d_f U_o} \frac{T + C_s}{B_s T^{1/2}} \tag{188}$$

and

$$\mathrm{Stk} = \frac{mU_o}{6\pi r d_f} \frac{T + C_s}{B_s T^{3/2}} \tag{189}$$

As in the case of selective and velocity characteristics, the mechanisms of diffusion and inertial deposition are oppositely dependent on the temperature, and, therefore, it can be expected that temperature characteristics will exhibit a minimum. The decreasing branch of this curve corresponds to inertial deposition and the increasing branch to diffusion deposition. Assuming that the sum of both parameters N_D + Stk is involved in the expression for the capture coefficient as in Davies' equation, Eq. (162), there can be written

$$\frac{d}{dT}(N_D + \mathrm{Stk}) = 0 \tag{190}$$

giving for the position of the minimum of the temperature characteristics T_m the equation

$$T_m = \frac{1}{k} E_k + \frac{1}{2} C_s + \sqrt{\frac{1}{k^2} E_k^2 + 7 \frac{C_s E_k}{k} + \frac{1}{4} C_s^2} \tag{191}$$

where $E_k = mU_o^2/2$ is the kinetic energy of the particle having the velocity U_o. Hence, the position of the minimum is dependent on the particle size and on the flow velocity. With increasing particle size and increasing velocity the minimum is shifted toward higher temperature, indicating that the inertial mechanism becomes more important.

Benarie and Quetier [275] analyzed the temperature dependence of the parameters N_D, N_R, and Stk and the temperature dependence of the corresponding capture coefficients E_D, E_R, and E_I using equations similar to Sutherland's expression, Eq. (187), for the viscosity-temperature dependence. They came to the same conclusions as Pich and Binek [274], namely, that the diffusion mechanism increases with the gas temperature, the direct interception mechanism is relatively independent of temperature, and the inertial mechanism decreases with increasing temperature.

Also, Strauss and Lancaster [276] analyzed the dependence of Stk and E_I on the temperature and concluded that the inertial mechanism decreases with increasing temperature. Furthermore, they stated that direct interception is not influenced by changes in external conditions and so will not change significantly at high temperatures and pressures. On the other hand, diffusion deposition tends to increase with increasing temperature but to decrease with increasing gas pressure. As the relative changes with gas pressure are much greater than those with temperature, diffusion deposition will also tend to decrease at high temperatures and pressures. Finally, they concluded that none of these three mechanisms will be more efficient at high temperature and pressure than that at low.

Engineering aspects of the filtration of hot gases have been discussed by Snyder and Pring [277] and Strauss and Thring [278]. First et al. [279] measured the efficiency of ceramic filters capable of withstanding temperatures up to 1093°C at temperatures of 21°C and 760°C. For fiber diameters $d_f = 20$ μm, efficiency values of E = 85% (t = 21°C) and E = 82% (t = 760°C) were found. For $d_f =$ 8 μm the corresponding values were E = 99% and E = 94%. Finally, for $d_f = 4$ μm the values E = 98% and E = 91% were found. For the experimental conditions ($d_f = 20$ μm, $d_p = 1$ μm, density $S_p = 6.4$ g/cm^3, $U_o = 177$ and 79 cm/sec) the values of the parameters N_D, N_R, and Stk are: $N_D = 6.69 \times 10^{-7}$, $N_R = 5 \times 10^{-2}$, Stk = 1.75 (t = 21°C); and $N_D = 9.61 \times 10^{-7}$, $N_R = 5 \times 10^{-2}$, and Stk = 0.714 (t = 760°C). It follows that under these conditions the inertial mechanism of particle deposition is most important, and the filter efficiency decreases in agreement with Eq. (189). The minimum of the temperature characteristics according to Eq. (191) is at $T_m \sim 10^8$K.

Hence, for $T < T_m$ the filter efficiency should decrease with increasing temperature. In this sense, the experiments of First et al. [279] qualitatively agree with theory.

Pich and Binek [274] reported measurements of temperature characteristics within the temperature interval 20 to 200°C. An increase of efficiency was observed throughout the investigated temperature interval. The efficiency increased from E = 94% at 20°C to E = 98.5% at 200°C. In these experiments monodisperse, cube-shaped, NaCℓ particles of edge 0.2 μm with density 2.16 g/cm^3 at velocity U_o = 0.6 cm/sec were employed. The kinetic energy of each particle was $E_k = 3 \times 10^{-15}$ erg, and Eq. (191) yields T_m = 235.67K. This means that, from -37.33°C upward, filter efficiency should increase due to the preponderant influence of the diffusion mechanism. To this extent these experiments are consistent with the theory.

Benarie and Quetier [275] measured the temperature characteristics of glass "fritte" and glass fiber filters in the temperature range 290 to 560K. Polydisperse silica dust having a determined cumulative size-weight distribution at a velocity of 10 cm/sec was used. For the first glass fritte filter with pore diameters from 90 to 150 μm, the minimum in the temperature characteristics was found, located at about 400K. From 290 to 400K the filter efficiency decreased, and from 400 to 560K the efficiency increased. For the second glass fritte filter with pore diameters from 20 to 40 μm, the filter efficiency decreased over the entire temperature range. For a glass fiber filter having all fibers parallel to the gas flow, the efficiency was found to be an increasing function of temperature over the entire temperature range. As the mean particle diameter and the mean fiber diameter of the filter are not given, it is difficult to interpret and compare their results with theoretical predictions. Nevertheless, the existence of a minimum in the temperature characteristics found experimentally is consistent with theory.

Pich [17], Petryanov et al. [230], and Benarie and Quetier [275] analyzed the dependence of the pressure drop of fibrous filters on the temperature of the filtered gas. All concluded that the pressure drop of filters operating in the continuum region should increase with increasing temperature in the same manner as does viscosity. For noncontinuum regions the dependence of the mean free path on the gas temperature must be taken into account.

Petryanov et al. [230] pointed out that for meteorological service using analytical fibrous filters it is important to know the filter properties at low temperatures. They measured the pressure drop of FPA-15 filters from +20 to -170°C in the pressure range 760 to 10 mmHg at constant velocity. They concluded that the decrease of temperature leads to a decrease in pressure drop, in agreement with theory. Benarie and Quetier [275] also measured the pressure drop of filters with gas velocities of 5 cm/sec and 10 cm/sec from ambient temperature to 600K. They concluded that the dependence $\log \Delta p = f(\log T)$ is a straight line with a slope 3/4, being parallel to the viscosity-temperature dependence in double logarithmic coordinates.

5. *Structure*

Under the classification structural characteristics are summarized the dependencies of the capture coefficient of the fiber or of filter efficiency on fiber or filter properties (mainly the filter structure). The most important of these characteristics are given below.

a. Dependence of Efficiency on the Geometric Size of the Filter

Efficiency does not depend on the size of the filtration area A, assuming that the filter is homogeneous. On the other hand, efficiency depends to a great extent on the filter thickness L. This dependence was established by Witzmann [280], who concluded that the filter penetration P decreases exponentially with the filter thickness L so that the filter efficiency can be expressed by

$$E = 1 - e^{-k_2 L} = 1 - e^{-L/k_L} \qquad (192)$$

where k_2 is a constant for the given filtration conditions and k_L is the Leers' constant (see Eq. 178).

b. Dependence of Efficiency on the Fiber Diameter

From the definitions of the parameters N_D, N_R, Stk, and others describing the intensity of individual filtration mechanisms, it follows that the values of these parameters—and, hence, the corresponding capture coefficients and the filter efficiency—decrease with increasing fiber diameter. This explains the universal desire to prepare fibers of the smallest diameter when preparing filtration materials. Other factors, e.g., shape of fibers, the state of their surface, and their chemical composition can, of course, also influence the deposition process. However, as concluded by Haupt [281], fiber cross-section and surface roughness have little influence on filter efficiency compared with the influence of fiber diameter.

c. Dependence of Efficiency on the Filter Porosity

From previously given equations for the capture coefficients and equations for filter efficiency, it follows that the efficiency increases with decreasing porosity, i.e., with increasing packing density β. Gallily [282] found somewhat different results in experiments with model filters (an array of wires of diameter 50 μm set perpendicularly within a metallic pipe) and monodisperse, electrically neutral aerosols of dioctylphthalate and varying the interwire distances 2b. The dependence of the capture coefficient of a wire in this model filter on the interwire distance was found to be a curve exhibiting a maximum at about 2b = 200 μm. Gallily tried to explain this unexpected phenomenon by the nonadherence to the wires of some impacting particles.

Grüner [283] investigated the structural characteristics of fibrous filters, viz. filter efficiency-filter porosity dependence, using a very interesting experimental approach. Experiments with pigment aerosols of mean particle diameter 0.56 μm and polyester fiber filters with d_f = 20 and 24 μm were conducted at a temperature of 140°C. At this temperature the decomposition of particles deposited on fibers occurred, thus completely avoiding the clogging of the filter. From the data Grüner concluded that the logarithm of filter penetration as a function of packing density β of the filter was a straight line with a negative slope. Hence, the filter efficiency increased with increasing packing density, which is in qualitative agreement with the theory.

The last conclusion seems to be generally accepted, but the second basic quantity, the pressure drop of the filter, increases rapidly with decreasing porosity as can be seen from the equations given for pressure drop or resistivity. This is the reason why highly porous substances find application in most gas filtration processes. (The porosity of fibrous aerosol filters is usually greater than 90%.) It should be kept in mind that this discussion relates only to high-porosity filters; for low-porosity filters a different analysis is required.

R. Penetration of Particles into Filters

Up to this point the paramount concern has been what proportion of the incoming particles is retained by a filter. There is, however, another important aspect of the filtration process, viz., where are particles deposited. The problem is solved for individual fibers by a calculation of the local capture coefficient, which describes the distribution of the deposited particles on the fiber surface. In a filter, even one assumed to be homogeneous, it can be expected that at various depths in the filter the amount of deposited particles will be different. Hence, the problem of the depth of penetration of particles into filters arises. Theoretically, it can also be expected that the depth of penetration of particles will increase with increasing fiber diameter, because the capture coefficient corresponding to different filtration mechanisms decreases with increasing fiber diameter. Studies of the depth of penetration of aerosol particles into filters are rather limited [284-290].

Sisefsky [284] investigated the depth of penetration of radioactive particles up to 2 μm in diameter in a glass fiber filter of thickness L = 0.2 mm and fiber diameter d_f = 0.6 μm. He devised the so-called stripping technique in which successive layers are separated from the investigated filter. In these experiments 225 cm^2 of the filter under investigation was pressed against a gummed cellulose-acetate film. After subsequent stripping, a thin layer of glass fibers was left on the sticky surface. The process was repeated until the whole filter was divided into twenty such layers, each about 10 μm thick. Each layer was radioautographed and its μ-activity measured. As the activity of different particles was known, their sizes could be approximately determined. The bigger particles were found more likely to penetrate deep into the filter than smaller ones, indicating that particles penetrate deeper the more momentum they have.

Jech [285], also using radioactive aerosols, concluded that in membrane filters having small pores a great part of the aerosol (at least 40%) is deposited in a surface layer of thickness 0.3 μm. In fibrous filters the depths of penetration are greater and the efficiencies of filtration smaller, this being in agreement with theory.

Sittkus and Backheuer [287] studied penetration using atmospheric aerosols labeled with radioactive thoron, finding that the depth of penetration into the filter as a function of filtration velocity exhibited a minimum. They suggested that this might be due to a combination of diffusion, inertial, and electrostatic mechanisms in the filtration process. An explanation—which would be very interesting from the theoretical point of view—is, however, lacking. There is probably a connection between the depth of penetration and the loading characteristics of the filter; this will be dealt with later. In the case of membrane filters, the particles are captured mostly on the filter surface; the loading thus has a "surface" character and the loading characteristic is usually linear. Depths of penetration into fibrous filters are greater; the loading may have, especially in the initial phase of filtration, a "volume" character, making the resulting loading characteristic nonlinear.

Lössner [288], using monodisperse SiO_2 particles of d_p = 0.55 μm labeled with thoron, concluded that three cases can be distinguished in the dependence of the quantity of deposit (arbitrary units) on the depth of penetration. In "surface retention" the majority of particles are deposited on the filter surface or in a very thin layer. Deeper into the filter the quantity of deposited particles decreases exponentially with the depth. And this latter dependence exhibits a maximum.

Madelaine and Parnianpour [289], using radioactive aerosols and the "stripping technique" described previously, concluded that the activity of deposited particles as a function of the depth of penetration exhibits a maximum. This maximum is more pronounced for more efficient filters and is shifted to the filter surface with increasing filter efficiency. The position of the maximum has been found to be from 45 to 25 μm from the filter surface, depending on the type of fibrous filter. For membrane filters the position of the maximum is 6 μm from the filter surface. Beyond the maximum, this function decreases exponentially.

A photographic study of the penetration of particles into filters was conducted by Biles and Ellison [290]. These investigators employed a method similar to that of Stevens and Hounam [286], consisting of the microscopic examination of sections cut by a microtome normal to the surface of the exposed filter. They published photomicrographs showing the distribution of deposited particles in fibrous and membrane filters.

S. Other Filtration Theories

Filtration theory is still incomplete. Present theories are based on the velocity fields around cylinders and in a system of cylinders. These velocity fields represent two extreme cases valid for very small Reynolds numbers (slow viscous flow, sometimes called creeping motion) and for large Reynolds numbers (potential flow). Between the extreme cases is an intermediate (transition) region of Reynolds numbers. For example, Whitby [175] assumed (without giving arguments for the classification) that the viscous flow region is limited to Re < 0.2, potential flow to Re > 150, and the transition region to 0.2 < Re < 150. As stated by Whitby, to be useful, filtration theory must be applicable also in the transition region of Re values. However, there are almost no theoretical studies that can be applied in this region. An exception is, for example, the paper by Yoshioka et al. [153], who attempted to solve the problem by construction of interpolation charts. Furthermore, most theories are usually valid only for high porosity filters ($\beta \ll 1$). The number of papers dealing with low porosity (dense) filters is very limited. There are no studies, to the knowledge of the author, of filters having intermediate values of porosity.

In existing filtration theory it is always assumed that the structural units of the filter are motionless (an exception is the paper by Radushkevich [232]. However, this assumption is not always necessary, as the theory and construction of rotary impaction filters shows as discussed later.

Finally, existing filtration theory is essentially based on the theory of transport phenomena. There are, however, other possible approaches. These attempts (phenomenological and statistical theories) are discussed in the following sections.

1. *Phenomenological*

Usually a filtration process is studied with the final aims being the derivation of equations for pressure drop and filter efficiency and an analysis of them to determine filter characteristics, filter quality, and suitability of application of the given filter under given filtration conditions. Another method for describing the filtration process was suggested by Radushkevich [291]. This approach does not deal with the details of the filtration process but tries to describe the process as seen from the "macroscopic point of view." The main objective of this approach is a relationship between the size distribution of particles entering the filter, the selectivity of the filter, and the size distribution of particles leaving the filter.

The penetration of particles having sizes in the interval r to r + dr through the filter is given by

$$P(r) = \frac{N(r)}{n(r)} \tag{193}$$

where n(r) is the concentration of particles having sizes between r and r + dr before the filter and N(r) the concentration of the same particles after the filter. The median penetration $\bar{P}$ is

$$\bar{P} = \frac{N_o}{n_o} \tag{194}$$

where n_o is the total number concentration (number of particles of any size per volume unit) before the filter and N_o the total number concentration after the filter. The quantities P(r) and $\bar{P}$ are related by

$$P(r) = \bar{P}\,\frac{f_2(r)}{f_1(r)} \tag{195}$$

where $f_1(r)$ is the normalized ($f_1(r) = n(r)/n_o$, so that $\int_0^\infty f_1(r)dr = 1$) size distribution function of particles before the filter and $f_2(r)$ that after the filter. Similar relations can be formulated for a discrete-particle size spectrum. If there are m filters arranged in series with penetrations $P_i(r)$ (i = 1, 2, $\cdots$, m) and mean penetration $\bar{P}_i$ (i = 1, 2, $\cdots$, m) then it can be shown that

$$P_1(r) \cdot P_2(r) \cdot \cdots \cdot P_m(r) = \bar{P}_1 \cdot \bar{P}_2 \cdot \cdots \cdot \bar{P}_m\,\frac{f_m(t)}{f_1(r)} \tag{196}$$

here $f_m(r)$ is the normalized size distribution of the outlet aerosol. The input spectrum $f_1(r)$ does not depend on the properties of the filter, while the outlet spectrum $f_2(r)$ depends on both $f_1(r)$ and P(r). Equation (195) permits solving three types of problems. First, if $f_1(r)$ and $f_2(r)$ are given, e.g., these spectra are determined experimentally, then P(r) can be determined from Eq. (195) and, since P(r) + E(r) = 1, it is possible to determine the selectivity of the filter. If the selectivity 1 - P(r) is determined theoretically or by way of experiments carried out with test aerosols and if the spectrum $f_1(r)$ is known, then using Eq. (195) it is possible to determine the spectrum after the filter $f_2(r)$. And, finally, if $P(r)$ and $f_2(r)$ are known, it is possible to determine $f_1(r)$ and to use the filter as an analyzer.

Obviously, in the case of a monodisperse aerosol $f_1(r) = f_2(r)$ and, consequently, $P(r) = \bar{P}$. Similarly, for a nonselective filter $P(r) = \bar{P}$ and, therefore, $f_1(r) = f_2(r)$ for a polydisperse aerosol also.

From a practical point of view the most important group of problems is the second, where the selectivity and inlet aerosol spectrum are known. Usually the inlet spectrum $f_1(r)$ exhibits a maximum (Gaussian distribution or log-normal distribution). Furthermore, the filter selectivity has a minimum so that the penetration $P(r)$ has a maximum. Then the following important cases can occur: (1) The inlet spectrum $f_1(r)$ is wide and the curve $P(r)$ is narrow with a distinct maximum. Phenomenological theory then shows that the outlet spectrum $f_2(r)$ also has a maximum located in the same region as the maximum of $P(r)$. (2) The reverse case is a narrow inlet spectrum with a distinct maximum while the curve $P(r)$ is wide. Then the outlet spectrum also has a maximum located in the same region as the maximum of the inlet spectrum. (3) Both functions $P(r)$ and $f_1(r)$ have distinct maxima located at a certain distance from each other. Then the outlet spectrum $f_2(r)$ also exhibits a maximum located between the maxima of the functions $f_1(r)$ and $P(r)$.

Hence, using the phenomenological theory of filtration, certain conclusions can be obtained regarding the changes in the aerosol spectrum without going into the details of the filtration process. However, it is obvious that the phenomenological approach in itself cannot determine the selectivity curve. On the other hand, this approach can be applied not only to filters but also to problems of particle deposition from the gas flow upon bodies of different geometrical shapes.

Another very important problem of the phenomenological description of filters is that of filter quality. This problem has recently been discussed by Juda and Chrosciel [1]. These authors correctly pointed out the need for a quantity that would characterize which of two filters is "better" or "worse" and that would permit comparison between different filters. In this connection, it was pointed out that the definition of filter quality given by Eq. (4) is insufficient, because this definition operates only with parameters (Δp and E) of the clean filter and does not take into account the behavior of the filter during filter clogging. It is also desirable that this as yet undefined filter quality should be dimensionless. However, the attempt of Juda and Chrosciel to construct such a quantity ignores existing filtration theory. For example, contentions that the filter penetration-velocity relationship is given with "sufficient accuracy" by $P = \exp(-cLU_o^{1/2})$, where c is a constant, have no justification as follows from the discussion of velocity characteristics.

2. *Statistical*

An attempt to develop a statistical theory of aerosol filtration based on the theory of probability was reported by Gutowski [292]. Gutowski divided the whole volume of the filter into ν^* elementary units such that $\nu^* = \varepsilon L/d_f$ and considered the probability x of the deposition of a particle in one elementary unit. He concluded that the quantities x and ν^* are related to filter efficiency E by

$$E = 1 - (1 - x)^{\nu^*} \tag{197}$$

For large values of ν^* the function $(1 - x)^{\nu^*}$ can be approximated by $(1 - x)^{\nu^*} \approx e^{-x\nu^*}$ so that Eq. (197) becomes

$$E = 1 - e^{-\varepsilon xL/d_f} \tag{198}$$

From comparison of this equation with Eq. (176) it follows that the probability x is proportional to the capture coefficient of the fiber composing the filter.

In order to evaluate the probability x, Gutowski distinguished two phases in the process of particle deposition. The first is a collision of the particle with a collecting surface, and the second is the sticking of the particle to the collecting surface. Denoting the probability of the particle colliding with the surface by x_c and the probability of the particle sticking to the collecting surface by x_s, it was assumed that $x = x_c \cdot x_s$. Furthermore, Gutowski introduced a collision coefficient k_c and a sticking coefficient k_s, which are related to x_c and x_s by $x_c = k_c$ for $0 \leq k_c < 1$ and $x_c = 1$ for $k_c \geq 1$, and, similarly, $x_s = k_s$ for $0 \leq k_s < 1$ and $x_s = 1$ for $k_s \geq 1$. Gutowski concluded (in an unknown manner) that each of these coefficients could be expressed as a function of a dimensionless parameter, so that $k_c = k_c(G_c)$ and $k_s = k_s(G_s)$. The parameter G_c—called the collision number by Gutowski—is a nondimensional complex of quantities such as gas velocity U_o, fiber diameter d_f, mean particle velocity in the direction to the collecting surface, hydrodynamic factor H*, and the acceleration of the particle. Similarly, G_s—called the sticking number by Gutowski—is a nondimensional complex involving such quantities as adhesion energy, a coefficient of static friction between particle and surface, a coefficient describing the contact of a particle with a surface, particle size and density, mean gas velocity, and the porosities of the particles and of the collecting surface. As the majority of these quantities are unknown, Gutowski includes them in an empirical coefficient dependent on the gas velocity, particle size, and density.

In order to calculate the parameter G_c for each filtration mechanism, it is necessary to calculate the particle acceleration and the mean particle velocity. For example, for a gravitational mechanism Gutowski identifies the particle acceleration with $U_o^2/H^{*2}d_f$ and the mean particle velocity with the steady-state sedimentation particle velocity. It is difficult to say how far these calculations are justified as only the final result was presented by Gutowski. Furthermore, Gutowski assumes (without giving arguments) that the function $k_c = k_c(G_c)$ is linear and claims to obtain approximately the same results for mechanisms of diffusion, inertial, and gravitational deposition as in existing filtration theory, which is called a "mechanical theory."

Finally, Gutowski discusses selective and velocity characteristics on the basis of his approach. He concludes that besides a minimum these curves should also exhibit a maximum at a certain particle size and velocity. After reaching a maximum these curves should decrease due to decreasing values of the sticking coefficient k_s with increasing velocity or particle size.

As already noted, it is difficult to evaluate the whole of Gutowski's approach as calculations have not been published in the open literature. Judging from the only summarizing paper presented by Gutowski at one symposium [292], there arc several weak points in the approach and some of the arguments are questionable and unconvincing. However, the approach is original, and this is its main merit at the present time.

3. *Low Porosity Filters*

The theory discussed so far has related to high-porosity filters where $\beta << 1$ and the distance between the axes of cylinders 2b is much greater than the fiber diameter 2a, i.e., the corresponding Fuchs' number $Fu = a/b$ is small ($Fu << 1$). In this case the fibers can be approximately treated as isolated cylinders, the interference effect being negligible or very small.

There is, however, an important opposite of low-porosity filters where $\varepsilon << 1$. The distance between the surfaces of cylinders 2h given by $2b = 2a + 2h$ is small in comparison with cylinder diameter 2a so that the Fu number is large and approaches unity ($Fu \sim 1$). For calculation of the velocity distribution, pressure drop, and capture coefficients due to individual deposition mechanisms, the model of parallel and equidistant cylinders is usually applied.

Keller [46] using a concept of lubrication theory and assuming a square filter structure ($\beta = (\pi/4)Fu^2$) derived for the pressure drop of low-porosity filters the equation

$$\Delta p = \frac{9\pi}{8\sqrt{2}} \frac{\eta U_o L}{a^2} \frac{Fu^2}{(1 - Fu)^{5/2}} \tag{199}$$

Hence, the resistivity X is given by

$$X = \frac{a^2 \Delta p}{\eta U_o L} = \frac{9}{2\sqrt{2}} \frac{\beta}{(1 - 2\sqrt{\beta/\pi})^{5/2}} = f(\beta) \tag{200}$$

and is independent of the Reynolds number.

Gupalo [23], for a triangular structure ($\beta = (\pi/2\sqrt{3})Fu^2$) of parallel equidistant cylinders, derived the equation

$$\Delta p = \frac{3\eta U_o L}{a^2} \frac{Fu^3}{(1 - Fu^2)^2} \left\{ 1 + \frac{3\sqrt{3}}{4} Fu + \frac{2Fu^2}{1 - (\sqrt{3}/2)Fu} - \frac{(\sqrt{3}/16)Fu^3}{[1 - (\sqrt{3}/2)Fu]^2} + \frac{6Fu}{(1 - Fu^2)^{1/2}} \operatorname{arccot}\left[0.268 \frac{(1 + Fu)}{(1 - Fu^2)^{1/2}} \right] \right\} \tag{201}$$

Here the corresponding resistivity is again independent of Re. Gupalo concluded that Eq. (201) is valid for $Re \ll 2Fu^2/(1 - Fu)^2$.

Calculation of particle deposition in low-porosity fibrous filters was first attempted by Fuchs [13]. Assuming that Poiseuille flow proceeds in the curved capillaries formed by the interstices of the fibers, Fuchs derived (for a triangular structure) approximate relations for the coefficients of absorption of the aerosol by the filter α_D, α_R, α_G, and α_I due to diffusion, direct interception, gravitational, and inertial deposition, respectively. These coefficients are given by Eq. (175), so the filter efficiency due to particle diffusion is given by $E = 1 - \exp(-\alpha_D)$, and so on. These equations, after introducing the dimensionless variables Fu, Pe, N_R, N_G, and Stk, become

$$\alpha_D = \frac{2^{5/3}(L/a)Fu^{7/3}}{3^{1/6}Pe^{2/3}(1 - Fu)^{2/3}} \tag{202a}$$

$$\alpha_R = \frac{\sqrt{3}\ N_R^2(L/a)Fu^3}{2(1 - Fu)^2} \tag{202b}$$

$$\alpha_G = \frac{N_G(L/a)Fu^2}{\sqrt{3}} \tag{202c}$$

$$\alpha_I = \frac{Stk(L/a)Fu^2}{\sqrt{3}(1 - Fu)^2} \tag{202d}$$

Equation (202a) was derived under the usual assumption of high Péclet numbers ($Pe >> 1$).

Another solution to the problem of particle diffusion in low-porosity filters was reported by Gupalo [23]. His equation for the fiber capture coefficient due to particle diffusion E_D in a low-porosity, triangular arrangement of parallel cylinders can be written

$$E_D = \frac{3.72Fu^{1/3}}{Pe^{2/3}} \left\{ \ln 3 + \frac{2Fu}{(1 - Fu^2)^{1/2}} \operatorname{arccot} \left[\frac{\sqrt{3} - 2Fu}{(1 - Fu^2)^{1/2}} \right] \right\} \tag{203}$$

Comparing Eq. (203) with Eq. (176) on the one side and Eq. (202a) on the other, it can be concluded that Fuchs' and Gupalo's theories yield $\alpha_D \sim (L/a)Pe^{-2/3} f(Fu)$. The function $f(Fu)$ is different due to the various approximations employed for the flow field.

4. *Rotary Filters*

In the theory discussed so far, the filter as a whole and its structural units (fibers) were considered motionless. Hence, the relative velocity between a particle and a collecting surface (cylinder) was, in principle, established by the velocity of the gas U_o approaching the cylinder. An increase of the velocity U_o was necessary in order to exploit the mechanism of inertial deposition of particles. However, a second possibility is to increase the relative velocity between the particle and the collecting surface by moving the collecting surface with sufficient velocity. This idea was tested by Pyne et al. [293] who have shown that particles can be removed from a gas stream by inertial deposition on a number of filaments mounted radially on a central disk rotated in a plane perpendicular to the direction of flow. Using the experimental data of Wong et al. [163] from which it follows that, for $Stk > 6$, the capture coefficient E_I is unity or very close to unity, they concluded that a filter rotating at 1900 rpm with an inner radius R_1 = 28 cm (corresponding linear velocity U_o, 5600 cm/sec) and filament diameter d_f = 460 μm should remove all particles of unit density above about d_p = 4 μm. In order to check the theoretical estimates, Pyne et al. [293] performed the experiments with a filter composed of 390 filaments, 10 cm long and 460 μm thick, mounted on a disk 56 cm in diameter. When the disk was rotated at 1900 rpm, just sufficient to ensure that the Stokes number for 4 μm particles of unit density was 6, and the air flow was 1.18 m^3/sec the measured filtration efficiency was 90 ± 5% for 4 μm particles of unit density. It was progressively greater for larger particles. Such rotary filters (sometimes called rotary filamentary or rotary impaction filters) have the advantage of being compact and self-cleaning, and they offer little resistance to large rates of air flow.

Soole [294] developed a general theory for the design of these filters. He has shown that there are five independent variables (two independent form factors, a nondimensional filament thickness, one dimension for fixing the scale, and a rate of revolution) from which all characteristics required for the filter design can be derived.

Soole [295] applied this general theory of rotary filter design to the requirement for the removal of all particles above a stated size from a given gas flow rate with a filter of a given diameter rotating at a given speed. He concluded that filters requiring the least power and producing the least pressure drop are those with the smallest radius ratio. Such filters must have a relatively large number of fine filaments. In practice, somewhat larger radius ratios with a smaller number of filaments may be preferable, despite a small increase in power consumption and pressure drop.

Finally, Soole and Meyer [296] reported the measurements of filtration efficiencies, pressure drop, and power consumptions for ten rotary filters operating with air flow rates of 4.72 m^3/sec. Five of the filters were designed to separate all droplets of 10,

15, 20, 30, and 40 μm (and above) at 500 rpm and five to separate all droplets of 10 μm (and above) at 1000 rpm. Efficiency of the filters was determined using polydisperse droplet sprays from radioactively-tagged glycerol solutions. They found that filtration efficiency decreased with decreasing filament thickness because of fragmentation of the larger entering drops caused by aerodynamic shear forces as they approached the filaments. They suggested that this effect could be minimized by designing the filters to have no greater values of the Stokes number than the necessary minimum to remove the desired droplets and by ensuring (using prefiltration, if necessary) that no drops larger than some (unknown) critical diameter were present in the incoming spray. The measured pressure drops of 6.5 to 46.1 mm of water gauge correlated with calculated mean velocities of flow through the filter interstices. Power consumption was found to be up to 37% less than predicted because of filament air screening.

V. NONSTATIONARY FILTRATION

The theory of stationary filtration as dealt with above is based on two basic assumptions: the collision efficiency of particles with collecting surfaces (fibers) is unity, i.e., a particle that once touches a fiber remains captured; and deposited particles have no further influence on the filtration process. In this case both basic parameters—filter efficiency E and pressure drop of the filter Δp—are time-independent and the process is stationary. This abstraction is approximately fulfilled in the initial stage of filtration at low concentration of incoming particles.

In reality, however, the filtration process is much more complicated. Particles, even if once deposited, can become separated from fibers due to the action of different forces and pass through the filter. Also, deposited particles change the geometrical shape of the individual collectors, thus producing structural changes in the filter. As a result both filtration efficiency and pressure drop are subject to change with time, and the process becomes unsteady. The causes of changes in E and Δp with time are called "secondary processes." This term was suggested by Radushkevich [297] who distinguished "primary processes" (diffusion, interception, inertial, and electrostatic deposition, etc.) from "secondary processes" (deposition of particles on one another, capillary phenomena, loss of electric charge, destruction of the filter).

However, the nonunity of the collision efficiency of particles with the fibers and the possible reentrainment of deposited particles can also lead to time variations of the filter efficiency. Obviously the adhesion forces acting between a particle and a fiber or between deposited particles play a very important role in the separation of deposited particles from the fibers. Furthermore, the deposition of particles on one another is a special case of filter clogging. Usually, in studying filtration the assumption is made that only E is varying with time and Δp is constant. However, sooner or later Δp must also vary with time.

Realizing these facts, the Radushkevich's classification may be modified in the following way, without claiming completeness, of course. Secondary processes in aerosol filtration include: (1) reentrainment of deposited particles from the fiber surfaces (the behavior of deposited particles and the problem of adhesion forces are related here), (2) clogging of filters, (3) capillary phenomena, (4) loss of electric charge by the filter, and (5) destruction of the filter. Secondary processes will be briefly discussed in this sequence.

A. Adhesion and Reentrainment of Particles

In the theory of stationary filtration it was assumed that, after contact with the surface of a fiber, particles remained fixed throughout the further course of the filtration process. The validity of this assumption depends on the strength of the bond between the particle and surface, called the force of adhesion. The adhesion force can be defined as the force required to dislodge an adhering particle from a given surface. Knowledge of adhesion forces is important not only for the quantiative description of filtration but also for other particle coagulation and industrial air-cleaning processes. For this reason a large number of theoretical and experimental papers are devoted to the problems of adhesion as reviewed by Corn [298], Morgan [299], Corn [300], Krupp [301], and Zimon [302]. According to Krupp [301], van der Waals and electrostatic forces are dominating attractive forces contributing to the adhesion of particles.

3. *Van der Waals Forces*

The theory of van der Waals forces has been treated by Margenau [303] and recently reviewed by Margenau and Kestner [304]. Three types are usually distinguished: the forces between molecules possessing permanent dipoles (the Keesom interaction); the forces caused by the polarizing action of one molecule on another (the Debye interaction); and nonpolar van der Waals forces, called also London or London-van der Waals forces (the London interaction). The London forces are also referred to as dispersion forces because London associated them with optical dispersion. The theory of London forces has been developed by London [305,306], Lennard-Jones [307], and Casimir and Polder [308]. They are usually dominant in molecular interaction according to Butler [309]. The attractive London force F between two molecules is $F = cS^{-7}$, where c is the constant of interaction and S the distance between the molecules.

In order to calculate the van der Waals forces between macroscopic bodies, essentially two approaches have been followed. The first, developed by Bradley [310] and especially by Hamaker [214] is based on the assumption of additivity of the energy of interaction of the London forces. As a result of this additivity, the London-van der Waals force F acting between two spheres of diameter d_1 and d_2 in a vacuum is found to be

$$F = \frac{\pi^2 q_o^2 c}{12S^2} \frac{d_1 d_2}{d_1 + d_2} \qquad (204)$$

where q_o is the number of atoms per cubic centimeter. The force acting between a sphere of diameter d_1 at a distance S from a flat surface is given by (see also Eq. 111)

$$F = \frac{\pi^2 q_o^2 c}{12S^2} d_1 \qquad (205)$$

The same approach for bodies of other geometrical shapes has been developed by Vold [311].

The second approach to the van der Waals interaction between macroscopic bodies was suggested by Lifshitz [312]. This macroscopic approach is based on the concept of the fluctuating electromagnetic field and was developed by Krupp [301]. It enables the derivation of van der Waals forces directly from the frequency-dependent dielectric constants of the adherents. Further development of the Lifshitz macroscopic approach has been reported by Langbein [313]. Both approaches have recently been reviewed by Gregory [314]. Direct measurement of van der Waals forces between macroscopic bodies has been reported by a number of authors, e.g., Deryaguin [315], Deryaguin et al. [316], Prosser and Kitchener [317], and Kitchener and Prosser [318].

2. *Electrostatic Forces*

As emphasized by Corn [300], knowledge of the electrostatic component of the adhesion force is rather limited and contradictions occur. Krupp [301] reported on the application of potential and boundary-layer theories of semiconductors for estimation of the electrostatic component of adhesion. Aleinikova et al. [319] investigated experimentally the electrostatic component of adhesion of dielectric particles (powdered PVC) to a metal surface (65-H steel). The charge of the PVC particles after tearing them from the steel surface increased about 100 times when compared with the charge of the particles adhering to the metal. The sign of the charge after the tearing could be either positive or negative, and the charge increase was proportional to the square of the particle radius, indicating that the adhesion was determined mainly by electrostatic forces.

In a further paper from this group, Deryaguin et al. [320] investigated the role of electrostatic forces in the adhesion of polymer particles to solid surfaces. Recently, Deryaguin and Smilga [321] developed an electronic theory of adhesion which enables a calculation to be made of the electrostatic component when certain physicochemical parameters are known. There are electrostatic components of the order of 10^8 to 10^9 dyne/cm^2. Adhesion forces can be varied by chemical modification of the surfaces (and sometimes of the bulk) of the contacting objects. Adaptation of theoretical equations to real systems, experimental methods for measuring adhesion forces, and the influence of other factors (particle and surface material, nature of contact, particle size and shape, temperature and surface roughness) were discussed partly by Corn [300] and later by Krupp [301].

Concerning the reentrainment of deposited particles from fibers, it can be said that the fate of deposited particles is dependent on the adhesion forces acting between the particles and the fibers, and also on the aerodynamic forces acting on the deposited particles. At low velocities the aerodynamic drag forces are too small to overcome the adhesion forces. Hence, to keep particles on fiber surfaces the drag forces must be smaller than the adhesion forces, and the energy of adhesion should be greater than the kinetic energy of the particle after collision with the fiber. In this case the particles stick to the fiber surface. The intensity of adhesion forces is dependent on the size, shape, chemical composition, electric charge, microgeometry of the particle and fiber surface, humidity, time of contact, and other factors. Usually, only experimental data for specific conditions or empirical relations are available as the phenomenon of particle separation from fibers is very complex. For example, at velocities U_o (in m/sec) $U_o < 0.3/d_f$ (d_f expressed in microns) silica particles remain on the glass fibers. There are also other complicating phenomena. For example, Gillespie and Langstroth [322] and Botterill and Aynsley

[323] observed that in collisions of dust particles with collecting surfaces electric charges arise on the surfaces. The value of the charge is dependent on the velocity of flow and on the number of collisions. Similar results are limited to specific conditions, and, therefore, the reader is referred to the original papers, which are arranged chronologically and listed under Refs. 324 to 331. Only papers published after 1967 are included, as in this year the review by Krupp [301] and the book by Zimon [302] appeared, summarizing and discussing the accumulated knowledge in the field.

B. Clogging of Filters

The clogging of filters by particles of different size, shape, and phase is also an important secondary process. The filter structure is changed by the deposited particles, and even a new layer consisting of deposited particles can be formed on the filter surface. These structural changes cause shifts in filter efficiency and of filter pressure drop. Structural changes are dependent on a number of factors, especially on the quantity of accumulated particles deposited within or on the filter. Quantity of the accumulated particles can be defined in several different ways. First, definition can be given by $Z_1 = M/A$, where M is the mass of deposited particles and A the face area of the filter. The second definition, due to Dorman [14] is given by $Z_2 = V'/A$, where V' is the volume of deposited particles. The advantages of this latter definition is that the loadings Z_2 of different filters can be compared for particles of different densities. The third definition, applied, e.g., by Yoshioka et al. [332], is the specific deposit $Z_3 = M/V_f$, where V_f is the volume of the filter. A fourth definition Z_4 is due to Billings [97]. The latter expresses an accumulation of solid particles in a fiber filter as the number of particles per square centimeter of filter face area. In order to describe the accumulation of particles on a fiber, the accumulation Z_5 becomes the number of particles deposited locally on a fiber per square centimeter of fiber cross-section normal to gas flow.

First to be discussed is the variation of filter pressure drop during the course of clogging.

1. *Changes of Pressure Drop*

Changes of the pressure drop in the course of clogging are of a special importance in the use of filters in industry and are closely connected with the "life" of the filter (sometimes so-called accelerated life-tests of the filters are carried out). The change of Δp due to deposited particles (clogging) depends mainly on the following factors. (1) Particle phase: In the filtration of liquid droplets from a gas Δp is subject to smaller changes than with solid particles, due to the different behavior of particles after contact with fibers. (2) Particle size: Finer particles usually cause a higher Δp than coarser particles. Another important factor is the depth of penetration of particles into the filter. (3) Filter structure: In some cases the filter structure determines whether particles are deposited within the filter structure or on its surface, which reflects the different ways in which Δp is changed with clogging. (4) Quantity of the deposited particles: The quantity Z_i (i = 1, 2, 3, 4) is called loading or accumulation, and the dependence $\Delta p = f(Z_i)$ is the loading characteristic of the filter pressure drop. This dependence has been studied by a large number of investigators.

Williams et al. [333] and Hemeon [334] assumed that there were no structural changes in a filter in the course of clogging and that deposition took place only on the filter surface, where a new filtration layer was formed by the deposited particles. The pressure drop of the loaded filter Δp is the sum of the filter pressure drop Δp_o and the pressure drop of a dust layer Δp_ℓ, so $\Delta p = \Delta p_o + \Delta p_\ell$ and

$$\Delta p = k_1 U_o Z_1 + \Delta p_o \tag{206}$$

where k_1 is a constant which can be approximated using the Kozeny-Carman equation, and Z_1 is the loading as defined above. According to Eq. (206) the loading characteristic is linear. This concept is approximately fulfilled for some membrane filters where the depths of penetration are very small. Some membrane filters show, in fact, a linear loading characteristic.

Snyder and Pring [277] found that relatively few aerosols behave in this way and that linear loading characteristics are found mainly for coarse fractions, while for finer fractions exponential or other relations are found. They concluded that the layer on the filter is subject to structural changes and that the porosity of this layer, consisting of deposited particles, changes.

Thomas [335] suggested an empirical equation for the time necessary for Δp to increase from the starting value Δp_o (clean filter) to Δp (loaded filter). However, the empirical "clogging coefficient" involved in this equation is dependent on the nature of the particles and their size.

Miczek [336] concluded that in the function $\Delta p = f(Z_1)$ two regions can be distinguished. In the first, Δp remains nearly constant or rises rapidly. In the second region, the increase of Δp with loading is more or less linear. In certain filters the first region is lacking and the dependence $\Delta p = f(Z_1)$ is linear from zero loading. As already given, this is found for some membrane filters and can be related to the depth of penetration of particles into filters.

Billings [97] found that the deposits formed during filtration of solid particles produce an increase in filter pressure drop. The deposited particle structures develop and grow out from the fibers, thus adding to the fiber drag. He assumed that the resulting drag on the deposit in the filter is proportional to the accumulation of particles Z_4 and the drag on a single particle. So, in the initial stage of clogging,

$$\Delta p = \Delta p_o + S_p \eta U_o r Z_4 \tag{207}$$

where S_p is the particle resistance coefficient. Using experimental data, Billings concluded that this coefficient can be expressed as $S_p = 33\beta$ in the range $0.007 < \beta < 0.035$. The coefficient S_p was independent of velocity in the range $13 < U_o < 140$ cm/sec and independent of the filter thickness L in the range $0.11 \leq L \leq 0.29$ cm.

Yoshioka et al. [332] measured the increase of the pressure drop due to clogging as a function of $C_i U_o t$, where C_i is the mass concentration of the inlet aerosol. Furthermore, they measured loadings characteristics $\Delta p = f(Z_3)$ and concluded that for $Z_3 > 1$ kg/m^3 this function can be approximated by

$$\Delta p = \Delta p_o(1 + 1.1Z_3) \tag{208}$$

They also tried to take into account the distribution of deposited particles in the filter.

Solbach [337] discussed the shape of the loading characteristic $\Delta p/U_o = f(Z_1)$ in connection with periodic cleaning of the filter and calculated how the flow rate decreases in the course of filter clogging at constant pressure drop.

Löffler [338] discussed the problem of "surface loading" (preferential accumulation of particles on the filter surface). The D'Arcy law for a layer consisting of particles deposited on filter surface can be expressed by

$$\Delta p_\ell = \frac{1}{k_\ell} \eta U_o L_\ell \tag{209}$$

where the index ℓ refers to a dust layer, so that k_ℓ is the permeability and L_ℓ the thickness of the layer of deposited particles. The efficiency of this layer E_ℓ can be described by

$$E_\ell = \frac{AS_p(1 - \varepsilon_\ell)L_\ell}{AU_o C_i t} \tag{210}$$

where, again, C_i is the mass concentration of the inlet aerosol. Combining Eqs. (209) and (210) results in

$$\Delta p_\ell = \frac{1}{k_\ell} \frac{\eta E_\ell C_i U_o^2 t}{S_p(1 - \varepsilon_\ell)} \tag{211}$$

where ε_ℓ is the porosity of the dust layer. Hence the pressure drop of the dust layer Δp_ℓ is linearly proportional to time t and particle concentration C_i and increases quadratically with U_o—assuming, of course, that the permeability k_ℓ is constant over the investigated period of time. Moreover, Löffler described an automatic experimental device for the measurement of changes of filter pressure drop and filter efficiency in the course of clogging, especially in the period after filter cleaning.

Another extreme case is when particles are deposited only inside the filter, i.e., the loading has a "volume" character. This case has been recently investigated by Juda and Chrosciel [339]. These investigators suggested a simple theoretical model to describe loading characteristics. In the course of "volume" loading, the packing density of the filter β is increasing according to

$$\beta = \beta_o + \frac{V'}{V_f} = \beta_o\left(1 + \frac{V'}{\beta_o V_f}\right) \tag{212}$$

where β_o is the packing density of the pure filter of volume V_f and V' is again the volume of deposited particles. Assuming uniform distribution of deposited particles on a cylindrical surface, they suggested that the hydrodynamical radius of the fiber of a clogged filter can be expressed by

$$a^2 = k_2 a_o^2\left(\frac{V'}{\beta_o V_f} + 1\right) \tag{213}$$

where $k_2 = (a/a_g)^2$, a_g is the geometric fiber radius of the clogged filter, and a_o is the fiber radius of the pure filter. Assuming k_2 = const and introducing Eqs. (212) and (213) into the Fuchs-Stechkina equation, Eq. (50), gives (the resulting Juda and Chrosciel's equation is dimensionally incorrect apparently due to a printing error)

$$\Delta p = \frac{\Delta p_o}{k_2} \frac{\ell n\ \beta_o + 2C}{\ell n\ \beta_o + 2C + \ell n[(V'/\beta_o V_f) + 1]} \tag{214}$$

where, again, C = 0.5 (Happel) and C = 0.75 (Kuwabara). However, this comparison of theory with experiments is rather unconvincing as nothing is described about the experiments and nothing is given about how the porosity of the deposit was determined. For this and other reasons the statement that this model is in good qualitative agreement with the real filtration process is unconvincing.

2. *Changes of the Fiber Capture Coefficient and Filter Efficiency*

When filtering solid particles, the distribution of particles on a fiber surface is not uniform; particles deposit one upon another (coprecipitation), forming so-called trees (dendrites) by which the new structure is built step by step. The parts of the fiber between the individual trees are relatively clean, i.e., incoming particles apparently prefer to deposit on already deposited particles instead of on the fiber. This phenomenon has been observed by several authors and described in detail by Leers [257], Billings [97], and Albrecht [340] (see Fig. 1). This effect results in increased filter efficiency while changes of the pressure drop may be relatively insignificant.

Radushkevich [297] made an attempt to describe this process theoretically. Making a number of simplifying assumptions, he derived the equation

$$P = P_o e^{-k_3 t} \tag{215}$$

where P_o is the penetration at time $t = 0$ (penetration of the clean filter) and k_3 is a constant under the given conditions. In a subsequent paper, Radushkevich [341] tried to describe theoretically the deposition of one particle upon another on the fiber and the distribution of aggregates consisting of particles deposited according to size. For the number of deposited particles on a fiber of length L_f at time t he derived the equation

$$n = \frac{C_1}{C_2}\left(e^{C_2 U_o n_o t} - 1\right) \tag{216}$$

where C_1 is the capture cross-section of the cylinder and C_2 the capture coefficient of the deposited particle. If there is no

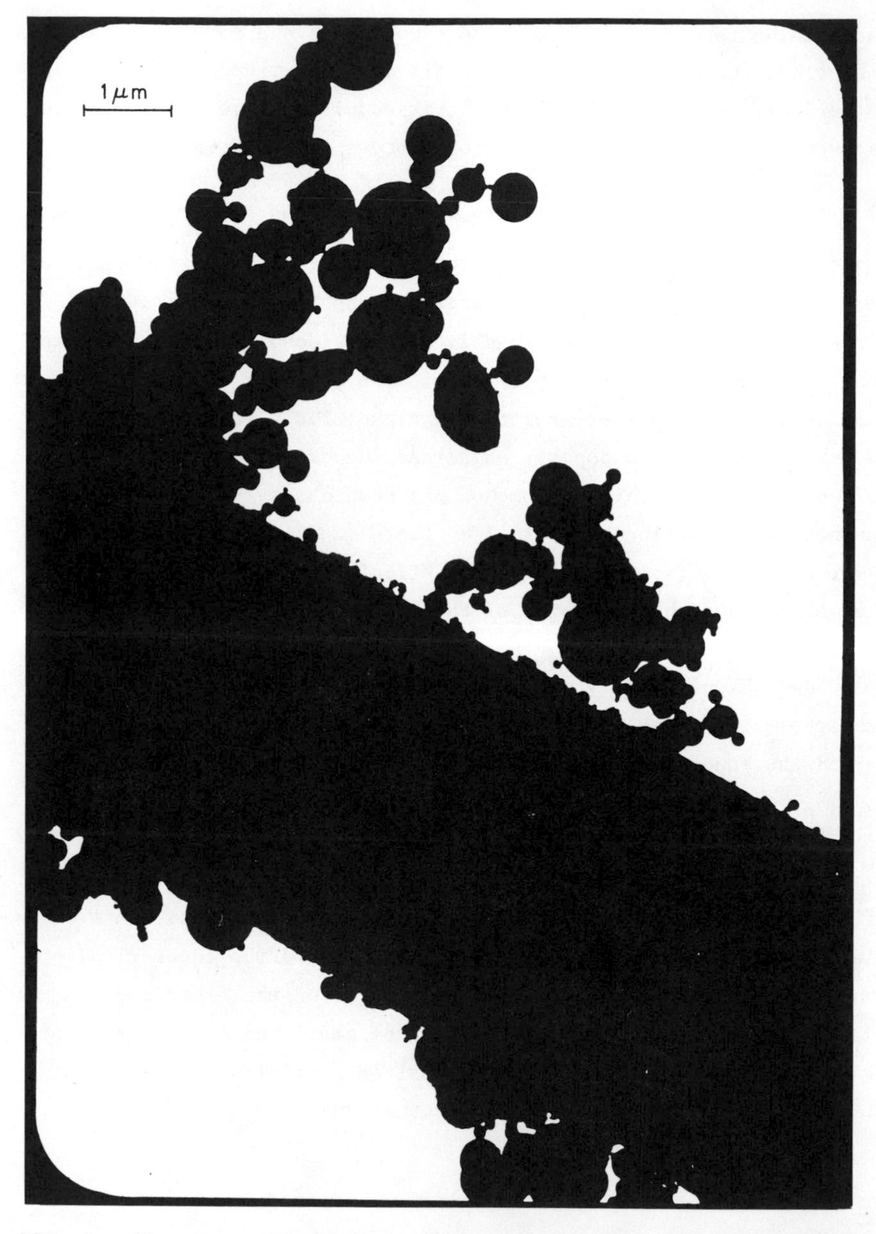

FIG. 1. Structure of the "trees" on a metal fiber of diameter d_f = 5 μm. The "trees" are formed by deposited particles of $K_2Cr_2O_7$ with diameter d_p = 0.1 to 1.2 μm at velocity U_o = 90 cm/sec. Electron microscopy. Photo by J. Albrecht.

coprecipitation, then $C_2 \to 0$, and Eq. (216) reduces to $n = C_1 U_o n_o t$. The same result is obtained from Eq. (216) for small values of t.

Kolganov and Radushkevich [342] checked Eq. (216) by experiments with model filters with $d_f = 16$ μm and using stearic acid particles with $0.2 < d_p < 0.4$ μm at $U_o = 10$ cm/sec and small inlet concentrations $n_o \sim 6$ to 7×10^4 cm^{-3}. They found that filter penetration decreases with time and that the pressure drop is almost constant during the first 30 hours (the total filtration time was 118 hours) and then increases considerably. The capture cross-section C_2 was found to decrease slightly with time, its value for the first 6 to 10 hours being $C_2 = 2.5 \times 10^{-11}$ cm^2.

Billings [97] in a detailed investigation of the effects of particle accumulation in aerosol filters assumed that the capture coefficient of the fiber E_f as a function of particle accumulation could be expressed by

$$E_f = E_f(0) + k_4 Z_5 \tag{217}$$

where $E_f(0)$ is the capture coefficient of the clean fiber and k_4 an accumulation coefficient which was assumed to be constant. In order to check this equation, single 10 μm glass fibers were exposed to a known concentration of 1.305 μm polystyrene particles in an aerosol tunnel. The exposed fibers were periodically removed, and the resulting deposit was photographed and counted. The accumulation coefficient k_4 was calculated from these data. The average value of this coefficient for five fibers tested at $U_o = 13.8$ cm/sec was found to be $k_4 = 1.36 \times 10^{-9}$ cm^2 for $Z_5 \leq 33 \times 10^6$ cm^{-2} of fiber area. For example, if $Z_5 = 20 \times 10^6$ cm^{-2} after time t, the fiber capture coefficient would increase by 2.72%.

Using Eq. (217) Billings derived an equation describing the changes of filter penetration due to clogging. His result is

$$P = P_o e^{-k_4 Z_4} \tag{218}$$

where P_o is the penetration of the clean filter. From experiments with fibrous filters, Billings concluded that $k_4 = 1.8 \times 10^{-9}$ cm^2

for $Z_4 \leq 5 \times 10^8\ cm^{-2}$, in good agreement with mean k_4 values obtained from experiments with single fibers. Finally, Billings concluded that the rate of particle accumulation in a fiber filter depends upon the filter efficiency and that the efficiency, in turn, depends in part upon the accumulated deposit at any time. He expressed the accumulation of material within the filter by an equation of continuity of material as

$$\frac{dZ_4}{dt} = n_o U_o [1 = P(Z_4)] \tag{219}$$

Using Eq. (218) this latter equation can be integrated using the condition that, at $t = 0$, $Z_4 = 0$. The result is

$$Z_4 = n_o U_o t + \frac{1}{k_4} \ell n \frac{1 - P_o}{1 - P} \tag{220}$$

Yoshioka et al. [332] measured the efficiency of a filter with $d_f = 18.6\ \mu m$ at $U_o = 102$ cm/sec as a function of $C_i U_o t$, where C_i is the mass concentration of the inlet aerosol. All the curves of $E = E(C_i U_o t)$ with filter thickness L as a parameter were found to be increasing functions. Using their experimental data—obtained with solid stearic acid particles with $d_p = 1.0\ \mu m$, glass fiber filters with $d_f = 12.6$ and $18.6\ \mu m$, and velocities $U_o = 50$, 102, and 204 cm/sec—they concluded that the capture coefficient of the fiber with accumulated particles could be expressed by

$$E_f = E_f(0)\ (1 + 5Z_3) \tag{221}$$

For the calculation of the efficiency of a filter with accumulated particles they also tried to take into account the distribution of deposited particles inside the filter.

Davies [343], proceeding from empirical equations relating to penetration, resistance, and time of filter operation, derived for the number of particles N deposited in a unit volume of the filter after time t the equation

$$N = \frac{n_o U_o t}{L} (1 - \bar{P}) \tag{222}$$

He related the mean penetration $\bar{P}$ over the time of operation to the initial P_o and final P(t) values.

C. Capillary Phenomena in Filters

Capillary phenomena play an important role, especially in the filtration of liquid particles. According to Radushkevich [297], capillary phenomena include: (1) flushing of drops on the fibers, their fusion, or their division into small drops, such phenomena having been observed by several authors, e.g., Petryanov and Rosenblum [344], Fairs [345], and Bondarenko [346]; (2) formation of fluid layers in places where the fibers are spliced; (3) capillary condensation of water vapor at points of contact of the particle with the fiber or at points of mutual contact of individual particles; (4) fusion of neighboring fibers due to capillary forces; and (5) melting of deposited particles of certain substances at an increased temperature, etc. The influence of capillary forces usually leads to a decrease of the filter efficiency.

Fairs [345] observed a considerable increase of filter penetration and a small increase of the filter pressure drop when filtering liquid particles.

Kolganov and Radushkevich [347] reported an experimental investigation of capillary phenomena in fibrous filters. They measured the selectivity of different filters (glass fiber filters, FPA-15 filters, and others) using solid polystyrol particles in the size range $0.1 < d_p < 0.6$ μm at velocities $U_o = 0.4$ and 5 cm/sec. After the selectivity measurement the filters were washed in water and dried to a constant weight. The selectivity measurement was then repeated. Using this procedure they concluded that the action of liquids (water, dibutylphthalate) leads to an increase of filter penetration, i.e., to a decrease of filter efficiency. Secondly, a decrease of selectivity was observed, but the general character of selectivity remained. Absorption of 60 mg/cm^2 of dibutylphthalate in a filter led to a greater increase of penetration than washing of the filter in water followed by drying. They interpreted these effects as due to an increase in the fiber diameter covered by a liquid layer and by the formation of liquid layers where the fibers intersected.

Mohrmann [348] measured the variation of filter penetration with time for cellulose-asbestos and glass fiber filters using monodisperse liquid DOP particles with d_p = 0.35, 0.5, 0.8, and 1.1 μm at velocity U_o = 9 cm/sec. He also found a considerable increase of filter penetration in the course of the filtration process.

D. Loss of Electric Charge and Filter Destruction

Electric charges on fibers usually influence filtration in a positive way, increasing the filter efficiency. However, in the filtration of certain aerosols (water and oil drops) a decrease of filter efficiency is observed which can also be explained by the neutralization of electric charges on the filter. Loss of electric charge also occurs during the flow of ionized gases, by radioactivity, and by X-rays. Another factor that influences the stability of the filter electric charge is humidity. For instance, Silverman et al. [195] found that a mechanically induced electrostatic charge disappears above 7.78 g of moisture per 453.28 g of dry air. Finally, a filter can be discharged, or at least a loss of charge is observed, when radioactive particles accumulate on it.

Complete destruction of a filter may be caused by a strong chemical reaction between the particles and fibers, by swelling of the fibers (cellulose fibers in the filtration of aqueous aerosols), by an abrupt increase of temperature, by dissolving the fibers in different solutions or organic liquids, etc.

VI. MEMBRANE FILTERS

There is a vast literature dealing with the preparation, structure, properties, and applications of membrane filters, hereafter referred to simply as MF. Although MF were known in the nineteenth century, systematic investigation of their preparation and properties only began early in this century. The historical aspect of the development has been discussed by Elford [349].

There are, to the knowledge of the author, three books dealing with MF. Chronologically, the first is by Jander and Zakowski [350] and covers mainly MF applications in chemical analysis. The second is a brochure by Tovarnickij and Glucharev [351] which focuses on the problem of MF preparation. And the third, by Daubner [352], discusses mainly the methods of filter preparation and their applications in hydrobacteriology. The latter book includes about 500 references relating to MF and their applications. Several reviews and papers of more general character have also been published. Chronologically, the first was by Bachmann [353] reviewing the MF, their action, and their application in science and technology; the second, by Manegold [354] discussing the physical properties of MF; and the third, by Ferry [355] reviewing preparation, properties, and MF applications. Further reviews have been published by Spurny [356] discussing the preparation, structure, mechanism of MF action, and MF applications, mainly in aerosol science. A last, short review was published by Pich [17], discussing mainly the theoretical aspects of MF action.

A. Preparation and Properties

The preparation of MF from different materials such as animal tissue, ceramics, glass, metals, alloys, porcelain, polyvinylchloride, nylon, and others have been reported. However, the most widely used material for MF preparation is a collodion. According to [351] there are two kinds of collodion: nitrocellulose solution in acetic acid, and nitrocellulose solution in an alcohol-ether mixture. From either basic substance MF are prepared by the evaporation of its solvent (gelling and drying of colloidal solutions). The result is a three-dimensional porous system, the skeleton of which consists of solid cellulose esters. As of today MF preparation is more a technological than scientific problem, the subject having been treated in detail by Fairs [345]. Only selected references [357 to 381], arranged chronologically, are given here with remarks included about some of them.

Starting with the pioneering work of Bechhold [357], MF preparation was pursued by Zsigmondy and his coworkers [360,361,364,365] who developed methods for producing nitrocellulose membranes applicable to a commercial sale. Membranfiltergesellschaft in Göttingen, West Germany, the first company to market MF commercially, began manufacturing the filters in 1927. Elford [368] developed a new series of graded collodion membranes and published detailed instructions for obtaining uniform MF of various pore sizes. These investigations have continued, with the result that such filters have been applied in biology (especially in virology) by Bauer and Hughes [370] and Elford [372]. Dianova and Voroshilova [369] described a method of MF preparation from photographic film that is both simple and suitable for laboratory MF preparation. Modern types of MF have been discussed by Gelman [379]. Following him, present membrane filters are made of cellulose triacetate or combinations of cellulose nitrate and acetate, with pore sizes ranging from 0.005 to 10 μm.

Richards et al. [380] described the preparation of metal MF (see Fig. 2). Recently, Desorbo and Cline [381] reported the preparation of metal membranes with uniform submicron-size pores. Such membranes were produced by selectively etching the rod phase of directionally solidified eutectics of NiAℓ-Cr and NiAℓ-Mo and by a two-state replication process. The resulting filters have uniform 0.5 μm diameter holes with number densities as high as 8.5×10^7 holes/cm^2.

As to physical properties, the tensile strength of nitrocellulose films was investigated by Jones and Miles [382]; the detection of defects in MF was treated by Loesche and Socransky [383]; electrical conductivity was studied by Manecke and Bonhoeffer [384] and Schmid and Schwarz [385]; and other electrical properties have been investigated by Spurny and Polydorova [386] and Havlicek et al. [387]. As a result dielectric constant values from 4.5 to 5, specific resistance of about 10^{10} ohm cm, and electric strength of about 100 kV/cm for nitrocellulose or PVC membranes have been reported [386,387], depending on the pore size. Present-day filters

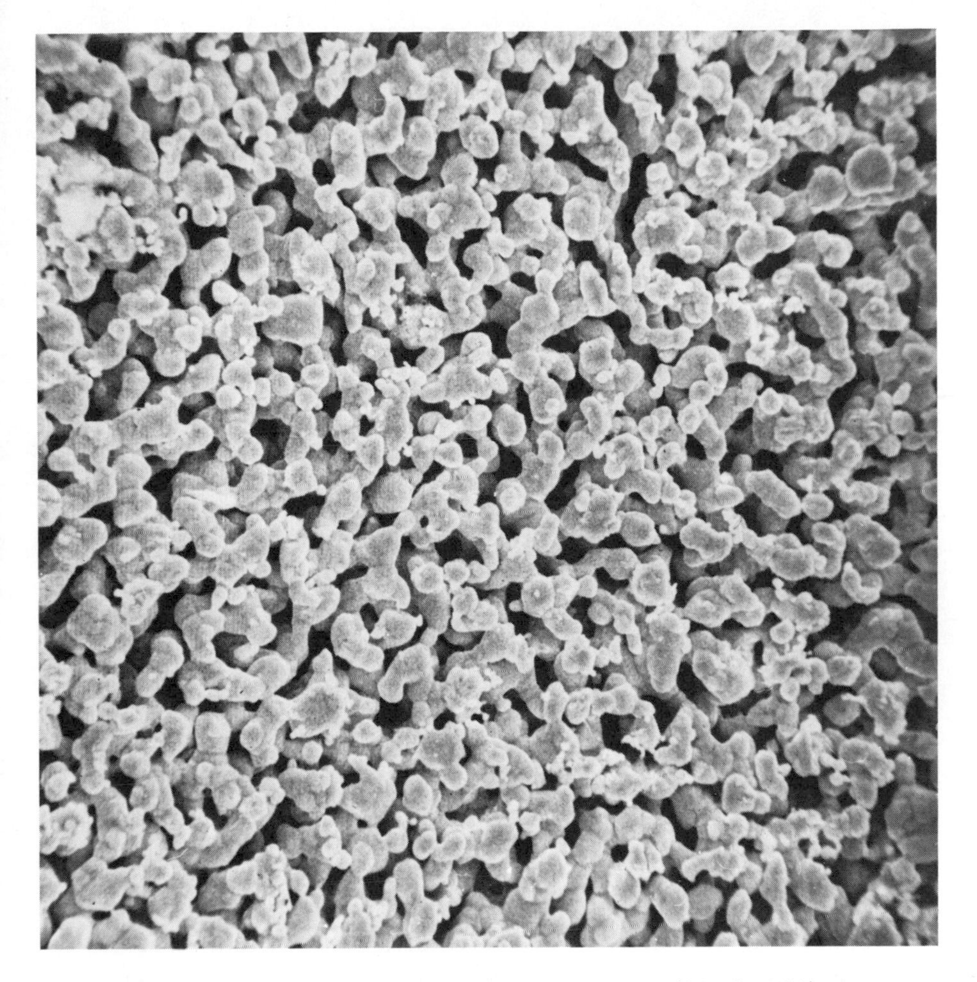

FIG. 2. Surface of the silver membrane filter (SELAS, USA), mean pore diameter 0.8 μm. Scanning electron microscopy, magnification × 1,000. Photo by R. Blaschke.

(MF without glycerine) have constant weight and a refractivity index of about 1.5; they are optically homogeneous and temperature resistant over the temperature range -20 to +80°C.

The most important chemical properties of MF are resistance to different compounds and solubility in organic solvents. For example, propyl alcohol, isopropyl alcohol, benzene, and toluene have no influence on MF; methyl alcohol and ethyl alcohol cause the swelling of MF; destruction of MF is caused by chloroform, aniline, and methylchloride; and MF are soluble in methylacetate, acetone, dioxane, and other solvents.

B. Structure

MF structure and morphology are usually described in terms of pore number per unit area N_p, pore size D_p, pore size distribution on a number basis, distribution of pore areas, filter porosity ε, and filter thickness L. Normally, the pore number is of the order of 10^7 cm^{-2}, filter porosity ε = 80 to 85%, and filter thickness L = 115 to 180 μm. A number of different methods have been elaborated for determining these quantities for a given filter. The most important of these methods are outlined in the following sections.

1. *Standard Sols*

It is possible to evaluate the minimum and maximum pore diameter by actually filtering a series of monodisperse systems according to Bechhold [388]. As the structure of MF is complex, this method is unreliable and tedious, and even refinements based on introducing different correction factors lead to no essential improvement.

2. *Mercury Porosimetry*

The mercury penetration method is based on the application of external pressure to overcome surface tension forces and to fill the pores of a filter with mercury. The technique has been applied to the measurement of MF pore radii by Honold and Skau [389].

3. *Two Immiscible Liquids*

This method, suggested by Erbe [390] and developed by others [391, 392], is based on the flow through the filter of liquids having low surface tension. Two immiscible liquids of a low interfacial tension (e.g., water and isobutyl alcohol giving 1.73 dyne/cm) are employed. The filter pores are first filled with one liquid and the second is forced through the filter by increasing the pressure.

The pore-size distribution is calculated from measurements of the pressure and the volume of the liquid passing through the filter.

4. *Permeability*

The permeability method is based on application of the laws governing the flow of fluids through porous media. For MF a capillary model of the structure is usually used, and, consequently, the Hagen-Poiseuille law is applied. Liquids as well as gases can be employed.

With liquids it is simple to calculate the pore radius by measuring the pressure drop, volume flow rate, porosity, and thickness of the filter (viscosity is usually known). The method was first applied to MF by Guerout [393] and developed further by Bjerrum and Manegold [394], Manegold et al. [395], and others. Such an approach is used by different MF manufacturers to characterize the mean pore diameter, the result being expressed in terms of the time required for a given volume of distilled water to flow through the MF at a given pressure. For example, Membranfiltergesellschaft characterizes its MF by the time required for the flow of 100 mℓ of distilled water through 100 cm^2 of surface at a pressure of 1 atm.

With gases, the situation is more complicated, as the laws governing gas flow through porous media are dependent on the value of the Knudsen number. Assuming a capillary model, the Hagen-Poiseuille equation ($Kn < 10^{-3}$), the Hagen-Poiseuille equation with slip correction ($Kn < 0.25$), the Adzumi equation ($Kn \sim 1$), and the Knudsen-Smoluchowski relationship (free molecule flow, $Kn > 10$) are applied, as discussed by Devienne [49] and Adzumi [396]. Lodge et al. [397] based their calculation of the mean pore diameter on Adzumi's equation, and Hampl and Spurny [398] used the Hagen-Poiseuille equation without, and with, slip correction and Adzumi's equation. Application of Adzumi's equation is obsolete because of its semi-empirical character, but no advanced approaches have been reported as yet.

5. *Electron Microscopy*

Investigation of MF structure using electron microscopy has been reported by a large number of authors [399-415]. Essentially two electron-microscopic methods have been applied. The first is the replica method used by Helmcke [403] and Kanig and D'Ans [404], and the second the method of ultrathin sections as described by Maier and Beutelspacher [402], Spandau and Kurz [406], Riley et al. [408], Fromme and Stöber [409], and Hampl and Pelzbauer [415]. The morphology of MF can be determined in the chosen filter plane using the method of ultrathin section, and, therefore, this method seems to be more suitable for quantitative evaluation than the replica of fracture surfaces. Particularly is this so if the stereometrical methods applied so far in metallography and in the morphology of polymers are employed, as shown by Hampl and Pelzbauer [415].

Results may be briefly summarized as follows: Surfaces of the majority of MF are likely to have very different structures [402, 403,408,409,411, and 415]. During production the filter material is poured onto a smooth base. The upper surface is thus in contact with the air during evaporation of the solvent, while the lower surface is in contact with the base. As a result, the lower surface contains only apertures mostly of a circular, or approximately circular, shape (see Figs. 3 and 4). Hampl and Pelzbauer [415] have shown that Czechoslovak MF (Synpor 1, 2, 3, 6, 8) having small pores are characterized by a well-developed surface film with approximately round pores, while with the filters having big pores the structure of the upper surface is practically the same as that of the lower. Furthermore, these latter investigators found that porosity values as well as mean pore sizes tend to increase from the upper surface toward the lower.

Certain investigators [402,409,415], working with MF from different manufacturers, have shown that the pore distribution in the region of small pores increases toward a maximum number and there-

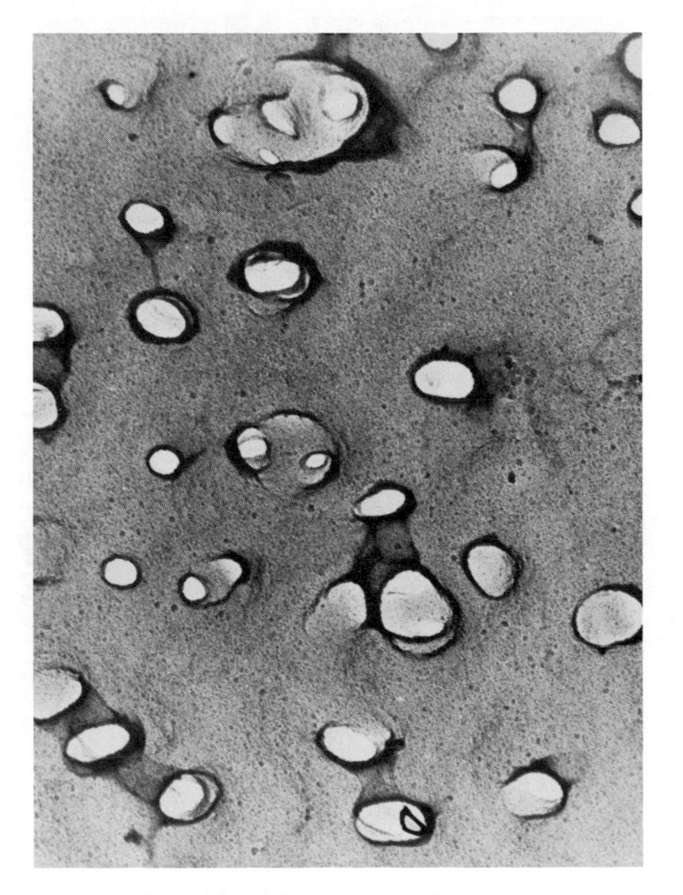

FIG. 3. Upper surface of HUFS membrane filter (Synthesia, Czechoslovakia), mean pore diameter 0.2 μm. Electron microscopy, magnification × 14,800. Carbon replica shadowed by gold palladium alloy. Photo by V. Hampl.

after gradually decreases. Comparison of results from different methods of measurement of the mean pore radius and pore distribution, as reported by Preusser [412], Hampl and Spurny [413], and Hampl et al. [414], show that values of mean pore diameter obtained by electron microscopy are greater than those obtained by indirect methods (permeability, Erbe's, and Hg porosimetry).

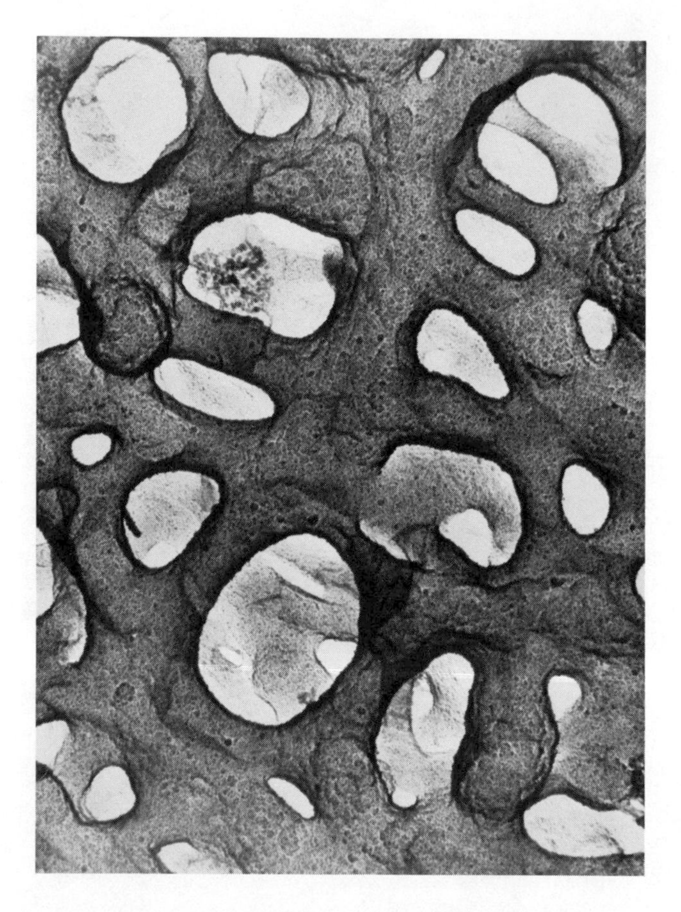

FIG. 4. Bottom surface of HUFS membrane filter (Synthesia, Czechoslovakia), mean pore diameter 0.2 μm. Electron microscopy, magnification × 14,800. Carbon replica shadowed by gold palladium alloy. Photo by V. Hampl.

C. Flow

For the investigation of fluid flow through MF, pressure drop Δp of MF, MF filtration mechanisms, the capillary model is applied. This model, according to Guerout [393], envisions a system of equidistant, parallel, circular capillaries of diameter D_p perpendicular to the filter surface and having a length equal to the filter thickness L.

If A is the area of the filter, A_p the area of one pore, and

N_p the number of pores per unit surface area, then the volume of pores V_p is $V_p = A_pN_pLA$. Since the volume of the filter $V_f = AL$ is related to the porosity ε by $\varepsilon = V_p/V_F$ there is obtained

$$\varepsilon = A_pN_p \tag{223}$$

If Q is the volume flow rate through the filter, then the superficial velocity U_o is given by $U_o = Q/A$. Similarly, if Q_p is the flow rate through one capillary, then $\bar{U} = Q_p/A_p$ is the mean flow velocity in the pore. For an incompressible fluid, $Q = Q_pN_pA$; and, using Eq. (223), there is obtained

$$\bar{U} = \frac{U_o}{\varepsilon} \tag{224}$$

the relation being formally equivalent to the Depuit-Forchheimer hypothesis [56].

Flow through MF has usually been studied by permeability methods with the primary goal being to determine the mean pore size. Flow in MF was studied by Machacova et al. [416]. Investigation of the flow of gases through plastic membranes was reported by Brubaker and Kammermayer [417].

The flow in a capillary in the continuum region is described by the Hagen-Poiseuille law, which, for a system of capillaries and an incompressible fluid, becomes

$$\Delta p = \frac{8\eta LU_o}{\pi a^4 N_p} \tag{225}$$

In the range of small Knudsen numbers, the slip correction must be taken into account, e.g., as discussed by Devienne [49]. Modifying the Hagen-Poiseuille law by including the slip correction to the capillary model gives

$$\Delta p = \frac{8\eta LU_o}{\pi a^4 N_p(1 + 3.992Kn)} \tag{226}$$

where diffusion reflection of the gas molecules from the pore surface has been assumed and, again, $Kn = \lambda/a$ is the Knudsen number. Finally, in the free molecule region ($Kn > 10$), modification of the Knudsen-Smoluchowski equation [49,56] results in

$$\Delta p = \frac{3LU_o p}{4a^3 N_p \sqrt{2\pi RT/M}} \tag{227}$$

where p is the gas pressure, R the gas constant, and M the molecular weight of the gas.

Relatively good agreement of these equations with experimental data has been found by Machacova et al. [416]. The semi-empirical Adzumi [396] equation has been applied [397,398] for the description of Δp in the transition region of Knudsen numbers ($Kn \sim 1$). No applications of new results in fluid dynamics to the problem of gas flow through MF have been reported to the knowledge of the author.

D. Deposition

Apparently, the first investigation of the mechanisms of particle retention in MF was by Duclaux and Errera [418]. Later, Cox and Hyde [419] investigated the physical factors involved in MF filtration. In these early papers attention was focused mainly on the sieve mechanism of particle retention. However, MF have the property of retaining particles much smaller than their pore size. Consequently, other mechanisms of particle retention must be taken into account. All published work relating to mechanisms use the capillary model of MF so that the basic structural unit is a cylindrical pore of diameter D_p arranged perpendicular to the surface of the filter. Deposition in each pore is assumed to be independent of the presence of other pores.

1. *Diffusion*

Neglecting the entrance length, in which the flow entering the capillary is stabilized, the problem is reduced to the diffusion of particles in laminar flow (very small Reynolds numbers at usual velocities) to the walls of the capillary. This problem has been investigated by a large number of authors and has been recently summarized by Wajsfelner [420] in connection with the theory of diffusion batteries. The mechanism is described by a dimensionless parameter

$$N_D = \frac{4DL}{D_p^2 \bar{U}} \tag{228}$$

where D is the diffusion coefficient of the particles. Defining E_D by $E_D = 1 - n_p/n_o$, where n_o is the inlet and n_p the outlet concentration of particles, then, according to Leveque [421] and Levich [422],

$$E_D = 2.57\ N_D^{2/3} \tag{229}$$

According to Fuchs [13] this equation is valid for $N_D < 0.03$. The equation derived by Twomey [423] can be applied for $N_D > 0.03$. It is

$$E_D = 1 - 0.819e^{-3.6568N_D} - 0.097e^{-22.304N_D} - 0.032e^{-56.95N_D} - 0.0157e^{-107.6N_D} \tag{230}$$

Equations (229) and (230) were used by Spurny and Pich [424] for calculating the intensity of diffusion deposition in a capillary model of MF. The same approach was reported by Megaw and Wiffen [425] who, using an equation similar to Eq. (230) taken from Fuchs [13], concluded that particle diffusion to the pore walls alone would ensure 100% efficiency for the filter up to a particle diameter of at least 0.05 μm (AA Millipore filters with ε = 80%, L = 170 μm, and velocity U_o = 40 cm/sec, i.e., $\bar{U}$ = 50 cm/sec were used). Megaw and Wiffen also calculated the penetration $P_D = n_p/n_o$ to a depth of 10 μm from the filter surface (L = 10 μm) and, having obtained very small values up to d_p = 0.02 μm, suggested that this would account for the fact observed by Kubie et al. [426] that MF retain all particles in a very shallow surface layer.

In Eqs. (229) and (230) and in other equations of similar type [420] the diffusion coefficient of particles D in a capillary is assumed to be the same as that in an unbounded fluid. This problem has been recently investigated by Uzelac and Cussler [427], measuring the diffusion of polystyrene latex particles of diameter 910 Å through Millipore filters at temperatures of 25 and 45°C. The temperature dependence of D and the variation of D with pore diameter were found consistent with the model of a rigid sphere surrounded by a continuum and moving in a cylindrical pore. The diffusion coefficients were roughly 10 times those predicted by the Stokes-Einstein equation.

2. *Direct Interception and Sieving*

If r is the particle radius, then the mechanism of direct interception is described by the parameter

$$N_R = \frac{2r}{D_p} \tag{231}$$

for $N_R < 1$. This parameter expresses the fact that a particle is not to be considered a mass point but that it is captured when it approaches the wall of the pore to a distance equal to its radius. No justified theoretical description of this mechanism has been reported as yet.

The sieve mechanism is a special case of direct interception and is described by the same parameter N_R, where $N_R > 1$. Obviously for $N_R > 1$ all particles are captured by the filter, which acts as a sieve in this case. This was realized by Manegold and Hoffmann [428], who concluded that, for $N_R > 1$, $P = n_p/n_o = 0$ (full retention) and incorrectly called the penetration $P = n_p/n_o$ "sieve constant." Statistical evaluation of this "sieve constant" has been reported by Ferry [429].

3. *Gravitation*

The problem of gravitational deposition in a circular tube under laminar flow was investigated by Natanson [430]. Denoting the efficiency of gravitational deposition by E_G, the resulting equation can be expressed by

$$E_G = \frac{2}{\pi}\left(2NG\sqrt{1 - N_G^{2/3}} + \arcsin N_G^{1/2} - N_G^{1/3}\sqrt{1 - N_G^{2/3}}\right) \quad (232)$$

where the dimensionless parameter of gravitational deposition is given by

$$N_G = \frac{3V_S L}{4D_p \bar{U}} \quad (233)$$

with V_s the steady-state sedimentation velocity of particles. For $N_G = 1$, Eq. (232) gives $E_G = 1$. Hence, for the condition $N_G = 1$, it follows that the critical length of capillary L_{cr} in which the entire aerosol would be deposited is given by the relation

$$L_{cr} = \frac{4D_p \bar{U}}{3V_s} \quad (234)$$

Thomas [431] analyzed the gravitational settling of particles in a horizontal cylindrical tube and derived an equation for the boundary line dividing the denuded area from that containing particles for any cross-section of the tube. Later, Davis [432] in experiments involving the visual examination of flow around pipe bends found satisfactory experimental agreement for the shape of the boundary line as predicted theoretically by Thomas [431]. The theoretical calculations of Thomas [431] were refined by Thomas and Knuth [433], commented on by Davis [434], and replied to by Thomas and Knuth [435].

4. *Inertial*

The impaction of particles around a pore on a filter surface was investigated by Pich [436], using an approximate model of the velocity field. Denoting by E_I the efficiency of impaction, the result was

$$E_I = 2E^* - E^{*2}$$

with

$$E^* = (1 + \alpha)^{-1}\left[4Stk\sqrt{\alpha} + 8Stk^2\alpha \exp\left(-\frac{1}{2Stk\sqrt{\alpha}}\right) - 8\alpha Stk^2\right] \quad (235)$$

where $\alpha = \varepsilon^{1/2}/(1 - \varepsilon^{1/2})$ and $Stk = U_o m/3\pi\eta d_p D_p$. From Eq. (235) it can be concluded that E_I is an increasing function of the Stokes number Stk; for $Stk \to 0$, i.e., for non-inertial particles, $E_I \to 0$; for $Stk \to \infty$, $E_I \to (1 + 2\alpha)/(1 + \alpha)^2$; and the relation $E_I = E_I(Stk^{1/2})$ plotted in semilogarithmic coordinates has the typical S-form characteristic for impaction systems [135].

5. *Electrostatic*

It is a common belief that the high efficiency of MF is due in part to their electrostatic charge. Such a statement is often offered by MF manufacturers without giving proof. Megaw and Wiffen [425], making experiments with Millipore AA filters (pore diameter D_p = 0.8 μm, porosity ε = 80%, and thickness L = 170 μm) and aerosols of potassium permanganate and radioactive tungsten (particle diameter 0.02 μm) at a face velocity U_o = 40 cm/sec, concluded that electrostatic effects play no role under these conditions. According to these investigators MF efficiency for small particles is predictable theoretically assuming that diffusion is the sole filtration mechanism. Removal of any electrostatic charge from the filter (e.g., by irradiating with electrons from a Kr^{85} source) does not alter its efficiency, and larger particles must be trapped by impaction in the filter pores.

On the other hand, Binek and Przyborowski [437] conducting experiments with MF of Czechoslovak make (PUFS type, D_p = 2 μm, ε = 80%, and L = 66 μm) at a face velocity U_o = 30 cm/sec and the NaCl particles in the diameter range 0.04 to 0.7 μm came to the opposite conclusion. A significant increase of efficiency due to a charge on the filter was observed both for uncharged or partly charged particles. The electrostatic charge on the filter was found to be of the order of 10^{-9} coulomb/cm^{-2} by an induction measurement.

As in the case of fibrous filters the total efficiency of MF is a function of all the partial efficiencies and parameters described previously. In order to find out this relationship, Pich [17] and Spurny and Pich [424] assumed that two mechanisms contribute in the filtration: inertial deposition on the filter surface, and diffusion deposition in the pores. In this case these two mechanisms do not act simultaneously but one after the other, i.e., impaction on the filter surface is followed by diffusion deposition inside the filter. This circumstance permits replacing the model filter of efficiency E by two filters having efficiencies E_D and E_I in series. Then

$$E = E_D + E_I - E_D E_I \tag{236}$$

This equation expresses the total efficiency by means of the partial efficiencies. These investigators [17,424] assumed that E_D was given by Eq. (229) or (230) and E_I by Eq. (235). An attempt to incorporate the mechanism of direct interception into Eq. (236) on empirical grounds was reported by Spurny and Madelaine [438].

Electron microscopy has been applied in the experimental demonstration of MF filtration mechanisms, using the replica or ultrathin section methods. Such research has been reported, in chronological order, by Henneberg and Haagen-Crodel [439], Beutelspacher [440], Moll [441], Spurny and Lodge [442], and Spurny et al. [443] who used scanning electron microscopy.

Another group of experimental investigations of MF action is represented by the papers of Refs. 444 to 446. D'Amico and Di Bari [444] reported the investigation of MF efficiency using a nephelometer and the Goetz aerosol spectrometer. They concluded that the maximum in the particle size distribution following the filter was shifted toward smaller particle sizes when compared with the peak in the size distribution before the filter. Petras [445] reported comparative studies of the filtration efficiency of MF. The efficiency of 17 types of MF manufactured by three companies was tested by electronic and microbiological analysis of particle suspension filtrates. The test particles, polystyrene latex, Colanyl Green, and bacterial cells, were suspended in 4% saline solution and taken as representative of different particle types with respect to size and shape. Results were summarized in two comprehensive tables presenting efficiency according to size and shape as well as to the maximum volume of the particles. Toth [446] measured the efficiency of MF (Czechoslovak make, AUFS type) for radon decay products. Efficiency was obtained by passing a monodisperse aerosol at constant velocity through two filters of equal thickness and measuring the α-radioactivity of both filters after an equal period of time with two different instruments. Toth concluded that the efficiency of AUFS filters made prior to 1964 was 93.3% and lower than that of filters made in 1966 (98.9%).

E. Characteristics

MF characteristics can be defined and classified in the same manner as for fibrous filters. Of these, the selective characteristic defined as the filter efficiency dependence on the particle sixe $E = E(r)$ is the most important. If it is assumed that only two mechanisms—inertial and diffusion deposition—are acting in the filtration, these may be characterized by efficiencies E_I and E_D, respectively. E_I, given by Eq. (235), is an increasing function of particle size; and E_D, given by Eq. (229) or (230), is a decreasing function of particle size. Hence, the total efficiency reaches a minimum at a certain particle size, and the selective characteristic has a minimum as in the case of fibrous filters. Using Eq. (236), the position of the minimum and the efficiency values at the minimum can be calculated as shown by Spurny and Pich [447].

Experimentally, this problem was first investigated by Fitzgerald and Detwiler [448]. These investigators measured the efficiency of American Millipore filters of AA type ($D_p = 0.8$ μm) and HA type ($D_p = 0.45$ μm) with particle size in the range 0.005 to 2.1 μm ($KMnO_4$ aerosol) at face velocities of 10, 20, and 40 cm/sec. They concluded that for both the AA as well as the HA type, there is an optimum particle size for total penetration through the filter (minimal efficiency). The efficiency for particles greater than 0.1 μm in diameter is very high and relatively independent of the face velocity. For particles with $d_p < 0.1$ μm the efficiency passes passes through a point of optimum penetration at approximately 0.02 μm.

Walkenhorst [449] measured the selectivity of German MF using Wolframoxid aerosols and did not find a minimum in the selectivity curve. He concluded that for particles $d_p = 0.1$ μm, the investigated filters are almost absolute; for smaller particles the efficiency decreased to about 94 to 97%—depending on the filter type—for $d_p = 0.01$ μm particles.

Megaw and Wiffen [425] measured the efficiency of type AA Millipore filters for particles of 0.02 μm diameter and a face velocity of 40 cm/sec using aerosols of potassium permanganate and radioactive tungsten. No significant penetration of the filter was detected, which contradicts the results of Fitzgerald and Detwiler. Megaw and Wiffen concluded that their results (decreasing filter efficiency with increasing particle size) can be explained if it is assumed that diffusion is the sole filtration mechanism. This is in agreement with the assumption of a minimum in the selectivity curve because, as stated by Megaw and Wiffen, other mechanisms must come into play and efficiency must again increase for larger particles ($d_p > 0.1$ μm). Spurny and Pich [447] measured the selectivity of Czechoslovak MF ($D_p = 1.8$ μm) using radioactive aerosols of NaC , AgI, H_2SO_4, $H_4P_2O_7$, and Hg at a face velocity of 25 cm/sec. A minimim in the selectivity curve was found at about 0.1 μm, the efficiency value being about 70% at the minimum.

Hence, from experiments there is disagreement concerning the shape of the selectivity curve. It should show a minimum since both the sieve effect and diffusion deposition lead necessarily to a minimum in the selectivity curve. Probably the position of the minimum is dependent on filtration velocity and pore size as in the case of fibrous filters.

Using equations from the kinetic theory of gases to describe the dependence of the gas viscosity and the mean free path of the gas molecules on the gas temperature and pressure, Eq. (229) or Eqs. (230), (235), and (236) lead to approximate descriptions of the temperature and pressure characteristics of membrane filters as pointed out by Spurny and Hrbek [450]. Qualitatively, the results of such an analysis are the same as in the case of fibrous filters.

VII. NUCLEPORE FILTERS

Price and Walker [451,452] and Fleischer et al. [453-456] developed a technique for producing small holes in certain materials as a result of studies of damage caused in solids by high energy radiation. A thin (10 μm) sheet of polycarbonate is irradiated with a collimated beam of fission fragments from uranium-235 and the damaged material removed in an etching bath. The size of the resulting pores is determined by the bath reagents and by the duration and temperature of the treatment.

Such sheets for use as filters are produced by the Nuclepore Corp. (formerly by the General Electric Co.) and called Nuclepore filters (see Fig. 5). They are transparent and have a thickness L of approximately 15 μm and a porosity between 5 and 20%; their pores are circular and perpendicular to the filter surface; and the number of pores on a unit surface N_p is usually between 10^5 and 10^7 cm^{-2} while pore diameters range from 0.3 to 8 μm depending on the filter type. The pore size distribution is narrow (geometric standard deviation being smaller than about 1.10), the filters are chemically resistant against many compounds, their density is 0.95 g/cm^3, and they do not lose their mechanical properties up to 140°C [453]. The preparation of filters with pore diameters as small as 50 Å and $N_p \sim 10^{11}$ cm^{-2} has been reported [454,456].

Flow in Nuclepore filters has been investigated by Kirsh and Spurny [457]. Pressure drop was measured at gas pressures from 7 to 745 mmHg. Qualitative agreement with the Knudsen-Smoluchowski equation, Eq. (227), was found with theoretical values being about 1.5 greater than the experimental ones. In order to predict Δp more precisely, an equation derived by Hasimoto [458] describing the hydrodynamic resistance of an isolated hole in an infinitesimally thin wall was employed. Denoting the resulting pressure drop by Δp_1, Hasimoto's equation becomes

$$\Delta p_1 = \frac{3\eta Q_p}{a^3} \tag{237}$$

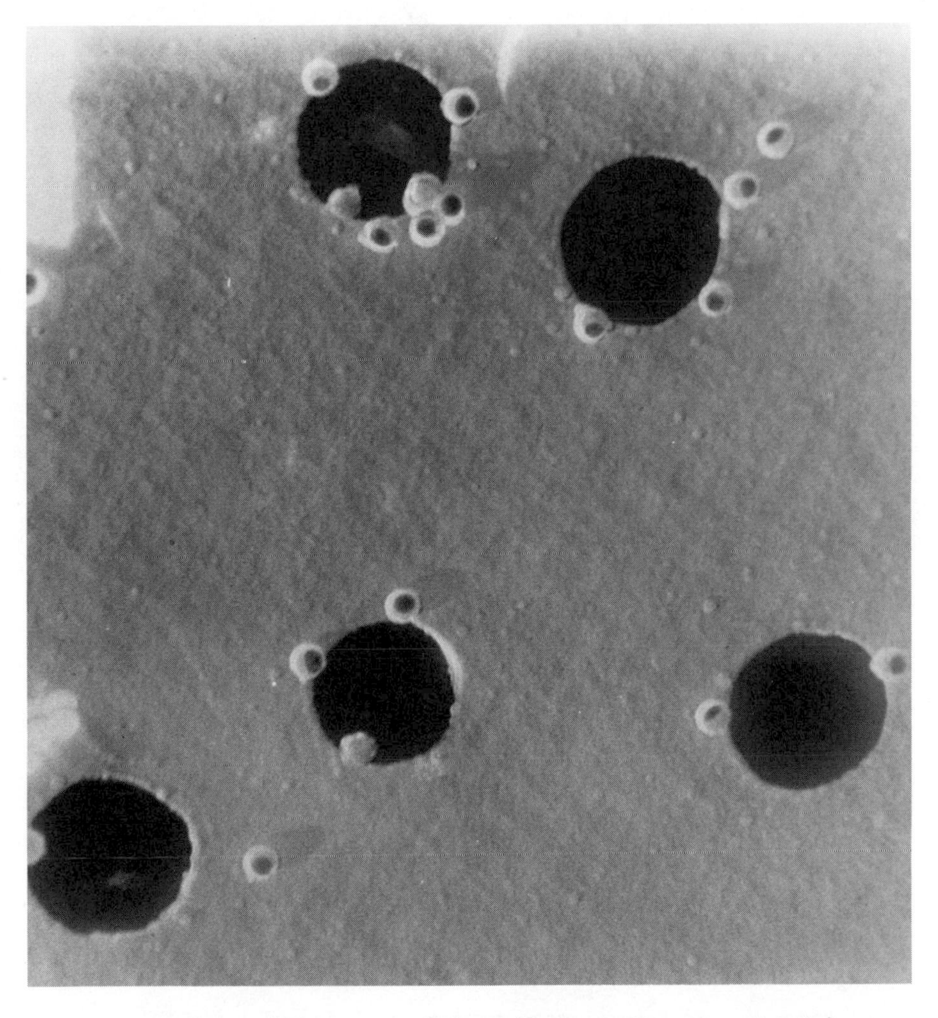

FIG. 5. Surface of Nuclepore filter (General Electric, U.S.A.), mean pore diameter 2 μm, with latex particles, mean particle diameter 0.357 μm. Electron microscopy, magnification × 17,600. SiO replica shadowed by chromium. Photo by E. R. Frank.

where $a = D_p/2$ is the radius of the hole. Summing this expression with the Hagen-Poiseuille law for one capillary (special case of Eq. 225), Kirsh and Spurny [457] obtained

$$\Delta p = \frac{3\eta L Q_p}{\pi a^4} \; 1 + \frac{3\pi a}{8L} \tag{238}$$

From this equation it follows that the hydrodynamic resistance of the filter surface can be neglected if $a/L << 1$. For $a/L = 0.5$, the pressure drop of the filter surface was experimentally found to be about one-third of the entire pressure drop, which is in agreement with Eq. (238). On the other hand, there are some inconsistencies in the Kirsh and Spurny paper [457]. Good agreement with the Knudsen-Smoluchowski equation for the pressure drop of Nuclepore filters has also been reported by Machacova et al. [416].

The filtration properties of Nuclepore filters have been studied by Spurny and Lodge [459] and Spurny et al. [460]. Particle deposition was described, again using the capillary model, as due to the three mechanisms of inertial deposition on the filter surface, diffusion deposition in the pores, and direct interception. Equation (235) was employed for the description of inertial deposition and Eq. (230), or similar equations known from the theory of diffusion batteries [420], for diffusion deposition. The effect of direct interception was accounted for empirically. They calculated a number of different characteristics of Nuclepore filters (e.g., selective and velocity characteristics) and claimed qualitative—in some cases even quantitative—agreement with experiments using radioactively labeled aerosols. They attempted to describe the clogging of Nuclepore filters and also discussed applications for aerosol sampling and aerosol measurement (see Fig. 6).

Frank et al. [461] discussed the use of Nuclepore filters in connection with the light and electron microscopic examination of aerosols. Another interesting application of Nuclepore filters reported by Seal [462] is in the isolation of cancer cells.

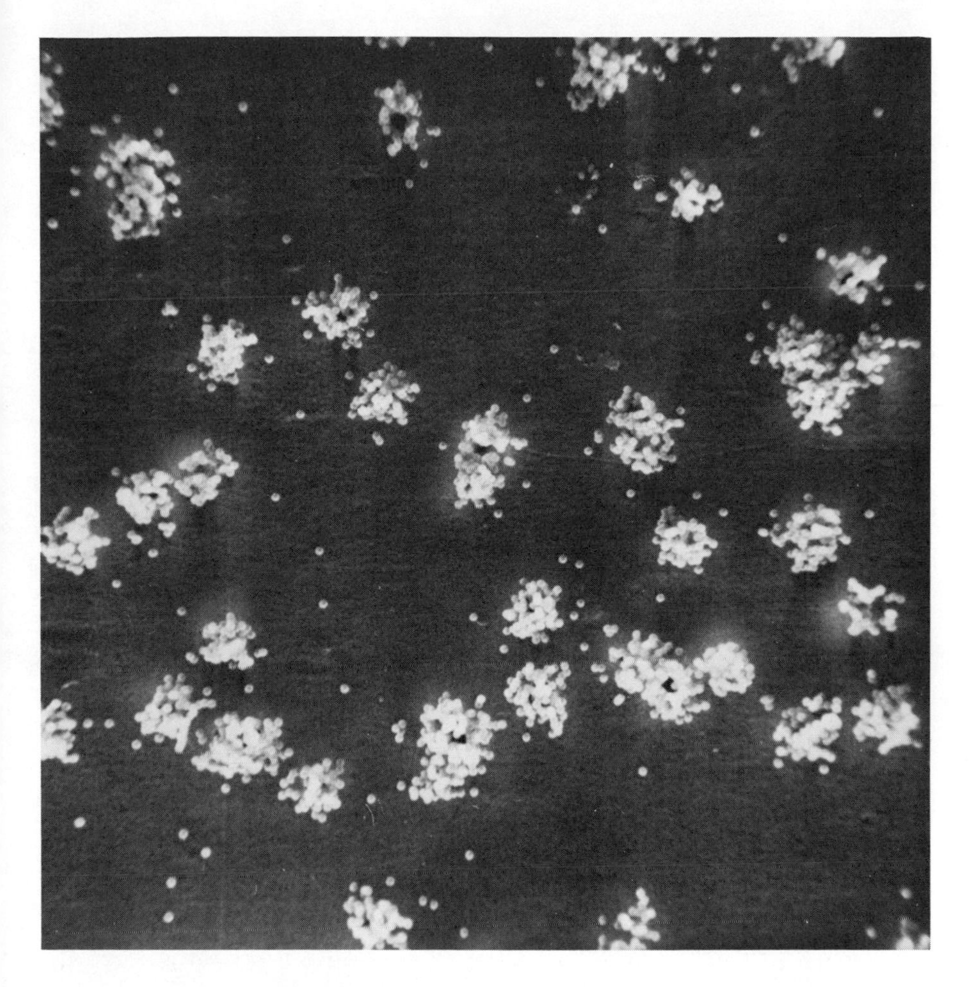

FIG. 6. Surface of Nuclepore filter (General Electric, U.S.A.), mean pore diameter 2 μm, with latex particle diameter 0.500 μm. Scanning electron microscopy, magnification × 2,400. Photo by R. Blaschke.

REFERENCES

1. J. Juda and S. Chrosciel, *Staub*, *30*, 522 (1970).

2. C. N. Davies, *The Ninth International Congress on Industrial Medicine*, London, September 13-17, 1949.

3. N. A. Fuchs, *The Mechanics of Aerosols* (in Russian), Izdat. Akad. Nauk SSSR, Moscow, 1955.

4. C. Y. Chen, *Chem. Rev.*, *55*, 595 (1955).

5. H. L. Green and W. R. Lane, *Particulate Clouds: Dusts, Smokes and Mists*, SPON, London, 1957.

6. H. Engelhard, *Z. Biol. Aerosol Forsch.*, *8*, 290 (1960).

7. Ch. H. Martius, *Z. Biol. Aerosol Forsch.*, *9*, 61 (1960); *10*, 63 (1961).

8. C. Kangro, *Staub*, *21*, 275 (1961).

9. H. Mohrman, *Z. Biol. Aerosol Forsch.*, *10*, 326 (1962); *11*, 516 (1964).

10. J. Ponroy, in *Seminaire de Physique des Aêrosols* (J. Bricard, ed.), Facultê des Sciences de Paris, Paris, 1962-1963.

11. C. N. Davies, *Recent Advances in Aerosol Research*, Pergamon, Oxford, 1964.

12. H. L. Green and W. R. Lane, *Particulate Clouds: Dusts, Smokes and Mists*, 2nd ed., SPON, London, 1964; translated into Russian, Izdat Khimyia, Moscow, 1969.

13. N. A. Fuchs, *The Mechanics of Aerosols*, Pergamon, Oxford, 1964.

14. R. G. Dorman, in *High-Efficiency Air Filtration* (P. A. F. White and S. E. Smith, eds.), Butterworth, London, 1964; translated into Russian, Atomizdat, Moscow, 1967.

15. S. Aiba, A. E. Humphrey, and N. F. Millis, *Biochemical Engineering*, Academic, New York, 1965.

16. R. G. Dorman, in *Aerosol Science* (C. N. Davies, ed.), Academic, New York, 1966.

17. J. Pich, in *Aerosol Science* (C. N. Davies, ed.), Academic, New York, 1966.

18. S. K. Friedlander, with the assistance of R. E. Pasceri, in *Biochemical and Biological Engineering* (N. Blakenbrough, ed.), Academic, London, 1967.

19. J. Pich, *J. Aerosol Sci.*, *1*, 17 (1970).

20. A. Fortier, *Mécanique des Suspensions*, Masson et Cie, Paris, 1967.

21. I. B. Stechkina, *Dokl. Akad. Nauk SSSR*, *167*, 1327 (1966) (in Russian).

22. I. P. Mazin, *The Physical Principles of Aircraft Icing* (in Russian), Gidrometeoirdat, Moscow, 1957.

23. Yu. P. Gupalo, *Dokl. Akad. Nauk SSSR*, *164*, 1339 (1965) (in Russian).

24. L. M. Levin, *Physics of Coarse Aerosols* (in Russian), Izdat. Akad. Nauk SSSR, Moscow, 1961.

25. L. Spielman and S. L. Goren, *Environ. Sci. Technol.*, *2*, 279 (1968).

26. S. Chandrasekhar, *Rev. Mod. Phys.*, *15*, 1 (1943).

27. S. V. Dawson, *Sc. D. Thesis*, Harvard University, Boston, Mass., 1969.

28. Yu. S. Sedunov, *Izv. Akad. Nauk SSSR, Ser. Geofiz.*, No. 7, 1093 (1964) (in Russian).

29. S. S. Duchin and B. V. Deryaguin, *Kolloid. Zh.*, *20*, 326 (1958) (in Russian).

30. G. Matherson, *Éléments pour une Théorie des Millieux Poreux*, Masson et Cie, Paris, 1967.

31. R. M. Werner and L. A. Clarenburg, *Ind. Eng. Chem. Process Des. Develop.*, *4*, 288 (1965).

32. R. M. Farrow, *Filtrat. Separat.*, 490 (1966).

33. A. Simon, V. Hampl, and J. Pich, *Collect. Czech. Chem. Commun.*, *34*, 3619 (1969).

34. S. Kuwabara, *J. Phys. Soc. Jap.*, *14*, 522 (1959).

35. Yu. M. Glushkov, *Izv. Akad. Nauk SSSR, Mekhanika Zhidkosti Gaza*, 176 (1968) (in Russian).

36. R. P. Labrecque, *Diss. Abstr.*, B 28 (2), 659 (1967) (Lawrence University, Wisconsin).

37. M. Benarie, in *Séminaire de Physique des Aérosols* (J. Bricard, ed.), Faculte des Sciences de Paris, Paris, 1967-1968.

38. M. Benarie, *Staub*, *29*, 74 (1969).

39. H. W. Piekaar and L. A. Clarenburg, *Chem. Eng. Sci.*, *22*, 1399 (1967).

40. E. L. Skau, E. Honold, and W. A. Boudreau, *Text. Res. J.*, *23*, 798 (1953).

41. H. Corte, *Filtrat. Separat.*, 396 (1966).

42. L. A. Clarenburg and J. F. van der Wal, *Ind. Eng. Chem. Process Des. Develop.*, *5*, 110 (1966).

43. R. R. Sullivan, *J. Appl. Phys.*, *12*, 503 (1941); *13*, 725 (1942).

44. J. L. Fowler and K. L. Hertel, *J. Appl. Phys.*, *11*, 496 (1940).

45. N. A. Fuchs and I. B. Stechkina, *Ann. Occup. Hyg.*, *6*, 27 (1963).

46. J. B. Keller, *J. Fluid Mech.*, *18*, 94 (1964).

47. S. Kuwabara, *J. Phys. Soc. Japan*, *14*, 527 (1959).

48. L. V. Radushkevich, *Zh. Fiz. Khim.*, *40*, 965 (1966) (in Russian).

49. M. Devienne, *Frottement et Échanges Thermiques dans les Gaz Rarefiés*, Gauthier-Villars, Paris, 1958.

50. F. Albrecht, *Z. Phys., 32*, 48 (1931).

51. W. Sell, *VDI Forschungsheft*, no. 347 (1931).

52. I. Langmuir, *OSRD Report*, no. 865 (1942).

53. I. Langmuir, *The Collected Works* (C. G. Suits, ed.), vol. 10, Pergamon, Oxford, 1961.

54. M. Muskat, *The Flow of Homogeneous Fluids through Porous Media,* McGraw-Hill, New York, 1937.

55. P. C. Carman, *The Flow of Gases through Porous Media*, Academic, New York, 1956.

56. A. E. Scheidegger, *The Physics of Flow through Porous Media*, University of Toronto, Toronto, 1957.

57. R. E. Collins, *Flow of Fluids through Porous Materials*, Reinhold, New York, 1961.

58. J. Pich and K. Spurny, *Zdrav. Tech. Vzduchotech, 7*, 114 (1964) (in Czech).

59. J. Kozeny, *Wasserkraft Wasserwirtsch., 22*, 67 (1927).

60. P. C. Carman, *Trans. Inst. Chem. Eng., 15*, 150 (1937).

61. E. J. Wiggins, W. B. Campbell, and O. Maas, *Can J. Res., 17*, 318 (1939).

62. R. R. Sullivan and K. L. Hertel, *J. Appl. Phys., 11*, 761 (1940).

63. R. R. Sullivan and K. L. Hertel, *Advances in Colloid Science*, vol. I, Interscience, New York, 1942.

64. L. A. Clarenburg and R. M. Werner, *Ind. Eng. Chem. Process Des. Develop., 4*, 293 (1965).

65. H. W. Piekaar and L. A. Clarenburg, *Chem. Eng. Sci., 22*, 1817 (1967).

66. L. A. Clarenburg and H. W. Piekaar, *Chem. Eng. Sci., 23*, 765 (1968).

67. L. A. Clarenburg and F. C. Schiereck, *Chem. Eng. Sci., 23*, 773 (1968).

68. H. C. Brinkman, *Appl. Sci. Res., A1*, 27 (1949).

69. C. R. Illingworth, in *Laminar Boundary Layers* (L. Rosenhead, ed.), Oxford University, London, 1963.

70. M. Van Dyke, *Perturbation Methods in Fluid Mechanics*, Academic, New York, 1964.

71. J. Happel and H. Brenner, *Low Reynolds Number Hydrodynamics*, Prentice-Hall, Engelwood Cliffs, 1965.

72. L. Bairstow, B. M. Cove, and E. D. Lang, *Phil. Trans. Roy. Soc.* (London), A, *223*, 383 (1923).

73. H. Faxen, *Nova Acta Reg. Soc. Sci. Upsal.*, 1-55, 1927.

74. H. Lamb, *Hydrodynamics*, 6th ed., Cambridge University, London 1932.

75. A. Thom, *Proc. Roy. Soc.*, A, *141*, 651 (1933).

76. C. N. Davies, *Proc. Phys. Soc.* (London), B, *63*, 288 (1950).

77. S. Tomotika and T. Aoi, *Quart. J. Mech. Appl. Math.*, *3*, 140 (1950).

78. I. Imai, *Proc. Roy. Soc.*, A, *208*, 487 (1951).

79. S. Tomotika, *Proc. Roy. Soc.*, A, *219*, 233 (1953).

80. S. Kaplun, *J. Math. Mech.*, *6*, 595 (1957).

81. K. Tamada and H. Fujikawa, *Quart. J. Mech. Appl. Math.*, *10*, 425 (1957).

82. K. Tamada and H. Fujikawa, *J. Phys. Soc. Jap.*, *14*, 202 (1959).

83. T. Miyagi, *J. Phys. Soc. Jap.*, *13*, 493 (1958).

84. J. Happel, *A.I.Ch.E. J.*, *5*, 174 (1959).

85. A. Kirsch and N. Fuchs, *J. Phys. Soc. Jap.*, *22*, 1251 (1967).

86. H. Hasimoto, *J. Fluid Mech.*, *5*, 317 (1959).

87. O. Emersleben, *Phys. Verh.*, *6*, 150 (1955).

88. O. Emersleben, *Anwendungen der Mathematik*, Greifswald, no. 3 (1958).

89. E. M. Sparrow and A. L. Loefler, *A.I.Ch.E. J.*, *5*, 325 (1959).

90. R. A. Leonard and R. Lemlich, *Chem. Eng. Sci.*, *20*, 790 (1965).

91. C. M. White, *Proc. Roy. Soc.*, A, *186*, 472 (1946).

92. C. M. White *Proc. Roy. Soc.*, A, *201*, 268 (1950).

93. R. K. Finn, *J. Appl. Phys.*, *24*, 771 (1953).

94. D. J. Tritton, *J. Fluid Mech.*, *6*, 547 (1959).

95. A. A. Kirsch and N. A. Fuchs, *Ann. Occup. Hyg.*, *10*, 23 (1967); also in *Kolloid. Zh.*, *29*, 682 (1967) (in Russian).

96. C. N. Davies, *Proc. Inst. Mech. Eng.*, 1B, 185 (1952).

97. C. E. Billings, *Ph.D. Thesis*, California Institute of Technology, California, 1966.

98. A. S. Iberall, *J. Res. Nat. Bur. Stand.*, *45*, 398 (1950).

99. S. T. Stern, H. W. Zeller, and A. I. Schekman, *J. Colloid Sci.*, *15*, 546 (1960).

100. L. Silverman and M. First, *Ind. Eng. Chem.*, *44*, 2777 (1952).

101. A. Blasewitz and B. Judson, *Chem. Eng. Progr.*, *51*, 6 (1955).

102. N. Kimura and K. Iinoya, *Kagaku Kogaku*, *23*, 792 (1959) (in Japanese).

103. K. T. Whitby, R. C. Jordan, and A. B. Algren, *ASHRAE J.*, *4*, 79 (1962).

104. I. B. Stechkina, *Inzh. Fiz. Zh.*, *7*, 128 (1964) (in Russian).

105. H. F. Johnstone and M. H. Roberts, *Ind. Eng. Chem.*, *41*, 2417 (1949).

106. W. E. Ranz, *Tech. Rept. No. 3*, University of Illinois, Urbana, 1951.

107. S. K. Friedlander, *A.I.Ch.E. J.*, *3*, 43 (1957).

108. G. L. Natanson, *Dokl. Akad. Nauk SSSR*, *112*, 100 (1957) (in Russian).

109. W. L. Torgeson, in *General Mills*, Rep. no. 1890 (1958).

110. J. Boussinesque, *Theorie Analytique de Chaleure*, vol. II, Paris, 1903.

111. W. K. Lewis and J. M. Smith, *OSRD Report*, no. 1251 (1942).

112. C. J. Stairmand, *Trans. Inst. Chem. Eng.*, *28*, 130 (1950).

113. L. G. Wilson and P. Cavanagh, *Atmos. Environ.*, *1*, 261 (1967).

114. L. G. Wilson and P. Cavanagh, *Atmos. Environ.*, *3*, 47 (1969).

115. B. Binek, *Atmos. Environ.*, *1*, 605 (1967).

116. T. Thalhammer, *Atmos. Environ.*, *2*, 535 (1968).

117. C. N. Davies, *Atmos. Environ.*, *3*, 595 (1969).

118. C. N. Davies, *Atmos. Environ.*, *3*, 695 (1969).

119. C. N. Davies, *Atmos. Environ.*, *4*, 592 (1970).

120. W. J. Megaw, *Atmos. Environ.*, *3*, 326 (1969).

121. L. G. Wilson and P. Cavanagh, *Atmos. Environ.*, *1*, 605 (1967).

122. L. G. Wilson and P. Cavanagh, *Atmos. Environ.*, *2*, 536 (1968).

123. L. G. Wilson and P. Cavanagh, *Atmos. Environ.*, *3*, 597 (1969).

124. L. G. Wilson and P. Cavanagh, *Atmos. Environ.*, *3*, 696 (1969).

125. L. G. Wilson and P. Cavanagh, *Atmos. Environ.*, *3*, 327 (1969).

126. R. Dobry and R. Finn, *Ind. Eng. Chem.*, *48*, 1540 (1956).

127. A. A. Kirsch and N. A. Fuchs, *Ann. Occup. Hyg.*, *11*, 299 (1968); also in *Kolloid Zh.*, *30*, 836 (1968) (in Russian).

128. L. V. Radushkevich and V. A. Kolganov, *J. Colloid Interface Sci.*, *29*, 55 (1969).

129. H. L. Sadoff and J. W. Almlof, *Ind. Eng. Chem.*, *48*, 2199 (1956).

130. W. H. Rodebush, in *Handbook on Aerosols*, U. S. Atomic Energy Commission, Washington, D. C., 1950.

131. G. L. Natanson, *Kolloid. Zh., 24*, 52 (1962) (in Russian).

132. T. Gillespie, *J. Colloid Sci., 10*, 299 (1955).

133. J. Pich, *Staub, 26*, 267 (1966).

134. I. B. Stechkina, A. A. Kirsch, and N. A. Fuchs, *Ann. Occup. Hyg., 12*, 1 (1969); also in *Kolloid. Zh., 31*, 121 (1969) (in Russian).

135. W. E. Ranz, *Principles of Inertial Impaction*, Engineering Research Bulletin B-66, Pennsylvania University, 1956.

136. M. N. Golovin and A. A. Putnam, *Ind. Eng. Chem. Fundam., 1*, 264 (1962).

137. A. C. Chamberlain, in *Airborne Microbes* (P. H. Gregory and J. L. Monteith, eds.), Oxford University, London, 1967.

138. F. R. S. Taylor, *Aeron. Res. Committee*, H.M.S.O., London, no. 2024 (1940).

139. M. Glauert, *Aeron. Res. Committee*, H.M.S.O., London, no. 2025 (1940).

140. I. Langmuir and K. B. Blodgett, *Gen. Elect. Res. Lab.*, Rep. no. RL 225 (1944-45).

141. I. P. Mazin, *Trudy CAO, 7*, 39 (1952) (in Russian).

142. R. J. Brun and H. W. Mergler, *NACA*, TN 2904 (1953).

143. C. N. Davies and C. V. Peetz, *Brit. J. Appl. Phys.*, Suppl. no. 3, 17 (1954).

144. R. Brun, W. Lewis, P. Perkins, and J. Serafini, *NACA*, Rep. 1215 (1955).

145. C. N. Davies and C. V. Peetz, *Proc. Roy. Soc.*, A, *234*, 269 (1956).

146. A. Robinson, *Comm. Pure Appl. Math., 9*, 69 (1956).

147. I. M. Yuryev, *Dokl. Akad. Nauk SSSR, 170*, 1035 (1966) (in Russian).

148. N. Yoshioka, H. Emi, and H. Sone, *Kagaku Kogaku, 31*, 1011 (1967) (in Japanese).

149. M. V. Subramanyam and N. R. Kuloor, *Ann. Occup. Hyg., 12*, 9 (1969).

150. W. L. Torgeson, in *General Mills*, Inc., Minneapolis 13, Min. (1961).

151. L. M. Levin and I. S. Sedunov, *Dokl. Akad. Nauk SSSR, 162*, 316 (1965) (in Russian).

152. N. Yoshioka, H. Emi, and M. Fukushima, *Kagaku Kogaku, 31*, 157 (1967) (in Japanese).

153. N. Yoshioka, H. Emi, H. Matsumura, and M. Yasunami, *Kagaku Kogaku, 33*, 381 (1969) (in Japanese).

154. M. K. Householder and V. W. Goldschmidt, *J. Colloid Interfac. Sci.*, *31*, 464 (1969).

155. J. A. Harrop and J. I. T. Stenhouse, *Chem. Eng. Sci.*, *24*, 1475 (1969).

156. I. B. Stechkina, A. A. Kirsch, and N. A. Fuchs, *Kolloid. Zh.*, *32*, 467 (1970) (in Russian).

157. H. Landahl and K. Hermann, *J. Colloid Sci.*, *4*, 103 (1949).

158. A. H. Yeomans, E. E. Rogers, and W. H. Ball, *J. Econ. Entomol.*, *42*, 591 (1949).

159. E. A. Ramskill and W. L. Anderson, *J. Colloid Sci.*, *6*, 416 (1951).

160. P. H. Gregory, *Ann. Appl. Biol.*, *38*, 357 (1951).

161. W. Ranz and J. Wong, *Ind. Eng. Chem.*, *44*, 1371 (1952).

162. J. B. Wong and H. F. Johnstone, *Tech. Rep. No. 11*, Eng. Expt. Station, University of Illinois, Urbana, 1953.

163. J. B. Wong, W. E. Ranz, and H. F. Johnstone, *J. Appl. Phys.*, *26*, 244 (1955); *27*, 161 (1956).

164. A. G. Amelin and M. I. Belyakov, *Kolloid. Zh.*, *18*, 385 (1956) (in Russian).

165. J. Lewis and R. Ruggieri, *NACA*, TN 4092 (1957).

166. V. Ignatiev, *Teploenergetika*, no. 3 (1958) (in Russian).

167. R. Jarman, *Chem. Eng. Sci.*, *10*, 268 (1959).

168. C. G. Haupt, *Ph.D. Thesis*, University of Birmingham, 1959.

169. J. Rosinski and C. Nagamoto, *Kolloid. Zh.*, *175*, 29 (1961).

170. M. J. Madock, *Ph.D. Thesis*, University of Birmingham, Birmingham, Alabama, 1962.

171. K. T. Whitby, D. A. Lundgren, A. R. McFarland, and R. C. Jordan, *APCA J.*, *11*, 503 (1961).

172. K. R. May and R. Clifford, *Ann. Occup. Hyg.*, *10*, 83 (1967).

173. J. R. Starr, *Ann. Occup. Hyg.*, *10*, 349 (1967).

174. J. I. T. Stenhouse, J. A. Harrop, and D. C. Freshwater, *J. Aerosol Sci.*, *1*, 41 (1970).

175. K. T. Whitby, *ASHRAE J.*, *7*, 56 (1965).

176. G. Asset and T. G. Hutchins, *Amer. Ind. Hyg. Assoc. J.*, *28*, 348 (1967).

177. J. Rosinski and T. Church, *Powder Technol.*, *1*, 272 (1967/68).

178. L. M. Levin, *Dokl. Akad. Nauk SSR*, *91*, 1329 (1953) (in Russian).

179. L. M. Levin, *Izv. Akad. Nauk SSSR, Ser. Geofiz.*, no. 3, 422 (1959) (in Russian).

180. G. L. Natanson, *Dokl. Akad. Nauk SSSR, 116*, 109 (1957) (in Russian).

181. V. M. Voloshchuk, *Izv. Akad. Nauk SSSR, Ser. Fiz. Atmosfery Okeana, 1* (12) (1965); *2* (2) (1966); *3* (8) (1967); *3* (9) (1967) (in Russian).

182. J. W. Thomas and R. E. Yoder, *Amer. Arch. Ind. Health, 13*, 550 (1956).

183. K. T. Whitby and Y. H. Liu, in *Aerosol Science* (C. N. Davies, ed.), Academic, New York, 1966.

184. R. J. H. Voorhoeve, *Rep. No. TDCK-45746*, 1966 (in Dutch).

185. G. Zebel, *Staub, 28*, 263 (1968).

186. G. L. Natanson, *Dokl. Akad. Nauk SSSR, 112*, 696 (1957) (in Russian).

187. H. F. Kraemer and H. F. Johnstone, *Ind. Eng. Chem., 47*, 2426 (1955).

188. W. L. Torgeson, in *General Mills*, Rep. No. 1919 (1960).

189. L. M. Levin, *Izv. Akad. Naúk SSSR, Ser. Geofiz.*, no. 7, 1073 (1959).

190. R. Cochet, *Rev. Génér. Électr.*, 62, 113 (1953).

191. G. S. Dawkins, *Tech. Rep. no. 15*, Eng. Expt. Station, Univ. of Illinois, Urbana, 1958.

192. L. V. Radushkevich, *Dokl. Akad. Nauk SSSR, 170*, 375 (1966) (in Russian).

193. G. G. Goyer, R. Gruen, and V. LaMer, *J. Phys. Chem., 58*, 137 (1954).

194. A. Rossano and L. Silverman, *Heat. Ventilat., 51*, 102 (1954).

195. L. Silverman, E. W. Conners, and D. M. Anderson, *Ind. Eng. Chem., 47*, 952 (1955).

196. D. A. Lundgren and K. T. Whitby, *Ind. Eng. Chem., 56*, 85 (1964).

197. D. A. Lundgren and K. T. Whitby, *Ind. Eng. Chem., Process Design Develop., 4*, 345 (1965).

198. B. I. Ogorodnikov and P. I. Basmanov, in *Radioactive Isotopes in the Atmosphere and Their Application in Meteorology* (in Russian), Atomizdat, Moscow, 1965.

199. N. Yoshioka, H. Emi, M. Hattori, and I. Tamori, *Kagaku Kogaku, 32*, 815 (1968) (in Japanese).

200. J. M. Thomas and E. J. Woodfin, *AIEE Trans., 78*, 276 (1959).

201. V. Havlicek, *Int. J. Air Wat. Pollut., 4*, 225 (1961).

202. R. D. Rivers, *ASHRAE J., 4*, 37 (1962).

203. K. Makino and K. Iinoya, *Kagaku Kogaku*, *33*, 684 (1969) (in Japanese).

204. K. Makino and K. Iinoya, *Kagaku Kogaku*, *33*, 701 (1969) (in Japanese).

205. W. Walkenhorst and G. Zebel, *Staub*, *24*, 444 (1964).

206. G. Zebel, *J. Colloid Sci.*, *20*, 522 (1965).

207. G. Zebel, *Staub*, *26*, 281 (1966).

208. G. Zebel, *Staub*, *29*, 62 (1969).

209. D. Hochrainer, G. M. Hidy, and G. Zebel, *J. Colloid Interfac. Sci.*, *30*, 553 (1969).

210. D. Hochrainer, *Staub*, *29*, 67 (1969).

211. W. Walkenhorst, *Staub*, *29*, 483 (1969).

212. W. Walkenhorst, *J. Aerosol Sci.*, *1*, 225 (1970).

213. N. N. Tunitskii and I. V. Petryanov, *Zh. Fiz. Khim.*, *17*, 408 (1943) (in Russian).

214. H. C. Hamaker, *Physica*, *4*, 1058 (1937).

215. L. A. Spielman and S. L. Goren, *Environ. Sci. Technol.*, *4*, 135 (1970).

216. C. Maxwell, *Phil. Trans. Roy. Soc.* (London), *170*, 231 (1879).

217. P. S. Epstein, *Phys. Rev.*, *23*, 710 (1924).

218. S. P. Bakanov and B. V. Deryaguin, *Dokl. Akad. Nauk SSSR*, *139*, 71 (1961) (in Russian).

219. S. Chapman and T. G. Cowling, *The Mathematical Theory of Non-Uniform Gases*, Cambridge University, London, 1952.

220. G. N. Patterson, *Molecular Flow of Gases*, Wiley, New York, 1956.

221. R. G. Dorman, in *Aerodynamic Capture of Particles* (E. G. Richardson, ed.), Pergamon, Oxford, 1960.

222. J. Pich and K. Spurny, *Rheol. Acta*, *6*, 12 (1967).

223. J. Pich, *Ann. Occup. Hyg.*, *9*, 23 (1966).

224. J. A. Wheat, *Can. J. Chem. Eng.*, *41*, 67 (1962).

225. J. Pich, *Staub*, *25*, 186 (1965); also in *Collect. Czech. Chem. Commun.*, *31*, 3721 (1966).

226. M. V. Velichko and L. V. Radushkevich, *Dokl. Akad. Nauk SSSR*, *154*, 415 (1964) (in Russian).

227. C. Y. Liu and R. Passamaneck, *Rarefied Gas Dynamics 5th Symposium*, *1*, 607 (1967).

228. J. Pich, *J. Colloid Interfac. Sci.*, *29*, 91 (1969).

229. H. Coudeville, P. Trepaud, and E. A. Brun, *Rarefied Gas Dynamics 4th Symposium, 1*, 444 (1965).

230. I. V. Petryanov, B. I. Ogorodnikov, and A. S. Suncov, in *Radioactive Isotopes in the Atmosphere and Their Application in Meteorology* (in Russian), Atomizdat, Moscow, 1965.

231. L. V. Radushkevich and M. V. Velichko, *Dokl. Akad. Nauk SSSR, 146*, 406 (1962) (in Russian).

232. L. V. Radushkevich, *J. Colloid Interfac. Sci., 34*, 205 (1970).

233. L. V. Radushkevich and V. A. Kolganov, *Izv. Akad. Nauk SSSR, Ser. Khim. Nauk, 1*, 23 (1962) (in Russian).

234. L. V. Radushkevich and V. A. Kolganov, in *Aerosols, Physical Chemistry and Applications* (K. Spurny, ed.), Prague, 1965.

235. J. R. Stalder, G. Goodwin, and M. O. Creager, *NACA*, Rep. 1032 (1951).

236. J. Pich, *Staub, 29*, 407 (1969).

237. S. K. Friedlander, *ARS J., 31*, 96 (1961).

238. S. K. Friedlander, *Ind. Eng. Chem., 50*, 1161 (1958).

239. G. L. Natanson and E. N. Ushakova, *Zh. Fiz. Khim., 35*, 463 (1961) (in Russian).

240. L. V. Radushkevich, *Zh. Fiz. Khim., 32*, 282 (1958) (in Russian).

241. L. V. Radushkevich, *Zh. Fiz. Khim., 35*, 467 (1961) (in Russian).

242. R. E. Pasceri and S. K. Friedlander, *Can. J. Chem. Eng., 38*, 212 (1960).

243. I. B. Stechkina and N. A. Fuchs, *Ann. Occup. Hyg., 9*, 59 (1966); also in *Kolloid. Zh., 29*, 260 (1967) (in Russian).

244. S. K. Friedlander, *J. Colloid Interfac. Sci., 23*, 157 (1967).

245. S. Aiba and T. Yasuda, *A.I.Ch.E. J., 8*, 704 (1962).

246. D. G. Thomas and C. E. Lapple, *A.I.Ch.E. J., 7*, 203 (1961).

247. H. L. Green and D. J. Thomas, *Proc. Inst. Mech. Eng., 1B*, 203 (1952).

248. D. Hasenclever, *Staub, 26*, 288 (1966).

249. S. G. Terjesen and G. B. Cherry, *Trans. Inst. Chem. Eng., 25*, 89 (1947).

250. A. E. Humphrey and E. L. Gaden, *Ind. Eng. Chem., 47*, 924 (1955).

251. E. L. Gaden and A. E. Humphrey, *Ind. Eng. Chem., 48*, 2172 (1956).

252. W. D. Maxon and E. L. Gaden, *Ind. Eng. Chem., 48*, 2177 (1956).

253. S. Aiba, S. Shimasaki, and S. Suzuki, *J. Gen. Appl. Microbiol.* (Tokyo), *7*, 192 (1961).

254. W. Zumach, *Atompraxis*, *3*, 10 (1957).

255. B. Shleien, T. P. Glavin, and A. G. Friend, *Science*, *147*, 290 (1965).

256. E. Landt, *Ges. Ing.*, *77*, 139 (1956).

257. R. Leers, *Staub*, *17*, 402 (1957).

258. E. Landt, *Staub*, *17*, 9 (1957).

259. J. Pich, *Zdrav. Tech. Vzduchotech.*, *4*, 119 (1961) (in Czech).

260. H. Freundlich, *Colloid and Capillary Chemistry*, Methuen, London, 1926.

261. J. J. Fitzgerald and C. G. Detwiller, *Amer. Ind. Hyg. Assoc. Quart.*, *18*, 47 (1957).

262. L. V. Radushkevich and V. A. Kolganov, *Zavod. Lab.*, *11*, 1365 (1964) (in Russian).

263. L. V. Radushkevich and V. A. Kolganov, *Kolloid. Zh.*, *27*, 95 (1965) (in Russian).

264. L. V. Radushkevich and V. A. Kolganov, *Zh. Fiz. Khim.*, *42*, 971 (1968).

265. V. A. Kolganov, L. V. Radushkevich, and V. G. Sazanova, *Zh. Prikl. Khim.*, *39*, 2725 (1966) (in Russian).

266. A. A. Kirsch, I. B. Stechkina, and N. A. Fuchs, *Kolloid. Zh.*, *31*, 227 (1969) (in Russian).

267. M. Benarie, *Staub*, *23*, 2 (1963).

268. W. J. Smith and N. F. Surprenant, *Proc. ASTM*, *53*, 1122 (1953).

269. H. Engelhardt and H. Mohrman, *Z. Naturforsch.*, B, *17*, 331 (1962).

270. C. L. Lindeken, R. L. Morgin, and K. F. Petrock, *Health Phys.*, *9*, 305 (1963).

271. J. Pich, paper presented at *45th National Colloid Symposium*, Atlanta, June 21-23, 1971.

272. L. H. Wilson, W. L. Sabitt, and M. Jakob, *J. Appl. Phys.*, *22*, 8 (1951).

273. J. A. Brink, W. F. Burggrabe, and L. E. Greenwell, *Chem. Eng. Progr.*, *62*, 60 (1966).

274. J. Pich and B. Binek, in *Aerosols, Physical Chemistry and Applications* (K. Spurny, ed.), Prague, 1965.

275. M. Benarie and J. P. Quetier, *I.R.Ch.A.*, Note Interieure no. 53 (1967).

276. W. Strauss and B. W. Lancaster, *Atmos. Environ.*, *2*, 135 (1968).

277. C. A. Snyder and R. T. Pring, *Ind. Eng. Chem.*, *47*, 960 (1955).

278. W. Strauss and M. W. Thring, *Trans. Inst. Chem. Eng.*, *41*, 248 (1963).

279. M. V. First, J. B. Graham, G. M. Butler, C. B. Walworth, and R. P. Warren, *Ind. Eng. Chem.*, *48*, 696 (1956).

280. H. Witzmann, *Z. Elektrochem.*, *46*, 313 (1940).

281. H. Haupt, *Chem. Tech.*, *18*, 99 (1966); *19*, 408 (1967).

282. I. Gallily, *J. Colloid Sci.*, *10*, 558 (1955); *12*, 161 (1957).

283. P. Grüner, *Staub*, *28*, 353 (1968).

284. J. Sisefsky, *Nature*, *182*, 1437 (1958).

285. C. Jech, *Staub*, *20*, 75 (1960).

286. D. C. Stevens and R. F. Hounam, *Ann. Occup. Hyg.*, *3*, 58 (1961).

287. A. Sittkus and K. Backheuer, *Staub*, *23*, 419 (1963).

288. V. Lössner, *Staub*, *24*, 217 (1964).

289. G. Madelaine and H. Parnianpour, *Ann. Ocup. Hyg.*, *10*, 31 (1967).

290. B. Biles and J. McK. Ellison, *Ann. Occup. Hyg.*, *11*, 13 (1968).

291. L. V. Radushkevich, *Izv. Akad. Nauk SSSR, Otd. Khim. Nauk*, *7*, 1190 (1962) (in Russian).

292. W. Gutowski, paper presented at Symposium *Physikalische Staubeigenschaften und ihr Einfluss auf den Entstaubungsprozess*, Zabrze (Poland), October 13-17, 1970.

293. H. W. Pyne, R. B. Wilson, and B. W. Soole, *Brit. J. Appl. Phys.*, *18*, 1177 (1967).

294. B. W. Soole, *Staub*, *28*, 274 (1968).

295. B. W. Soole, *Staub*, *30*, 112 (1970).

296. B. W. Soole and H. C. W. Meyer, *J. Aerosol Sci.*, *1*, 147 (1970).

297. L. V. Radushkevich, *Izv. Akad. Nauk SSSR, Otd. Khim. Nauk*, no. 3, 407 (1963) (in Russian).

298. M. Corn, *J. Air Poll. Control Assoc.*, *11*, 523, 566 (1961).

299. B. B. Morgan, *BCURA Monthly Bulletin*, *25*, 125 (1961).

300. M. Corn, in *Aerosol Science* (C. N. Davies, ed.), Academic, New York, 1966.

301. H. Krupp, *Advances in Colloid and Interface Science*, *1*, 111 (1967).

302. A. D. Zimon, *Adhesion of Dust and Powders* (in Russian), Khimiya, Moscow, 1967; translated into English, Plenum, New York, 1969.

303. H. Margenau, *Rev. Mod. Phys.*, *11*, 1 (1939).

304. H. Margenau and N. R. Kestner, *Theory of Intermolecular Forces*, Pergamon, New York, 1969.

305. F. London, *Z. Phys.*, *63*, 245 (1930).

306. F. London, *Trans. Faraday Soc.*, *33*, 8 (1937).

307. J. E. Lennard-Jones, *Proc. Phys. Soc.* (London), *43*, 461 (1931).

308. H. B. G. Casimir and D. Polder, *Phys. Rev.*, *73*, 360 (1948).

309. J. A. V. Butler, *Ann. Rep. Chem. Soc.* (London), *34*, 75 (1937).

310. R. S. Bradley, *Trans. Faraday Soc.*, *32*, 1088 (1936).

311. M. J. Vold, *J. Colloid Sci.*, *9*, 451 (1954).

312. E. M. Lifshitz, *Zh. Eksp. Teor. Fiz.*, *29*, 94 (1955) (in Russian).

313. D. Langbein, *J. Adhes.*, 237 (1969).

314. J. Gregory, *Advances in Colloid and Interface Sci.*, *2*, 396 (1970).

315. B. V. Deryaguin, *Discuss. Faraday Soc.*, *18*, 24 (1954).

316. B. V. Deryaguin, I. I. Abrikosova, and E. M. Lifshitz, *Quart. Rev. Chem. Soc.*, *10*, 295 (1956).

317. A. P. Prosser and J. A. Kitchener, *Nature*, *178*, 1339 (1956).

318. J. A. Kitchener and A. P. Prosser, *Proc. Roy. Soc.*, A, *242*, 403 (1957).

319. I. N. Aleinikova, B. V. Deryaguin, and Yu. P. Toporov, *Kolloid. Zh.*, *30*, 177 (1968) (in Russian).

320. B. V. Deryaguin, I. N. Aleinikova, and Yu. P. Toporov, *Powder Technol.*, *2*, 154 (1969).

321. B. V. Deryaguin and V. Smilga, *Adhesion*, Rep. Int. Eng. Conf., 1966 (Pub. 1969), pp. 152-63.

322. T. Gillespie and G. O. Langstroth, *Can. J. Chem.*, *30*, 1056 (1952).

323. S. M. Botterill and E. Aynsley, *Brit. Chem. Eng.*, *12*, 1593, 1899 (1967).

324. A. D. Zimon, *Kolloid. Zh.*, *29*, 883 (1967) (in Russian).

325. H. Krupp, *Chem. Ing. Tech.*, *39*, 374 (1967).

326. T. Gillespie and W. J. Settineri, *J. Colloid Interfac. Sci.*, *24*, 199 (1967).

327. F. Löffler, *Staub*, *28*, 11 (1968).

328. D. Tabor and R. H. S. Winterton, *Nature*, *219*, 1120 (1968).

329. W. Schnabel, *Staub*, *28*, 448 (1968).

330. J. Visser, *Rep. Progr. Appl. Chem.*, *53*, 714 (1968).

331. A. D. Zimon and K. A. Lazarev, *Kolloid. Zh.*, *31*, 59 (1969) (in Russian).

332. N. Yoshioka, H. Emi, M. Yasunami, and H. Sato, *Kagaku Kogaku*, *33*, 1013 (1969) (in Japanese).

333. C. E. Williams, T. Hatch, and L. Greenburg, *Heat. Pip. Air Condit.*, *12*, 259 (1940).

334. W. C. Hemeon, *Heat. Ventilat.*, *37*, 191 (1940).

335. D. J. Thomas, *Inst. Heat. Vent. Eng.*, *20*, 35 (1952).

336. G. Miczek, *Zdrav. Tech. Vzduchotech.*, *4*, 1 (1961) (in Czech).

337. W. Solbach, *Staub*, *29*, 24 (1969).

338. F. Löffler, *Staub*, *30*, 518 (1970).

339. J. Juda and S. Chrosciel, *Staub*, *30*, 196 (1970).

340. J. Albrecht, *Ph.D. Thesis*, Institute of Air Technique, Prague, 1968 (in Czech).

341. L. V. Radushkevich, *Kolloid. Zh.*, *26*, 235 (1964) (in Russian).

342. V. A. Kolganov and L. V. Radushkevich, *Kolloid. Zh.*, *29*, 518 (1967) (in Russian).

343. C. N. Davies, *J. Aerosol Sci.*, *1*, 35 (1970).

344. I. V. Petryanov and N. A. Rosenblum, *Dokl. Akad. Nauk SSSR*, *61*, 661 (1948) (in Russian).

345. G. Fairs, *Trans. Inst. Chem. Eng.*, *36*, 476 (1958).

346. V. S. Bondarenko, *Zh. Fiz. Khim.*, *35*, 2775 (1961) (in Russian).

347. V. A. Kolganov and L. V. Radushkevich, *Izv. Akad. Nauk SSSR, Ser. Khim.*, no. 6, 1208 (1967) (in Russian).

348. H. Mohrmann, *Staub*, *30*, 317 (1970).

349. W. J. Elford, *J. Roy. Micr. Soc.*, *48*, 36 (1928).

350. G. Jander and J. Zakowski, *Membran-Cella-und Ultrafeinfilter*, Akademische Verlagsgesellschaft, Leipzig, 1929.

351. V. I. Tovarnickij and G. P. Glucharev, *Ultrafilters and Ultrafiltration* (in Russian), Medgiz, Moscow, 1951.

352. I. Daubner, *Membrane Filters and Their Application in Hydrobacteriology* (in Slovak), Slovak Academy of Sciences, Bratislava, 1960.

353. W. Bachmann, *Angew. Chem.*, *2*, 616 (1919).

354. E. Manegold, *Kolloid. Z.*, *61*, 140 (1932).

355. J. D. Ferry, *Chem. Rev.*, *18*, 373 (1936).

356. K. Spurny, *Z. Biol. Aerosol Forsch.*, *12*, 369 (1965); *13*, 44 (1966); *13*, 398 (1967).

357. H. Bechhold, *Z. Phys. Chem.*, *60*, 257 (1907).

358. S. L. Bigelow and A. Gemberling, *J. Chem. Soc.*, *29*, 1576, 1675 (1907).

359. N. Brown, *Biochem. J.*, *9*, 320, 591 (1915).

360. R. Zsigmondy and W. Bachmann, *Z. Anorg. Chem.*, *103*, 119 (1918).

361. R. Zsigmondy, *Z. Anal. Chem.*, *58*, 241 (1919).

362. A. H. Eggerth, *Amer. J. Biol. Chem.*, *28*, 201 (1921).

363. F. E. Bartell and M. Van Loo, *J. Phys. Chem.*, *28*, 161 (1924).

364. R. Zsigmondy, *Biochem. Z.*, *171*, 198 (1926).

365. R. Zsigmondy, *Angew. Chem.*, *39*, 398 (1926).

366. H. Bechhold and N. Silbereisen, *Biochem. Z.*, *199*, 1 (1928).

367. A. P. Krüger and R. C. Ritter, *Amer. J. Gen. Physiol.*, *13*, 409 (1930).

368. W. J. Elford, *J. Pathol. Bacteriol.*, *34*, 505 (1931).

369. E. V. Dianova and A. A. Voroshilova, *Mikrobiologiya*, *1*, 3 (1932) (in Russian).

370. J. H. Bauer and T. P. Hughes, *Amer. J. Gen. Physiol.*, *18*, 143 (1934).

371. P. Grabar, *Bull. Soc. Chim. Biol.*, *17*, 965, 1245 (1935).

372. W. J. Elford, *Trans. Faraday Soc.*, *33*, 1094 (1937).

373. P. Grabar and J. A. de Loureiro, *Ann. Inst. Pasteur*, *63*, 159 (1939).

374. R. E. Montonna and L. T. Jilk, *J. Phys. Chem.*, *45*, 1374 (1941).

375. A. E. Rukina and W. I. Bijosowa, *Mikrobiologiya, 21,* 60 (1952) (in Russian).

376. S. G. Mokruschin and V. J. Borisichina, *Zh. Prikl. Khim.*, *25*, 1182 (1952) (in Russian).

377. S. Jacobs, *Nature*, *175*, 133 (1955).

378. P. H. Carnell and H. G. Cassidy, *J. Polym. Sci.*, *55*, 233 (1961).

379. Ch. Gelman, *Analyt. Chem.*, *37*, 29A (1965).

380. R. T. Richards, D. T. Donovan, and J. R. Hall, *Amer. Ind. Hyg. Assoc. J.*, *28*, 590 (1967).

381. W. Desorbo and H. E. Cline, *J. Appl. Phys.*, *41*, 2099 (1970).

382. G. G. Jones and F. D. Miles, *J. Soc. Chem. Inc.*, *52*, 251 (1933).

383. W. J. Loesche and S. S. Socransky, *Science*, *138*, 139 (1962).

384. G. Manecke and K. F. Bonhoeffer, *Z. Elektrochem.*, *55*, 475 (1951).

385. G. Schmid and H. Schwarz, *Z. Elektrochem.*, *55*, 4, 8 (1951).

386. K. Spurny and M. Polydorova, *Collect. Czech. Chem. Commun.*, *26*, 921 (1961).

387. V. Havlicek, M. Polydorova, and K. Spurny, *Collect. Czech. Chem. Commun.*, *26*, 932 (1961).

388. H. Bechhold, *Z. Phys. Chem.*, *64*, 328 (1908).

389. E. Honold and E. L. Skau, *Science*, *120*, 805 (1954).

390. F. Erbe, *Kolloid. Z.*, *59*, 33 (1932); *63*, 277 (1933).

391. M. Pisa, *Kolloid. Z.*, *63*, 139 (1933).

392. P. Grabar and S. Nikitine, *J. Chim. Phys.*, *33*, 721 (1936).

393. M. Guerout, *C. R. Acad. Sci.*, *75*, 1809 (1872).

394. N. Bjerrum and E. Manegold, *Kolloid. Z.*, *42*, 97 (1927).

395. E. Manegold, S. Komagata, and E. Albrecht, *Kolloid. Z.*, *66*, 142 (1936).

396. H. Adzumi, *Bull. Chem. Soc. Jap.*, *12*, 292 (1937).

397. J. P. Lodge, J. B. Pate, and H. A. Huitt, *J. Amer. Hyg. Assoc.*, *24*, 380 (1963).

398. V. Hampl and K. Spurny, *Collect. Czech. Chem. Commun.*, *31*, 1152 (1966).

399. G. Hansmann and H. Pietsch, *Naturwiss.*, *36*, 250 (1949).

400. J. G. Helmcke, *Zbl. Bakt.*, *159*, 308 (1953).

401. K. H. Maier, *Naturwiss.*, *40*, 605 (1953).

402. K. H. Maier and H. Beutelspacher, *Kolloid. Z.*, *135*, 10 (1954).

403. J. G. Helmcke, *Kolloid. Z.*, *135*, 29, 101, 106 (1954).

404. G. Kanig and A. M. D'Ans, *Kolloid. Z.*, *149*, 1 (1956).

405. F. M. O'Leary, G. E. Hess, and F. E. Kubalski, *Amer. J. Bact. Proc.*, A, *28*, 31 (1956).

406. H. Spandau and R. Kurz, *Kolloid. Z.*, *150*, 109 (1957).

407. D. Schyma, H. Gärtner, and G. Moll, *Zbl. Bakt.*, *181*, 22 (1961).

408. R. Riley, J. O. Gardner, and U. Merten, *Science*, *143*, 801 (1964).

409. H. G. Fromme and W. Stöber, *Staub*, *26*, 270 (1966).

410. E. Frank, W. H. Fischer, and J. P. Lodge, *Staub*, *26*, 244 (1966).

411. V. Hampl, *Collect. Czech. Chem. Commun.*, *32*, 1983 (1967).

412. H. J. Preusser, *Kolloid-Z. Z. Polym.*, *218*, 129 (1967).

413. V. Hampl and K. Spurny, *Collect. Czech. Chem. Commun.*, *32*, 4181 (1967).

414. V. Hampl, Z. Pelzbauer, and K. Spurny, *Staub*, *29*, 245 (1969).

415. V. Hampl and Z. Pelzbauer, *Collect. Czech. Chem. Commun.*, *34*, 185 (1969).

416. J. Machacova, J. Hrbek, V. Hampl, and K. Spurny, *Collect. Czech. Chem. Commun.*, *35*, 2087 (1970).

417. D. W. Brubaker and K. Kammermayer, *Ind. Eng. Chem.*, *45*, 1148 (1953).

418. J. Duclaux and J. Errera, *Rev. Gen. Colloides*, *2*, 130 (1924).

419. H. Cox and R. R. Hyde, *Amer. J. Hyg.*, *16*, 667 (1932).

420. R. Wajsfelner, in *Seminaire de Physique des Aérosols* (J. Bricard, ed.), Faculté des Sciences de Paris, Paris, 1967-1968.

421. M. Leveque, *Ann. Mines Carbur.*, *13*, 276 (1928).

422. V. G. Levich, *Physicochemical Hydrodynamics* (in Russian), Izdat. Fiz. Mat. Lit., Moscow, 1959; translated into English, Prentice-Hall, Englewood Cliffs, 1962.

423. S. Twomey, *Bull. Obs. Puy de Dome*, 173 (1962).

424. K. Spurny and J. Pich, *Collect. Czech. Chem. Commun.*, *28*, 2886 (1963); *30*, 2276 (1965).

425. W. J. Megaw and R. D. Wiffen, *Air Water Pollut.*, *7*, 501 (1963).

426. G. Kubie, C. Jech, and K. Spurny, *Collect. Czech. Chem. Commun.*, *26*, 1065 (1961).

427. B. M. Uzelac and E. L. Cussler, *J. Colloid Interfac. Sci.*, *32*, 487 (1970).

428. E. Manegold and R. Hofmann, *Kolloid. Z.*, *51*, 220 (1930).

429. J. D. Ferry, *Amer. J. Gen. Physiol.*, *20*, 95 (1936).

430. G. L. Natanson, quoted without giving a reference in N. A. Fuchs, *Mechanics of Aerosols*, Pergamon, Oxford, 1964.

431. J. W. Thomas, *J. Air Pollut. Control Ass.*, *8*, 32 (1958).

432. R. E. Davis, *Air Water Poll.*, *8*, 177 (1964).

433. J. W. Thomas and R. H. Knuth, *Air Water Pollut.*, *10*, 569 (1966).

434. R. E. Davis, *Atmos. Environ.*, *1*, 69 (1967).

435. J. W. Thomas and R. H. Knuth, *Atmos. Environ.*, *1*, 70 (1967).

436. J. Pich, *Staub*, *24*, 60 (1964); also in *Collect. Czech. Chem. Commun.*, *29*, 2223 (1964).

437. B. Binek and S. Przyborowski, *Staub*, *25*, 533 (1965).

438. K. Spurny and G. Madelaine, *Collect. Czech. Chem. Commun.*, 36 (1971).

439. G. Henneberg and B. Haagen-Crodel, *Zbl. Bakt.*, *161*, 314 (1954).

440. H. Beutelspacher, *Kolloid. Z., 137*, 32 (1954).

441. G. Moll, *Kolloid-Z. Z. Polym.*, *203*, 20 (1965).

442. K. Spurny and J. P. Lodge, *Staub*, *28*, 503 (1968); also in *Collect. Czech. Chem. Commun.*, *33*, 3679 (1968).

443. K. Spurny, R. Blaschke, and G. Pfefferkorn, *Collect. Czech. Chem. Commun.*, *36*, 950 (1971).

444. G. D'Amico and I. L. Di Bari, *Rassegna Chim.*, *17*, 270 (1965) (in Italian).

445. E. Petras, *Kolloid-Z. Z. Polym.*, *218*, 136 (1967).

446. A. Toth, *Magy. Fiz. Foly.*, *15*, 435 (1967) (in Hungarian).

447. K. Spurny and J. Pich, *Air Water Poll.*, *8*, 193 (1964).

448. J. J. Fitzgerald and C. G. Detwiller, *Arch. Ind. Health*, *15*, 3 (1957).

449. W. Walkenhorst, *Staub*, *19*, 69 (1959).

450. K. Spurny and J. Hrbek, *Staub*, *29*, 70 (1969).

451. P. B. Price and R. M. Walker, *J. Appl. Phys.*, *33*, 2625, 3400, 3407 (1962).

452. P. B. Price and R. M. Walker, *Phys. Rev. Let.*, *8*, 217 (1962).

453. R. L. Fleischer, P. B. Price, and R. M. Walker, *Rev. Sci. Instrum.*, *34*, 510 (1963).

454. R. L. Fleischer, P. B. Price, R. M. Walker, and E. L. Hubbard, *Phys. Rev.*, A, *133*, 1443 (1964).

455. R. L. Fleischer, P. B. Price, and E. M. Symes, *Science*, *143*, 249 (1964).

456. R. L. Fleischer, P. B. Price, and R. M. Walker, *Science*, *149*, 383 (1965).

457. A. A. Kirsh and K. R. Spurny, *Zh. Prikl. Mekh. Tekh. Fiz.*, *3*, 109 (1969) (in Russian).

458. H. Hasimoto, *J. Phys. Soc. Jap.*, *13*, 633 (1958).

459. K. Spurny and J. P. Lodge, *Staub*, *28*, 179, 503 (1968); also in *Collect. Czech. Chem. Commun.*, *33*, 3679, 3931, 4385 (1968).

460. K. R. Spurny, J. P. Lodge, E. R. Frank, and D. C. Sheesly, *Environ. Sci. Technol.*, *3*, 453, 464 (1969).

461. E. R. Frank, K. R. Spurny, D. C. Sheesly, and J. P. Lodge, *J. Microsc.*, *9*, 735 (1970).

462. S. H. Seal, *Cancer*, *17*, 637 (1964).

Chapter 2

LIQUID FILTRATION THEORY AND FILTRATION PRETREATMENT

Richard J. Akers
Anthony S. Ward

Department of Chemical Engineering
Loughborough University of Technology
Loughborough, Leicestershire, England

I.	Introduction	170
	A. Types of Filtration Processes	171
	B. Flow through Packed Beds	173
II.	Filtration Process Theory	179
	A. Incompressible Cake Filtration	179
	B. Compressible Cake Filtration	186
	C. Units and Dimensions	190
	D. Depth Filtration	191
	E. Rapid Filtration Mechanism	193
III.	Modification of Cake Properties	197
	A. Coagulation	198
	B. Flocculation	214
	C. Use of Flocculants and Coagulants	227
	D. Assessment of Flocculants and Coagulants	231
IV.	Filter Aids	236
	A. Diatomaceous Earths	236
	B. Perlite	239
	C. Carbon, Cellulose, and Other Materials	239
	D. Selection of Filter Aids	240
V.	Miscellaneous Pretreatment Techniques	242
	A. Freezing	242
	B. Heating	242
	C. Aeration	243
	D. Mechanical and Ultrasonic Vibration	243

E. Electrical and Magnetic Treatments 243
F. Radiation 244
G. Conclusions 244
Notation 244
References 247

I. INTRODUCTION

The term solid-liquid filtration covers all those processes where a liquid containing suspended solid is freed of some or all of the solid when the suspension is drawn through a porous medium. The device in which this porous medium is retained is known as a filter. The main features of a filter are shown in Fig. 1.

Not all the features shown in Fig. 1 are distinguishable in every filter. For example, in many cartridge filters using paper media the paper is self-supporting, whereas in a deep bed filter the "filter cake" is distributed throughout the depth of the medium rather than lying on top of it. However, it will be found on close inspection that the function of each component shown is exercised in some way or other. The choice of a filter design for a given process depends on many factors, among which are the properties of the

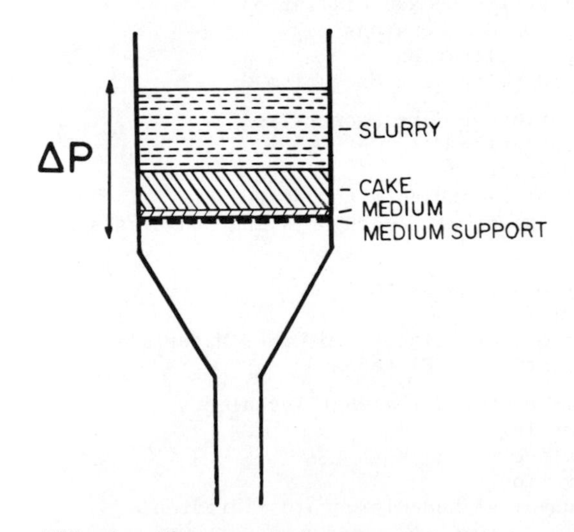

FIG. 1. The parts of a general filter.

solid particles to be removed, i.e., particle size and shape distribution and state of aggregation; the properties of the fluid, i.e., viscosity, density, and interaction with structural materials; the quantity of material to be handled; whether the process should be batchwise, continuous, or either; the dryness of the cake produced, if applicable; the concentration of solids in the suspension; the value of the materials being processed; whether the material to be retained is the solid, the liquid, or both; and whether the product need be washed.

An important factor in the design of a filter is the source of the driving force, which may be gravity, suction, positive pressure, or the application of centrifugal force. This choice will in turn depend upon consideration of the factors listed above.

A. Types of Filtration Processes

Filtration processes are conventionally divided into classes for convenience when considering the mathematical analysis of the factors involved. However, in virtually all real filtration processes more than one of these mechanisms may take part, and either the mathematical analysis is adjusted to account for this or simplifying assumptions are made. These simplifying assumptions are often justified by the dominance of mechanisms.

1. *Medium Filtration*

In medium filtration the particles are retained because they are larger than the holes in the filter medium. In this sense the filter is behaving as a sieve. This relatively uncommon mechanism is usually thought of in relation to the screening of large particles, but it also applies to the retention of fine particles on membrane filters and woven metal cloths.

If it is required that *absolutely* no particles greater than a given size should pass, medium filtration must be resorted to, although the differential pressure loading and maximum mass loading characteristics of medium filters are far from ideal. In any other type of filter there is only a probability that a certain size of particle will pass, although in a good filter design this probability is practically zero.

2. *Depth Filtration*

In depth filtration the separation process occurs *within* the medium only, the particles being smaller—often very much smaller—than the pores of the medium. If the medium is considered as a multitude of tortuous channels, for filtration to occur the particles present must impact on the walls of the channel and then be retained there by some force. For the particles to impact on the channel walls they must leave the fluid streamlines, and the rate at which this is achieved will depend on the balance of inertial and drag forces experienced by the particles. The depth medium may be either a bed of granular material or a porous solid. Examples of the former include deep-bed sand filters and precoat filters where the medium is a bed of diatomaceous earth or similar material supported on a coarse screen; and of the latter, felt and sintered metal filters. As particles become deposited in a depth filter the retentivity becomes greatest at the upstream side of the medium, leading eventually to blockage at that point. For long medium life, depth-type filters today are often constructed with a graded porosity through the medium, the pores being finest at the downstream side. With particulate media this may be done by using different particle size ranges of material. Because of the low loading of solid that can be tolerated before blockage occurs, both medium and depth filters are suitable only for the removal of small quantities of solids.

3. *Cake Filtration*

In cake filtration the solid material accumulates on the surface of the medium, so that, after a short initial period, filtration is through the bed of deposited solid. This process will continue until the pressure drop across the cake exceeds the maximum permitted by economic or technical considerations or until the space available is filled. This method of filtration is the most widely employed in the process industries and is very well suited to the filtration of concentrated suspensions and the recovery of large quantities of solid. The most important factor in cake filtration is the permeability or resistance of the filter cake, and this may be controlled, more or less, by altering the particle size distribution of the material, sometimes by adding another solid to it, and also by altering the state of aggregation of the solid. This topic will be discussed in more detail below (Sec. III). Because of its importance, a major part of this chapter will be taken up with a discussion of the flow of liquids through packed beds and its application to cake filtration.

As stated above, real filtration processes are composite in nature, several or all of the filtration mechanisms occurring simultaneously or consecutively. For example, in cake filtration, the very important initial layer of the cake must be retained on the surface of the medium by medium filtration, or depth filtration must occur until the sizes of the pores within the medium are reduced to such a point that medium filtration and, subsequently, cake filtration occurs.

B. Flow through Packed Beds

From a study of the flow of liquids through beds of sand, Darcy proposed the empirical relation, now known as Darcy's law [1],

$$u = K' \frac{\Delta P}{L} \tag{1}$$

where u is the overall fluid velocity [= $(1/A)(dV/d\theta)$], L the thickness of the bed, ΔP the pressure drop across the bed, K' a constant characteristic of the bed and fluid properties, V the volume of fluid flowing in time θ, and A the cross-sectional area of the bed.

Shortly before Darcy's work was published, Poiseuille [2] presented the equation for the flow of liquid through a capillary of circular cross section

$$\frac{dV}{d\theta} = \frac{\Delta P \pi r^4}{8\mu L} \tag{2}$$

where r is the capillary radius, and μ the viscosity of the fluid.

The Poiseuille equation states that the rate of flow of fluid through a capillary is inversely proportional to the viscosity of that fluid. If this relationship is assumed for packed beds, the Darcy equation may be rewritten

$$u = \frac{1}{A}\frac{dV}{d\theta} = \frac{K\ \Delta P}{\mu L} \tag{3}$$

which defines K, the permeability of the bed. This equation has been found to be obeyed when the rate of flow of fluid through the pores in the bed is of such a velocity to give laminar flow. The reciprocal of permeability is defined as the resistance of the bed, denoted R. The unit of K when expressed in dimensions of 1 $cm^3/(sec\ cm^2)$ for a liquid with viscosity 1 cP and at a ΔP of 1 atm is known as the Darcy.

Two other important properties of a packed bed are the porosity ε, defined as the fraction of volume of the bed not occupied by solid material (also known as the fractional voidage or voidage), and the specific surface area of the bed S, the surface area of the packed bed per unit volume of bed. It will be seen that the volume fraction of bed occupied by the solids is $1 - \varepsilon$.

Consider a bed of unit length and unit surface area containing n capillaries perpendicular to the face. By Eq. (2)

$$\frac{dV}{d\theta} = \frac{n\pi r^4 \Delta P}{8\mu L} \tag{4}$$

But for this particular bed

$$\varepsilon = n\pi r^2 \tag{5}$$

and

$$S = 2n\pi r \tag{6}$$

Hence

$$\frac{dV}{d\theta} = \frac{\varepsilon^3 \Delta P}{2S^2 L\mu} \tag{7}$$

However, for beds of irregular particles or irregular and interconnecting channels, it is not easy to define the radius of the pores and, hence, to derive an equivalent equation to that for circular pipes. By combining these terms Kozeny [3] defined the equivalent diameter δ of the pore space in a random bed as the void volume per unit of internal surface by

$$\frac{\varepsilon}{S} = \delta \tag{8}$$

where S is the specific surface area of the bed. For beds in which only point contacts occur (i.e., a negligible amount of surface area is lost)

$$S = S_p(1 - \varepsilon) \tag{9}$$

where S_p is the specific surface area of a particulate substance. Hence

$$\delta = \frac{\varepsilon}{S_p(1 - \varepsilon)} \tag{10}$$

This quantity is related to the hydraulic mean diameter, which is defined as

$$4 \times \frac{\text{Cross-sectional area}}{\text{Wetted perimeter}} = 4 \times \frac{\text{Volume of voids}}{\text{Surface area in bed}} = 4\delta \tag{11}$$

and for a circular pipe the hydraulic mean diameter d is

$$4\,\frac{\pi r^2}{2\pi r} = 2r = d \tag{12}$$

i.e., the hydraulic mean diameter and actual diameter are equal.

Kozeny proposed that the equivalent diameter δ of a random bed be substituted into the Poiseuille equation for a circular tube, to give

$$\bar{u} = \frac{\delta^2 \Delta P}{K'' \mu L'} \tag{13}$$

where $\bar{u}$ is the mean velocity of fluid through the pores, K" a dimensionless constant characteristic of the structure of the bed, and L' the length of the pores. Making the reasonable assumption that $L' \propto L$ and using the Dupuit [4] relation that

$$\bar{u} = \frac{u}{\varepsilon} \tag{14}$$

which, as has been shown by Pirie [5], is approximately true for random packings but not, as discussed by Coulson and Richardson [6], for a cubic array of spheres; and substituting these approximations into Eq. (13) gives

$$u = \frac{1}{A}\frac{dV}{d\theta} = \frac{1}{K''}\frac{\varepsilon^3}{S^2}\frac{1}{\mu}\frac{\Delta P}{L} \tag{15a}$$

$$K'' = \frac{1}{u}\frac{\varepsilon^3}{S_p^2(1-\varepsilon)^2}\frac{\Delta P}{\mu L} \tag{15b}$$

known as the Kozeny-Carman equation [7]. The constant K" is dependent on particle shape and size and the packing of the bed; for beds of moderate porosity it lies in the range 3.5 to 6.0. Typical values are shown in Fig. 2.

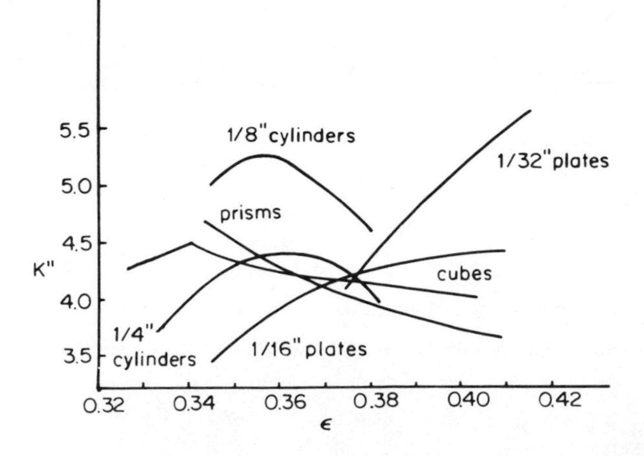

FIG. 2. Variation of K" with ε for various shapes. Reproduced, by permission, from Ref. 6.

Many workers have attempted to correlate permeability and porosity. In the older literature the two terms are used indiscriminately and are still confused by some authors. However, experiment and theory show clearly that permeability is a function of porosity and other variables such as the specific surface of the packed bed. Most attempts at correlation have compared different packings and sizes of one shape of particle, e.g., Raschig rings, and as such are not general. Scheidegger [8] considers it hardly credible that general correlations should exist. The proportionality between flow rate and pressure drop as stated in Darcy's law does not apply at high flow rates. This, by analogy with the flow of liquid through cylindrical channels, has been equated with the concept of a critical "Reynolds number." The designation of a Reynolds number to a system requires that some linear dimension equivalent to diameter must be defined, and this has been one of the main stumbling blocks of hydraulic radius theories of fluid flow. Deviations from linearity occur when the inertia term in the Navier-Stokes equation begins to become significant. This is known to happen in curved tubes long before the onset of turbulence. Friction-factor charts have been published for packed bed systems, but, in view of the great range of published values of the critical Reynolds number, they must be used with caution. However, this general problem is not serious in filtration, as the flow rates encountered in almost all filtration processes give behavior consistent with Darcy's law.

The earlier attempts at developing a theory of flow through packed beds made use of capillary models. Starting with a bundle of straight parallel capillaries and the Hagen-Poiseuille equation, the capillary models proceed to consider orthogonal capillary groups, capillaries of different diameters in series, and branched capillary models. Capillary models rely on a simplified structure for the bed, and as such the equations produced are very sensitive to the assumed capillary diameter distribution. However, by the use of an empirical tortuosity factor they can be made to fit any porous medium.

Kozeny developed the hydraulic diameter theory, in which the porous medium is considered as an assembly of channels of varying cross section but of definite length. The Navier-Stokes equation is then solved for a section normal to the direction of flow. The permeability is related to a hydraulic radius defined by the relationship

$$r = \frac{F(\varepsilon)\ K^{1/2}}{c} \tag{16}$$

where $F(\varepsilon)$ is the porosity factor, K the permeability, and c a dimensionless constant. The concept of hydraulic radius is related to the fact that the dimensions of permeability are area, i.e., length^2. This dimensional argument proceeds to consider that the permeability is related to the diameter of some equivalent pore defined by the hydraulic radius. Kozeny's theory goes on to give an equation, which was modified by Carman and known by both of their names, in which the permeability of the bed is expressed as a function of porosity, a form of specific surface area, and the empirical constant K". Although this equation is one of the most generally applicable solutions to the problem and certainly the most widely used, it is open to serious criticism.

The equation is frequently used to measure surface area even though the results obtained often differ considerably from those given by the well-established method of gas adsorption and also from particle size measurement. It is also observed that the measured permeability is a function of the packing of the bed in a way that the equation does not predict.

Other theories of flow through packed beds have included hydraulic drag theories and statistical theories. Drag theories, initiated by Iberall [9], consider the pressure drop across the bed in terms of the drag exerted on pore walls by the fluid passing through them. Statistical theories have considered the passage of fluid through a bed as describing a random walk. Relating to disordered systems, statistical theories assume that the porous medium is isotropic and comprised of an ensemble of macroscopically identical systems. Scheidegger [8] employed the techniques devised by Einstein to solve the relationship of Brownian diffusion to permeability by introducing the concept of dispersivity, i.e., the sideways motion undergone by the fluid. More recently Scarlett [10], using a sterological description of geometry for an irregular pore, has obtained an equation for laminar flow that gives a value of the Kozeny-Carman constant of $\pi^2/2$, very close to the observed value of 5 for beds of irregular particles.

II. FILTRATION PROCESS THEORY

A. Incompressible Cake Filtration

The process of cake filtration involves the separation of fine solid particles, usually of such a size that they will not easily settle, from the liquid in which they are dispersed by collecting the solid particles on the surface of a medium which is permeable only to the liquid. As the process continues, additional particles are collected on the surface of the initial layer of solids, thereby building up a cake.

At any instant of time during the build-up of the cake, the liquid will be flowing through solids already deposited. The relationships governing this mechanism have already been discussed in detail. The first is Darcy's Law, rewritten here as

$$\frac{1}{A}\frac{dV}{d\theta} = \frac{K}{\mu}\frac{\Delta P}{L} \tag{17}$$

where V is the volume liquid passing in time θ, A the superficial area of the cake perpendicular to the direction of flow, ΔP the pressure drop in the liquid flowing across the cake of thickness L, μ the viscosity of the liquid, and K a proportionality constant. This constant, called the permeability, is defined by this equation. An alternative expression is the Kozeny-Carman equation, written here as

$$\frac{1}{A}\frac{dV}{d\theta} = \frac{\varepsilon^3}{K''\mu(1 - \varepsilon)^2}\frac{\Delta P}{S_p^2 L} \tag{18}$$

where the additional terms are: ε the voidage or porosity of the cake, i.e., void volume divided by total volume; S_p the specific area of the particles, and K" a numerical constant normally taken as having the value 5. Inspection of these equations will show that the flow rate of a given liquid obtained for a given pressure difference will depend on the permeability K and the porosity. (It can be assumed that the specific surface of the particles S_p will remain unchanged.)

If the values of permeability and porosity change as the pressure difference is increased, then the cake is termed "compressible." If these parameters remain constant, then the cake is termed "incompressible."

To describe the continuous process of cake filtration it is necessary to obtain a relation between the mass of solids in a given suspension and the mass of cake produced on subsequent filtration of the suspension. This is achieved by means of a mass balance on the solids. Referring to Fig. 3, the filtrate volume V is produced by the separation of the mass of solids represented by the cake having thickness L, area A, and porosity ε from the suspension, which has a solids mass fraction c. The mass of solids in the cake is equal to the mass of solids in the suspension. This is expressed by

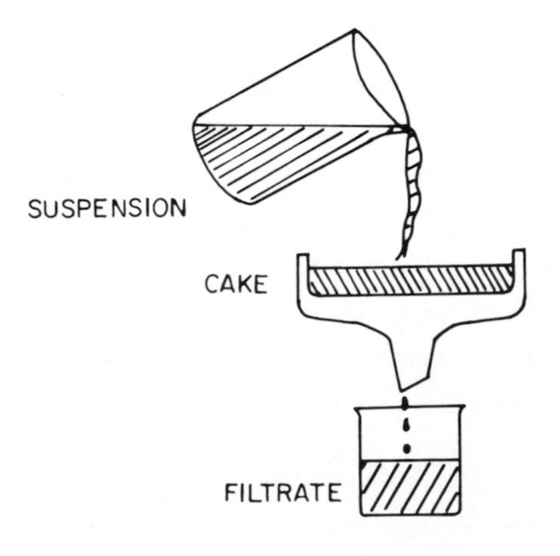

FIG. 3. The filtration process.

$LA(1 - \varepsilon)\rho_s = (V + \varepsilon AL)\rho_f[c/(1 - c)]$, so

$$L = \frac{V\rho_f \frac{c}{1 - c}}{A\left[(1 - \varepsilon)\rho_s - \varepsilon\rho_f \frac{c}{1 - c}\right]} \tag{19}$$

where ρ_f and ρ_s are liquid and solid densities, respectively.

Substituting for L in Darcy's Law gives

$$\frac{dV}{d\theta} = \frac{KA^2}{\mu V} \frac{\left[(1 - \varepsilon)\rho_s - \varepsilon\rho_f \frac{c}{1 - c}\right]}{\rho_f \frac{c}{1 - c}} \Delta P \tag{20}$$

From the definition of an incompressible cake, permeability and porosity are taken as constant, and the equation may be integrated to give

$$\frac{V^2}{2} = \frac{KA^2}{\mu} \frac{\left[(1 - \varepsilon)\rho_s - \varepsilon\rho_f \frac{c}{1 - c}\right]}{\rho_f \frac{c}{1 - c}} \Delta P\theta \tag{21}$$

on, in terms of resistance,

$$\frac{V^2}{2} = \frac{A^2}{R\mu} \frac{\left[(1 - \varepsilon)\rho_s - \varepsilon\rho_f \frac{c}{1 - c}\right]}{\rho_f \frac{c}{1 - c}} \Delta P\theta \tag{22}$$

A single constant is used to represent the system parameters, i.e.,

$$C = \frac{\mu\rho_f \frac{c}{1 - c}}{2\left[(1 - \varepsilon)\rho_s - \varepsilon\rho_f \frac{c}{1 - c}\right]} \tag{23}$$

giving

$$V^2 = \frac{KA^2\Delta P\theta}{C} \quad \text{and} \quad V^2 = \frac{A^2\Delta P\theta}{RC} \tag{24}$$

1. *Medium Resistance*

The resistance to flow of the medium may be significant and, therefore, should be taken into account. The liquid hold-up in the medium is taken as insignificant and the flow regime in the pores is assumed laminar, so that a straight proportionality relation between flow and pressure difference may be used. It is then possible to postulate a thickness of cake L_o, and an associated volume of filtrate V_o, which is equivalent in resistance to the medium. Equation (24) can now be written

$$\frac{dV}{d\theta} = \frac{KA^2 \Delta P}{2C(V + V_o)} \tag{25}$$

and integrated to give

$$\frac{V^2}{2} + V_o V = \frac{KA^2 \Delta P \theta}{2C} \tag{26a}$$

or

$$\frac{\theta}{V} = \frac{C}{KA^2 \Delta P} (V + 2V_o) \tag{26b}$$

2. *Equivalent Forms of Basic Filtration Equations*

The above equations may be written in several different, but nevertheless equivalent, forms, each of which has been favored at one time or another by various workers in the field. Some of these forms are as follows:

$$\frac{\theta}{V} = \frac{C}{KA^2 \Delta P} (V + 2V_o) \tag{27}$$

where

$$C = \frac{\mu \rho_f c}{2[(1 - \varepsilon)\rho_s (1 - c) - \varepsilon \rho_f c]} \tag{27a}$$

$$\frac{\theta}{L} = \frac{B}{K\ P} (L + 2L_o) \tag{28}$$

where

$$B = \frac{\mu}{2c\rho_f} [(1 - \varepsilon)\rho_s (1 - c) - \varepsilon \rho_f c] \tag{28a}$$

$$\frac{\theta}{V} = \frac{R\mu v}{2A^2 \Delta P} (V + 2V_o) \tag{29}$$

where

$$R = \frac{1}{K} \quad \text{and} \quad v = \frac{\rho_f c}{(1 - c)(1 - \varepsilon)\rho_s - c\varepsilon\rho_f} \tag{29a}$$

$$\frac{\theta}{L} = \frac{R}{2v\Delta P}(L + 2L_o) \tag{30}$$

$$\frac{\theta}{V} = \frac{\alpha D}{A^2 \Delta P}(V + 2V_o) \tag{31}$$

where

$$\frac{1}{D} = \frac{2}{\mu}\left[\frac{1}{\rho_f}\left(\frac{1 - c}{c}\right) - \frac{1}{\rho_s}\left(\frac{\varepsilon}{1 - \varepsilon}\right)\right] \tag{31a}$$

$$\frac{\theta}{L} = \frac{E\alpha}{\Delta P}(L + 2L_o) \tag{32}$$

where

$$E = \frac{\mu(1 - \varepsilon)\rho_s}{2}\left[\left(\frac{1 - c}{c}\right)(1 - \varepsilon)\frac{\rho_s}{\rho_f} - \varepsilon\right] \tag{32a}$$

In the above equations, v is the volume of cake deposited by unit volume of filtrate. Clearly

$$L = \frac{Vv}{A} \quad \text{and} \quad v = \frac{\rho_f c}{(1 - \varepsilon)(1 - c)\rho_s - \varepsilon\rho_f c} \tag{33}$$

The relationships between the constants are

$$C = \frac{\mu v}{2} \qquad B = \frac{\mu}{2v} \qquad R = \alpha(1 - \varepsilon)\rho_s \tag{34}$$

with α being a specific resistance defined by writing Darcy's Law in the form

$$\frac{1}{A}\frac{dV}{d\theta} = \frac{1}{\mu\alpha(1 - \varepsilon)\rho_s}\frac{\Delta P}{L} \tag{35}$$

where $(1 - \varepsilon)\rho_s L$ is the weight of dry cake of thickness L. These equations have been developed by integrating between the limits $\theta = \theta$ and $\theta = 0$, $V = V$ and $V = 0$, and $L = L$ and $L = 0$.

For constant pressure filtration a plot of θ/V vs. V (or θ/L vs. L) will give a slope and intercept from which the unknowns K, α or R, and V_o or L_o may be evaluated.

The choice of filtration equation depends on the type of investigation being undertaken and the relative ease of measurement of the various parameters. Obviously, Eqs. (28), (30), and (32) do not require a measure of the filter area, although a measure of cake thickness is required. The latter may be easily determined in a simple laboratory vacuum leaf test but may be difficult even to estimate in pressure filtration. If the temperature of the liquid is constant throughout the filtration then the viscosity symbol can be combined with the other terms in a single constant as in Eqs. (27) and (28). If temperature changes are significant, then use of Eqs. (29) and (30) would allow the consequent viscosity effects to be investigated. It is obvious that there are other ways in which these equations may be written, and the assiduous investigator will develop a form which most closely suits the problem in hand.

Disregarding V_o as having a negligible value, Eq. (29) may be rearranged to give

$$\frac{V^2}{A^2\theta} = \frac{2\Delta P}{R\mu v} \tag{36}$$

and

$$\frac{V}{A\theta} = \left(\frac{2\Delta P}{R\mu v\theta}\right)^{1/2} \tag{37}$$

The left-hand expression is the volume of filtrate per unit area per unit time of filtration and is often called the "form filtration rate."

If Eq. (36) is multiplied by v, the volume of cake deposited by unit volume of filtrate, there is obtained

$$\frac{Vv}{A\theta} = \left(\frac{2v\Delta P}{R\mu\theta}\right)^{1/2} \tag{38}$$

which is an expression for the volume of cake deposited per unit area per unit time of filtration. Alternatively, the basic equation may be written with w, the weight of dry cake per unit volume of filtrate, instead of v, giving

$$\frac{Vw}{A\theta} = \left(\frac{2w\ \Delta P}{R\mu\theta}\right) \tag{39}$$

as a "cake-weight form filtration rate." This approach has its advocates especially in vacuum filtration; its use is further discussed in Refs. 11 and 13.

A plot of log $Vw/A\theta$ vs. log θ should give a slope of -0.5 for an ideal cake. The disadvantage of this method of analysis is that values for cake or medium resistance are not obtained; also, because a log-log plot is used, it is relatively insensitive to change in the various parameters. The use of the "cake-weight form filtration rate" in the calculation of rotary vacuum-filter yields is discussed at some length by Gale [13].

The applicability of these filtration equations has been investigated by many workers. A notable contribution was made by Ruth and coworkers in a series of papers [12]. One of their investigations concerned the reproducibility of filter behavior. A diatomaceous earth, known as Filter-Cel, was chosen as the solid, and, using a filter press with a single frame about 9 in. in diameter, filtrations were carried out at constant pressures up to 60 psi. The data obtained were plotted in the form V^2 vs. θ, as in Fig. 4. Each curve is straight in the center portion but curves at the ends. The final curve is due to the fact that the press is filling and the effective filtration area is being reduced. Ruth reasoned that the initial curve of each plot was related to the resistance of the cloth and that the reduction in the value of V^2 for a given θ in subsequent filtrations related to the change in cloth resistance. It was noticed that later tests of a series showed better agreement with each other than earlier ones. Ruth was able to straighten the initial curve by adding—by trial and error—a constant V_o to V giving $(V + V_o)^2$ vs. θ and hence, he deduced that $dv/d\theta \propto K/2(V + V_o)$, which is the same form as the equation employed in the derivation of the filtration equations.

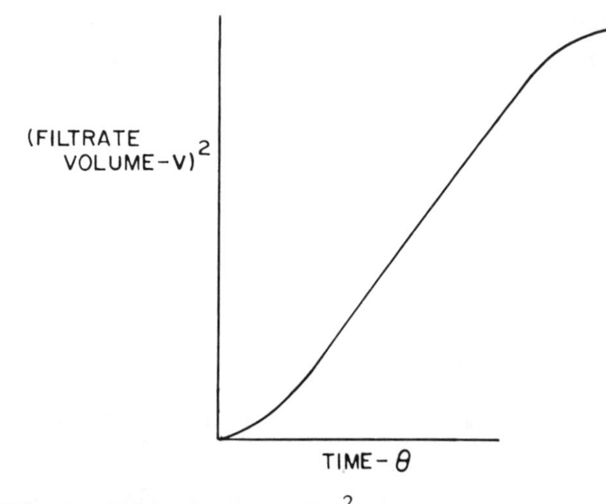

FIG. 4. Typical curve of V^2 vs. θ. (After Ruth.)

B. Compressible Cake Filtration

A truly incompressible cake will have a constant permeability and porosity which is independent of the pressure difference or the liquid flow rate. However, almost all filter cakes show some changes in permeability or porosity as filtration conditions are altered. Porosity differences are most easily seen as changes in moisture content of the cake. Clearly in a completely incompressible cake it is impossible to reduce the moisture content by increasing the filtration pressure.

The change in permeability produced by changing filtration pressure can be seen best by plotting on log-log graph paper the values obtained. A straight line with a negative slope will usually result. This implies a relationship of the form

$$K \propto K_o(\Delta P)^{-s} \tag{40}$$

where s is a compressibility coefficient and K_o is the initial permeability. Obviously, s has a zero value for totally incompressible systems; the larger the value, the more compressible the material.

A moderate amount of compressibility is advantageous because it allows a reasonably low moisture content to be achieved without an unduly low filtration rate making the process uneconomic. In contrast, a highly compressible sludge will have a high resistance to the lowest of pressure differences, giving poor filtration rates and high moisture contents and rendering as unacceptable the idea of using a filtration process. In these cases resort must be made to altering the sludge-producing process or using filter aids or other pretreatment techniques in order to reduce compressibility.

An approach to filtration which is applicable to compressible sludges is due largely to Tiller [16-20]. The basic flow equation used is the Kozeny-Carman equation,

$$\frac{1}{A}\frac{dV}{d\theta} = \frac{\varepsilon^3}{K''\mu(1 - \varepsilon)^2}\frac{\Delta P}{S_p^2 L} \tag{41}$$

derived earlier. This relationship has been developed from a model of a packed bed, and the porosity and specific surface terms have been introduced to eliminate other, less easily measurable parameters. If K" is considered to be constant and the specific surface of the particles can be related to the particle size distribution, which is a constant for a given set of particles, the porosity becomes the single variable parameter in the equation.

If the cake is considered as a series of thin slices with their boundaries parallel to the septum, of thickness dx and porosity ε_x, the equation—using q for $(1/A)(dV/d\theta)$ in order to simplify the notation—can be written to describe the flow through any one slice, e.g.,

$$q = \frac{\varepsilon_x^3}{K''\mu(1 - \varepsilon_x)^2}\frac{\Delta P}{S_p^2\,dx} \tag{42}$$

In order to integrate this equation it is first necessary to know how porosity varies within the filter cake. It has been observed that ε_x is least in the region close to the septum, i.e., where the hydraulic pressure p_x is low, and greatest at the interface between cake and suspension where p_x is large. Considering a plane within the cake parallel to the septum, the porosity changes from point to point depending on the orientation of the particles and the areas of particle-particle contact; and in the region close to the wall of the vessel the porosity will differ from that in the main body of the cake due to the frictional drag effect of the walls. To make some sense of this complicated situation, it is assumed that all particles are in point contact and that the vessel used is sufficiently large for wall-friction effects to be ignored.

Therefore the total pressure p on the cake can be shown to give rise to two pressures within the cake, a hydraulic pressure and p_s, a so-called compressive stress pressure in the solids, so that

$$p = p_x + p_s \tag{43}$$

A local filtration resistance α_x can be defined by an equation of the form

$$q = \frac{dp_x}{\mu\alpha_x(1 - \varepsilon_x)\rho_s dx} \tag{44}$$

which by comparison with the earlier equation shows α_x to be a function of local porosity ε_x. A local permeability K_x may be defined similarly. The relationship between porosity and pressure must be established experimentally. This is done by using a compression-permeability cell which is a device in which the cake is consolidated under a mechanical loading of p_s, achieved by placing suitable weights on a porous solid plug positioned over the cake. The permeability as a function of resistance is then determined by allowing the filtrate to pass through the cake under a low hydrostatic head. In this apparatus, the permeability of the cake is determined and assumed to be equal to that of a filter cake where the apparent compressive pressure, $p_s = p - p_x$, is the same as the consolidation pressure used in the cell.

The results of work using the above technique shows that in general

$$\alpha_x = \alpha_o p_s^n \tag{45}$$

$$\varepsilon_x = \varepsilon_o p_s^{-\lambda} \tag{46}$$

for p_s values greater than a minimum p_i, which is about 1.0 psi, where α_o is the initial filtration resistance and ε_x the initial porosity.

Several authors [21-27] have developed approximate equations to predict the distribution of porosity within a cake. For example,

$$\frac{p_s}{p} = \left(1 - \frac{x}{L}\right)^{1/1-n-\beta} \qquad \text{for } p_s \geq p_i \tag{47}$$

where β is the exponent in the relation

$$1 - \varepsilon_x = Bp_s^{\beta} \tag{48}$$

B being a constant that varies from 0 to 0.25.

Tiller also observed that in cakes where there is a change in local porosity there will also be a change in local flow rate, q_x. An expression for this is

$$q_x = \frac{-dp_s}{\mu\alpha_x(1 - \varepsilon_x)\rho_s dx} \tag{49}$$

To simplify this, $(1 - \varepsilon_x)\rho_s$ dx is replaced by dw_x, the mass of solids per unit area in thickness dx, so that

$$\frac{dp_s}{\alpha_x} = -\mu q_x \, dw_x \tag{50}$$

If the hydraulic pressure at the septum-cake interface is set at P_1, then the compressive pressure at that point is $P - P_1$. Similarly, w is zero at the septum-cake interface, and w = w at the cake suspension plane. Thus

$$\frac{1}{\mu}\int_0^{P-P_1} \frac{dp_s}{\alpha_x} = \int_0^w q_x \, dw_x \tag{51}$$

Now q_x is assumed to vary with x and w_x and the right-hand side of the above equation may be multiplied by $q_1 w$, where q_1 is the value of q at the cake-medium interface. In this way a factor J is defined by

$$Jq_1w = q_1w \int_o^w \frac{q_x}{q_1} \frac{dw_x}{w} = q_1w \int_o^1 \frac{q_x}{q_1} d\left(\frac{w_x}{w}\right) \tag{52}$$

and

$$J = \int_o^1 \frac{q_x}{q_1} d\left(\frac{w_x}{w}\right) \tag{53}$$

A new filtration resistance α_T is defined by

$$\alpha_T = \frac{(P - P_1)}{\mu q_1 w} = (p - p_1) \frac{\int_o^1 \frac{q_x}{q_1} d\left(\frac{w_x}{w}\right)}{\int_o^{P-P_1} \frac{dp_s}{\alpha_x}} \tag{54}$$

The average filtration resistance, α, defined earlier, is expressed by

$$\alpha = \frac{P - P_1}{\int_o^{P-P_1} \frac{dp_s}{\alpha_x}} \tag{55}$$

so it is seen that

$$\alpha_T = J\alpha \tag{56}$$

For dilute slurries J has a value approaching unity, will decrease in value for increasing slurry concentration, and will change with pressure. From the basis set out here a number of further equations may be developed and used in the calculation of filtration problems. Examples will be given in the chapters on industrial filtration.

C. Units and Dimensions

The basic dimensions of force F, mass M, length L, and time T, are related by Newton's second law of motion, Force = mass × acceleration, so that the dimensions of force are MLT^{-2}.

Pressure is defined as force per unit area, and its dimensions are $ML^{-1}T^{-2}$. Viscosity is the ratio of shear stress to velocity gradient and therefore has the dimensions $ML^{-1}T^{-1}$. If the appropriate dimensions are substituted for the symbols in the Darcy's Law equation, the dimensions of permeability K can be obtained. These are L^2. In petroleum production engineering the unit of permeability is the Darcy, defined under the conditions of 1 atm pressure difference over a bed of 1 cm depth with a flow of 1 cm/sec of a liquid of viscosity of 1 poise. The resistance R has dimensions of L^{-2} and the specific resistance α, LM^{-1}. Constants B and C have dimensions of $ML^{-1}T^{-1}$, i.e., those of viscosity, and constants D and E, $M^2L^{-4}T^{-1}$, i.e., those of viscosity x density.

In a system that uses force, F, as a defining term the appropriate dimensions can be obtained by substituting $FL^{-1}T^2$ for M.

D. Depth Filtration

Depth filtration is a method of clarifying liquids by allowing them to percolate through a bed of particles or fibers. In the context of solid-liquid separation it is most commonly applied to water purification and tertiary effluent treatment, and for these purposes sand is the most common filter medium. Because of the build-up of material within the bed which leads to blocking of the available pores, depth filtration is suitable only for liquids containing no more than 500 mg/liter of solid and usually much less than this. For example, in water treatment the solids concentration presented to the filter may be as low as 0.01 mg/liter, the solids being in the form of very fine particles.

The particles to be removed are usually much smaller than the pores in the bed. A typical algae cell occurring in water is about 10 to 30 μm in diameter compared with the 500-μm diameter sand particles making up the usual filter bed.

Depth filtration is also practiced with media other than sand; for example, sintered-steel felt filters are used in critical hydraulic systems.

In water treatment processes, depth filters are divided into two classes, which, although apparently are not very different in construction, differ widely in terms of the mechanism of solids removal.

1. *Slow Filters*

These filter at the rate of approximately 60 gal per day per square foot of filter surface. Because of their slow filtration rate, they need to be very large and are, consequently, very expensive in terms of land, some filters being several acres in extent. The water is filtered through a bed of sand 3 to 4 ft in thickness supported on 1 ft of gravel. Filtration occurs mainly at the top surface of the bed, the particles being removed as the water passes through a superficial layer of material which is biological in origin. Because of this there is little penetration of retained particles into the bed, which may be cleaned by removing the top layer of sand every month or so or by washing the top layer in situ from above. Filters of this type are frequently operated without chemical pretreatment of the water.

2. *Rapid Filters*

These filters are designed for throughputs of about 2,000 to 2,500 gal per day per square foot, i.e., 30 times that of a slow filter. The thickness of sand used is generally less than in slow filters and the sand has a larger particle size and narrower particle size distribution. These properties of the sand bed permit the considerable increases in permeability necessary for rapid filtration. The removed solids are retained within the bed, and, because of the much greater flow rate and hence rate of solids deposition, the filter cycle time is hours rather than days. Such filters are cleaned by back flushing to fluidize the bed, this sometimes being aided by air scouring. Because of the fluidization process and subsequent particle sedimentation, the finer particles are concentrated in the upper layers of the bed. This is not desirable for down-flow operation and materials of different densities are now used in some filters to overcome these particle size classification effects. The construction and operation of rapid filters is discussed in detail in a chapter in Part II of this book, but the mechanism of particle removal is briefly described below.

E. Rapid Filtration Mechanism

For a rapid filter to remove particles from a liquid, these particles must come into contact with a particle of the bed and then adhere to that particle. The mechanisms by means of which impaction occurs have been studied for both liquid and gas filtration processes and are reasonably well understood. Flow is laminar under the conditions prevailing in a rapid filter, there being a velocity gradient from zero at the walls to a maximum near the center of the pore. Because of the irregular shape of the pores, the flow is more complex than in a circular channel and is characterized by a modified Reynolds number, the Blake number, defined as

$$B = \frac{6d_m u}{(1 - \varepsilon)\mu} \tag{57}$$

where d_m is the diameter of a particle in the filter-medium bed and u the flow rate per unit area of filter. The term B typically has a value of 0.2. The mechanisms of particle capture in depth filters are interception, inertia, diffusion, sedimentation, and hydrodynamic effects; these are shown diagramatically in Fig. 5.

1. *Interception*

In the absence of other forces, particles will move through a bed with the streamlines of the fluid. As flow in the bed is laminar, the streamlines will diverge to flow around a medium particle and reconverge behind it. If the position of a particle in a streamline is such that the distance between its center and the center of the medium particle at any point during its passage around the latter is less than their collision diameter, i.e., $(d_p + d_m)/2$ where d_p is

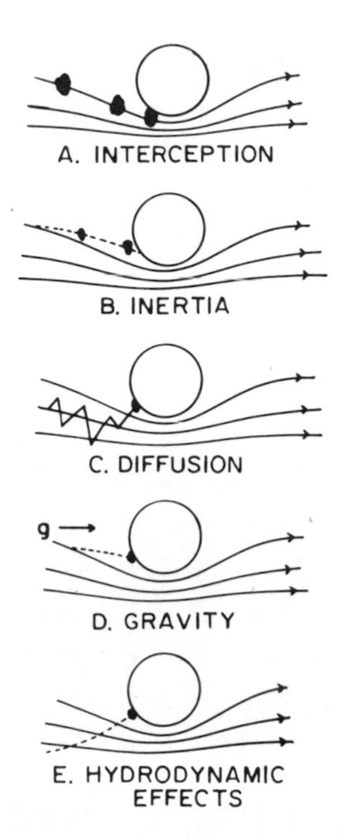

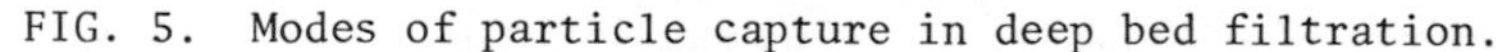
FIG. 5. Modes of particle capture in deep bed filtration.

the Stokes diameter of a particle, interception will occur. The probability of this occurring is related to the ratio of these diameters d_p/d_m.

2. *Inertia*

If a particle has a greater density than the liquid in which it is suspended, it will experience inertial forces that cause it to cross the fluid streamlines as they diverge before a medium particle. Davies [28] has shown that the inertial effect is related to a dimensionless group, the Stokes number, by

$$St = \frac{(\rho_s - \rho_f) d_p^2 \bar{u}}{18 \mu d_m} \tag{58}$$

where $\bar{u}$ is the mean velocity of a fluid through a pore. The greater the Stokes number the greater is the effect of inertial forces on the particle. From theoretical and experimental work on the capture of particles on fibers in gas filters, Harrop and Stenhouse [29] have shown how the capture efficiency may be related to St and the ratio d_p/d_m.

3. *Diffusion*

Small suspended particles will experience random diffusional movement due to collisional energy transfer by the molecules of the suspending fluid. The diffusion coefficient D associated with this Brownian diffusion is given by

$$D = \frac{\underline{k}T}{3\pi\mu d_p} \tag{59}$$

where $\underline{k}$ is Boltzmann's constant and T the absolute temperature. Brownian diffusion is significant for aqueous suspensions at ambient temperature for particles with diameters of 1 μm and less.

4. *Sedimentation*

The rate of gravitational sedimentation of particles from 2 to 10 μm in diameter has been shown by Ison and Ives [30] to be important in rapid filtration. The significance of this effect may be characterized by a dimensionless group relating the Stokes sedimentation velocity to the approach velocity of the particles:

$$S = \frac{(\rho_s - \rho_f)\, d_p^2 g}{18\mu\bar{u}} \tag{60}$$

where S is the sedimentation coefficient and g the gravitational force.

5. *Hydrodynamic Action*

When particles are suspended in a liquid in which a shear gradient exists they will experience forces causing them to cross the fluid streamlines. The resultant motion of particles in fluids moving within cylindrical pores is not completely understood because of the interaction of several effects.

As inertial impaction, hydrodynamic action, and sedimentation are enhanced by increasing particle size and as diffusion is enhanced by decreasing particle size, it might be expected that the efficiency of a depth filtration process would show some minimum effectiveness at a particle size intermediate between those for these mechanisms. This minimum is well known in gas filtration and has been demonstrated in rapid filtration by Yao [31], who in his own experiments found it to occur with 1 μm plastic spheres. As rapid filters generally show decreasing efficiency with increasing flow rate, it is to be concluded that diffusional impaction, the rate of which will be proportional to the residence time of the particles within the bed, is the dominant mechanism and that inertial impaction is relatively unimportant. The proportionality between the rate of diffusional impaction and residence time is expressed by the Peclet number P:

$$P = \frac{d_p \bar{u}}{D} \tag{61}$$

Once the particles have impacted with the medium, they must be retained there by an attractive force. Ives and Gregory [32] have studied the capture of particles within rapid filters in relation to the forces considered in the DLVO theory of colloid stability (see Sec. III. A). It appears from this and similar studies that van der Waals' forces are sufficient to account for the observed retention. It is also thought that polyelectrolyte bridging between particles and the filter medium may occur. From experiments on detergency and related phenomena it is known, that because of the interaction of particle forces and hydrodynamic effects, it may be very difficult to remove particles from a surface by applying fluid shear. This is overcome in the cleaning of rapid filters by fluidizing the bed, which causes attrition between the medium particles and removes any matter adhering to their surfaces.

III. MODIFICATION OF CAKE PROPERTIES

Often in the cake filtration of fine materials, the first layer of cake deposited will blind the medium to such a degree that fluid flow virtually ceases. Even where such an extreme state of affairs does not exist the economics of the process may be greatly enhanced by increasing the permeability of the cake and, hence, of the filtration rate. The permeability of a filter cake may be changed by altering the size distribution and the packing of the particles making up the cake. The size distribution may be altered by adding materials having a different particle size distribution to the slurry prior to filtration. Such materials are known as filter aids; their selection, properties, and use are described in Sec. IV. An alternative way of changing the permeability of a cake is by altering the state of aggregation of the particles, causing them to form into relatively large open flocs, each floc consisting of many particles. These flocs—which usually have a very irregular shape—will pack as large particles, giving rise to an open filter cake of greatly increased permeability.

There are two practical ways of bringing about changes in the state of aggregation of a suspension of particles, and La Mer and Healey [33] have proposed that they be distinguished by the terms coagulation and flocculation. The proposal is that the term coagulation be reserved for those phenomena brought about by reducing the zeta potential of a particle suspended in an electrolyte by changing the nature and concentrations of the ions present. Flocculation is reserved for those processes where certain types of long chain polymer or polyelectrolytes cause the particles to aggregate by forming "bridges" between them. Although in any actual pretreatment process both of these effects may be induced by the same reagent, it is usually clear which of them is dominant. The terms coagulation and flocculation have been used indiscriminately in the literature of colloid and surface chemistry, probably because flocculation in La Mer's sense has received much less attention than coagulation which has been the subject of much study for over a century. The terms coagulation and flocculation are also used in a different sense in the water treatment industry, coagulation being used there to describe the addition of reagents to the water to induce the reaction, and flocculation to describe the subsequent slow agitation during which floc growth occurs.

Chemical pretreatment, and to a lesser extent the use of filter aids, will lead to a more compressible filter cake which will in turn produce less-desirable filtration characteristics, and to a more bulky cake which may, for example, require the use of a larger plate and frame press. In any particular process, the balance of increased permeability and increased compressibility must be optimized. This consideration will be much more significant for high-pressure process, e.g., for filter presses rather than rotary vacuum filters where large bulk and high compressibility are less important.

Flocculation and coagulation, increasingly combined in the same overall pretreatment, are used in deep bed filtration processes such as the production of potable water. The particular problem in such processes is the removal of small quantities of very fine particles, mainly of an organic colloidal nature. It is found that pretreatment gives a considerably enhanced filter life as well as increasing the efficiency of the filtration process.

A. Coagulation

1. *Colloid Stability*

The phenomenon described as coagulation by La Mer is the adhesion of particles by forces of molecular and atomic origin. Although the theories of colloid stability are usually considered in relation to particles smaller in size than those encountered in most solid-liquid filtration processes, the basic mechanism of coagulation is the same. According to the Derjaguin-Landau-Vervey-Overbeek (DLVO) [34,35] theory of colloid stability, the presence or absence of coagulation depends upon the balance between the attractive or dispersive van der Waals' forces and the repulsive electrical double-layer forces.

Van der Waals postulated the existence of attractive forces between molecules and atoms to account for some of the deviations from ideal behavior exhibited by real gases. Although they may exist between electrically neutral atoms and molecules, these forces are electrostatic in nature, originating in dipole-dipole, dipole-induced dipole, and interactions of fluctuating fields in nonpolar molecules. The latter forces, which account for most of the van der Waals' attraction, were first explained in 1930 by London [36] who showed that they were of very short range, falling off inversely as the sixth power of the distance between two atoms. However, for macroscopic bodies the fall-off is much less; e.g., for two spheres it is proportional to the inverse distance squared. Various workers have postulated theories of these forces that agree in their general prediction and also with direct measurements of force magnitudes. The van der Waals' forces are a property of all matter, their magnitude depending on the nature of the material and of the intervening medium, the shape of the bodies, and the distance between their surfaces. In general, it is not possible in solid-liquid systems to control them to any real extent, so the tendency to coagulate is controlled by altering the magnitude of the repulsive electrical double-layer interactions. These will be described in Sec. III.A.2.

The overall effect of the attractive and repulse forces is best expressed in the form of potential energy, U, diagrams. Figure 6a shows such a diagram for a system not exhibiting coagulation. Curve I is the van der Waals' energy of the system, showing that it is an attractive force by its increasingly negative value at decreasing interparticle distances r. At very small values of r, the curve rises steeply as the electron fields of the particles overlap and the particles begin to be forced into one another. Curve II shows the repulsive electrical double-layer force and curve III the resultant. The presence of the maximum in the resultant shows that

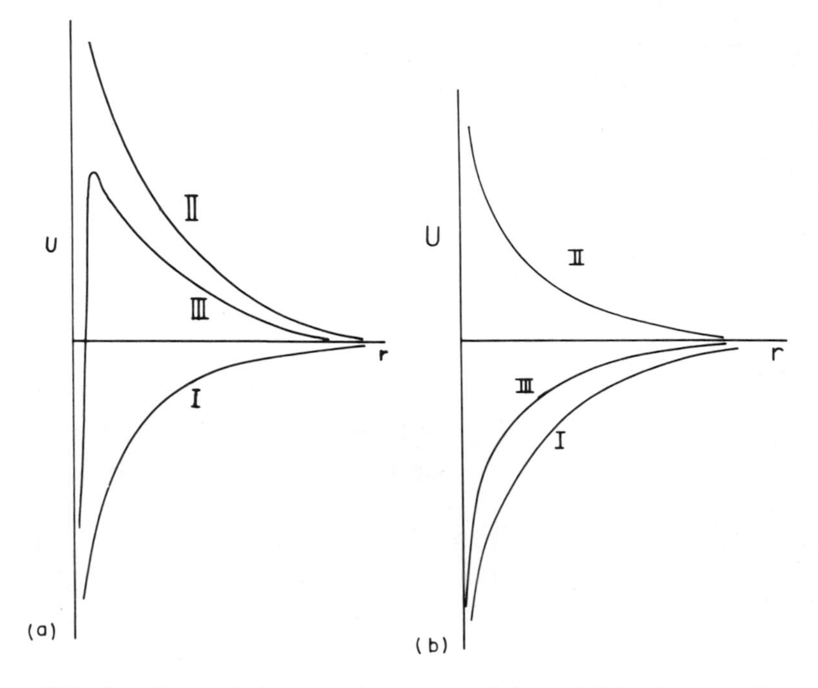

FIG. 6. Potential energy curves: (a) stabilized particles; (b) coagulated particles.

an energy barrier is present sufficient to prevent the particles approaching closely enough to pass over into the stable energy minimum. Hence, the system will not coagulate. Figure 6b shows the situation when the repulsive double-layer energy is less, and the resultant curve III shows no maximum. In this case, the particles will coagulate if they should happen to come into sufficiently close contact. Another type of interaction found with larger particles of the size of interest in filtration processes is that where a secondary minimum in the potential energy curve occurs. An example of this is shown in Fig. 7.

Although the curve contains a maximum at r_o capable of inhibiting coagulation, there is a secondary minimum at a greater distance of separation r_1. If this minimum is deep enough it will give rise to a weak form of coagulation known as secondary minimum coagulation. Although this form of coagulation is important in the rheological

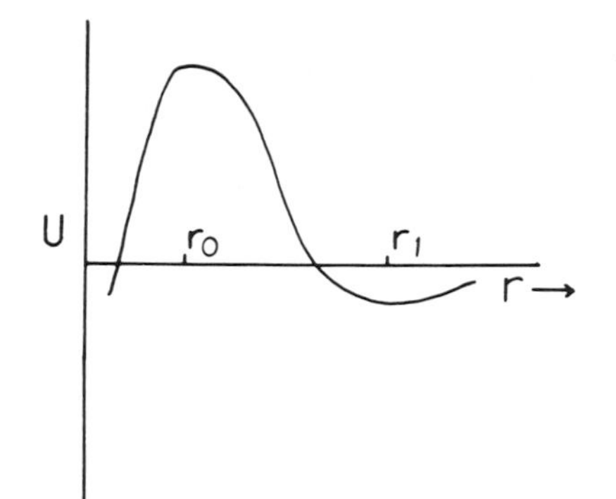

FIG. 7. Potential energy curve showing a secondary minimum.

properties of suspensions and the stability of emulsions it may not be stable enough to be of great significance in pretreatment processes. It is associated with particles greater than 1,000 Å in diameter and may occur in the presence of nonionic surfactants.

The basic rate equation for coagulation was published in 1916 by von Smoluchowski [37]. This treatment assumes that when particles collide they will coagulate if their potential energy of interaction is suitable. For particles too small to experience a significant gravitational force, the rate at which they collide will be governed by their rate of Brownian diffusion. This theory of perikinetic coagulation yields the result that the rate of disappearance of primary particles is given by

$$\frac{-d\nu}{dt} = 8\pi DR\nu^2 \tag{62}$$

where D is the Brownian diffusion coefficient of the particles, ν the particle concentration, and R the collision diameter of the particles. For monosized, spherical particles where R = 2a, a being the particle radius, the diffusion coefficient D is given by

$$D = \frac{k\bar{T}}{6\pi\mu a} \tag{63}$$

Hence, in terms of the molar concentration of particles c, it may be written that

$$\frac{-dc}{dt} = \frac{8N_o kT}{3\mu} c^2 \tag{64}$$

where N_o is Avogadros number.

In order to treat the overall collision reaction, the rate of collision of these primary pairs with other particles must be described, and so equations have been produced in series form which predict that the overall reaction is second order with respect to particle concentration. Hence, the coagulation of very dilute suspensions of particles such as those found in water treatment may be very slow. Figure 8 shows the numbers of primary particles and aggregates of particles up to four as a function of time in rapid coagulation. The rate is plotted on a reduced time basis where T, the time of coagulation, is defined as

$$T = \frac{1}{4\pi D_1 R \nu_o} \tag{65}$$

with $2D_1R = D_{ij}R_{ij}$, $R_{ij} = r_i + r_j$, and $D_{ij} = D_i + D_j$.

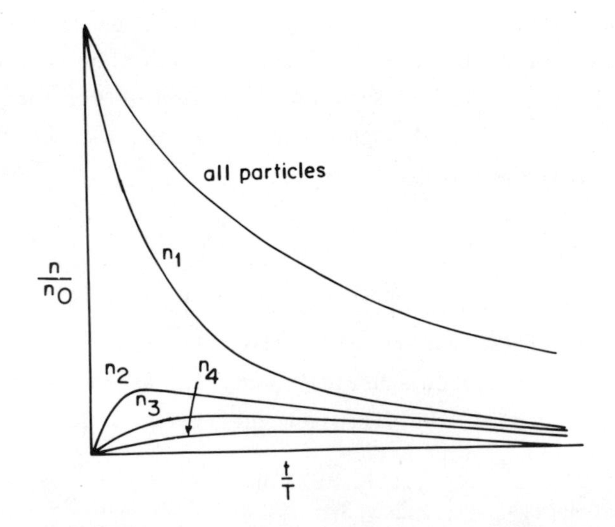

FIG. 8. The number of particles as a function of time in rapid coagulation. Reproduced, with permission, from *Colloid Science*, vol. 1, ed. Kruyt, Elsevier, 1952.

The von Smoluchowski treatment assumes that the potential energy curve is an infinitely deep well with no maximum. If a maximum insufficient to inhibit coagulation completely is present, only those particles colliding with enough energy to pass over this barrier will coagulate. This situation has been treated by Fuchs [38] in his theory of slow coagulation. Müller [39] has shown how, using the basic von Smoluchowski mechanism, the fact can be accounted for that in a mixed system of small and large particles the smaller particles appear to be removed more rapidly. This is because collisions between large and small particles do not effectively change the concentration of large particles. The impression is that the small particles are being caught by the large ones, however. Figure 9 shows how the rate of removal of small particles is affected by the large particles being 10, 20, and 100 times larger in diameter than the smaller ones. The upper line is the coagulation rate of the smaller particles alone. This effect is very important in the removal of low concentrations of very fine material, as, e.g., is frequently required in water treatment processes.

The above coagulation processes have assumed that the particles are undergoing Brownian diffusion only. In the presence of

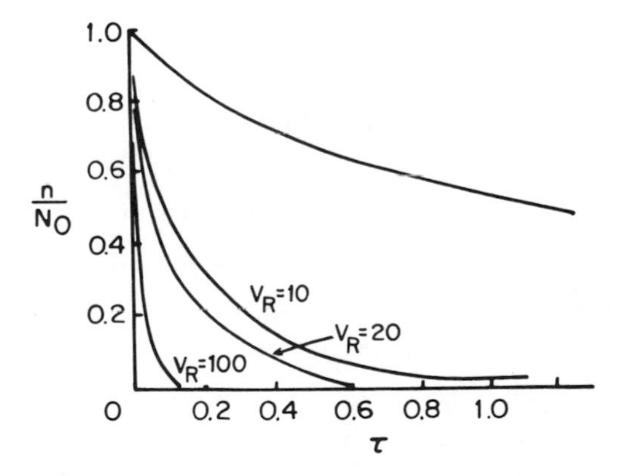

FIG. 9. The rapid coagulation of small particles in the presence of larger ones. n = number of small particles; $n_o = N_o$ = number of small with respect to large particles at t = 0; V_R = ratio of diameters of large and small particles; $\tau = t/T_{smol}$.

sedimentation or shear gradients significant changes in the coagulation rate are observed. This type of coagulation is referred to as orthokinetic. In a shear gradient, the ratio of the coagulation rate to that of the same-size particles in slow coagulation is given by

$$\frac{J}{I} = \mu \frac{(R_{ij})^3 \, du/dz}{2\underline{k}T} \tag{66}$$

where J is the collision rate in a shear gradient du/dz and I the Brownian collision rate. The form of the equation shows that the rate is very dependent upon R_{ij}, the collision diameter of species i with species j. The effect of sedimentation will only be observed in dilute suspensions when the system is polydisperse. This is because in dilute monodisperse systems the particles will all sediment at the same velocity, whereas in polydisperse systems the larger particles will catch up with the smaller ones and increase the overall collision rate. There is an optimum particle-size range for this effect.

In more concentrated systems the behavior will be complicated by phenomena such as hindered settling and fine particles being carried upwards by displacement flow. As the particle size in a system grows due to perikinetic coagulation, particles will begin to sediment, and the coagulation rate will increase due to the resulting change to orthokinetic coagulation. In this sense the coagulation reaction is autocatalytic.

When the particle concentration is low, as in potable water treatment, Eq. (57) might predict that the coagulation rate would be unacceptably slow. In these situations the system is subjected to shear to induce orthokinetic coagulation. Many types of coagulator have been designed, the energy to provide the velocity gradient being induced mechanically or hydraulically. A system used commonly is a series of stirred tanks, the rate of stirring decreasing from one tank to the next because as the coagula increase in size the required velocity gradient decreases (Eq. 61) and also because large coagula would be liable to rupture by excessive shear. Other systems combine the coagulator and gravity clarifier in one vessel, the mass being mechanically stirred in the coagulating section.

The design of coagulators is based on the premise of Camp and Stein [40] who introduced the concept of mean shear rate $\bar{G}$ into flocculator design and derived the relationship

$$\bar{G} = \left(\frac{P}{V\mu}\right)^{1/2} \quad (67)$$

where P is the power expended in agitating the fluid, V the volume of the coagulator, and μ the fluid viscosity. By deriving the form of P for different types of coagulator, the dimensionless group $\bar{G}\bar{t}$ can be used to scale up coagulator designs and compare different types, $\bar{t}$ being the mean residence time of the fluid in the coagulator. Although this approach has limitations, for example, it does not take account of residence time and shear rate distributions which may effect overall coagulator efficiency, it is a convenient concept for coagulator design and has been widely used.

2. *Electric Double Layer and Zeta Potential*

When a solid is in contact with an electrolytic (e.g., an aqueous) liquid, there is generally an electrical potential difference at the interface. This may be due to adsorption of ions by the solid, dissolution of ions from the solid, or ionization of molecules at the surface of the solid. As a result of this potential difference, ions of opposite charge (counter-ions) in the liquid phase will be preferentially attracted toward the interface and ions of like charge (co-ions) repelled from it. The solid will become surrounded by an electrical double layer as a result of this charge separation in the liquid phase due to the presence of the interface. Gouy [41] and Chapman [42] proposed a double-layer theory for a flat infinite surface and point charges in a homogeneous medium. The potential ψ at any distance from the surface depends upon competition between thermal diffusion of the ions moving them randomly and the attractive effect exerted by the surface charge. This theory leads to ionic concentration distributions and potential gradients as shown in Fig. 10.

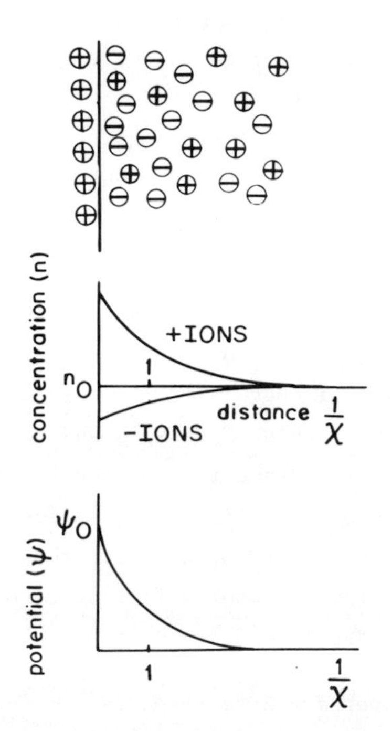

FIG. 10. Ion and potential distributions in the diffuse double layer.

The Debeye-Hückel constant χ in Fig. 10 has the dimensions of length^{-1} and has the value of

$$\chi = \left(\frac{2e^2N_oz^2c}{\varepsilon \underline{k}T}\right)^{1/2} \tag{68}$$

where c is the electrolyte concentration, z the valency of the ions, e the electronic charge, and ε the dielectric constant of the medium. Because of the assumptions made, particularly that the ions behave as point charges, the Gouy and Chapman model is at variance with experiment. Stern [43] proposed a semi-empirical theory in which the ionic double layer was considered as having two parts, an inner layer of ions sufficiently strongly attached to the surface to overcome effects due to thermal agitation and an outer diffuse layer to which the Gouy-Chapman treatment was applied. The effective potential exhibited by a particle was that at the interface between the two layers. This potential is known as the zeta (ζ) potential and is what is measured in electrokinetic experiments. The Stern model is shown in Fig. 11. Double-layer theories are very complex, and the Stern model with its assumption of a well-defined plane of shear is only approximate but does enable coagulation phenomena to be understood readily.

When an electrical field is applied to an electrolyte containing particles, they will be attracted to the electrode bearing a charge opposite in sign to that of their zeta potential. This phenomenon, known as electrophoresis, is one of the electrokinetic phenomena employed in measuring zeta potential. Electrokinetic phenomena may be classified as follows:

1. Electrophoresis—the movement of a charged surface relative to a stationary liquid under the influence of an applied electric field.

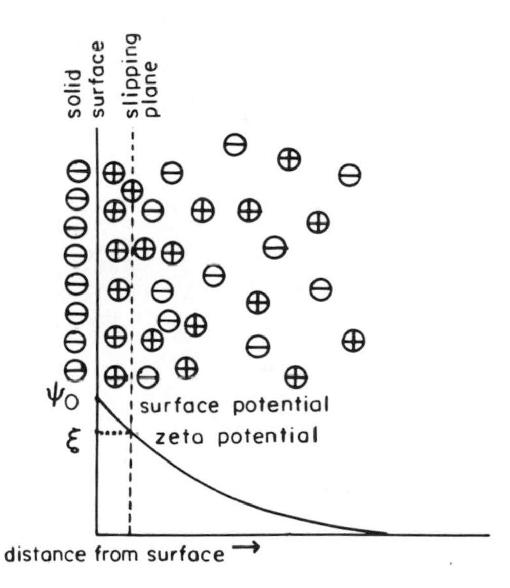

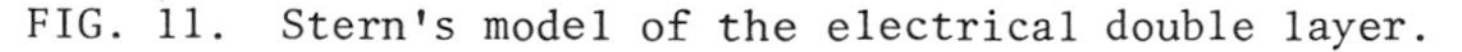

FIG. 11. Stern's model of the electrical double layer.

2. Electroosmosis—the movement of liquid through a capillary or porous plug having charged surfaces under the influence of an applied field. If the experiment is performed under static conditions, the pressure produced is known as the electroosmotic pressure.
3. Streaming potential—the potential difference generated across a capillary or porous plug when liquid flows through it (the converse of osmosis).
4. Sedimentation potential—the electrical potential generated in a liquid when particles sediment through it (the converse of electrophoresis).

For measurements on particle systems, electrophoresis is the most widely used and convenient phenomenon to observe, although there are situations in which it may be more convenient to measure electroosmosis or streaming potential.

The best-known technique for measuring particle electrophoresis is that of microelectrophoresis. In this the rate of movement of individual particles through the electrolyte is observed by either direct or reflected light microscopy. The suspension is placed in either a flat or cylindrical cell and a potential difference applied across it. Because there is also a velocity gradient within the liquid, it is necessary to measure the particle velocity at 0.146 of the internal diameter from the wall of a cylindrical cell or about 0.2 of the thickness from the wall of a flat cell, if corrections are not to be made. To obtain a satisfactory measurement the average of several measurements in each direction must be obtained. Most workers make their own apparatus but versions are commercially available, e.g., Zeta-Meter (U.S.A.) and Rank (U.K.). Figure 12 shows an apparatus for microelectrophoresis.

Microelectrophoresis is a very time-consuming and fatiguing technique, and these objections are to some extent overcome with the mass transport technique of Sennett and Olivier [44]. This instrument is shown in Fig. 13. The latter consists of a cell containing an electrode and having an orifice at one end. This cell is immersed in a bath or inserted into another cell containing a second electrode. Both compartments are filled with the suspension to be studied, the measuring cell is weighed, the apparatus assembled, a known current

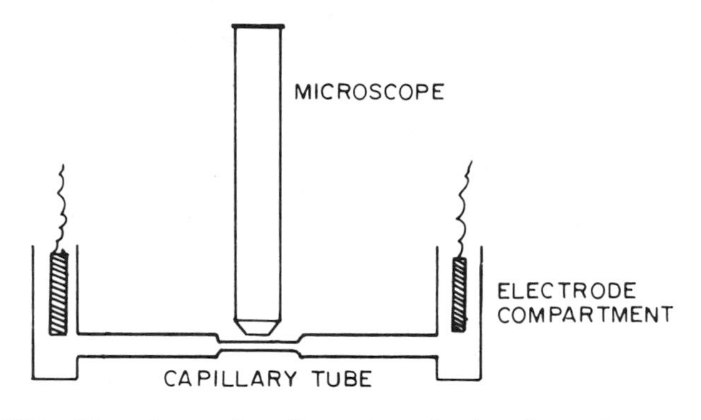

FIG. 12. Apparatus for microelectrophoresis.

passed for a given time, and the cell reweighed. The change in weight will be equal to the weight of particles that have passed through the orifice due to electrophoresis less the weight of fluid displaced. This technique is much less tedious than microelectrophoresis, but considerable care has to be taken. Differences in weight are small; and for them to be sufficient, relatively concentrated particle suspensions have to be used. The theory of the electrokinetic behavior of concentrated suspensions in which the double layers are close enough to interact is not fully understood, and more comparison data are needed between the microelectrophoresis technique, which deals with very dilute suspensions, and

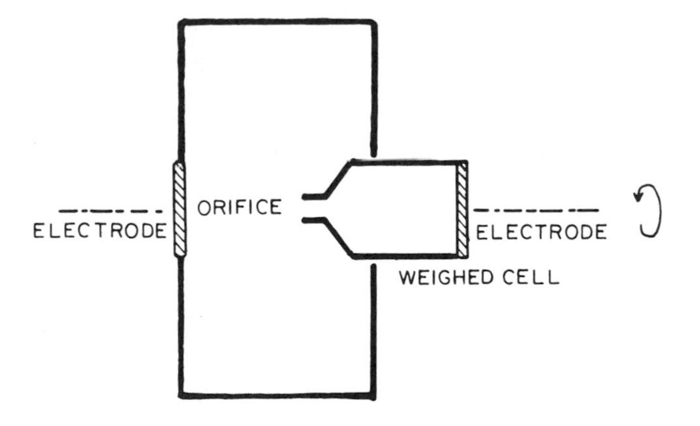

FIG. 13. Apparatus for electrophoretic mass transport analysis.

electro-osmosis, which uses packed beds of particles. A practical problem with the mass transport apparatus is changes in weight due to absorption of water by the material of the cell. This may be overcome by either keeping the cell in water or, as in the author's laboratory, by coating it with a polyurethane varnish. In a commercially produced form of the apparatus (Micromeritics Corp.) the cell is rotated to retain particle suspension, thus enabling measurements to be made on particles that would be large enough to sediment in a microelectrophoresis apparatus.

From the electrophoretic mobility it is possible to calculate the zeta potential. For curved surfaces it is necessary to consider the shape of the surface, and this is achieved through the parameter $\chi\underline{a}$, where χ is the Debeye-Huckel reciprocal thickness and $\underline{a}$ the diameter of a spherical particle. For small $\chi\underline{a}$ values, each particle will behave as a point charge, and for large $\chi\underline{a}$ as a plane surface.

For small values of $\chi\underline{a}$, a particle may be considered as a charged Stokes law sphere, and the equation

$$U_E = \frac{\xi E}{1.5\mu} \tag{69}$$

may be obtained, where U_E is the electrophoretic mobility. This equation will not apply to particle sizes and values of $\chi\underline{a}$ found in filtration problems other than those involving nonaqueous systems where the double layer is usually much thicker. For large $\chi\underline{a}$ values, the relationship between electrophoretic mobility and zeta potential

$$U_E = \frac{\xi E}{\mu} \tag{70}$$

is found; it is the equation generally applicable to particle systems. These relationships are complicated by effects due to surface conductance, double-layer relaxation, and the influence of the electric field strength adjacent to the surface on the dielectric constant and viscosity of the liquid. The relaxation effect may be significant for $0.1 < \chi\underline{a} < 300$, these values often being found in systems of particles where $\underline{a}$ is of the order of micrometers but errors may be corrected for by using the method of Wiersema et al. [45].

3. *Ions and the Use of Coagulants*

The effect of electrolyte concentration and charge on the stability of hydrophobic colloids was recognized by Schülze [46] in 1882 and Hardy [47] in 1900, while six years later Burton [48] related them to the electrophoretic properties of colloidal particles. The generalization of these observations, known as the Schülze-Hardy rule, states that the coagulation value, i.e., the minimum ionic concentration necessary to produce coagulation, is approximately proportional to the sixth power of the reciprocal of the counter-ion charge. The sixth-power law is an upper limit, values as low as the square of the reciprocal charge being found in some cases. The effect of charge on coagulation is matched by its effect on electrophoretic mobility and, hence, on zeta potential.

Figure 14 shows the effects of the counter-ions K^+ (as KCL), Ca^{2+} (as $Ca(NO_3)_2$), Al^{3+} (as $Al(NO_3)_3$), and Th^{4+} (as $Th(NO_3)_4$) on the zeta potential of glass as a function of ionic concentration. It is evident that the decrease in zeta potential is much more marked in the case of multivalent ions and that some ions, e.g., Al^{3+} and Th^{4+}, are even capable of reversing the zeta potential. The ability of multivalent ions to depress the zeta potential may be explained by double-layer compression where the ions, due to their greater charge concentration, enable most of the charge of the double layer to be brought within the plane of shear between

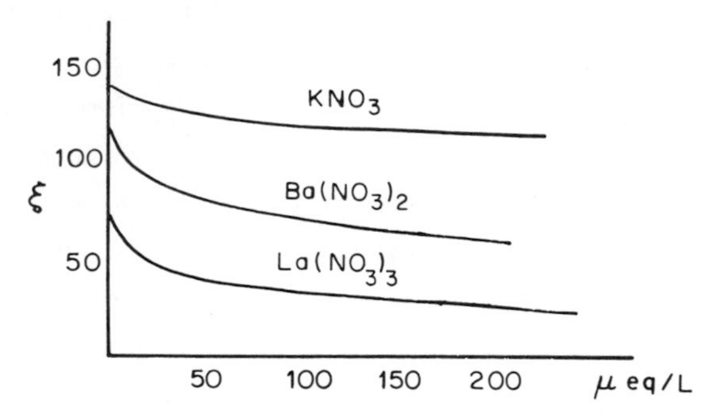

FIG. 14. Effect of ions on the ξ potential of glass.

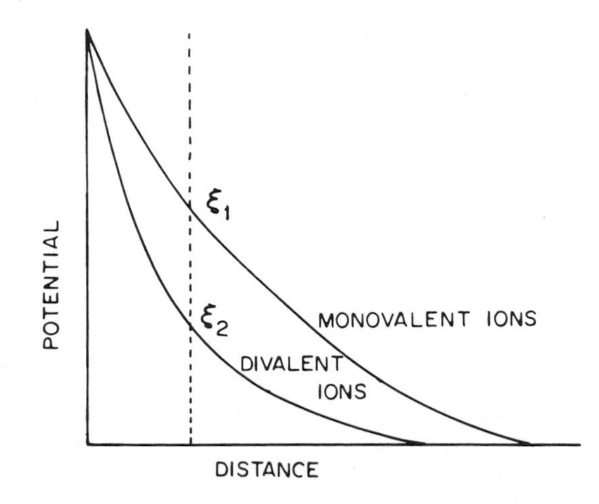

FIG. 15. Double-layer compression by multivalent ions.

the Stern and diffuse layers although the surface potentials may be the same. This is shown in Fig. 15. Charge reversal may be explained by nonspecific adsorption of ions so that the amount of charge adsorbed is greater than that present on the solid surface. This may happen as with $A\ell^{3+}$ and Th^{4+} where hydrolysis products of the ions could adsorb by hydrogen bonding, and with polyelectrolytes where nonspecific van der Waals' adsorption may occur.

From the above it may be seen that the most effective ions for coagulation are multivalent cations. Of these, the ones used in practice are Al^{3+}, Fe^{2+}, Fe^{3+}, and Ca^{2+}. Aluminum is usually obtained as aluminum sulfate, often incorrectly referred to as alum (the alums are double sulfates of aluminum with other metals) and supplied as either a solid or solution, or sodium aluminate, a freely soluble powder. Iron salts are cheaper and may be more readily available than aluminum salts, being the byproducts of many processes, e.g., pickle liquors and TiO_2 manufacture. Fe^{2+} is obtainable as ferrous sulfate (copperas) although it is frequently used as chlorinated copperas, prepared by saturating a 10 to 15% solution of ferrous sulfate with one part of chlorine to every seven of ferrous sulfate. Ferric ions may also be obtained from ferric chloride, which may be hazardous to handle, or ferric sulfate. Most iron salts are corrosive, some being very much so, and money saved on the cost of materials may well be offset by increases in initial plant and maintenance costs. Calcium in the form of limestone or slaked lime is mainly used in conjunction with other coagulants in water treatment processes.

Coagulating salts are almost universally used in water treatment, where the problem is usually the removal of very small concentrations of suspended solid and of organic materials such as the humic acids which impart odor, taste, and color to the water. Because of the low particle concentrations, present coagulation rates may be unacceptably slow due to their dependence on the square of the number of particles present (Eqs. 62 and 66). Used under conditions of correct pH the common coagulating salts produce hydrolysis products which contribute to the particle population and hence enhance the coagulation rate. There are optimum conditions for the hydrolysis of these salts; for example, Al^{3+} forms soluble aluminates at high pH and will not hydrolyze at all below a critical pH and ion concentration. These effects have been discussed in detail by O'Melia [19]. Hence the use of any particular salt depends on the economic optimation of pH modification and salt demand. The unstable salt hydrolysis intermediates are able to adsorb the undesirable organic materials, anions such as phosphate, and also heavy metal ions. As well as potable water treatment these properties are made use of in physicochemical waste treatment processes, in which biological oxidation is replaced by a combination of solids removal by conventional separation procedures following salt coagulation (frequently with the use of polymeric flocculants to enhance separation) and adsorption to remove soluble components not adsorbed on the coagula. These processes have been developed to reduce discharged phosphate levels that may lead to eutrification problems at the discharge point.

An undesirable side effect of the use of hydrolyzing coagulation salts is that the bulk of the dried sludge produced during a process may consist of salt hydrolysis products, and the economics of sludge disposal have become a serious factor in waterworks operation in large urban areas.

Measurement of zeta potential in itself is no substitute for direct studies of coagulation behavior (see Sec. III. D.), but it may explain anomalous behavior due to phenomena such as charge reversal; it is also of value in selecting the best coagulant for a particular operation and, in particular, for optimizing the dose of that coagulant. There is often an optimum coagulant dose. Exceeding that dose leads to an impaired coagulation reaction in addition to the extra cost of reagent.

The presence of hydrophilic colloids adsorbed onto the surface of the particles may prevent coagulation in situations where it might otherwise be expected. Theories of stabilization by adsorbed layers have recently been reviewed by Vincent [50] but are not yet at a predictive stage. They are a particular problem in certain situations such as "white water" treatment in paper mills where considerable quantities of starch may be present, and also in the dewatering of activated and digested sludges in sewage treatment processes. In these latter situations the bulk of the hydrated adsorbed layers may be such as to make solids concentration difficult beyond quite a modest level.

B. Flocculation

1. *Nature of the Flocculation Reaction*

Natural polyelectrolytes have long been used as flocculants, although it is only recently that the term polyelectrolyte has come into common use. Packham [51] pointed out that the crushed nuts of an Indian tree have been used for centuries for clearing potable water and that isinglass, a protein extract of fish swim bladders, particularly of the sturgeon, has been used since ancient times for clarifying beer by the process known in England as "fining." The action of these materials depends on the presence of water-soluble polymers, which are usually proteins or polysaccharides. Among other polymers used to induce flocculation are soluble starches, gelatine, guar gum, tannins, and sodium alginate. More recently these naturally occurring materials have been increasingly displaced by synthetic polyelectrolytes, usually derivatives of polyacrylic acid. In spite of their higher unit cost, the synthetic materials have the advantages of much lower dosage rates and the production of firmer flocs which are more able to withstand the shear produced by the agitation inherent in many processes.

Effects related to the nature of the flocculating agent, the particles being flocculated, and operating conditions on the flocculation reaction are best thought of in terms of the overall flocculation reaction mechanism. Ruehrwein and Ward [52] proposed that the principal effect of polyelectrolyte flocculants was to form "bridges" from one particle to the next, as shown in Fig. 16. This mechanism requires that the polymer chain be adsorbed from solution onto one particle and that, when another particle comes within close enough range for the extended polymer chain to be adsorbed onto it, a physical bridge will be formed between the two particles. This elementary floc then grows by bridging with other particles until an optimum floc size is reached. As discussed later, the flocculation reaction unlike the coagulation reaction, is irreversible, hence it is

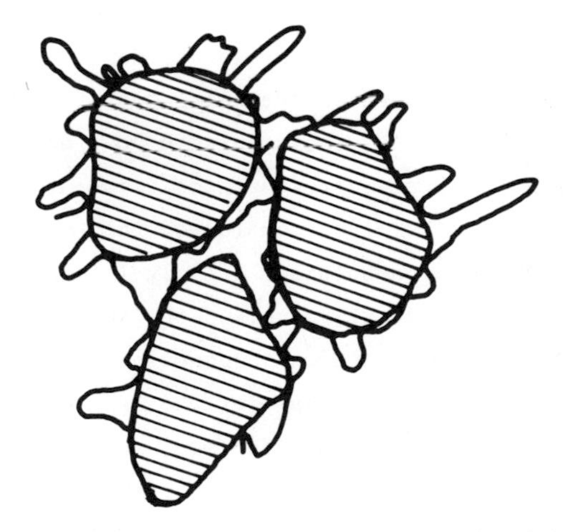

FIG. 16. Particles flocculated by polymer bridging.

misleading to think of the flocs as reaching an equilibrium size. The results of much subsequent work have confirmed this bridging hypothesis.

In general, the coagulation reaction brought about by reduction of zeta potential is not important in polyelectrolyte flocculation. This is shown both by the efficiency of neutral polyelectrolytes as flocculants and also by the high efficiencies exhibited in some systems where the flocculant carries the incorrect electrical charge with respect to that of the material being flocculated. The main effect of highly charged polyelectrolytes on electrical double-layer properties is to increase, not decrease, the zeta potential. For example, low molecular weight, i.e., 10,000 to 50,000, polyacrylates are used as dispersants, not flocculants. Because of their considerable amount of noncharged organic skeleton they are able to adsorb nonspecifically, and being charged, give rise to an increase in zeta potential or even to charge reversal. On ground chalk whiting in water, for example, sodium polyacrylates will give zeta potentials of about -60 mV, and calculation of the potential energy curve for the system shows that this is sufficient to account for the stabilization observed [53]. Charge effects may be more important with some lower molecular weight flocculating polyelectrolytes such as the polyethylene imines. In these materials the polymer chain length is insufficient for bridging to occur to any considerable extent, and alternative mechanisms such as zeta potential reduction must be invoked. Investigation has shown this to be true.

Flocculation is best represented in terms of a sequence of elementary reactions. It will be considered that a small quantity of a relatively concentrated solution of polyelectrolyte flocculant is to be added to a much larger volume of the suspension of the solid to be flocculated. The reaction sequence proposed is dispersion of the flocculant in the liquid phase of the suspension, diffusion of flocculant to the solid-liquid interface, adsorption of flocculant onto the solid surfaces, collision of a particle bearing adsorbed polymer with another particle, adsorption of free polymer chain onto a second particle forming a bridge, and subsequent collision and adsorption reactions leading to build-up of floc.

a. Dispersion in the Suspension

Polyelectrolytes in solution exhibit high viscosities and low diffusion rates because of their high molecular weights. This first step in the reaction sequence is analogous to a diffusion rate controlling reaction in solid-fluid catalysis. As the adsorption reaction is usually very much faster than the diffusion reaction, it is necessary to disperse the phases one into the other mechanically. In practice, this is usually accomplished by having a short, vigorous premixing stage during the process to enable dispersion to be carried out in the shortest practical time. Techniques used to achieve this rapid mixing will be discussed in Sec. III. C.

b. Diffusion to the Solid-Liquid Interface

The factors influencing this stage of the reaction are identical to those in section a above.

c. Adsorption at the Solid Surface

The bridging hypothesis requires that the polyelectrolyte chain be adsorbed on the solid surface at only a few points of attachment, with the bulk of the chain projecting into the liquid phase away from the solid-liquid interface. This model of adsorption of a flexible chain carrying many possible points of attachment has been treated in detail by Silberberg [54] and others. Among other things, Silberberg shows that this model of adsorption gives rise to an adsorption isotherm of the Langmuir type, although not for the same reasons as in the Langmuir case; that the amount of adsorbed material increases with the chain length, i.e., with the molecular weight of the adsorbed molecules; and that the number of points of attachment of the chain to the solid surface increases until an equilibrium compressed layer is achieved. This is shown diagramatically in Fig. 17. It also follows from this theory that the nature of the

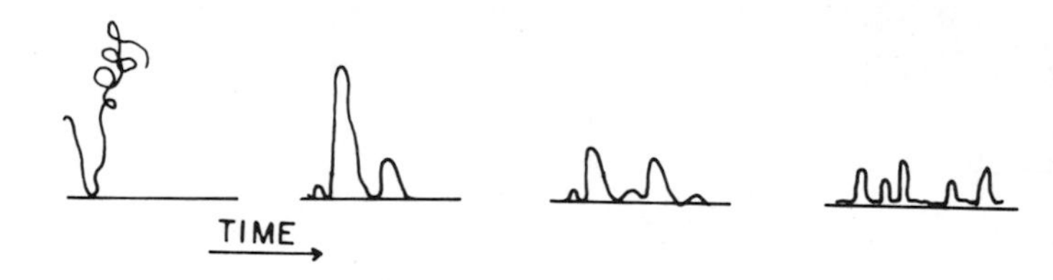

FIG. 17. Compression of adsorbed polyelectrolyte layer.

adsorption process is dependent on the configuration of the polyelectrolyte in solution which in its turn depends upon the structure of the polyelectrolyte, on the solvent power of the solvent for the polyelectrolyte, on the solution pH, and on the concentrations of other ions present.

Because of the difficulties of obtaining materials of narrow molecular weight range and the lack of a really good analytical technique, little precise data is available concerning the adsorption of polyacrylamide derivatives. However, what is known is generally consistent with the Silberberg model. The initial adsorption is rapid and irreversible as predicted. Although the energy of adsorption of each link may be quite low and the adsorption at that particular point reversible, the probability of a large number of links leaving the surface at any one time becomes so low as to be discounted. It has been claimed that the high molecular weight components are selectively adsorbed from a solution of wide molecular weight range. Silberberg claims that this is more a question of the time taken to reach equilibrium and that when equilibrium is reached there is no molecular weight selectivity.

Bonding forces between the polymer and the solid surface depend on the nature of both materials and the conditions within the solution, but the most commonly invoked adsorption mechanism is hydrogen bonding with evidence of ionic charge effects and nonspecific van der Waals' adsorption in some systems. Although the energies involved in the hydrogen and nonspecific bonding mechanisms are very low for each active site, and this is confirmed by the very low temperature dependence of the adsorption isotherms, the overall energy over many sites is sufficient to account for the phenomena observed.

d. Collision of One Polymer-Coated Particle with Another

It would be expected that the rate of flocculation of particles having loops of polymer at their surface would follow the Smoluchowski type of rate equations for coagulation reactions. These equations were all derived for coagulating systems, but as they apply to irreversible processes they might be expected to be obeyed more strictly by flocculating systems, having made allowances for the increased collision diameter of the particles due to the presence of the polymer loops extending from their surfaces.

e. Subsequent Collisions Leading to Flocs

The points discussed in section d above would also apply to the final stages of the reaction. Sutherland [55] has described a computer simulation of floc structure on the basis of the Smuluchowski model. The irregularly shaped flocs of very high porosity obtained are consistent with those observed in actual flocculation reactions.

3. *Effects of Operating Variables*

A detailed study of the flocculation of polyacrylamides has been made by Linke and Booth [56]; and, while systems differ widely in their characteristics, their conclusions are generally valid and illustrate well the behavior of flocculating systems.

a. Polyelectrolyte Concentration

Practically all polymer is adsorbed at optimum dosage. The optimum dosage is proportional to the solid surface area over a wide range of particle sizes and solution concentrations. La Mer and Healey [33] claim that optimum flocculation occurs when half the area of solid is covered with polyelectrolyte. To simplify the argument this is the optimum coverage between the maximum number of extended loops available to form bridges and the maximum uncovered area of solid available to permit the loops from other particles to adsorb and form bridges. Any detailed consideration of this point would involve making assumptions about the nature of the adsorption reaction, but it is in any case academic as the dosage of polyelectrolyte required to bring this coverage about would be grossly uneconomic.

At high concentrations the degree of flocculation decreases, and this may reach such a point that the particles are completely protected by the presence of an adsorbed polymer layer (see Fig. 18).

The linear relationship between surface area and dosage of polyelectrolyte is consistent with the model of adsorption previously discussed.

Departures from linearity at very low solids concentrations might be expected on the grounds that collision is less likely to occur before the polymer layer adsorbed on the particle surfaces becomes compressed and that the length of the loops available for bridging are much shorter.

b. Agitation

The effects of agitation are complex. As previously mentioned, agitation is necessary for the dispersal of the flocculant and for

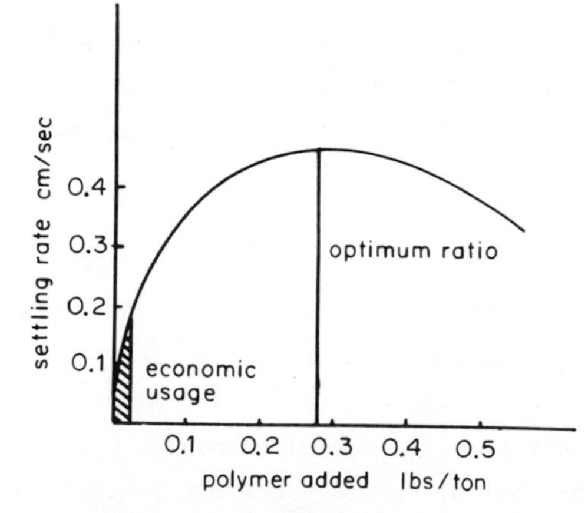

FIG. 18. Effect of polyelectrolyte concentration on flocculation reaction. Reproduced, with permission, from Ref. 56.

ensuring that flocculant is available at the solid surfaces. If the initial agitation is insufficient, the flocculant will be adsorbed in excess on some of the solids, leaving the remainder completely uncovered. The presence of shear gradients due to agitation will also give an increased collision rate and, hence, flocculation rate.

In the presence of vigorous agitation, the flocs will be ruptured and the size of flocs produced will be reduced. Linke and Booth [54] considered that during floc rupture the bridging polymer is stripped from the solid surface, thus leaving vacant sites to adsorb more polymer if it becomes available. It is reasonable to consider that there must be an equilibrium between the number of polymer molecules stripped from the surface and the number ruptured, the position of this equilibrium depending on the number of positions at which adsorption has occurred. Continued agitation will lead to a decrease in the amount of flocculated material and an increase in the amount of "haze" present. This is shown in Fig. 19.

c. pH and Ionic Strength

Solution pH will have two main effects on the flocculation reaction. One is on the nature and intensity of charge present on the solid surface; the other is on the nature and intensity of charge present on the polymer chain if this possesses sites capable of acquiring or losing charge.

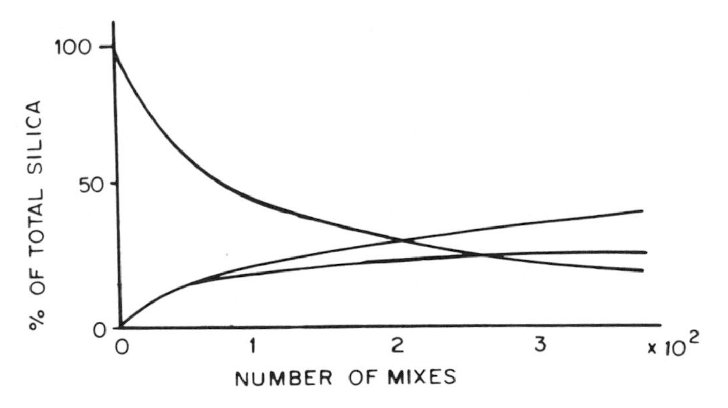

FIG. 19. Effects of agitation on the flocculation of silica by polyacrylamide. Reproduced, with permission, from Ref. 56.

The presence of a high surface charge will inhibit the adsorption of a polyelectrolyte bearing charge of the same sign and will enhance the adsorption of that bearing an opposite charge. If the charge is sufficiently high it may inhibit the collisions necessary to form flocs, although this effect would decrease with increasing polymer molecular weight as the need for very close approach would be lessened. For the system silica and polyacrylamide, it has been shown that the optimum polymer dose decreases as the pH moves away from the isoelectric point, i.e., the pH corresponding to zero net surface charge. Figure 20 is a plot of optimum flocculant dose over a range of pH values for the system silica-polyacrylamide.

Linke and Booth [56] explained this by postulating that the lower the polymer concentration, the slower the rate of protective surface covering. Hence a longer time is available for suitably energetic collisions leading to coagulation to occur.

If a polyelectrolyte possesses ionizable sites, e.g., an anionic or cationic polyacrylamide, the presence of charge at these points will cause them to be mutually repulsed, leading to an extended rigid polymer chain. In the absence of charge effects, a polymer molecule in solution will tend to exist as a randomly coiled chain. This might be expected to lessen the ease with which it will

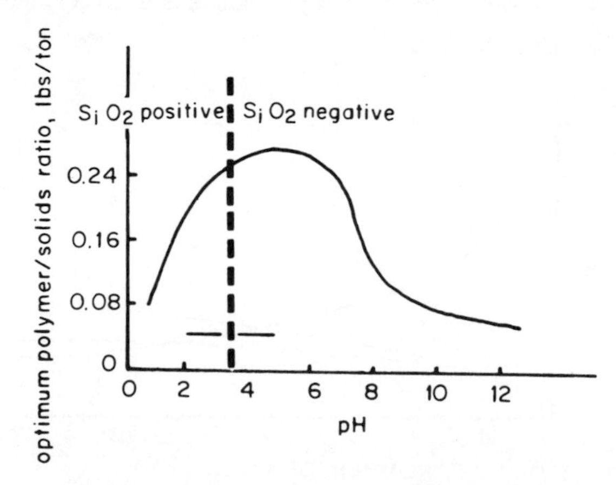

FIG. 20. Effect of pH on the flocculation of silica by polyacrylamide. Reproduced, with permission, from Ref. 56.

be adsorbed at a solid interface, and also to lessen the chances of bridging occurring once adsorption has taken place. However, according to the Silberberg adsorption model, the adsorption of a polymer coiled in solution still occurs by extended multipoint adsorption on the solid surface.

The charge intensities on both a solid surface and a polymer chain will also be affected by the concentrations of other ions present. Ions—particularly multivalent ions—will in general lead to a decrease in the potential at the solid-liquid interface (see Sec. III. A), which in turn will bring about a decrease in the mutual repulsion and, hence, an improvement in the efficiency of the flocculation reaction. The effect of salts on the overall reaction is also due to their contribution to the charges present and, hence, the configuration of the polymer chain and to the interaction between the polymer chain and its solvent.

d. Polyelectrolyte Molecular Weight

Experience shows that there is an optimum polymer molecular weight, i.e., chain length, for efficient flocculation. If the molecular weight is too low the polymer chains will not be long enough for bridging to occur; if too high the material becomes difficult to dissolve, and manufacture becomes more expensive. For polyacrylamide derivatives, the practical range covers average molecular weights from 100,000 to 10,000,000. Figure 21 shows how the
lar weights from 100,000 to 10,000,000. Figure 21 shows how the
efficiency of the flocculation reaction and the amount of material adsorbed increases with increasing polymer molecular weight. The latter is consistent with the Silberberg adsorption model. In general, increasing molecular weight leads to an increasing flocculation efficiency and floc strength. This is offset by the increased cost of polyelectrolyte due to the greater quantities required and possibly to an increased unit cost.

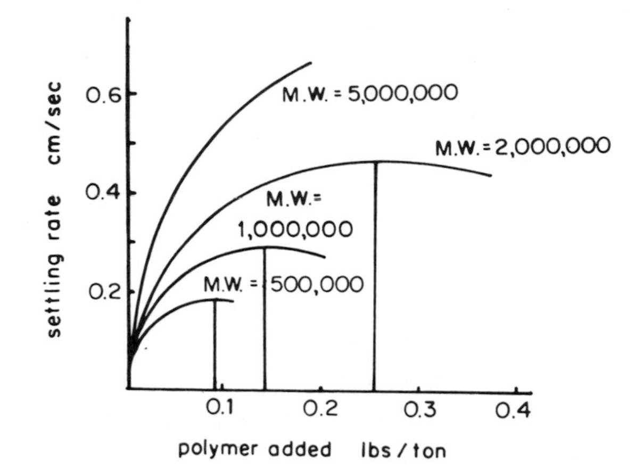

FIG. 21. Effect of molecular weight of polyelectrolyte on the flocculation reaction. Reproduced, with permission, from Ref. 56.

e. Solids Concentration

The strength of flocs is found to increase with solids concentration, although, as described in section a above, the optimum flocculant dosage is proportional to surface area except at low solids concentration when the particle collision rate is very low. In processes involving systems of this latter type, as in potable water treatment, the combined use of flocculants and coagulants brings about an increase in the amount of solids present. This will be discussed in Sec. III. C.

3. *Naturally Occurring Polyelectrolytes*

Guar gum, obtained from the seed of a small legume *cyamopsis tetragonolobus* is a widely used industrial flocculant. It is a polysaccharide consisting principally of a straight chain mannan with galactose branches and an average molecular weight of about 200,000. Being nonionic its action is not seriously affected by wide variations in pH. Under conditions of low pH or high temperatures the efficiency of guar gum deteriorates markedly, probably due to hydrolysis of the gum. It is also sensitive to enzymic degradation on storage, although this may be inhibited by the addition of preservatives such as citric or oxalic acid.

Glue and gelatine have been widely used as flocculants in the mineral processing industries, as they have the advantages of being inexpensive and easily available. Gelatine is unique as a flocculant in that it will flocculate almost any suspension. This could in part be due to the multiplicity of ionic groupings, their zwitterionic character, and the many hydrogen bonding sites occurring in protein materials. Because of these many active sites the configuration in solution of the gelatine molecule may be profoundly altered by changes in pH and electrolyte concentration. These may be correlated with the viscosity. Kragh and Langston [57] have made a detailed study of the flocculation of quartz by gelatine which typically has a molecular weight of the order of 300,000. The efficiency of a gelatine flocculant may also be increased by cross-linking with a reagent such as formaldehyde until the point is reached where the gelatine almost ceases to be soluble in water.

Starches, which have also been widely used in the mineral processing industries, have been studied by several groups of workers. The properties of naturally occurring polyelectrolyte flocculants differ widely with differing sources of material, this being particularly so in the case of starch. The main factors seem to be the ratio of amyloses to amylopectin in the starch and the degree of cross-linking present.

Sodium alginate, extracted from seaweed, has been and still is used as a flocculant. It was particularly useful in potable water treatment, where, until the recent introduction of low residual monomer polyacrylamides, its lack of toxicity was a considerable advantage. In hard water it has the disadvantage of being precipitated as calcium alginate. Another group of flocculants widely used in the raw water, sewage, and general effluent fields are the tannins. They have the disadvantage of being susceptible to degradation on storage. They are most effective under conditions of acid pH.

In general the natural polyelectrolytes used as flocculants are much less effective on a weight-dosage basis than synthetic materials; this requires that they should be inexpensive and widely obtainable, and that the use of relatively larger quantities should not be disadvantageous in the total process.

4. *Synthetic Polyelectrolytes*

The most widely used class of synthetic polyelectrolyte is polyacrylamide and its derivatives. These are made by the catalytic polymerization of either acrylamide or acrylonitrile. Polyacrylamide has the structure

$$-\left[\begin{array}{l} CH_2 - \underset{\underset{\displaystyle NH_2}{|}}{\underset{|}{\overset{}{CH}}} \!\!\!\!\!\!\!\!\!\!\!\! \end{array}\right]_n$$

$$-\left[CH_2 - CH(C{=}O)NH_2 - \right]_n$$

where for flocculant materials 1500 < n < 150,000. Polyacrylamide may be used as a neutral polyelectrolyte or in the form of anionic and cationic derivatives. Hydrolysis of the polyacrylamide gives polyacrylic acid, an anionic material which ionizes under conditions of high pH as shown below:

$$-\left[CH_2 - CH(C{=}O)O^{\ominus} - \right]_n$$

In commercial flocculants between 5 and 30% of the amide groups are hydrolyzed. Similarly, a cationic flocculant may be produced by reduction and quarternization of the amide groups or copolymerization of acrylamide and quarternary acrylamides, such as

$$\left[CH_2 - CH(C{=}O)\overset{\oplus}{N}(CH_3)_3 - \right]_n$$

Neutral acrylamide flocculants are effective over a wider range of conditions and with a greater variety of materials than the anionic and cationic forms; however, the latter are more efficient where conditions of pH, and the degree of surface charge on the solid favors their use.

The residual monomer left in polyacrylamides is toxic. In the last few years several countries have given permission for the use of these materials in the potable water and food industries, provided that the residual monomer concentration is reduced to an acceptable level by extraction. The principal manufacturers market special materials for this use.

Studies [58] of the level of residual acrylamide monomer in surface waters have been made, and a W.H.O. report [59] has been devoted to the subject. It is a subject that is likely to receive continued attention from public health authorities.

Many other synthetic polyelectrolytes are described in the patent literature, some being copolymers of acrylamide with other materials. Some synthetic polyelectrolyte flocculants are of a lower molecular weight, e.g., polyethylene imines, than might be expected for effective bridging agents; it is likely their main effect is the encouragement of coagulation by zeta potential reduction.

C. Use of Flocculants and Coagulants

Section III.B presented the sequence of steps in the flocculation reaction, therefore the design of any flocculation process should attempt to achieve optimum conditions for these steps. In general, this means providing for a period of rapid mixing followed by a longer period of slow agitation to bring about the conditions for orthokinetic flocculation without excessive shearing of the flocs produced. The shear rupture problem is more severe with flocculating than coagulating systems because the flocculation reaction is irreversible. In a coagulating system, shear would lead to a smaller equilibrium size of floc.

Flocculating polyelectrolytes are generally used in the form of dilute (0.1 to 0.01% weight by volume) solutions, although for convenience of handling these may be prepared from more concentrated bulk solutions. Synthetic polyelectrolytes are usually supplied as low-density, fluffy powders. The viscosities of their solutions impose limits on the maximum concentrations that can conveniently be prepared. Depending on molecular weight, these are usually in the range 0.1 to 1% by volume. The solid powder is added slowly to the liquid to ensure thorough wetting and to prevent the formation of large lumps of gel which take a great deal of time to disperse.

A convenient way of handling batch quantities and preparing laboratory solutions is to drop the solid slowly into the vortex created by a stirrer as indicated in Fig. 22, which shows a complete dosing system based on this principle. In this apparatus, flocculant in hopper A is dosed by the weigh hopper B into a mixing chamber where it is slowly fed into a vortex in the incoming water. This mixture then flows into the main chamber and is stirred with a paddle mixer until solution is complete.

In doing this it is important not to allow the solid to fall onto the stirrer shaft, otherwise it will collect there as a large lump. The mixture is then stirred until solution is judged to be

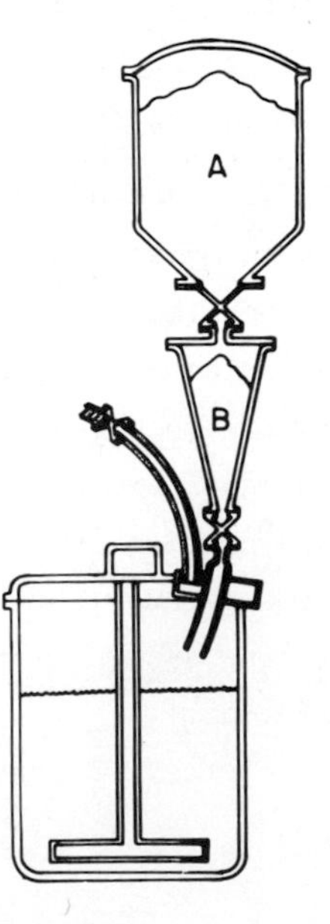

FIG. 22. A complete polyelectrolyte addition system.

complete, although this may be difficult to assess even in a transparent-wall vessel. Complete solution may take several hours. Manufacturers warn that excessive agitation of polyelectrolyte solutions may lead to degradation of the polymer, although the evidence of how important this is in practice is not very clear. There are also proprietary devices available for dispersing the polymer into a stream of liquid. One is shown in Fig. 23. This device is troublesome if there are irregularities in the water flow rate, as these may lead to powder in the feed line becoming wet and blocking the flow.

Proprietary mixing systems [60] have been developed. This is a subject that has received insufficient attention by both researchers and plant designers. It is only recently that attention has been given by flocculant manufacturers to the considerable amounts of insoluble material that may be present in some flocculants.

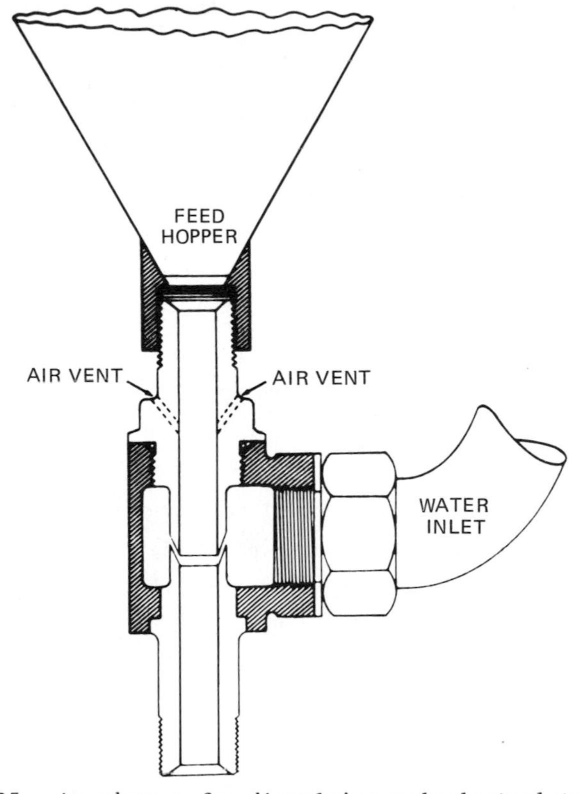

FIG. 23. An eductor for dissolving polyelectrolyte.

Having obtained a solution of convenient strength of polyelectrolyte, it is necessary to add it to the suspension to be treated so as to ensure rapid and thorough mixing. The flocculant solution may be added to a relatively small stirred holding tank through which the suspension passes. This method of addition is particularly convenient in thickener operation. Alternatively, the mixing may be induced by the geometry of the system through which the suspension is flowing. It is often found that more efficient flocculation is obtained if the flocculant is added at several points. This may be done in either an open launder or a closed pipe, and the mixing achieved either by baffles or by sharp changes in direction in the pipe, as shown in Fig. 24. From this mixing zone the suspension is then led to a further vessel where slow stirring can be maintained or the suspension can be allowed to flocculate during flow along the pipe or launder.

Because of their low viscosities, the addition and mixing of coagulants is easier than for flocculants. The coagulation solution is often added to the suspension in a small stirred holding tank before the mixture is passed to the vessel in which coagulation takes place. As the reaction is reversible, the residence time and rate of mixing in the mixing tank are not so critical compared with flocculation, the size of the resultant flocs being dependent on the amount of shear rather than on the previous history of the suspension.

In general, flocs produced by synthetic polyelectrolyte reagents are much more resistant to rupture than those produced by coagulants, and they give rise to more open filter cakes. Also, problems with media blinding are less likely with polyelectrolyte flocculants.

Now that synthetic polyelectrolytes pure enough for use in water treatment processes are becoming available, increasing use is being made of them as coagulant aids in which they are combined with ionic coagulants such as Ac^{3+} and Fe^{3+}; at $3 < pH < 5$, coagulation due to polymeric hydrated ferric oxide occurs; and at $pH > 5$, precipitation of $Fe(OH)_3$ occurs. Under certain conditions adsorption of the hydrolyzed complex will cause charge reversal sufficient to stabilize the system. The precipitation of $Fe(OH)_3$ will lead to an increase in the available particle concentration and, hence, in the

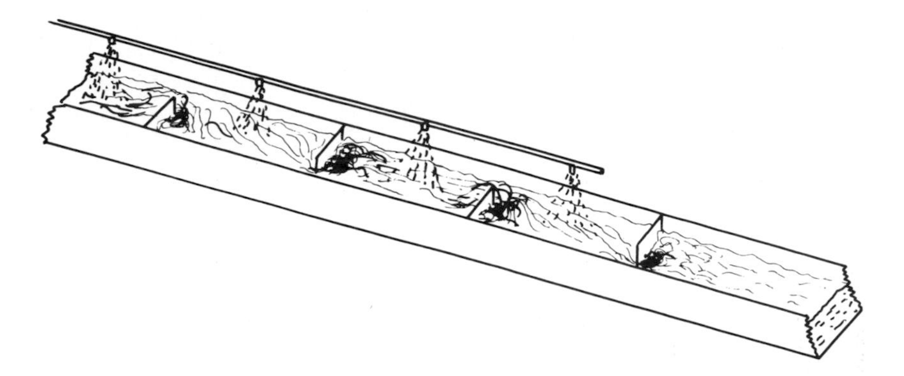

FIG. 24. Multipoint addition of polyelectrolyte to a launder.

collision frequency and coagulation rate. Subsequent addition of a flocculating polyelectrolyte will then flocculate both the SiO_2 and $Fe(OH)_3$ precipitate. In this way the use of polyelectrolytes is possible for systems in which the solids concentration would be too low to flocculate effectively because the adsorbed flocculant would become compressed before flocculation had occurred. A set of useful equivalents for use in pretreatment processes are given in Table 1.

TABLE 1 Useful Equivalents

1 mg flocculant/100 g solid = 0.0079 lb/ton (U.S.) = 0.0088 lb/ton (long)
1.0 mℓ 0.05% flocculant in 500 ml = 1 ppm
1 ppm flucculant per 1% solids = 0.20 lb/ton solids (U.S.) = 0.224 lb/ton (long)
Settling rate in ft/hr = 0.1 in. in 30 sec
Settling rate of 2 ft/hr ≃ 1 cm/sec

D. Assessment of Flocculants and Coagulants

Settling rates, filtration rates, sediment volumes, and refiltration rates have all been advocated as methods of assessing the efficiency of pretreatment processes. Although these phenomena are interrelated, they do not measure the same property of the system. In order to evaluate a pretreatment technique for use in filtration, it is necessary ultimately to test it in actual filtering situations, although other methods of testing may be useful in the preliminary evaluation.

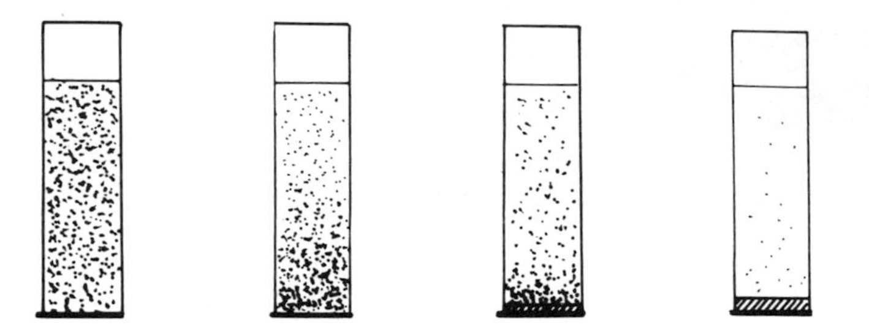

FIG. 25. Sedimentation of a nonflocculated suspension.

The measurement of setting rate is the most widely used of the preliminary testing methods. During the settling of a nonflocculated suspension, a compacted sediment will be observed at the base of the container above which will be a layer of suspension that decreases upward in concentration. If the particle size range is considerable, the fine particles will remain in the upper layers because they take longer to settle. The sediment will occupy a volume approaching that of the close packed volume of the particles when sedimentation is complete. After some time this sediment may become quite firm and difficult to redisperse, as shown in Fig. 25.

The behavior of a flocculated suspension is quite different. Because of their size, flocs of particles will settle much more rapidly than individual particles, and a clear upper mud line will be observed. Sedimentation will continue until the sediment is unable to compress further. These types of behavior are shown diagramatically in Fig. 26. The sediment obtained will have a considerably

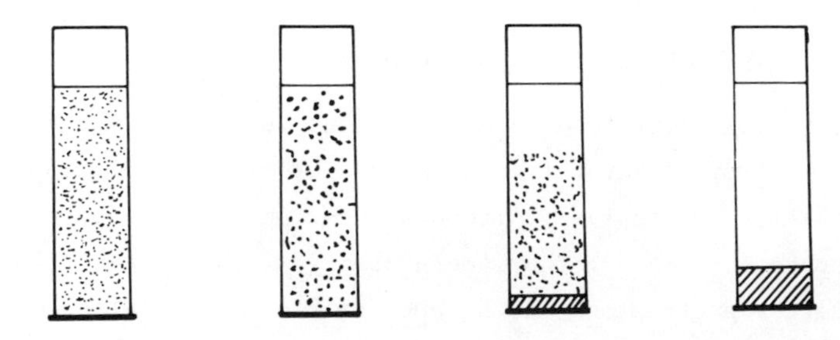

FIG. 26. Sedimentation of a flocculating suspension.

greater volume than the corresponding nonflocculated sediment, because the particles are in a structural array that prevents them from packing closely. As a result this of this structure the sediment may be very compressible. Figure 27 shows a settling rate plot for such a sediment.

Settling rate tests are a very convenient way of screening suitable flocculants because of their simplicity, and they aid in obtaining the optimum dosage. The tests may be carried out in a variety of ways, the main thing being to ensure that the technique is as reproducible as possible. A 500 ml measuring cylinder makes a convenient sedimentation vessel. It is desirable that cylinders of equal diameter and having parallel sides be selected for the tests, as wall effects may be considerable. The cylinder is filled with suspension to the top mark and the appropriate quantity of flocculant added, conveniently as a 0.05% solution made up that day from a concentrated stock solution, and at the same time a timer is started. The contents of the cylinder are then thoroughly mixed. This may be done either with an agitator such as a perforated metal disc attached

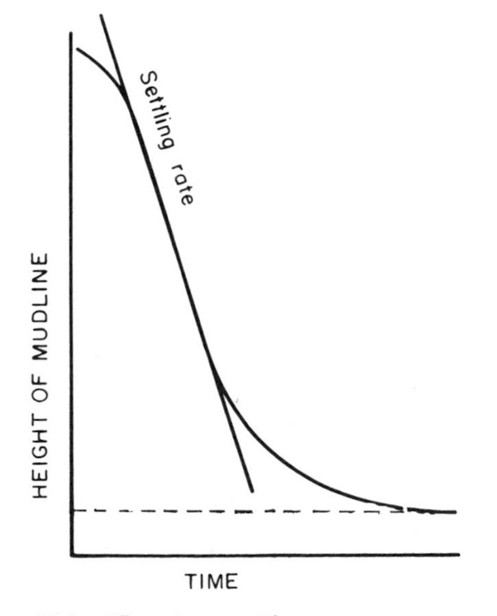

FIG. 27. A settling rate curve.

to the end of a rod which is pushed up and down the cylinder, or by inverting the cylinder after closing with a stopper or the palm of the hand. Whatever mixing technique is chosen it must be thorough, rapid, and reproducible. After mixing, the cylinder is allowed to stand, and the height of the mud line is noted at regular intervals until no further sedimentation occurs. From the linear portion of the settling curve (Fig. 27) the settling rate may be evaluated and from the height of the final mud line, the sediment volume.

The choice of flocculating agent is very empirical. The economics of the process may dictate whether, for example, a relatively inefficient natural flocculant such as starch is used or a more efficient but also more expensive polyacrylamide. If the material to be flocculated is of very constant composition or if variables such as pH are not expected to change, it would probably be better to use a specific flocculant. On the other hand, if conditions are only poorly controlled, a less specific but also less efficient flocculant is probably best. General knowledge of the surface properties of the solid to be flocculated and the pH and salt concentration of the suspension may also help, and zeta potential measurement may be useful. As a general rule for polyacrylamides, it has been found that anionic materials are best for mineral suspensions and cationic reagents for organic suspensions. Having selected a possible flocculant on the basis of its ability to flocculate the suspension, paying particular attention to the removal of residual haze, it is then necessary to find the optimum dose by carrying out a series of tests covering the practical dosage range. Once this has been done, the appropriate filter testing procedures, as described in Chap. 4, should be used to confirm that the optimum procedure as determined by settling tests also applies to the filtration technique under consideration.

A very convenient test for filtrability testing of flocculated and coagulated pulps is the capillary suction test of Baskerville and Gale [55d], which was designed for sewage applications. A reasonable correlation of test results (i.e., within ±30%) has been found [62] with rotary vacuum filter performance, the materials used covering filtration yields over a 25-fold range.

La Mer and Healey [33] considered that refiltration rate was a valuable test of flocculation and that the refiltration rate equation—refiltration rate proportional to $\phi^4(1 - \phi^4)$ where ϕ is the proportion of solid surface area covered by the polymer layer—was unique for a system that had been subjected to polyelectrolyte flocculation. Other workers [63] have shown that this relationship is not uniquely true and that no advantage is to be obtained from using this more complicated testing procedure.

Ives [64] has reported that depth filters 15 cm in diameter give the same pressure head losses and other characteristics as full-scale filters. This ability to scale the results directly from model filter beds formed in 6 in. pipes facilitates the performance of meaningful tests of pretreatment techniques in depth filtration processes. Indeed, Conley [65] has suggested that a small model filter be run in parallel with the main filter to help the operators optimize the correct pretreatment dose.

In testing coagulants and flocculants for water treatment, where the solids concentration is too low for studying settling rates, the water and reagents are placed in beakers and slowly stirred. The change in the state of aggregation of the particles may be followed by observing changes in the light scattering or light absorbing properties of the suspension, both of which would decrease with increasing aggregation, or by observing the rate of settling.

The scale-up of sedimentation test data for thickener design is still based on empirical experiments using pilot plant for those situations where the pulp is compressible and the underflow concentration is dependent on the thickener depth, sediment depth, and detention time. These latter factors are not taken account of in the widely used flux-curve design procedures.

IV. FILTER AIDS

As an alternative to altering the particle size distribution of a suspension by shifting the state of aggregation of the component particles, it is possible to bring about this change by adding solid material having a different size distribution. Such materials are known as filter aids. As these materials are also used as filter media in precoat filtration processes the term "body aid" may be used to avoid any confusion. The use of the word "body" signifies that the material is added to the main body of suspension being filtered. Although the use of a body aid is frequently combined with precoat filtration, the two processes and their mechanisms must be differentiated.

The requirements of a filter aid are that it should have suitable particle size and shape characteristics for achieving the desired cake permeability, that it should be chemically inert under the conditions used, and that it should be inexpensive. This latter requirement rules out many potentially useful substances and also the use of sophisticated classifying processes in aid manufacture. Table 2 indicates the important properties of the principal filter aid materials.

A. Diatomaceous Earths

These are the sedimentary deposits of fossilized diatoms that had sunk to the bottom of prehistoric seas and lakes. For deposits to be commercially exploitable as filter materials, it is necessary that they be free of excessive contamination with both soluble and insoluble impurities. Diatom skeletons exhibit a great variety of shapes, and it is this property that enables them to be used in producing both filter cakes and precoats of high permeability. Figure 28 is a scanning electron micrograph of a diatomite filter aid showing the remains of a diatom skeleton. Mined diatomaceous earth is refined by a combination of grinding, screening, drying, and, frequently, calcining operations. The effect of calcinations is to reduce the number of very fine particles present with a subsequent

TABLE 2 The Properties of Filter Aid Materials

Material	Chemical composition	Advantages	Disadvantages
Diatomaceous earth	Silica	Wide size range available Fines reduced by calcination Can be used for very fine filtration	Slightly soluble in dilute acids and alkalies
Expanded perlite	Silica and aluminosilicates	Wide size range available Not capable of finest retention of diatomites	More soluble than diatomites in acids and alkalies May give highly compressible cakes
Asbestos	Aluminosilicate	Usually used in conjunction with diatomites Very good retnetion on coarse screens	Chemical properties similar to perlite
Cellulose	Cellulose	Used mainly as a coarse precoat High purity Excellent chemical resistance—slightly soluble in dilute and strong alkalies, none in dilute acids	Expensive
Carbon	Carbon	May be used for filtering very strongly alkaline solutions	Available in coarser grades only Expensive

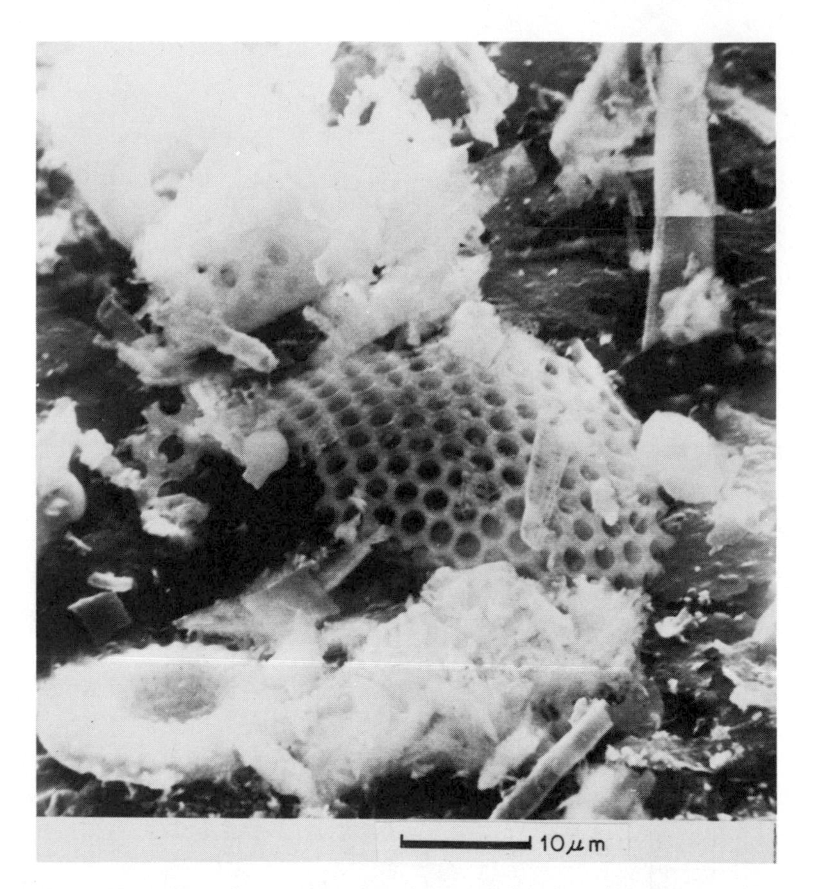

FIG. 28. Scanning electron micrograph of a diatomite filter aid.

improvement in cake permeability. This reduction in the size range is probably achieved by a fusion of small particles. A narrower size distribution is obtained by flux calcination in the presence of alkali; while this gives a more desirable product, it has the disadvantage of a considerably increased soluble alkali metal content. If this increase in alkali metal is undesirable, nonfluc calcined materials are available. As a diatomaceous earth consists mainly of silica, i.e., from 85 to more than 90% silica, it is soluble in both strong acids and, particularly, strong alkalies.

B. Perlite

Perlite is an igneous rock formed by the quenching of molten volcanic lava in water. Because of the rapid cooling that it has experienced, it is a supercooled material with a glassy structure. Occluded within the small "pearls" that make up the rock is several percent of water. If crushed and screened and then rapidly heated to its softening point, the occluded water in perlite will vaporize and cause the "pearls" to swell to hollow spheres many times their original size. This process is similar in principal to that used in preparing some breakfast cereals. The thin walled hollow spheres are then crushed (c.f. crushed eggshells) and classified to give the final product. A scanning electron micrograph (Fig. 29) shows the irregularly shaped particles obtained.

The greater amounts of Al_2O_3, Na_2O, and K_2O found in perlite as compared to diatomite makes it even more soluble and limits the pH range in which it may be used from 4 to 9. Perlites cost about the same as diatomites, but because of their lower bulk density, e.g., 3 to 6 lb/ft^3 as compared with 8 to 12 lb/ft^3 for diatomites, they are more economical to use. With the exceptions of the extreme ends of the particle size range, it is possible to obtain perlites corresponding to most grades of diatomite. It has been claimed that in some applications their cost advantages may be offset by significantly greater cake compressibility.

C. Carbon, Cellulose, and Other Materials

Carbon is an expensive material for use as a filter aid but is invaluable for the filtration of strongly alkaline liquids which will attack diatomites, perlites, and other aluminosilicates such as asbestos. The particle sizes available correspond to the larger particle-size diatomite and perlite materials. Asbestos, with its fibrous structure may also be used as a filter aid, usually in conjunction with diatomite or perlite. These mixtures are probably used more as precoats than as aids, on account of their excellent retention on coarse screens. Cellulose filter aids are expensive

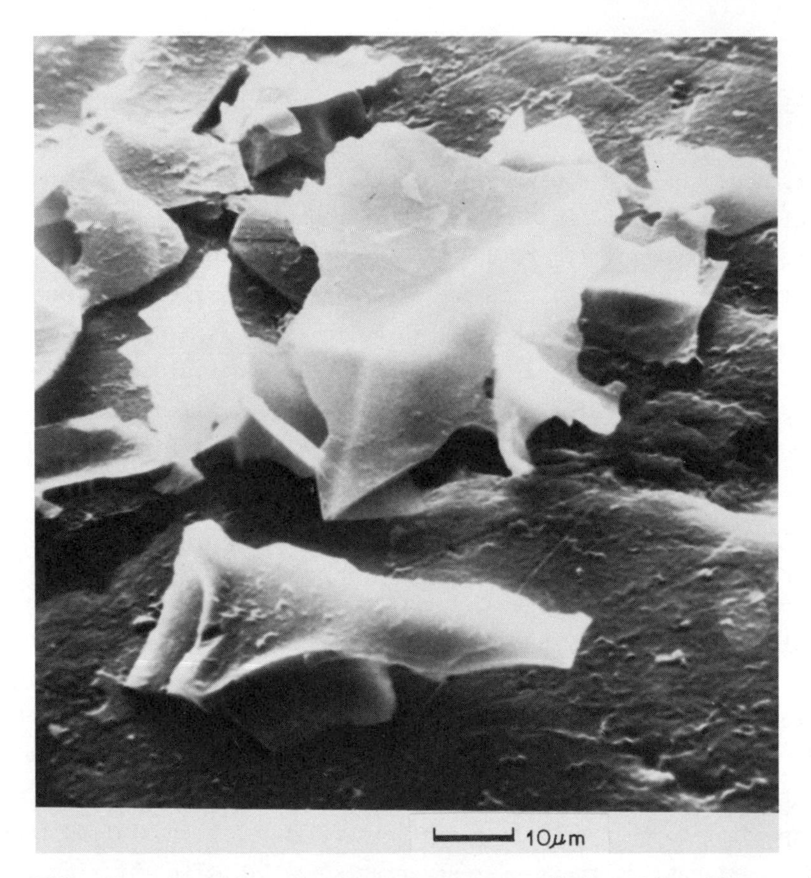

FIG. 29. Scanning electron micrograph of a perlite filter aid.

materials but may be used where their purity and resistance to dilute acid attack is required. Any available solid material, e.g., sawdust, may be used as an aid, provided that its presence does not impair the product of the filtration process.

D. Selection of Filter Aids

In general, a material having a wide particle-size distribution will produce a packed bed of low porosity and low permeability. The aim of adding a body aid is to disturb the packing of the bed to such a point that a considerable increase in porosity and permeability occurs. The addition of an aid having a particle size small with respect to the material being filtered would tend to block the existing pores in the packing, giving rise to decreased permeability and porosity but an increased ability to retain fine particles. This might be advisable in a situation where it was necessary to clear a liquid containing a bimodal distribution of coarse and fine particles. This, however, is not a common circumstance. If the aid particles are very much larger than the solid being filtered, they will have little effect on the overall bed structure. For maximum effect, an aid needs to have a narrow particle-size range lying within or close to that of the solid being filtered. The manufacturers of aids commonly quote the water permeabilities and relative flow rates through packed beds of aid. This is useful information in selecting aids for precoat use but is of little value in the selection of a body aid as it is the properties of the whole system of retained solids that are required. Having selected a suitable particle size range of aid it is then necessary to determine the optimum dosage. Figure 30 is a typical plot of filter aid concentration

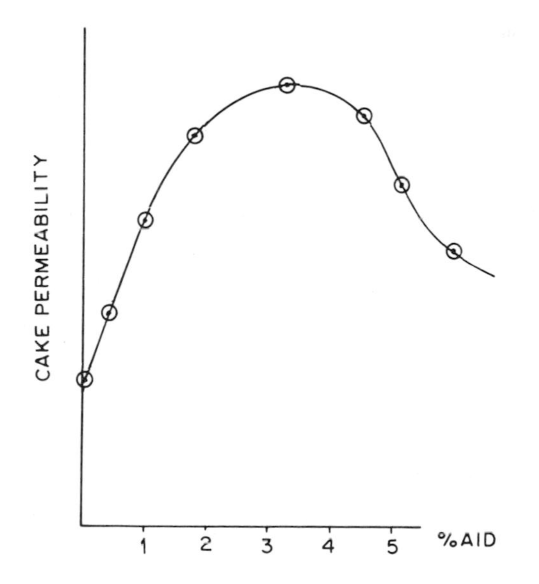

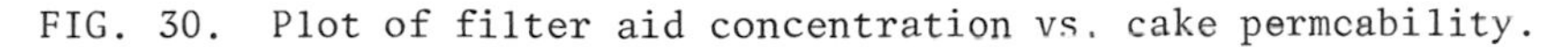
FIG. 30. Plot of filter aid concentration vs. cake permeability.

vs. permeability showing an optimum dose level. For a practical aid, the optimum dose will usually lie within the range 1 to 10% by weight with respect to the retained solids.

As the effect of the aid is to increase cake permeability, the associated increase in porosity may lead to a considerably increased bulk of cake which may be undesirable in those filtration operations where the cake volume is limited, e.g., in plate and frame presses.

V. MISCELLANEOUS PRETREATMENT TECHNIQUES

A. Freezing

Freezing and thawing has been successfully applied to dewatering of the difficult organic sludges produced during water treatment. Normally these materials have to be lagooned as they are impossible to concentrate further by filtration or sedimentation. Doe and colleagues [66,67], at the Fylde Water Board, England, made a detailed study of this process in which they found that a hard granular material could be produced if very strict control of the operating conditions was maintained. The purification plant to which the process was applied had an output of 18,500,000 gal per day, producing 14,000 gal per day of sludge. The costs of processing this sludge was 0.075 cents per 1,000 gal of pure water more than conventional lagooning, say $2 per 1,000 gal of sludge. One of the difficulties encountered was the design and maintenance of tanks that would withstand repeated freezing and thawing of water in them. The process is grossly uneconomical for systems with as high a solid content as sewage effluents, although work is still being continued, investigating systems such as direct freezing with liquified petroleum gas [68].

B. Heating

Sewage sludges have been successfully treated by "pressure cooking" at about 200°C and 150 psi. Heat treatment processes have recently been reviewed [69].

C. Aeration

Injection of air has been shown to enhance coagulation in many cases, but with some materials, particularly digested sewage sludges, inhibition has been found. The effect of aeration is probably to increase the collision and, hence, overall coagulation rate. The cases in which inhibition occurred are those in which oxidation of the system modified somewhat the surface properties.

D. Mechanical and Ultrasonic Vibration

Ultrasonic vibration has been successfully applied in ore benefication, sewage treatment process, and colliery washing plants [70,71, 72]. However, there is much dispute as to the efficiency of the process, some workers claiming no benefit at all. Coagulation is a reversible process, and at any particular shear rate an equilibrium will exist between shear-induced coagulation and dispersion. This would be very dependent on the geometry of the system, the frequency of the applied vibration, and the properties of the suspension. These same restrictions would apply to any other form of mechanical agitation.

E. Electrical and Magnetic Treatments

Electrolysis with aluminum or iron electrodes has been shown to have a coagulating affect, due to the release of Al^{3+} or Fe^{3+}. These act as coagulants themselves and, under conditions of correct pH, would hydrolyze the precipitate and increase the overall coagulation rate by increasing the particle concentration. The application of a magnetic field will enhance the coagulation of magnetic powders. This is probably due to enhancing the collision rate by causing mutual attraction and also, possibly, due to enhancing orthokinetic coagulation by causing the particles to move.

F. Radiation

In the period 1960 to 1964 the United States Atomic Energy Commission published a series of reports [73] on the effect of ionizing radiation on sewage sludges and montmorillonite clays. From the increase in coagulation rate observed, it was concluded that the technique would be valuable if the isotope costs at that time could be reduced by a factor of six. A reappraisal of the economics of this process might be appropriate.

G. Conclusions

Apart from the freezing technique, the economics of which are marginal for sludge treatment, the other forms of treatment described in this section are only in an early stage of development and their future is speculative. Ultrasonic and mechanical agitation techniques and aeration are all means of adding mechanical work to the system to increase the collision rate. The very intensive local accelerations present in systems exposed to ultrasonic energy may enable the particles to overcome greater potential energy barriers to coagulation than they might otherwise. The effects of heating, electrical, and irradiation techniques are probably due to changes induced in the surfaces of the materials.

NOTATION

- a the diameter of a particle
- $\underline{a}$ diameter of spherical particles
- A cross-sectional area of a packed bed or filter cake
- B a constant
- B Blake number
- c a dimensionless constant
- c molar concentration of particles in a suspension
- c solid mass fraction
- c the concentration of an electrolyte
- C a constant

d	the hydraulic diameter of a pore
d_p	the Stokes diameter of a particle
d_m	a characteristic dimension
D	a diffusion coefficient
e	unit electronic charge
E	electric field gradient
$F(\varepsilon)$	porosity factor
g	gravitational force
$\bar{G}$	mean shear rate in a coagulator
I	particle collision rate in perikinetic coagulation
J	particle collision rate in orthokinetic coagulation
$\underline{k}$	Boltzmann constant
K	permeability
K'	a constant
K''	Kozeny-Carman constant
K_o	initial permeability
K_x	local permeability at depth x in cake
L	thickness of a packed bed or filter cake
L'	length of a pore
L_o	thickness of a filter cake hydraulically equivalent to the medium
N_o	Avogadro number
p	power expended in agitating a field
p_i	minimum pressure
p_s	compressive stress pressure
p_x	hydraulic pressure at depth x in cake
P	Peclet number, power applied to a coagulator
P	pressure
P	total pressure on the cake
P_1	pressure at septum-cake interface
ΔP	pressure drop across a packed bed or filter cake
q	flow rate
q_1	flow rate at cake-medium interface

q_x flow rate at depth x in filter cake
r interparticle distance
r radius of a capillary
R resistance of a porous bed
R collision diameter of a pair of particles
s a compressibility coefficient
S specific surface area of a packed bed
S_p specific surface area of a particulate substance
S_t Stokes number
t time
$\bar{t}$ mean residence time
T absolute temperature
T coagulation time
u overall fluid velocity through a packed bed
$\bar{u}$ mean velocity of fluid through a pore
U potential energy
U_E electrophoretic mobility
v volume of cake deposited by unit volume of filtrate
V volume of coagulator
V volume of fluid flowing in time θ
V_o a constant
V_o volume of filtrate corresponding to cake thickness L_o
w weight of dry cake corresponding to unit volume of filtrate
dw_x mass of solids per unit area in cake at thickness x
z valency of an ion
z vertical coordinate
α particle diameter
α specific cake resistance
α_o initial filtration resistance
α_x specific cake resistance at depth x
α_T a new filtration resistance
β a constant
δ a form of hydraulic diameter = d/4
ε dielectric constant
ε porosity of a packed bed or filter cake

ε_o	initial porosity
ε_x	porosity of a packed bed or filter cake at depth x
ξ	zeta potential
θ	time
μ	viscosity of a fluid
ν	number of particles per unit volume
ρ_s	density of solid
ρ_f	density of fluid
ϕ	proportion of solid surface covered by polymer
ψ	potential
χ	Debeye-Hückel double-layer reciprocal thickness

REFERENCES

1. H. P. G. Darcy, *Les Fontaines Publiques de la Ville de Dijon*, Dalamont, 1856.
2. J. L. M. Poiseuille, *Compt. Rend*, *11*, 961 (1840); *12*, 1041 (1841); *15*, 1167 (1842); *Ann. Chim. Phys.*, *21* (3), 76 (1842).
3. J. Kozeny, *Sber. Akad. Wiss. Wien.* (Abt. IIa) 136, 271 (1927).
4. A. J. E. J. Dupuit, *Etudes theoriques et pratiques sur le mouvement des eaux,* 1863.
5. J. M. Pirie, in J. M. Coulson, *Trans. Inst. Chem. Eng.*, *27*, 237 (1949).
6. J. M. Coulson and J. F. Richardson, *Chemical Engineering*, 2nd ed., Pergamon, London, 1968.
7. P. C. Carman, *Trans. Inst. Chem. Eng.*, *15,* 150 (1936).
8. A. E. Scheidegger, *Physics of Flow Through Porous Media*, University of Toronto, Toronto, 1957.
9. A. S. Iberall, *J. Res. Nat. Bur. Stand.*, *45*, 398 (1950).
10. B. Scarlett, Paper read to Inst. Chem. Eng. Symposium, *Filtration*, 1968.
11. D. A. Dahlstrom, M. E. O'K. Trowbridge, and D. Bradley, in *Solid-Liquid Separation*, (Poole and Doyle, eds.), H.M.S.O. London, 1966.
12. B. F. Ruth, G. H. Montillon, and R. E. Montonna, *Ind. Eng. Chem.*, *25* (1), 67 (1933); *25* (2), 153 (1933).
13. B. F. Ruth, ibid., *27*, 708 (1946); *27*, 806 (1946); *38*, 564 (1946).
14. R. S. Gale, *J. Wat. Pollut. Contr.*, 70 (1), 1 (1971).

15. R. S. Gale and R. C. Baskerville, ibid., *69* (5), 514 (1970).

16. F. M. Tiller, *Chem. Eng. Progr.*, *49*, 467 (1953).

17. Idem, ibid., *51*, 282 (1955).

18. F. M. Tiller, *A.I.Ch.E. J.*, *4*, 170 (1958).

19. F. M. Tiller and H. Cooper, *A.I.Ch.E. J.*, *6*, 595 (1960).

20. F. M. Tiller, *A.I.Ch.E. J.*, *10*, 61 (1964).

21. F. M. Tiller and H. Cooper, *A.I.Ch.E. J.*, *8*, 445 (1962).

22. M. Shirato and S. Ikamura, *Chem. Eng.* (Japan), *19*, 104, 111 (1955).

23. Idem. ibid., *20*, 98, 678 (1956).

24. Idem, ibid., *23*, 11, 226 (1959).

25. F. B. Hutto, Jr., *Chem. Eng. Progr.*, *53*, 328 (1957).

26. H. P. Grace, *Chem. Eng. Progr.*, *49*, 303, 367 (1953).

27. D. R. Boylan and F. A. Kottwitz, *A.I.Ch.E. J.*, *4*, 175 (1958).

28. C. N. Davies, *Proc. Inst. Mech. Eng.*, *B*, *1*, 185 (1952).

29. J. A. Harrop and J. I. T. Stenhouse, *Chem. Eng. Sci.*, *24*, 1475 (1969).

30. C. R. Ison and K. J. Ives *Chem. Eng. Sci.*, *24*, 717 (1969).

31. K. M. Yao, *Ph.d. Thesis*, University of North Carolina, Chappel Hill, 1968.

32. K. J. Ives and J. Gregory, *Proc. Soc. Water Trtmt. Exam.*, *15*, 93 (1966).

33. V. I. La Mer and T. Healey, in *Solid-Liquid Separation* (Poole and Doyle, eds.), H.M.S.O., London, 1966.

34. B. V. Derjaguin and L. Landau, *Acta. Phys.-chim.*, U.S.S.R., *14*, 633 (1941).

35. E. J. Verwey and J. Th. G. Overbeek, *Theory of the Stability of Lyophobic Colloids*, Elsevier, Amsterdam, 1948.

36. F. London, *Z. Phys.*, *63*, 245 (1930).

37. M. von Smoluchowski, *Z. Phys.*, *17*, 557, 585 (1916); *Z. Phys. Chem.*, *92*, 129 (1917).

38. N. Fuchs, *Z. Phys.*, *89*, 736 (1934).

39. H. Müller, *Kolloid. Z.*, *38*, 1 (1926); *Kolloiden Beihefte*, *26*, 257 (1928).

(a) T. R. Camp and P. C. Stein, *J. Boston Soc. Civ. Eng.*, *30*, 219 (1943).

40. T. R. Camp and P. C. Stein, *J. Boston Soc. Civ. Eng.*, *30*, 219 (1943).

41. G. Gouy, *J. Phys.* (Paris), *9* (4), 457 (1910); *Ann. Phys.* (Paris), *7* (9), 129 (1917).

42. D. L. Chapman, *Phil. Mag.*, *25* (6), 475 (1913).

43. O. Stern, *Z. Electrochem*, *30*, 508 (1924).

44. P. Sennett and J. P. Olivier, *Ind. Eng. Chem.*, *57*, 33 (1965).

45. P. H. Wiersema, A. L. Loeb, and J. Th. G. Overbeek, *J. Colloid Interfac. Sci.*, *22*, 78 (1966).

46. H. Schulze, *J. Prakt. Chem.*, *25* (2), 433 (1882); *27*, 320 (1882).

47. W. B. Hardy, *Proc. Roy. Soc.*, *66*, 110 (1900); *Z. Phys. Chem.* *33*, 385 (1900).

48. E. F. Burton, *Phil. Mag.*, *11* (6), 425 (1906); *12*, 472 (1906); *17*, 583 (1909).

49. C. R. O'Melia, in *Physico-Chemical Processes for Water Quality Control* (W. J. Weber, Jr., ed.), Interscience, New York, 1972.

50. B. Vincent, *Advan. Colloid Interfac. Sci.*, *4*, 193 (1974).

51. R. F. Packham, *Proc. Soc. Water Trtmt. Exam.*, *16*, 18 (1967).

52. R. A. Ruehrwein and D. W. Ward, *Soil Sci.*, *73*, 485 (1952).

53. R. J. Akers and P. W. Riley, unpublished results.

54. A. Silberberg, *J. Phys. Chem.* *66*, 1884 (1962); *J. Chem. Phys.* *46*, 1005 (1967); *48*, 2835 (1968).

55. D. N. Sutherland, *J. Colloid Interfac. Sci.*, *25*, 373 (1967).

56. W. F. Linke and R. B. Booth, *Trans. Amer. Inst. Min. (metal.) Eng.*, *217*, 364 (1959).

57. A. M. Kragh and W. B. Langston, *Brit. Glue Gelat. Res. Ass.*, Report C21 (1960).

58. B. T. Croll, G. M. Arkell, and R. P. J. Hodge, *Water Res.*, *8*, 989 (1974).

59. *Health Aspects Relating to the Use of Polyelectrolytes in Water Treatment for Community Supply*, Technical Paper no. 5, W.H.O., The Hague.

60. A. Elphick and L. J. Lake, *Effl. Water Trtmt J.*, *14*, 378 (1974).

61. R. C. Baskerville and R. S. Gale, *J. Inst. Water Pllut. Contr.*, *2*, 3 (1968).

62. R. S. Gale, *Filtrat. Separat.*, *9*, 341 (1972).

63. R. W. Slater and J. A. Kitchener, *Diss. Farad. Soc.*, *42*, 267 (1966).

64. K. J. Ives, *Water Res.*, *4*, 206 (1970).

65. W. R. Conley, *J. Amer. Water Works Ass.*, *57*, 1333 (1966).

66. P. W. Doe, D. Benn, and L. R. Bays, *J. Inst. Water Eng.*, *19*, 251 (1965).

67. P. W. Doe, ibid., *12*, 409 (1948).

68. M. J. Smyth, I. M. Stokes, R. Murdoch, and J. B. Lewis, *The Application of Chemical Engineering to the Treatment of Sewage and Industrial Liquid Effluents*, Institution of Chemical Engineers, London, Symposium Series, no. 41, 1975.

69. *Process Design Manual for Sludge Treatment and Disposal,* U. S. Environmental Protection Agency Report, EPA 625/1 74-006, 1974.

70. W. A. Lyon, *Sewage Ind. Waste*, *23*, 1084 (1951).

71. W. Husmann, *Gesundheitzung*, *73*, 127 (1952).

72. T. Imaizumi and T. Inone, *Nihon Koggo Kaishi*, *76*, 93 (1960).

73. *Nuclear Science Abstracts*, *15*, nos. 7656, 400 (1961); *16*, no. 15047 (1962); *17*, nos. 2007, 16057 (1963); *18*, no. 5362 (1964).

Chapter 3

FILTER MEDIA

A. Rushton
P. V. R. Griffiths

Chemical Engineering Department
The University of Manchester
Institute of Science and Technology
Manchester, England

I. Introduction 252

II. Media Classification and Requirements 254

III. Criteria of Choice 256

IV. Measurement of Pore Size and Particle Retention 257

V. Permeability of Clean Media 260
A. Woven Multifilament Filter Media 262
B. Nonwoven Filter Media 272
C. Monofilament Filter Media 274

VI. Particle Deposition Mechanisms 288

VII. Flow Resistance of Used Media 291

VIII. Effect of Media Structure on Cake Resistance 296

IX. Media Bridging 299

Notation 302

References 306

I. INTRODUCTION

A distinguishing feature of the unit operation of filtration is the variety of machinery available for the separation of solid particles from fluids. This variety has been created by the multiplicity of filtration problems, where suspensions of high or low particle concentrations must be separated from gases or liquids, which in turn may be corrosive, highly viscous, etc. Despite this complexity, process information on the particles and fluid can be used to narrow the wide field of choice; thus the particle size, the filterability and compressibility of the filter cake, the solids concentration, etc. may lead to decisions on the suitability of a particular form of filter and to the optimum operating conditions. After this specification of machine type, the remaining and, perhaps, more intractable problem must be considered: selection of the most suitable medium. Indeed, the successful performance of a filter station depends largely upon the filter medium and its capability to separate the particles and fluid without concomitant medium blinding and deterioration.

Most authors dealing with media selection are agreed that the multitude of media available makes the specification of the optimum medium a particularly difficult task. With the exception, perhaps, of very coarse particulate matter, where perforated-plate separators would most likely be used, there usually exists a large number of possible solutions to the problem of medium choice. A second exception here would pertain to extremely fine particles, where membrane filters would be necessary.

Of course, not all filter media would be considered for a particular type of machine, and this fact does reduce the field somewhat in certain cases. However, the inherent difficulties associated with media selection are illustrated by reference to woven fabrics. Here an extremely large number of variables exist, such as weave pattern and materials of construction, all of which may influence the successful application of the fabric and, perhaps, the feasibility.

Faced with a new separation problem, the engineer must consult the manufacturer's literature for answers to such questions as media strength, stability, and chemical resistance. A useful, qualitative review of the principal forms of filter media is given by Dickey [1], dealing with questions of strength, etc., and also containing much practical information on the performance of media. French [2] reviews woven and nonwoven filter fabrics and discusses the various modes of fabric construction; information is supplied on chemical and physical properties of cotton, glass, paper, and man-made fibers. Kovacs [3] describes media composed of wire cloth, metal edge filters, fused metallic fibers, ceramics, and impregnated cellulose. Again, certain physical properties are tabulated. The spectrum of filter media is covered by Shoemaker [4], who includes practical suggestions on media selection and classification.

The suitability of a medium for a particular separation is usually decided by practical trials, since there is a relative dearth of information to facilitate the prediction of filter action from measured properties of the medium and solid-fluid mixture. Dobie [5] reports some useful experiences in the field of woven fabrics and indicates that practical trial weighs very heavily in the successful outcome of media selection.

In recent years, the lack of fundamental understanding on the role of the medium has received some attention; the results of various workers are reviewed below, where the effect of media construction on permeability to fluids, filtration mechanisms, particle retentivity, etc. is mentioned, along with process conditions necessary for efficient separations by media pore-bridging.

As may be seen in the section below dealing with media classification, a wide choice of constructional materials is available, most of which are fashioned in many modes of construction as woven cloths, sheets, and cartridges. In such a short space it is, of course, impossible to reproduce all the technical information that is available on each combination of media type and material. However, an attempt has been made to report the general principles of media performance that have been established by research workers using media of various classification.

II. MEDIA CLASSIFICATION AND REQUIREMENTS

A useful classification of filter media has been presented [6] on the basis of rigidity, and a partial list of the more commonly accepted media is presented in this form in Table 1. The latter also gives an indication of the minimum particle size normally retained by various media; such specifications are so sensitive to process conditions and particle concentrations that the figures quoted must be viewed as rough guides only.

Table 1 contains interesting information at the extreme ends of the particle-size spectrum met with industrially; thus it suggests that above a certain size (>100 μm) the coarseness of the

TABLE 1 Media Classification

	Type	Example	Minimum particle trapped (μm)
1.	Edge filter	Wire-wound tube Scalloped washers	5-25
2.	Porous solids	Ceramic Stoneware Sintered-metal cartridge	1 1 3
3.	Metallic sheets	Perforated plates Woven wire	100 >5
4.	Porous plates	Plastic pads and sheets Membranes	3 0.005
5.	Woven fabrics	Woven cloths of natural and synthetic fibers	10
6.	Cartridges	Spools of yarn or graded fiber	2
7.	Nonwoven sheets	Felts, laps, etc. Paper Sheets and mats	10 2 0.5
8.	Loose solids	Beds of sand or diatomaceous earth	Submicron

particles indicates the use of machines based on perforated plates, while membranes must be used for the extremely fine (<0.005 μm) particles.

A further broad classification of media may be made on the basis of the basic mechanism involved in the separation of particles and fluids. Edge filters, perforated plates, plain-weave metal wire or monofilament cloths, and certain grades of paper generally effect the separation by a sieve-like action, where the particles are retained on the surface of the medium and pore penetration does not occur in a successful separation. Apart from particles that, by virtue of size, shape, or adhesive properties, become lodged in the pores of such filters, particles are generally sieved out or lost as bleeds. As may be envisaged, such filters are normally used with relatively coarse suspensions, since the creation of the necessary pore size to remove extremely fine particles would impose high resistance to fluid flow and lead to slow, uneconomic filtrations.

Small particles may be removed from a flowing fluid by internal deposition in the second class of medium which is characterized by felts, mats, pads, multifilament yarn cloths, and ceramics. Removal of particles may occur by several mechanisms including internal sieving, electrostatic attachment to internal fibers, and so on. Generally, the filtration models used in describing the performance of deep-bed sand filters may be considered to describe the action of felts and multifilament yarns, which will have application to suspensions of small particles, when the mean size of the latter is often smaller than the mean size of the filter pore.

Except where previous experience dictates the solution to a specific problem, practical filtration trials are used to test the suitability of a medium in meeting the following requirements:

1. An efficient retention of particulate matter with clear filtrate
2. Absence of medium blinding of a sudden or progressive character
3. Good cake-discharge properties

4. Adequate cleaning potential either by back-flushing or laundering
5. Physical strength and resistance to chemical attack
6. An economic filtration rate
7. Resistance to microorganisms

Quite often, certain of the above requirements are relaxed if, by so doing, other more important media properties can be made available, e.g., allowing a slight preliminary bleed of particles (recovered by recycling the filtrate) through media that have good nonblinding characteristics.

III. CRITERIA OF CHOICE

Purchas [6] suggests three criteria by which a filter medium may be judged:

1. What size of particle will be retained by the medium?
2. What is the permeability of the clean medium?
3. What is the solids-holding capacity of the medium and the resistance to fluid flow of the used medium?

Along with these general questions, information is necessary on the physical state (particle concentration, temperature, etc.) of the suspension and its chemical constitution. In some cases, the desirable component in the slurry is the liquid, which may be required in clarified form; here the particles may be regarded as valueless and the choice of a deep-bed element of large solids-holding capacity may be indicated. On the other hand, where the solids are valuable, a sieve-like mechanism is favored, so that information of the pore size of the medium may be of more direct use in cloth selection. Again, the picture is not entirely clear, since increases in particle concentration and fluid velocity can result in good solids removal, even in cases where the mean particle size of the solids is much less than the pore, or aperture, of the medium.

The permeability of the clean medium will have importance in determining power requirements and in deciding the initial flow rate of fluid through the septum. It has been observed [7] that initial flow rates can influence the structure of the filter cake, resulting in changes in the specific resistance of the deposit. This phenomenon is dealt with in some detail in a later section. The third question is, perhaps, the most serious one since, having chosen a medium that adequately retains the particles at minimum power consumption, failure to release such solids by back-flushing, or the onset of a large increase in resistance to fluid flow, will have serious consequences in a filtration process. Apart from disposable-cartridge filters, a medium that after one usage does not respond to a cleaning cycle or cake-discharge operation must be considered a process failure, despite the fact that adequate filterability and pressure drops were experienced in the short period of medium life.

IV. MEASUREMENT OF PORE SIZE AND PARTICLE RETENTION

Although it has been suggested above that it is doubtful whether information on the pore size of a medium will be of much use in finalizing media selection, there is obviously great value in having such information since, in the limit, the pore structure of the medium will determine the feasibility of a separation. As will be discussed below, there is a maximum pore size in the medium across which it is impossible to bridge with a certain particle size, irrespective of the slurry concentration and particle shape. Therefore, the pore size of a medium—particularly for filters of the edge, perforated, simple wire, or monofilament type—is of use in deciding the upper limit of aperture size required by a particular process. In filters composed of random fibers, sintered or porous elements, staple or natural fiber cloths, the mean pore size will have less significance and use in predicting media behavior.

In certain cases (perforated plate and simple weaves) the geometry of the septum allows the use of direct microscopic measurement of the aperture or pore size. In more random situations, or where complex weave patterns (twills, sateens) produce a distribution of pore size, other techniques such as a bubble point test or a permeability test are used. In the latter the pressure drop concomitant with the flow of a fluid, usually water or air, is measured and the data are used to calculate a pore size by means of a theoretical relationship such as the Kozeny-Carman equation. In the bubble point test [8] a sample of medium is submerged in a wetting liquid and the air pressure necessary to force air bubbles through the fabric is recorded. The radius of the pore (r_{bp}) is then calculated from:

$$r_{bp} = \frac{2\sigma}{P} \tag{1}$$

where σ is the surface tension of the fluid and P the bubble pressure. Certain difficulties arise in the interpretation of results when the test is applied to systems possessing a distribution of pore sizes. It has been necessary, in such cases, to use the pressure at which the third bubble of air appears on the surface of the specimen. Careful measurements [9] have shown that when applied to woven media the three methods show a reasonable degree of correlation for woven monofilaments, and the equations

$$r_c = 1.26r_p \tag{2}$$

$$r_{bp} = 1.58r_p \tag{3}$$

have been shown to relate r_p, the pore radius calculated from permeability measurements, to the bubble pressure r_{bp} and the pore radius determined by microscopic inspection of the cloth construction (r_c).

When applied to multifilament yarns, no simple relationships are available since the permeability of the yarns is not accounted for by inspection or a bubble test; the latter tends to measure the larger interyarn pores and gives pore-size results that are larger than inferred by permeability tests. Thus tests of the bubble-point type are of direct use in evaluating simple, sieve-like media but are less useful in correlating pore sizes from different methods in deep-bed elements.

The presence of interfiber pores must be taken into account in any attempts to interpret the behavior of woven material composed of multifilament yarns. At the present time, no method is available for direct measurements of interfiber pores since the technique of mercury intrusion [10], useful in media of a rigid structure, would tend to destroy the pore structure in nonrigid media. However, by assuming that the Kozeny-Carman equation (Eq. 25, see later) applies to flow through and around yarns, it is possible to calculate the average pore size in multifilament media; in all cases the interfiber pore radius r_f is less than the measured, overall permeability radius which, in turn, is less than the calculated interyarn radius r_y, i.e., $r_f < r_p < r_y$.

While measurements of the above type cannot be used in rigorous predictions of the particle stopping power of the medium or its capacity for particle retention, a knowledge of pore size may be used in a limited way in predicting the course of particle deposition and in quality control tests on media. Zievers [11] reports a fast, commercially accurate method of determining the size of particle retained by paper, felts, and nonwoven fabrics; this work also deals with the relationships between permeability of clean media to air and liquids and the retentivity of particles.

Media efficiency tests are made using dilute suspensions of particles, which are passed through the medium under inspection; the concentration of particles in the fluid before and after passing through the medium is measured. Great care is necessary in such work, where attention must be given to adequate particle sampling and dispersion.

In the liquid field, a wide variety of "standard" powders have been used for testing. Generally, the various industries tend to use a test mixture typical of the products of interest, e.g., yeast in the brewing industry.

Air filtration tests are, perhaps, more stringent since smaller particles must be stopped by the filter. The basic mechanisms of particle removal here are: direct interception, gravitational settling, inertial impact, diffusion, and electrostatic attraction.

In a full test range, it is often necessary to use several types of test particles. Examples of the latter are: sodium chloride (0.03-5.0 μm); methylene blue (0.01-1.5 μm); dioctylphthalate (0.3 μm). The methylene blue test has been included in British Standard 2831. Care must be taken in reporting filter efficiencies to record the basis upon which the efficiency is calculated. Efficiencies are generally expressed in terms of the percentage of material removed from the impinging fluid stream, and the percentage may be calculated on a weight basis or a particle number basis. Efficiencies calculated on the latter basis will generally be lower, and such information has little value unless a complete description of the particle size distribution of the solids is available.

V. PERMEABILITY OF CLEAN FILTER MEDIA

As was pointed out in Sec. 3.III the permeability of the clean medium has a direct influence on the pressure losses occurring during filtration, and also on the fluid flow conditions during the building up of the first layers of filter cake. Generally, low media resistance, if blinding is absent, will be associated with low cake resistances, since an open, porous structure tends to be stabilized by higher fluid flows [7]. Grace [12] reports that where successful cake filtration is obtained, the medium resistance should be a low percentage (< 10%) of the mean filter-cake resistance. Tiller [13] points out the importance of cloth-resistance variations on flow conditions within the cake and, subsequently, on the level of cake resistance occurring during filtration. Apparently a good medium choice should yield an equivalent thickness of cake L in the range, $0.02 < L < 0.15$.

It is important, therefore, to have quantitative information on the permeability level of a medium and, more importantly, on the effect of media structure on flow resistance. It would be a great convenience if the many filter media could be described by an easily measured parameter which could form the basis of a system of media indexing. This index might conceivably be linked to particle properties, so that the user could make a reasonably accurate estimate of the probability of blinding, bleeding, and overall system resistance. An index of this type has been postulated for use with woven cloths of monofilament and multifilament yarns [14].

Although many publications report the permeability of media, the variety of units—and the relatively few attempts to relate permeability and media structure—results in the available data being of little use in promoting an understanding of the behavior of the medium as a filter. Most recent data are reported in terms of the basic unit, the Darcy. The latter is defined by: A 1-cm thick sample of a medium possessing a permeability of 1 darcy will pass 1 cm^3/sec of fluid through a filter area of cm^2 under a differential pressure of 1 atm; the fluid will have a viscosity of 1 Cp, and the flow conditions must be streamline. The latter conditions are usually interpreted as producing a linear relationship between the pressure drop and fluid velocity.

Although much information is available in this form, the constraint of low Reynolds numbers, i.e., streamline flow, makes the use of such units of limited applicability to many users of coarser media. Heertjes [15] has reported some information of the onset of turbulence in woven media. For streamline conditions the Reynolds number ($\rho u_p d_p/\mu$) is proportional to ΔP, the pressure drop across the filter, i.e.,

$$Re = \left(\frac{\rho d_p^3}{32L\mu^2}\right)\Delta P \tag{4}$$

and is proportional to the 0.55th power of the pressure drop for turbulent flow

$$Re = \left(\frac{\rho d_p^3}{32L\mu^2}\right)\Delta P^{0.55} \tag{5}$$

where d_p is the pore diameter, L the thickness of the filter medium, ρ the fluid density, and μ the fluid viscosity. A transition region for certain monofilament woven filters and perforated plates lies in in the range $3 < Re < 7$ [15].

It will be realized that deposition of solids may cause flow transitions from turbulent to streamline in coarse filters. Recent reports [14] in deposition studies have concentrated on those systems where the flow required is streamline before and after deposition, thus avoiding the complications associated with turbulence effects.

As this influence of the medium is felt to its greatest extent during the initial period of the filtration process, it is convenient to subdivide the deposition into two stages: (1) the period prior to the deposition of solids, and (2) the impingement of solids on the medium, with the subsequent formation of filter cake.

It is the purpose of this section to give as broad a picture as possible of the various attempts that have been made to evaluate the permeability of clean media and, in particular, to highlight the relationship between media structure and behavior permeability. The behavior of the medium subsequent to the deposition of solids will be dealt with in a later section. Here, most of the results discussed refer to woven media, although certain conclusions and results have direct application to random structures.

A. Woven Multifilament Filter Media

The chief difficulty encountered when dealing with multifilament media is the highly complex geometry of the fibers and yarns that make up the cloth. Even in a fabric of apparently simple weave and construction, such as a plain-weave, continuous-filament cloth, most of the flow takes place in the highly tortuous channels present in the yarns. The fact that these flow channels were of too complicated a nature for complete hydrodynamic analysis led to many of the early workers [16,17] simplifying the problem by considering the case of a monofilament weave, as represented by a metallic screen. They were then able to view the fabric as a flow barrier, consisting of a large number of square orifices or nozzles forming the flow channels. Robertson [17] obtained a good correlation between the two dimensionless variables, log C_D and log Re, for air flow through plain-weave, metallic meshes. The discharge coefficient C_D and the Reynolds number were defined as

$$C_D = \frac{Q}{A_2\left[\frac{2\ \Delta P}{\rho(1 - A_3^2)}\right]^{1/2}} \tag{6a}$$

$$Re = \frac{u_p d_p \rho}{\mu} \tag{6b}$$

where Q is the volumetric flow rate through an orifice, ΔP the pressure drop across the medium, A_2 the orifice projected open area, A_3 the ratio of the orifice projected open area to the upstream flow channel area, d_p the width of a square orifice, ρ the fluid density, and μ the fluid viscosity.

Robertson's [18] attempt to extend this approach to loosely woven cloths of a more complex nature failed badly; the author attributed this failure to the effect of weave on pore construction. Grace [19] suggested that the failure resulted from the fact that in a multifilament cloth—especially one that was very tightly woven—flow is possible both in and around the yarns. While the orifice analogy approach must be viewed as a failure, the information it provided of the effect of weave pattern, etc., on pore construction proved extremely useful in the case of flow through monofilament fabrics. This work is described in more detail below in the section dealing with monofilament filter media.

Provided that laminar flow conditions prevail, the flow of fluid through a filter medium may be described by the Darcy equation

$$\frac{1}{A}\frac{dV}{dt} = u = \frac{B\ \Delta P}{\mu L} \tag{7}$$

where A is the area of the filter medium, dV/dt the volumetric flow rate, and B the specific permeability of the filter medium.

The outstanding problem is to link B, the specific permeability of the medium, with its geometric structure. For the flow of air through fibrous pads, Davies [20] reported that the correlation

$$\frac{\mu u L}{\Delta P} = B = \frac{d_f^2}{64(1 - \varepsilon)^{1.5}[1 + 56(1 - \varepsilon)^3]} \tag{8}$$

(where d_f is the fiber diameter and ε the porosity of the fibrous pad) represented the results in the range $0.6 < \varepsilon < 1.0$, provided that the Reynolds number of flow with respect to the fiber is less than unity. Turbulence develops within the interstices of the filter at high air velocities and Eq. (8) is no longer valid. Agreement with the general form of the Davies equation has been reported by several other workers [21-24]. Recognizing that Eq. (8) was evidently successful in predicting the resistance to water flow of fibrous pads, Rushton et al. [25] attempted to extend its use to woven multifilament media. Figure 1 shows a logarithmic plot of the two dimensionless variables $\Delta P d_f^2/\mu u L$ vs. $1 - \varepsilon$. A correlation was produced only by those cloths of a tightly woven nature, namely, plain-weave, continuous filaments where interfiber flow predominated, although the resistance of the cloths was several times greater than that predicted by the Davies equation. The authors attributed this to the fact that the tightness of the weave gave rise to zones of compression at the points of contact between the yarns, thus creating a porosity that is far less than the average for the complete assembly of the cloth. For these cloths the authors suggested the relationship

$$B = \frac{d_f^2 \times 10^{-5}}{2(1 - \varepsilon)^{6.5}} \tag{9}$$

Also included in the study were a number of twill weave cloths composed of staple fibers. As the surface of the constituent yarns were of a hairy nature, a high degree of particle retention could be achieved with a relatively open weave. Therefore, these cloths are characterized by large interyarn pores, with a high proportion of the flow taking place around the yarns. The resultant data for these cloths showed a high degree of scatter which was not improved significantly by the substitution of yarn diameter d_y, for fiber

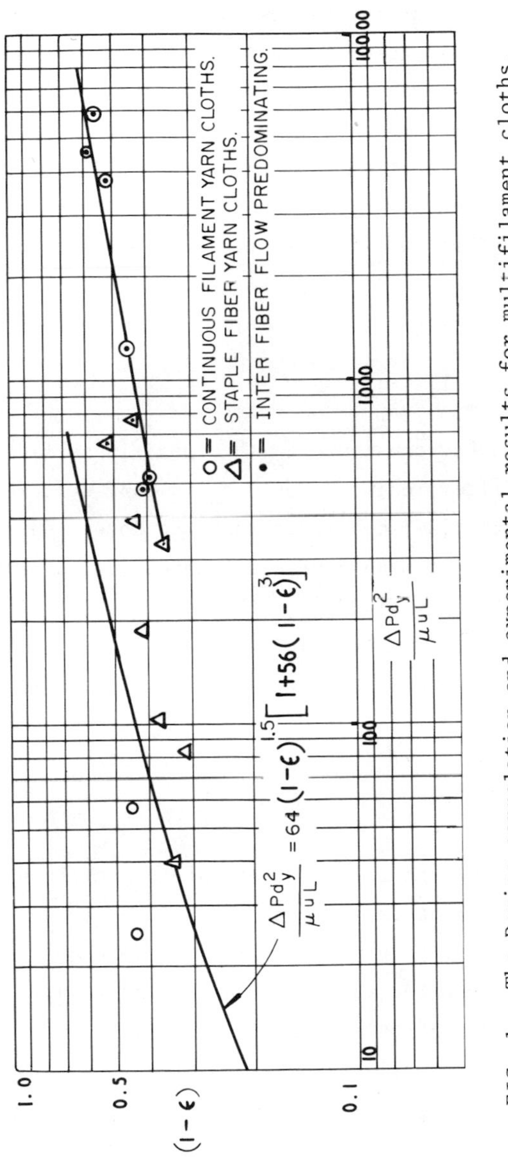

FIG. 1. The Davies correlation and experimental results for multifilament cloths.

diameter d_f in Eq. (8). However, the authors state that a roughness factor, which takes into account the hairy nature of the yarns, must be determined before the data may be correlated on a yarn diameter basis. Therefore, use of the Davies equation is questionable in the absence of prior knowledge of the division of flow within the cloth. In the case of monofilament media—where the problem of division of flow does not occur—the Davies equation has been used with limited success [25] (see the section on monofilament filter media).

The best-known equation for describing the specific permeability of a media in terms of its structural properties is the Kozeny-Carman equation:

$$\frac{\mu u L}{\Delta P} = B = \frac{1}{K_o S_o^2} \frac{\varepsilon^3}{(1 - \varepsilon)^2} \tag{10}$$

where K_o is the Kozeny constant, S_o the specific surface, and ε the porosity of the filter medium. Despite its many well-known limitations [10], McGregor [26] concluded that the Kozeny-Carman theory, based on the concept of mean hydraulic radius, could provide a satisfactory description of fluid flow through woven textile media, provided that the flow was laminar and that the permeability of the yarns themselves was taken into account. Equation (10) may be rewritten as

$$K_o = \frac{1}{B S_o^2} \frac{\varepsilon^3}{(1 - \varepsilon)^2} \tag{11}$$

and K_o should be constant for an assemblage of particles or fibers of given specific surface. McGregor summarized the work of several authors; Fig. 2 shows the experimentally determined values of the Kozeny constant K_o plotted as a function of the total porosity of assemblies of fibers and cylinders.

McGregor made the following observations:

1. Assemblies of fibers or cylinders give lower values of K_o when flow is parallel to their long axes than when it is perpendicular to them. Sullivan and Hertel [27] gives $K_o = 6.04$ for flow normal to the long axes of parallel-laid glass fibers and $K_o = 3.07$ for flow along the axes.

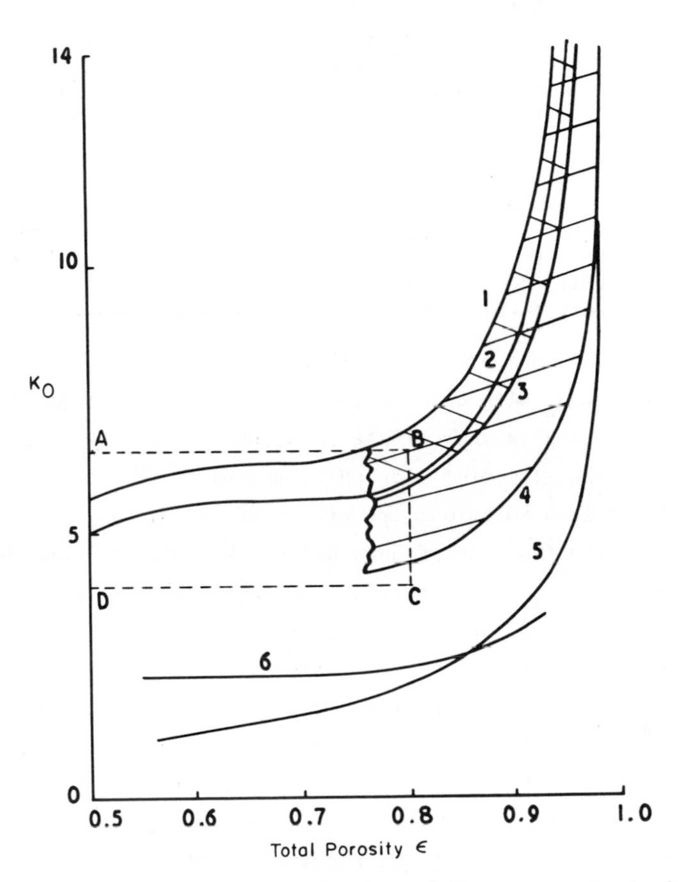

FIG. 2. Experimental variation of Kozeny constant with total porosity for assemblies of fibers and cylinders. Values of Kozeny constant for measurements at lower porosities fall in rectangle ABCD [64,65]. Area marked \\ is for fibers that tend to lie parallel and normal to flow [40]. Area marked // is for more random assemblies of fibers [19,23,41]. Curve 1. Davies [20]; random assemblies of fibers. Curve 2. Ingmanson et al. [24]; random assemblies of fibers. Curve 3. Lord [43]; random assemblies of wool fibers. Curve 4. Lord [43]; random assemblies of cotton fibers. Curve 5. Sullivan [42]; cylinders parallel to flow. Curve 6. Sullivan [29]; cotton fibers parallel to flow.

2. Low values of K_o are associated with nonuniform packing of fibers because chaneling of a high percentage of the fluid occurs in the larger pores.
3. For random wads of fibers and assemblies of fibers arranged perpendicular to the flow of air and water, K_o is fairly constant in the porosity range $0.5 < \varepsilon < 0.8$, having a value in the range $4.0 < K_o < 6.5$ with 5.5 as an average value.
4. For values of porosity greater than 0.8, the permeability is usually less than that predicted by the Kozeny-Carman equation. High values of K_o are obtained; K_o is no longer constant but becomes a function of porosity. This can be attributed to the change in flow mechanism from one of capillary flow to one of free flow around cylinders [10].

McGregor also summarized the theoretical investigations into the variation of K_o with total porosity. These theoretical considerations were based on approximate solutions of the Navier-Stokes equations for the flow of fluids past assemblies of finite bodies; the results of such solutions are presented graphically in Fig. 3. Inspection of the two sets of data allowed McGregor to consider the validity of the various theoretical approaches when compared with the experimental evidence. The validity of Emersleben's [28] solution for flow parallel to the long axes of a square array of cylinders becomes less as the porosity decreases because it is based on the assumption that the cylinders are widely spaced. Consequently, the agreement between Emersleben's theory and experimental measurements is best for high porosities; such agreement as exists for lower values of porosity is with results for flow parallel to the long axes of fibers [29], as would be expected. The results of Sparrow and Loeffler [30] and Happel and Brenner [31] for flow along the axes of cylinders give results that are substantially higher than those calculated from experimental data for similar conditions. The discrepancy is probably due to the difficulty in producing an even distribution of the filament packing, low values of K_o being associated with nonuniform packing of the fibers.

Results obtained by Happel and Brenner [31] for the flow of fluid normal to an arrangement of cylinders show good agreement with the experimental results for flow perpendicular to the long axes of

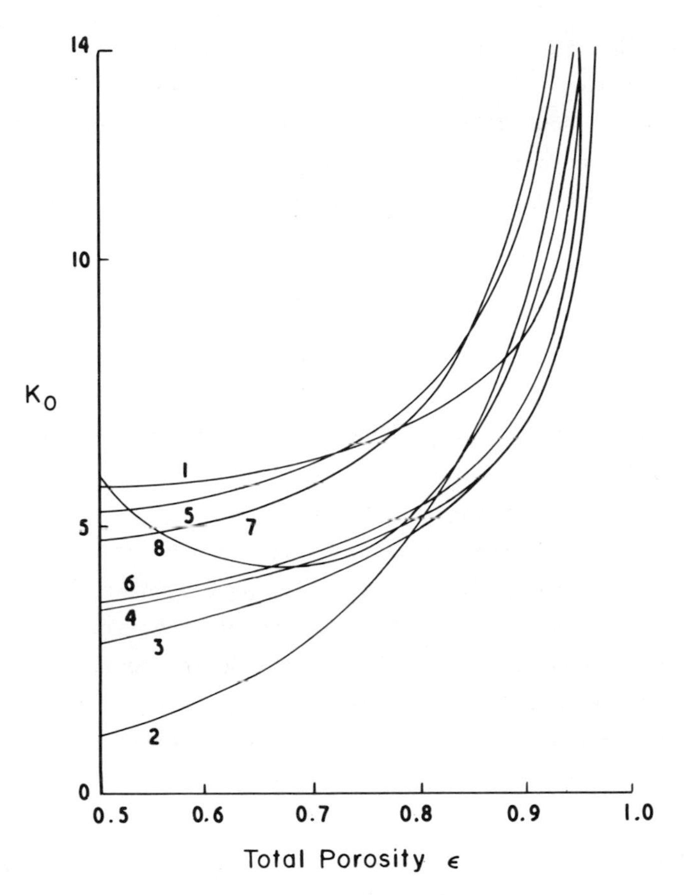

FIG. 3. The theoretical variation of Kozeny constant with total porosity. Curve 1. McGregor [26]; cylinders normal to flow. Curve 2. Emersleben [28]; cylinders parallel to flow. Curve 3. Sparrow and Loeffler [30]; cylinders parallel to flow (square array). Curve 4. Sparrow and Loeffler [30]; cylinders parallel to flow (triangular array). Curve 5. Happel and Brenner [31]; cylinders normal to flow. Curve 6. Happel and Brenner [31]; cylinders parallel to flow. Curve 7. Happel and Brenner [31]; spheres. Curve 8. Brinkman [32]; spheres.

fibers. Curve 7 in Fig. 3, for flow through an assemblage of spheres, is of the same form; while Brinkman's [32] solution for flow through a package of impermeable spheres shows reasonable agreement with experiment for flow normal to a bed of fibers, if the porosity of the bed is greater than about 0.6. McGregor [26] suggested that in order to describe the flow through an assembly of permeable yarns, such as is found in a woven textile cloth, it is reasonable to adopt Brinkman's [33] approximate solution to the problem of calculating the fluid drag on an assembly of permeable spheres for such a purpose, and the Brinkman equation becomes

$$\beta = \left(\frac{B}{B_1}\right) = 1 + 1.80\left(\frac{B_o}{B_1}\right) + 2.68\left(\frac{B_o}{B_1}\right)^{1/2}, \text{ for } \frac{B_o}{d_y^2} < 0.0017 \qquad (12)$$

where B is the total specific permeability of the package, B_o the specific permeability of the yarns of the package, and B_1 the specific permeability of the package if the yarns were impermeable cylinders.

In order to test the validity of Eq. (12), McGregor used Denton's [34] data for the air permeability of a cross-wound, cotton-yarn package. Denton found that at low package densities the flow through such a package could be satisfactorily described by the Kozeny-Carman equation, Eq. (10), if the Kozeny constant K_o was assumed to have a value of 5.0 and the yarns were considered to be impermeable cylinders. Less satisfactory agreement was found at higher package densities, as the proportion of fluid flowing via the yarns increased and the specific permeability became greater than that predicted by the Kozeny-Carman equation. However, McGregor showed that the permeability of the cotton-yarn package at high package density, as predicted by Eq. (12), was substantially in agreement with the experimental variation of permeability with package density found by Denton. These results are shown in Fig. 4. The data in Fig. 5 illustrate how the percentage of the total flow that passes via the yarns of the package (i.e., interfiber flow) increases as the interyarn porosity decreases. The actual volume flow rate of liquid is probably smaller because the overall permeability of the package decreases. The applicability of the Kozeny expression to situations of mixed particulates and preferential flow paths [35] has been reported. Despite the poor representation of permeability in such situations, the Kozeny expression is still universally used for estimations of flow rates in media.

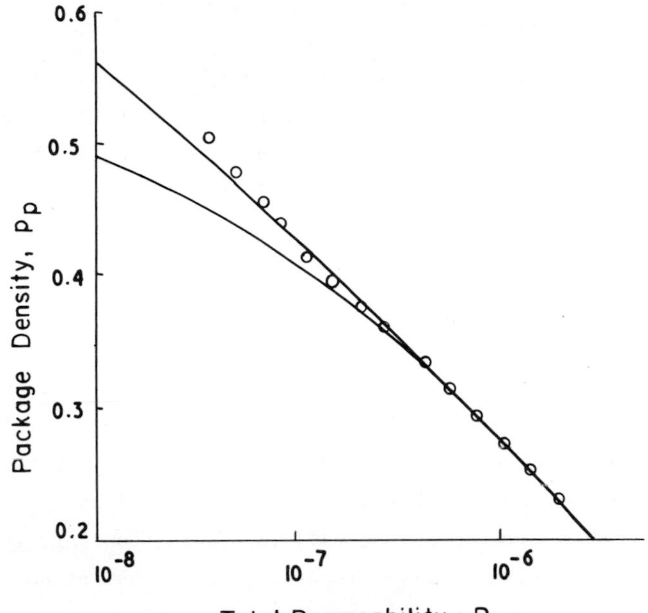

FIG. 4. The effect of package density on the specific permeability of a cross-wound, cotton-yarn package. The lower line is calculated from the Kozeny-Carman equation for flow around impermeable yarns. The upper line is calculated taking yarn permeability into account [26].

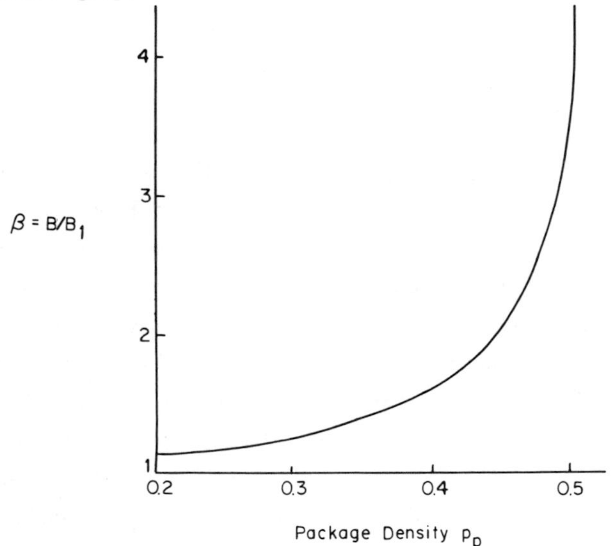

FIG. 5. The effect of package density on the division of flow through a cross-wound, cotton-yarn package.

B. Nonwoven Filter Media

In the previous section mention was made of the disparity that existed between the large volume of literature related to the filtration performance of woven media and the relatively small amount devoted to the link between the structure and permeability of these materials. This is also the case with nonwoven filter media; whereas the manufacture and filtration characteristics have been extensively discussed, particularly by Wrotnowski [36-38], few attempts have been made to correlate pressure drop-flow data in terms of the structural properties of the medium. Attempts that have been made have consisted of a semi-empirical approach based on Darcy's Law [39] and use of a randomly packed bed analogy [40]. In this latter approach, use is made of the Davies equation mentioned in the previous section. A recent method [41] for fluid flow through high porosity fibrous beds is also discussed, though as yet no attempt has been made to apply this technique to the felted materials more commonly found in filtration practice.

For felted materials and airflow, Cunningham et al. [39] stated that the pressure drop ΔP could be represented by the viscous flow equation

$$\Delta P = k\mu u \tag{13}$$

which is an amended form of the Darcy equation, where u is the superficial velocity based on the cross-sectional flow area of the sample holder, and μ the fluid viscosity. The proportionality constant k is peculiar to each material, as it is dependent on the geometrical configuration of the cloth (length and shape of flow channels, porosity, etc.). Furthermore, Cunningham and his coworkers found that for certain felted materials (wool, rayon, and cotton) k was approximately equal to $(4.29 \times 10^6)W_c$, where W_c is the cloth weight in grams per square centimeter. This correlation was recommended as an approximation of the air-flow resistance of wool, rayon, or cotton-felted materials of structure similar to those tested. For a limited number of materials, flow data on water and oils of varying viscosity was obtained. Flow was again found to be generally represented by the viscous flow equation, though an attempt to correlate the data by means of a friction factor-Reynolds number type plot failed, the authors attributing this to wetting effects and experimental errors.

The use of the Davies equation, Eq. (8), to predict the flow resistance of woven, multifilament media was severely handicapped by the fact that flow was possible both in and around the yarns [25], though this problem does not arise in the case of nonwoven media as the constituent fibers are impermeable. The possibility of using Eq. (8) to express the permeability of a felted material in terms of its fiber diameter and cloth porosity has been examined by Griffiths [40]. For water flow through uncalendered wool felts, both the experimentally determined resistances and those predicted by Eq. (8) were found to be in agreement, though the limited number of cloths studied prevented firm conclusions being made.

Kyan et al. [41] have recently proposed a pore model for the flow of a single-phase fluid through a bed of random fibers. This model incorporates an effective pore number to account for the influence of dead space on flow; also, a deflection number is used to characterize the effect of fiber deflection on pressure drop. A generalized friction factor-Reynolds number equation is presented, which successfully correlated the data obtained from the flow of several fluids (viscosity range 1-22 Cp) through fibrous beds composed of glass, nylon, and Dacron fibers of 8-28 μm diameter. The authors also showed that the Kozeny constant K_0 was not only a function of the fiber orientation and porosity—as had previously been found by other workers [20,27,42,43]—but was also strongly dependent on both the Reynolds number and the deflection number. On this basis they concluded that the usual one-term Kozeny-Carman equation, Eq. (10), was not applicable for the flow of single-phase fluids through fibrous beds.

Although this technique is evidently successful in predicting the flow resistance of high porosity beds (the porosity range studied was 0.682-0.919), it remains to be seen whether this method can be applied to the lower porosity felts found in common usage.

C. Monofilament Filter Media

The methods used to predict the permeability of monofilament media are similar to those discussed in the preceding two sections, namely, use of an orifice analogy, a randomly packed bed analogy, and use of drag theory.

In view of the fact that the yarns of a monofilament fabric are not only impermeable but also fairly smooth and cylindrical, it is not surprising that the orifice analogy and drag theory approaches have been the most successful in predicting the resistance of these materials to fluid flow. Indeed, when the present state of the art in the monofilament field is compared with that in both the multifilament and nonwoven fields, it may be safely concluded that the link between the permeability and the structural properties of a filter medium has only been successfully established in the case of these monofilament fabrics. It is interesting to note that Backer [44] considered a full understanding of the nature of fluid through monofilament fabrics to be essential before any meaningful attack could be made on the problem of flow through multifilament cloths.

1. *Orifice Analogy*

Much of the early work in this field resulted from investigations into the permeability of multifilament media. In order to simplify the problem, the yarns of a loosely woven multifilament cloth were assumed to be impermeable; thus the cloth could be considered to be a flow barrier, with the interyarn pores forming the flow channels.

Simplifying the problem even further, metallic screens and meshes were taken to represent idealized fabrics, with the orifices or nozzles taking the place of the flow channels. Robertson [17] obtained a good correlation between the Reynolds number and discharge coefficient (see Section 3.V.A) for airflow through plain-weave metallic meshes, though the technique failed badly when

Robertson [18] applied it to loosely woven multifilament cloths of differing weave pattern.

Backer [44] suggested that this failure was due to the fact that the pores of a multifilament cloth, especially twill-type pores, were of too complicated a nature to be characterized by the simple projected open area used in the Robertson Reynolds number discharge-coefficient correlation. Backer considered that the factor governing the flow rate to the greatest extent was the minimum pore cross-sectional area. By using scale models and mechanical integration, Backer was able to determine these minimum pore areas for the four pore types possible in a textile fabric. These pore types are shown in Fig. 6. Although some difficulty was encountered when the dimensions were scaled down to realistic fabric proportions, Backer was able to show that use of a minimum pore area greatly

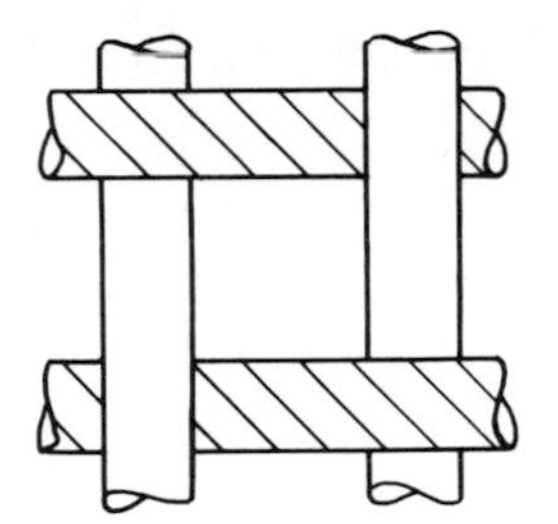

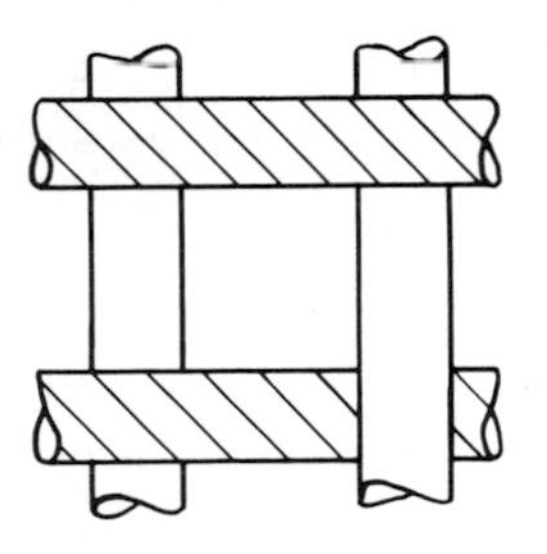

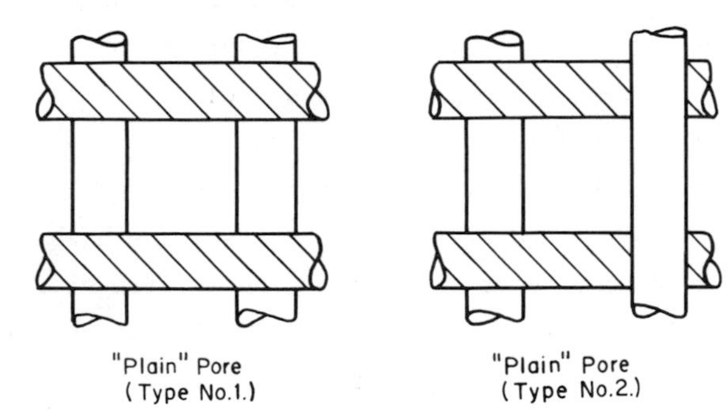

FIG. 6. Four types of pore patterns. \\ denotes warp yarn.

reduced the scatter produced by calculations based on the projected open area.

Penner and Robertson [45] took the study a stage further when they suggested that, ultimately, the rate of fluid flowing through a fabric pore was essentially dependent on the cross-sectional area of the stream-tube or jet and the velocity of the flowing fluid. A visual study was performed, in which a tracer technique was used to study the flow of oil through Plexiglass models of the four pore types depicted in Fig. 6. By means of this technique they were able to determine the minimum jet cross-sectional area (M.J.A.) at a known Reynolds number; furthermore, these values of the M.J.A. could be compared to the associated values of the minimum pore cross-sectional area (M.P.A.) and the projected open area (P.A.). Their results confirmed those of earlier workers; a sateen-type pore was found to have 30% more jet area than a plain-type pore, a similar relationship having been found by Backer [44] for the minimum pore areas of these two pore types. The most important result was that, for a given value of the Reynolds number, the ratio M.J.A./P.A. varied, whereas the ratio M.J.A./M.P.A. was approximately constant and equal to 0.8. As the ratio M.J.A./P.A. varied among the four pore types, the authors concluded that to compare fabrics of differing weave on the basis of projected open area was erroneous, as the need for geometric similarity was ignored.

The work mentioned above, though useful for the information it provided on pore shape, area, and construction, was directed mainly toward the multifilament field. This situation obtained until Pedersen [46] published a paper devoted exclusively to flow in monofilament fabrics. Pedersen approached the problem in a similar fashion to that of the early workers; the fabric was considered to be a collection of irregularly shaped holes through which flow could take place, although, by analogy with flow in irregularly shaped ducts, each orifice was characterized by an "effective area" and an "effective diameter." For a monofilament fabric the "effective diameter" D was taken to be 4 (area for flow/wetted perimeter). Thus

$D = 4A_o/W$, with A_o being the area available for flow where flow is most constricted and W the perimeter of the orifice where the flow is most constricted. Pedersen then incorporated the five controlled variables in a simple monofilament fabric into the two significant variables A_o and W, which characterize the orifice. The three controlled variables are: the end count (ec), in other words, the number of warp yarns per unit length, and the warp yarn diameter d_2; the pick count (pc), i.e., the number of weft yarns (filling) per unit length, and the weft yarn diameter d_1; and the weave.

The discharge coefficient used by Pedersen was defined as

$$C_D = \left[\frac{\rho u^2(1 - \alpha_o^2)}{2\ \Delta P\ \alpha_o^2}\right]^{1/2} \tag{14}$$

where ΔP is the pressure difference across the medium, u the superficial or approach velocity, and ρ the fluid density. The effective fraction open area α_o is given by

$$\alpha_o = A_o(ec)(pc) \tag{15}$$

where A_o is the area available for flow where flow is most constricted.

The discharge coefficient C_D was anticipated to be a function of the Reynolds number within the fabric, i.e.,

$$C_D = f\left(\frac{Du_p\rho}{\mu}\right) \tag{16}$$

where D is the "effective diameter" of an orifice, u_p the mean velocity in the orifice, ρ the fluid density, and μ the fluid viscosity.

Since $D = 4A_o/W$ and the effective fluid velocity in the fabric $u_p = u/\alpha_o$, Eq. (16) becomes

$$C_D = f\left[\frac{4\rho u}{W(ec)(pc)\mu}\right] \tag{17}$$

where W is the wetted perimeter of the orifice.

In order that a completely rigorous determination of the geometric configuration of any fabric could be made, Pedersen had to make assumptions that filling yarns are straight, weaving is perfect, yarns are cylindrical, warp yarns are perfectly straight between filling yarns, and flow through an orifice is independent of the presence of other orifices.

Although the five conditions listed above are never fully met in practice, the deviations from this imaginary state were considered to be small. Making these assumptions enabled Pedersen to determine A_o (and thus α_o) and W analytically for both a plain and a twill pore.

For a plain pore

$$\alpha_p = \frac{\phi}{2} \ln\left(\frac{1 + \sqrt{1 + \phi^2}}{\phi}\right) + \frac{\sqrt{1 + \phi^2}}{2\phi} - (ec)d_2 - (ec)(pc)d_1^2 \sqrt{\left(\frac{\frac{1}{ec} - d_2}{d_1}\right)^2 + 1} \tag{18}$$

and

$$W_p = 2\sqrt{\left(\frac{1}{ec} - d_2\right)^2 + d_1^2}\; E\left\{\left[1 + \left(\frac{d_1}{\frac{1}{ec} - d_2}\right)^2\right]^{-1/2}, \pi/2\right\} + \frac{2}{pc}\sqrt{1 + (d_1 + d_2)^2 (pc)^2} \tag{19}$$

For a twill pore

$$\alpha_T = \frac{\phi}{2} \ln\left(\frac{1 + \sqrt{1 + \phi^2}}{\phi}\right) + \frac{\sqrt{1 + \phi^2}}{2\phi} - (ec)d_2 - \frac{(ec)(pc)}{2}\left[d_1^2 \sqrt{\left(\frac{\frac{1}{ec} - d_2}{d_1}\right)^2 + 1} + d_1\left(\frac{1}{ec} - d_2\right)\right] \tag{20}$$

and

$$W_T = \sqrt{\left(\frac{1}{ec} - d_2\right)^2 + d_1^2}\; E\left\{\left[1 + \left(\frac{d_1}{\frac{1}{ec} - d_2}\right)^2\right]^{-1/2}, \pi/2\right\}$$

$$+ \left(\frac{1}{ec} - d_2\right) + \frac{1}{pc} + \frac{1}{pc}\sqrt{1 + (d_1 + d_2)^2(pc)^2} \tag{21}$$

where ϕ is the dimensionless group $[(ec)(d_1 + d_2)]^{-1}$.

Pedersen tested his theory by plotting log C_D and log Re for airflow through a wide range of plain 1/1 and 2/2 twill monofilaments. The correlation obtained was excellent, the usual error being in the range ±12%. It is interesting to note at this stage that Pedersen found that some of the data for the twill fabrics would have fallen ten times above his correlation if the projected area, rather than the effective open area α_o, had been used.

Rushton and Griffiths [47] extended Pedersen's technique to include water flow through 1/1 plain, 2/1, 2/2 twills, and 5/1 sateen-weave monofilaments. Pedersen confined his to 1/1 plain and 2/2 twills; these cloths consist solely of plain and twill-type pores, respectively. The situation becomes more difficult in the case of 2/1 and 5/1 weaves, as two different types of pores exist side by side in these fabrics. For these cloths, a unit cell was defined which, regardless of where it was placed on the fabric, always contained the two (or three) types of pores present in the same ratio. Thus, the unit cell depicted in Fig. 7 always contains six twill pores and three plain pores independent of position on the fabric. In this case

$$\alpha_o = \frac{2}{3}\alpha_T + \frac{1}{3}\alpha_p \tag{22}$$

where α_o can be taken to be an "average" effective fraction open area, α_T as an effective fractional open area of the twill-type pore (Fig. 6), and α_p as an effective fractional open area of the plain-type pore (Fig. 6).

Similarly, for the orifice perimeter

$$W = \frac{2}{3}W_T + \frac{1}{3}W_p \tag{23}$$

where W can be taken to be an "average" wetted perimeter, W_T as the wetted perimeter of the twill-type pore (Fig. 6), and W_p as the

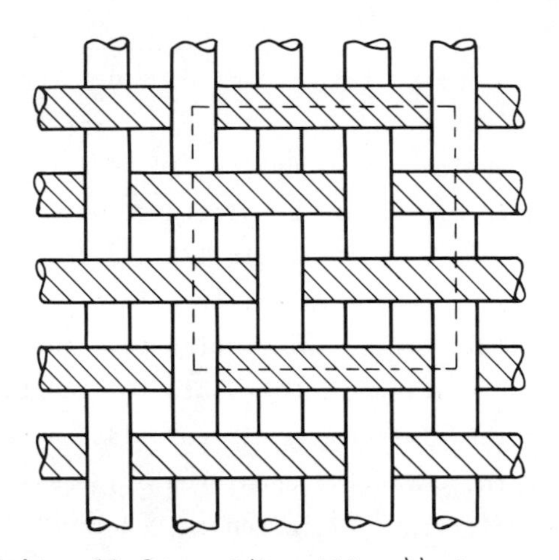

FIG. 7. Unit cell for a 2/1 twill. \\ denotes warp yarn.

wetted perimeter of the plain-type pore (Fig. 6).

A similar technique was used by Backer [44] to define the average minimum pore area of various twills. Figure 8 shows a plot of experimentally determined values of log C_D against log Re for water flow through several plain- and twill-weave fabrics. In the range $1 < Re < 10$ the equation

$$C_D = 0.17(Re)^{0.41} \tag{24}$$

may be used to predict the permeability to water flow of plain- and twill-weave fabrics with a maximum error of ±18%. The plain-weave cloths generally show a better degree of correlation than the twills. This is to be expected as the plain-weave cloths were as near as perfectly woven as possible, whereas the twill-weave cloths were somewhat irregular. Figure 9 shows the result of basing the correlation on projected fraction open area as opposed to effective fraction open area. Up to 100% deviations occur in the case of the twill weaves, though the plain-weave cloths are still correlated fairly well.

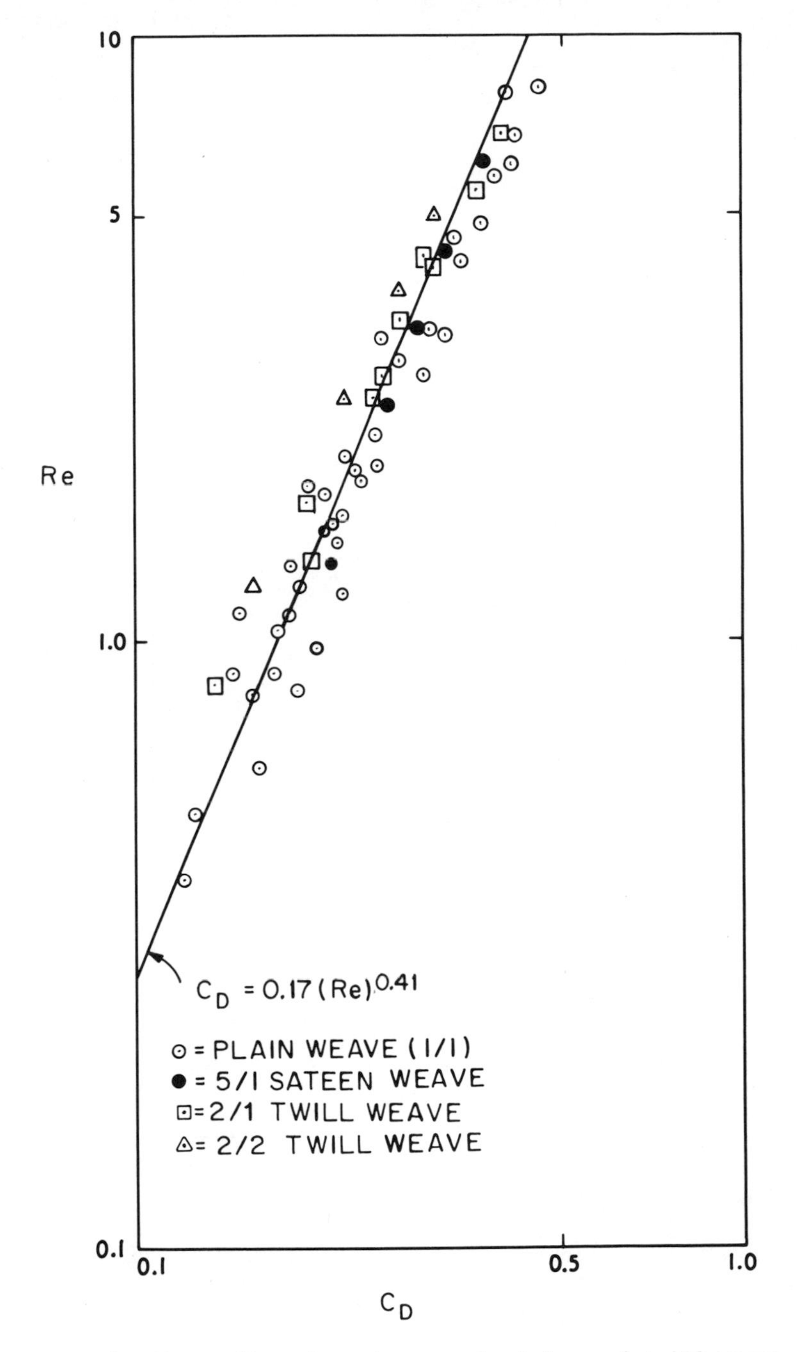

FIG. 8. Water flow through several plain- and twill-weave monofilament fabrics.

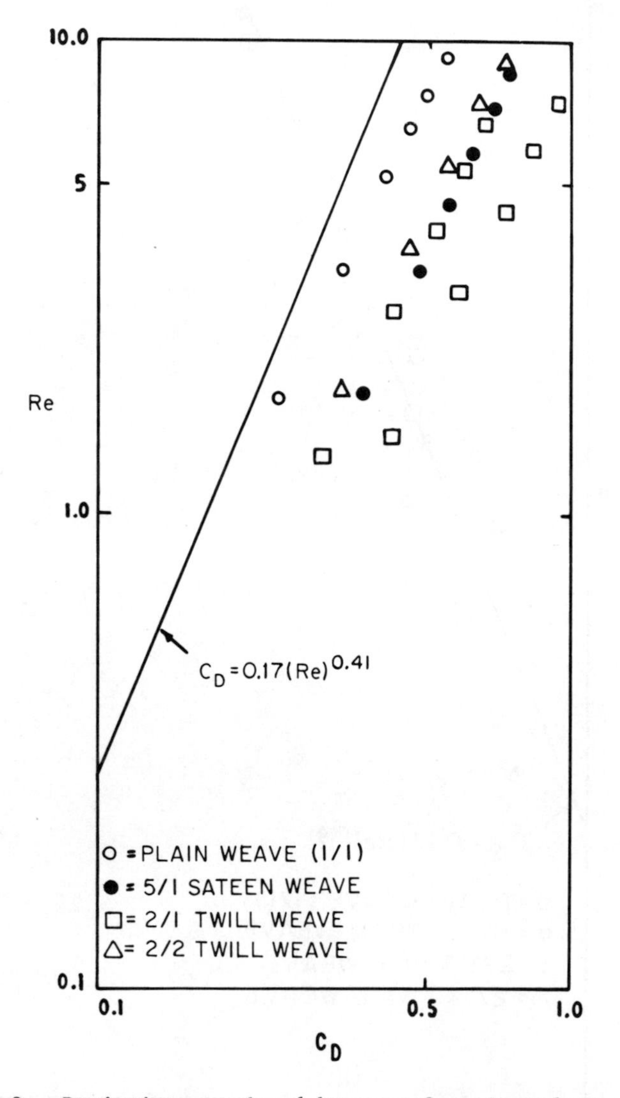

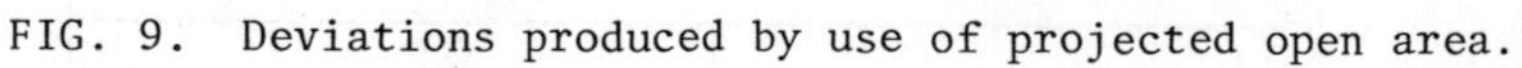
FIG. 9. Deviations produced by use of projected open area.

2. *Randomly Packed Bed Analogy*

It will be recalled that in the previous two sections, use was made of the Davies equation, Eq. (8), in attempts to predict the permeability to fluid of both woven multifilament and nonwoven filter media. The attempt failed badly in the case of woven multifilament cloths [25], because, in these particular materials, division of flow is possible both in and around the yarns. In the case of nonwoven media, the felted structure of which was more likely to approximate a randomly packed bed, the Davies equation was found to correlate the data quite well [40], although the limited number of cloths studied prevented any firm conclusions being drawn. The attempts [25,47] to predict the permeability of monofilament media by means of the Davies equation fall neatly between the two cases mentioned above, as no flow is possible through the yarns, and the cloth construction is of a highly symmetrical nature.

At this stage it is instructive to examine the derivation of Eq. (8) in some detail, as one of the main assumptions has particular relevance to monofilament media. The Kozeny-Carman equation may be written

$$B = \frac{\mu u L}{P} = \frac{1}{K_o S_o^2} \frac{\varepsilon^3}{(1 - \varepsilon)^2} \tag{25}$$

where B is the specific permeability of the medium, ΔP the pressure difference across the medium, L the thickness of the medium, ε the porosity of the medium, S_o the specific surface, K_o the Kozeny constant, u the superficial or approach velocity, and μ the fluid viscosity. Equation (25) may be simplified if the cloth yarns are regarded as smooth cylinders of diameter d_y, in which case $S_o = 4/d_y$ and the equation becomes

$$B = \frac{d_y^2}{16K_o} \frac{\varepsilon^3}{(1 - \varepsilon)^2} \tag{26}$$

McGregor [26] has shown that in the porosity range $0.5 < \varepsilon < 0.8$, K_o has an average value of 5.5, but at higher values of porosity

K_o increases rapidly. For airflow through random beds of fibrous materials of widely varying physical dimensions, Davies [20] showed that the relationship between K_o and ε may be represented by the equation

$$K_o = \frac{4\varepsilon^3}{(1 - \varepsilon)^{0.5}}[1 + 56(1 - \varepsilon)^3] \tag{27}$$

By utilizing Eqs. (26) and (27) the Davies equation may be written

$$\frac{\Delta P \, d_y^2}{\mu u L} = 64(1 - \varepsilon)^{1.5}[1 + 56(1 - \varepsilon)^3] \tag{28}$$

Rushton et al. [25] found that Eq. (28) satisfactorily predicted the water permeability of several plain-weave monofilament cloths; thus the assumption that $S_o = 4/d_y$ was valid. Extending the study to include twill- and sateen-weave monofilaments, Rushton and Griffiths [47] found that a high degree of scatter was produced by several of these fabrics, the results being depicted graphically in Fig. 10. No reasonable explanation can be given at the moment as to why plain- and twill-weave monofilament fabrics, of comparable porosity and yarn diameter, should differ so markedly in their degree of correlation. For this reason the attempt to extend the Davies equation to more complex monofilament cloths must be viewed as a failure.

3. *Drag Theory*

An alternative approach to those outlined in the previous two sections may be found in the "drag" theory of permeability, originally postulated by Emersleben [28] to provide a physical explanation of the permeability of porous media. In this approach the pore walls are treated as obstacles to the flow of a viscous fluid. The drag of the fluid on each portion of the walls is estimated using simplified forms of the Navier-Stokes equations, the sum of all the drags being taken to equal the flow resistance of the porous medium. Similar approaches have been made by several other workers [30-33,48-50].

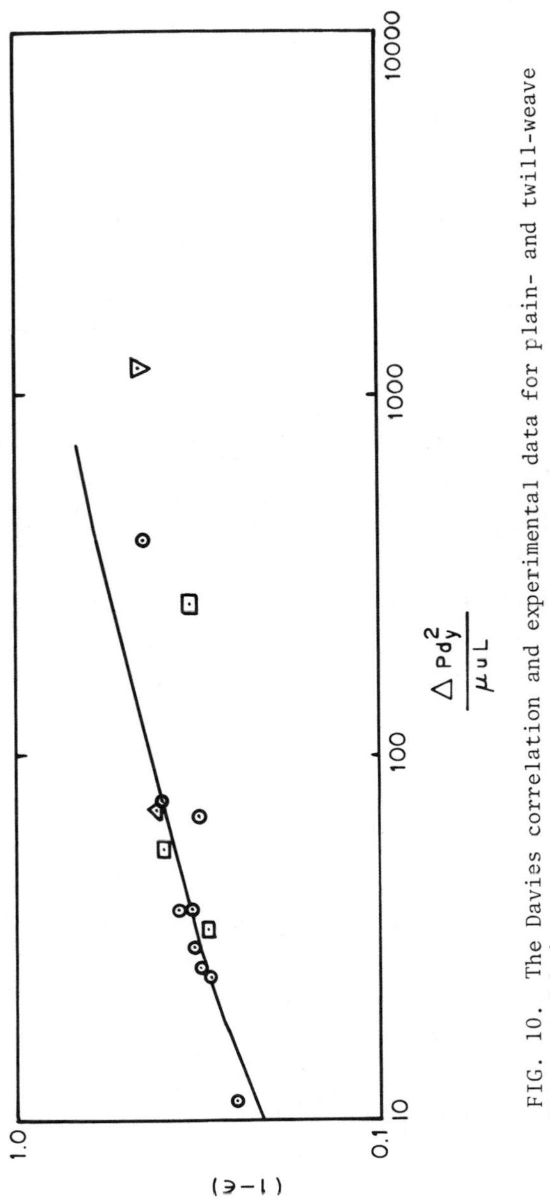

FIG. 10. The Davies correlation and experimental data for plain- and twill-weave monofilament fabrics.

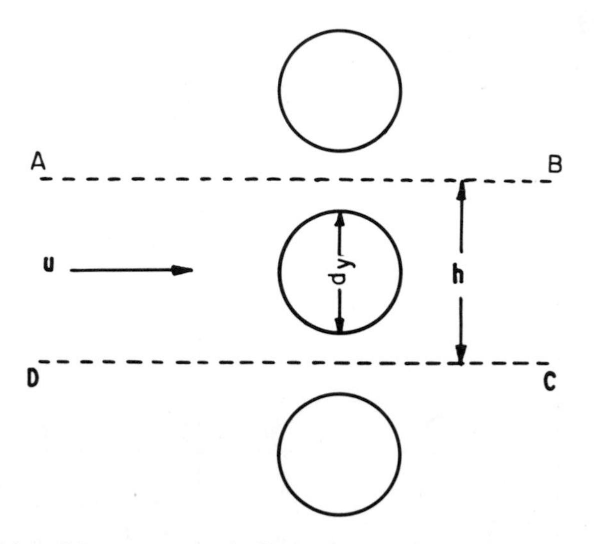

FIG. 11. Flow past an infinite row of circular cylinders. There is no flow perpendicular to the medial planes AB and CD.

Hutson [51] has recently modified Iberall's [48] model for fibrous porous materials to predict successfully the resistance of gauze filters to water flow. In Hutson's model (see Fig. 11) the woven material is replaced by two perpendicular rows of equal parallel cylinders. Using an approximate solution of the Stokes equations, Hutson obtained the expression

$$B = \frac{\mu u L}{\Delta P} = d_y^2 \frac{(1 - x)}{x\ f(x)} \tag{29}$$

for the permeability of a gauze having the same geometry in the two directions parallel to the two sets of wires, i.e., plain-weave gauzes. The author states that a simple extension of the method gives the resistance for gauzes composed of two sets of wires of different thicknesses and separations, e.g., twills.

In the above expression $x = d_y h^{-1}$ is the single parameter used to define the row geometry, where the diameter of the wires is taken to be d_y and the separation of the central axes to be h. The function f(x) is defined as $\underline{f}/\mu u = f(x)$, where $\underline{f}$ is the drag force per unit length, and for $x < 0.6$ it may be obtained from the expression

$$\frac{8\pi}{f(x)} = 3.31571 - 2 \ln 10x + 1.6449x^2 - 0.67644x^4 \qquad (30)$$

Hutson compared the theoretical values of B predicted by Eq. (29) with experimental results for a number of plain weave wire gauzes and nylon cloths [25]. A graphical comparison of theoretical and experimental results for the latter is given by Fig. 12 which shows that the resistance of plain-weave wire gauzes and nylon cloths to water flow may be predicted adequately by Hutson's equation.

In conclusion, it would appear that of the three approaches outlined above, the use of the Davies equation will be the least

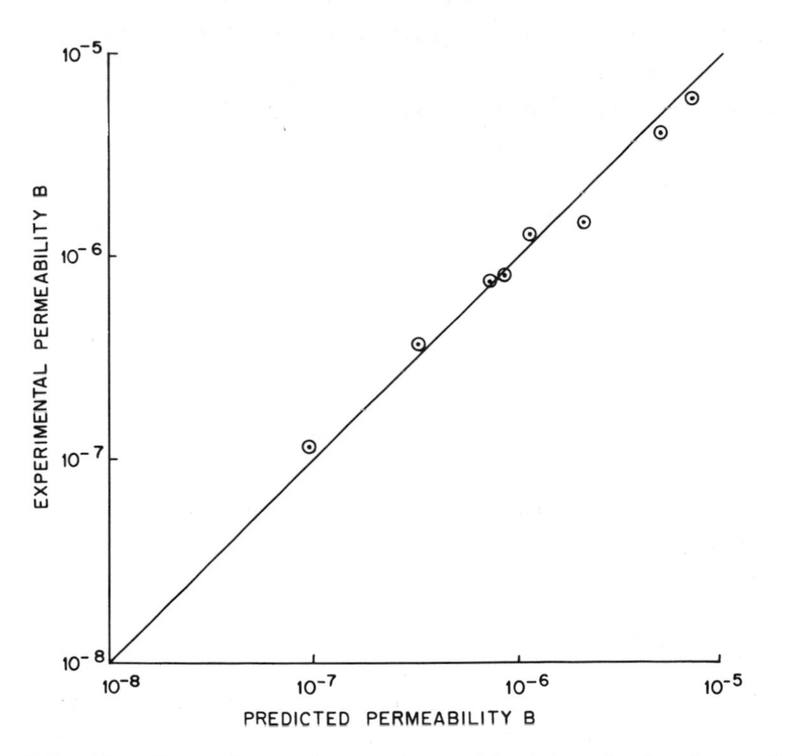

FIG. 12. Comparison of experimentally determined values of permeability with those predicted by Eq. (29).

satisfactory in predicting the permeability of monofilament fabrics. With the exception of yarn diameter, the parameters that completely characterize the cloth construction (weave, yarn separation, diameter, etc.) are not taken into account in Eq. (8). Although the porosity term ε may be taken to represent the degree of yarn separation, its determination depends on an accurate measurement of the cloth thickness, which is extremely difficult to obtain. Another disadvantage of the Davies approach is that its use appears to be limited to plain-weave cloths.

These difficulties are not met in the orifice analogy and drag theory approaches. Both methods take into account yarn diameter and separation. The Pederson approach goes even further, in that the weave is included, i.e., different yarn diameter and separations (twills) in both the warp and weft direction are accounted for. Also both methods obviate the necessity for a thickness determination.

Of the two methods, the Pedersen approach is more developed and comprehensive than that of Hutson, and for this reason it is recommended for use in the design of monofilament fabrics. Pedersen's technique can be used to design cloths of a desired stopping power with maximum permeability; the latter may be estimated without recourse to experiment.

VI. PARTICLE DEPOSITION MECHANISMS

During the deposition of solid matter on, or within, a filter medium, a principal interest is that of the instantaneous resistance of the medium to fluid flow. Certain relationships between the filtrate volume V, and time t may be obtained from the general differential expression

$$\frac{d^2t}{dV^2} = K\left(\frac{dt}{dV}\right)^n \tag{31}$$

in which the form of the (t,V) relation depends upon the value of the exponent n assumed for the integration. Thus when n is zero, the particle size or concentration level precludes penetration of particles into the pores of the medium, and a sieve-like action ensues. Conditions are termed "cake-filtration" where particles are retained on the surface of the medium. Integration of Eq. (31) with n = 0 yields the well-known expression for cake-law conditions

$$\frac{t}{V} = K_1 V + K_2 \qquad (32)$$

where the parameter K_2 includes the resistance of the medium R and is given by

$$K_2 = \frac{\mu R}{A \Delta P} \qquad (33)$$

where μ is the fluid viscosity, A the filtration area, and ΔP the pressure drop over the system. Much has been written on the true value of R which must be used in filtration computations.

Apparently, the clean resistance of the filter medium R_o is augmented by the partial penetration of particles into surface pores, and it is recommended that the combined resistance of the medium plus surface particles should be used in predictive work with Eq. (32). Recent developments on the effects of particle depositions on R_o are reported in Sec. 3.VII.

Deposition mechanisms other than sieving may be described with values of $n \neq 0$. It must be noted, however, that agreement between theoretical and practical data on filtration rates does not prove conclusively that the model of the deposition mechanism is, in fact, equivalent to the physical processes taking place within the unit. Gonsalves [52] showed that the same practical data may be predicted from different theoretical models. Examples of the latter are: n = 2, complete pore blocking; n = 1.5, partial pore blocking; and n = 1.0, an exponent detected in experiments and describable by a combined model.

It is probable that several mechanisms of particle deposition occur during the filtration cycle. In the early moments of cake formation, particles may be sieved out on the surface of the medium, deposit within the latter, blind certain pores partially or completely, or may bleed through the medium. The particular combination of mechanisms occurring will depend upon the relative size of particles and pores and also upon the concentration and particle approach velocity [53]. After the passage of enough filtrate for the surface of the medium to be covered by particles, further deposition will occur by "cake-law" sieving-out on the initial deposit. At this stage, however, the possibility of the migration of ultrafine particles, present in the initial slurry or removed from previously deposited agglomerates by fluid shear, must be considered. In fact, the possibility of a filtration cycle following Eq. (32) as written is relatively rare, particularly in the early stages of the deposition. Most experimental studies are characterized by curved (t/V,V) plots, indicating gradual build-up of first deposits or pore penetration.

In "standard-law" filtration conditions ($n = 1.5$), a gradual deposit of particles on the pore channels in the medium is postulated, and, where the particles are much smaller than the pore diameter, attachment to the walls may proceed by an adsorptive mechanism. Where the attractive forces are of an electrokinetic nature, the capacity of the medium may be enhanced by modifications of the zeta potential of the system using polyelectrolyte additives [54].

The deposition process in deep-bed filters is described by equations of the type

$$-\frac{dc}{dL} = \mu k_f c \tag{34}$$

where k_f is the so-called filtration coefficient, which is not a constant but varies over the bed length L, c the concentration, and μ the fluid viscosity. Ives [55] postulated that the deposition of material in the bed increases the adsorbing surface but simultaneously an increase in velocity occurs, thereby decreasing the filtration coefficient.

Several mathematical models have been put forward in which the filtration coefficient is considered to be a function of the specific deposit within the packing and the interstitial velocity [56-58].

The fact that several deposition mechanisms may occur simultaneously was recognized by Smith [59] who reported eight physical effects encountered during filtrations using multifilament woven media. This work pointed to the fact that it is rare for a single mechanism to predominate during particle deposition, especially in the early stages of cake build-up or in the filtration of dilute suspensions. It is interesting to note that Smith concluded that high-twist, multifilament yarns are preferable to the low-twist variety, since a predominance of flow through the yarn will result in an accumulation of fine particles in the interfiber pores. Depositions within the latter lead to cloth blinding or, at least, situations where cloth laundering is difficult. Recently, attempts have been made to quantify the effect of divided flow through and around multifilament woven media [14].

VII. FLOW RESISTANCE OF USED MEDIA

Information is required on the resistance of the medium immediately before use in quantitative calculations of the time required for a particular filtration. In most situations, a successful medium will be used many times before it is discarded for reasons of wear, pore blocking, etc.; and, although attempts will have been made to clean the medium after each usage, it is inevitable that a certain amount of deterioration in the ability of the cloth to pass fluids will occur.

Hatschek [60] published a study of the throttling of the free area in a medium which may occur upon the deposition of solids. The effect is most pronounced in the deposition of the first layer of particles, and the study suggested that the structure of subsequent deposits may be influenced by the initial throttling process. Hatschek showed that the resistance of a cloth plus a thin layer of cake is very much greater than would be expected from the sum of the resistance of the clean cloth and the resistance of the cake; quantitative and reproducible measurements are made difficult by the fact that the degree of pore plugging is related to the shape of the particle, which tends to orientate itself in the flowing fluid so that it presents minimum resistance to fluid drag forces.

Later experience indicated that both the initial filtration pressure and the solids concentration had a great influence on the initial resistance. Smith [59] stated that the desirable conditions for cake filtration could not occur above a certain critical applied pressure, as the particles cannot form a stable bridge but collapse and plug the medium. Similarly, if the slurry is below a certain concentration, bridging cannot occur, as particles too frequently arrive individually. The conditions for pore bridging are referred to again in a later section. Lack of understanding concerning these initial mechanisms during the first period of separation results largely from the extremely short time periods involved.

Grace [12] developed the standard blocking law to include the properties of the filter medium by introducing a term into the basic filtration equation to account for the volume of a pore occupied by deposited particles. The resulting equation may be written as

$$\frac{t}{V} = \left[\frac{c/(1 - \varepsilon_p)}{\pi N A \ell_e r_o^2}\right] t + \left(\frac{8\ell_e}{\pi N r_o^4}\right)\left(\frac{\mu}{A\ \Delta P}\right) \tag{35}$$

where the concentration c is expressed as the volume of solid particles removed per volume of solution filtered, r_o the effective pore radius, ε_p the packing porosity of the solids, N the number of pores per unit area of filter medium, A the surface of the filter, ℓ_e the effective pore length, μ the fluid viscosity, and ΔP the pressure difference across the filter medium. Grace was able to calculate the "clogging value" (= $N\ell_e r_o^2$) for various filter media under similar filtration conditions and to show that a definite correlation exists between the clogging value and the modal value of the interfiber pore radius, as measured by the mercury intrusion method. The clogging value then is a measure of the volume of pore constrictions that plug during filtration.

While investigating the performance of filter media, Grace observed an initial bleeding of particles followed by an initial filtration period indescribable by any existing theory, then a period obeying the standard law, and finally a period of cake filtration.

The latter occurred even though no visible cake was evident. All these data indicate that the degree of particle retention, assumed constant by Hermans and Bredee [61], may change considerably in any one deposition.

Heertjes and Haas [53] suggested that the initial deposition stage had serious consequences on the complete cycle of events in filtration, since the average supernatent cake resistance was determined by the underlying layers. Further, these authors disputed the claims of Hermans and Bredee, stating that there was no abrupt change from one filtration mechanism to another during the cycle, but rather a gradual change in the form of pore blocking. This work demonstrated the importance of solids concentration in determining the course of the deposition; blocking as well as cake filtration may be realized in the same system by changes in particle concentration. With very small concentrations, particles follow the fluid streamlines, entering pores and thereby giving rise to high degrees of blocking. Increase in concentration results in more particles arriving simultaneously at the pore, and the degree of blocking is reduced. Above a certain concentration, enough particles arrive to create bridging action or cake filtration conditions. These effects are depicted in Fig. 13.

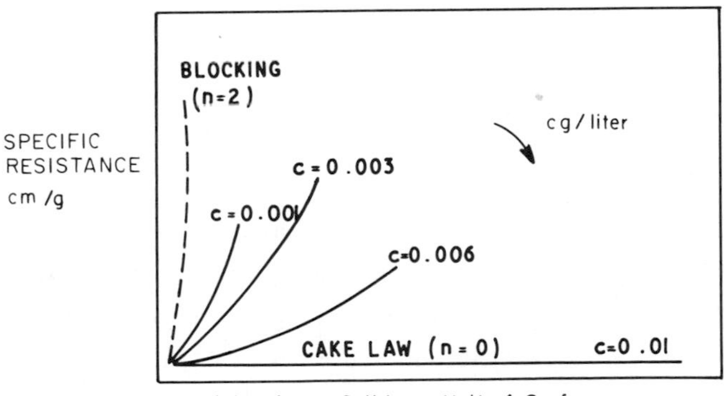

FIG. 13. Effect of concentration on filtration mechanism and deposition mechanism.

Heertjes and Haas also pointed out that if a graph of R_T (total resistance of the medium plus solid deposit) vs. V (volume of filtrate) was extrapolated to zero volume, the initial resistance is not always the resistance of the clean cloth, and is generally much higher. Further, the effect was related to the concentration of the mixture being filtered in that, the higher the concentration of the particles, the nearer the extrapolated resistance approached the initial "clean" medium resistance. This demonstrates again that increases in particle concentration reduce the possibility of particle penetration into the medium and thereby eliminates the large increases in R_T associated with dilute suspensions.

In a later paper [62] these mechanisms were expressed mathematically by

$$R_T = R_o\left(\frac{OaN}{OaN - w}\right)^S \tag{36}$$

where R_o is the clean medium resistance, w the weight of solids per unit area of filter, and a the weight of a particle. The parameters O and S increase with increases in slurry concentration. Again, as the weight of solids per unit surface of filter w increases, the value of R_T increases. Heertjes also dealt with the problems of experimental reproducibility in such work and recorded the influence of vibration and lack of homogeneity [63].

Another relationship [15] between the resistivity of thin deposits and slurry concentration takes the form

$$\frac{r - r_\infty}{r_t - r_\infty} = \exp\left\{-m'\left[\frac{(c - c_t)}{aN}\right]\right\} \tag{37}$$

where r is the specific cake resistance at the concentration c, r_t the specific cake resistance at the concentration c_t at which blocking action changes to cake filtration, r_∞ the specific cake resistance at infinite concentration, and m' a proportionality constant.

Velocity-effect studies on thin cake deposits have shown that high velocities tend to stabilize loose cake structures of low inherent resistance. This effect is more noticeable for cakes, where r is naturally high. Thus velocity may be of importance at low concentrations, the effect being described by the equation

$$r = r_a \exp\left(\frac{u_o}{b}\right)^y \tag{38}$$

where r_a is the cake resistance for zero initial filtrate velocity, u_o the filtration velocity, b a parameter dependent on particle size, and y an exponent.

A report by Dmitrieva and Pakshver [64] which appears to have received little attention points out the importance of particle sedimentation velocity in pore plugging. These investigators showed that blocking filtration tends to develop when liquid viscosities increase and the concentration and particle size of the suspended solids diminish. These quantities are related by a dimensionless ratio π involving the velocity of flow through the medium and the sedimentation velocity of the particles v_s, which is obtained from the Stokes relationship for a sedimenting particle

$$v_s = \frac{\bar{d}^2 g(\rho_s - \rho)}{18\mu} \tag{39}$$

The parameter π is given by

$$\pi = \frac{18\ \Delta P}{R_o \bar{d}^2 (\rho_s - \rho) g} \tag{40}$$

where $\bar{d}$ is the mean particle diameter, ρ_s and ρ are the solid and liquid densities, respectively, and g is the acceleration of gravity. In Eq. (40) the velocity of flow has been replaced by the ratio of the pressure drop ΔP to the medium resistance R_o.

Experimental data indicate that pore plugging occurs for π values greater than 1,000 and cake filtration for π values less than 100; between these limits, i.e., $100 < \pi < 1{,}000$, an intermediate type of deposition occurs. These data apply to filtrations on horizontal surfaces.

The relationship between the effective resistance of a partly clogged filter medium to the resistance of the medium at the beginning of the cycle has been considered by Kehat et al. [65]. Two equations were reported. For complete blocking, the expression is

$$R_T = K_a R_o - R_b - R_c \tag{41a}$$

and for standard blocking

$$R_T = IR_o \tag{41b}$$

where R_o is the effective resistance at the beginning of the cycle, R_b the reduction in R_o due to bleeding, R_c the reduction in R_o due to bridging, and I and K_a are constants dependent on particle properties and filtration conditions.

Many practical studies have been reported on the relationship between R_o and R_T, particularly for plain, woven, monofilament cloths [14]. Here relationships of the following type have been developed for various deposits:

$$R_T = \psi R_o \tag{42}$$

where $\psi = \{1 + m_1/[\bar{d}/(d_p + d_y)]\}^{n'}$ and $\bar{d}$ is the mean particle diameter, d_p the width of the pore aperture, d_y the yarn diameter, m_1 a proportionality constant, and n' an exponent.

For example, with sand and glass beads:

	m_1	n'
Glass	2.25	1.65
Sand	0.07	2.94

English [66] presents a useful analysis of the process of media blinding and develops mathematical relationships for the rate of blinding of screens.

VIII. EFFECT OF MEDIA STRUCTURE ON CAKE RESISTANCE

It was noted in the previous section that the initial filtration velocity has an influence on the deposited cake structure and, therefore, on the cake resistance. The initial flow rate will be determined by the medium resistance and pressure drop in the system according to

$$u_o = \frac{\Delta P}{\mu R_o} \tag{43}$$

where u_o is the initial filtration velocity, ΔP the pressure difference over the medium, μ the fluid viscosity, and R_o the resistance of the clean medium. Substitution of the above relationship into Eq. (38) gives

$$r = r_a \exp\left\{\left(\frac{K_3}{R_o \mu}\right)^y\right\} \tag{44}$$

where r is the true specific cake resistance, K_3 a constant, and y an exponent. Thus the specific resistance of filter cakes formed under similar process conditions will be affected by the medium resistance.

All such effects have been shown to be demonstrable only in those systems where cake-law filtration conditions are obtained [67]. Data presented in Fig. 14 show that a correlation between r and medium resistance follows when n = 0 characterizes the deposition. Again the correlation was improved by using the deposited resistance R_T rather than the clean value R_o.

However, no correlation has been suggested for blocking or "standard law" situations. It may be that for n ≠0, the resulting

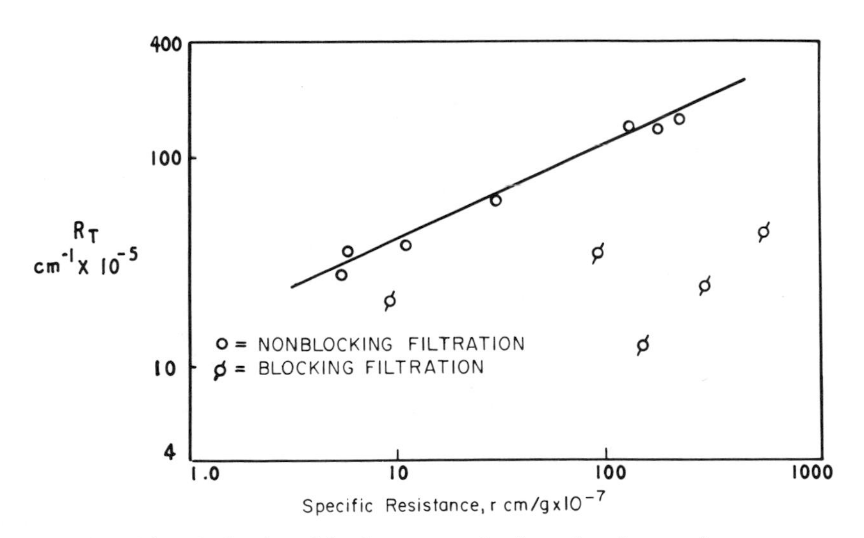

FIG. 14. Relationship between cloth and cake resistance.

r value cannot be related to the medium resistance, clean or used. In these cases the cloth resistance depends upon the course of the initial stage in the filtration process. Those situations in which cloths are most severely plugged will produce the largest effects on r; thus a large change $(R_T - R_0)$ is associated with a large r for substantially the same process conditions.

Further evidence has been collected which demonstrates the importance of media structure in determining the course of a deposition [67]. With multifilament cloths of varying β factor, or flow division through and around yarns, flow conditions in the medium immediately prior to particle deposition alter the subsequent filtration mechanism. This is shown in Table 2 where the β and n parameters are reported for the filtration of a slurry with a mean particle diameter of 0.7 μm.

The small particle size used in this work gave rise to the possibility of particle penetration into the interfiber pores. Where n = 1.5, the deposition model is that of a relatively coarse channel, being gradually reduced in diameter by wall deposits. Table 2 indicates that n = 1.5 situations arise when β is low, i.e., when the interyarn pore sizes are large relative to the interfiber pores. The increase in β results in a larger percentage of flow passing through the interfiber pores with a change in n to unity. These results indicate that the overall permeability of the medium must

TABLE 2 Effect of Media Structure on Filtration Mechanism

Mean pore in cloth (μm)	β	n
7	1.3	1.5
17	2.0	1.5
15	2.6	1.5
8	16.0	1.0
5	16.1	1.0
6	20	1.0

be considered when attempts are made, by the use of high-twist yarns, to reduce particle penetration into the interfiber pores [59]. The suggestion that high-twist yarns are more suitable in preventing particle penetration is true if the overall permeability is reasonably low. A high-twist yarn that is closely woven, resulting in a high β factor, will produce a preponderance of flow through the yarns and give the same overall result as a low-twist yarn.

The application of laboratory test results to industrial practice is often made difficult because of changes in the medium when used in a filtration plant. For example, it has been demonstrated [68] that the permeability of clean media may change considerably when subjected to centrifugal force. Changes in medium resistance in upward filtering systems, e.g., rotary vacuum filtration, may lead to loss of solids since insufficient fluid force is available to support the growing filter cake against gravitational and drag forces caused by the agitator and drum motion [69]. Experimental evidence has been collected [70] which demonstrates the important influence of medium resistance in the course of a filtration when conducted in an upward direction. Quite often, loss of solids may lead to erroneous estimates of the filter cake resistance.

IX. MEDIA BRIDGING

Most process difficulties associated with blinded media would be avoided if a mathematical treatment were available from which could be predicted the concentration conditions necessary for bridging the medium pores and thereby creating cake-filtration conditions. However, few quantitative data are available which permit such calculations, except for relatively simple media and a restricted number of particle types.

The effect of concentration on the percentage of particles bleeding through media has been studied quantitatively for a few systems [14]. The general trend is reported in Table 3 where cloth A is a multifilament cloth of loose weave and cloth B of tight weave, the information referring to the filtration of calcium carbonate

TABLE 3 Concentration Effects in Particle Bleeding

Suspension concentration (Calcium carbonate) (g/100 g)	Weight passing	
	Cloth A (β = 2; $\bar{d}/2r_p$ = 1.1) (%)	Cloth B (β = 20; $\bar{d}/2r_p$ = 3) (%)
1.0	23.3	1.25
2.0	9.0	0.55
5.0	0.9	0.03

from water. The low β factor in cloth A produces monofilament characteristics. At low concentration, and where the particle to pore-size ratio approaches unity, large bleeds are found. Industrial users are often heard to express the belief that cloths that exhibit bleeding do not produce blinding. Close examination of these combined effects casts some doubt on this view since, while it is true that the lower bleeds of cloth B are associated with much lower filtration rates, the latter are reproducible upon laundering or backwashing the cloth. In contrast, the interfiber pores in cloth A do tend to accumulate finer particles since wash fluid will channel through the larger interyarn pores. The filtration rates obtained here gradually decline after several usages.

The principal feature of such results is the suggestion that a critical concentration may exist above which particle bleeding will be negligible. Quantitative studies on single capillaries and samples of monofilament and multifilament cloths have shown that such concentrations do exist. The general pattern of behavior is depicted in Fig. 15. For a particular particle size filtered at various concentrations through cloths having various pore sizes, a locus exists which separates conditions of bleeding from those producing cake filtration or bridging. At high concentrations the locus is horizontal, suggesting an upper pore size above which bridging is impossible irrespective of concentration.

This phenomenon was examined by Hixson [71] who used glass capillaries and highly concentrated quartz suspensions in establishing

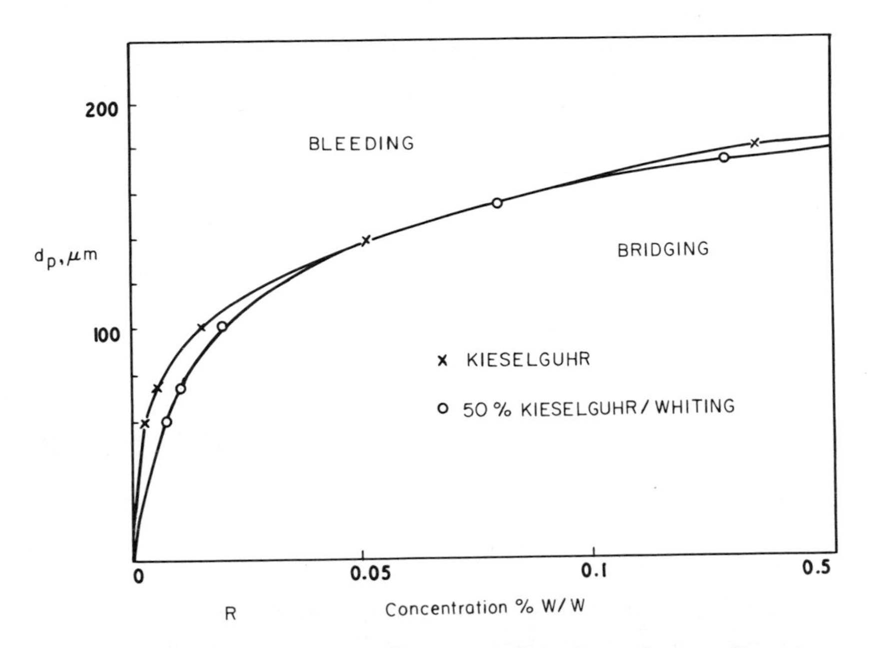

FIG. 15. Bridging concentration as a function of pore diameter.

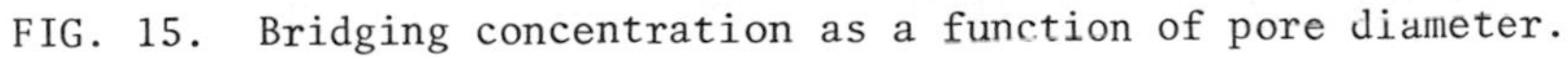

the relationship relating the limiting pore size to particle size

$$d_p = k_1(\bar{d})^{1/4} \quad \text{for } c > 20\% \tag{45}$$

where d_p is the pore size and $\bar{d}$ the mean particle size. The parameter k_1 was found to have a value of 175. Recent work [72] has demonstrated that the parameter k_1 depends on the nature of the particles, typical values being given in Table 4.

TABLE 4 Practical Values of Bridging Parameter k_1

Material	k_1
Kieselguhr	500
Calcium carbonate	438
Magnesium carbonate	249
Sand	175

TABLE 5 Practical Values of Bridging Parameters m and k_2

Material	k_2	m
Kieselguhr	21	0.26
Magnesium carbonate	13	0.35
Calcium carbonate	10	1.04
Calcium carbonate plus kieselguhr	33	0.38

It will be noticed in Fig. 15 that, as the concentration of solids falls, the pore size required for bridging also decreases. Such effects may be reported in the form

$$d_p = k_2\bar{d}(c)^m \tag{46}$$

where c is the slurry concentration and m an exponent. Using filter cloths composed of monofilament yarns, bridging concentrations occur in the range $0.002 < c < 2$ and typical k_2 and m values are as listed in Table 5.

It is interesting to note that in the situation of body-feed filter aid, the bridging conditions are nearer those of the pure filter aid than the filtered solids.

Further work has shown that the pore structure of the medium, e.g., plain, twill, sateen, etc., is important in determining the conditions of bridging. It is obvious that such empirical work, while of interest in identifying important variables in the process, will be enhanced when a fuller understanding of particle-type (shape, agglomerate, etc.) effects, along with pore shape effects, is established.

NOTATION

a	Weight of a particle	gm
A	Surface area of filter	cm^2
A_o	Effective orifice area	cm^2

Symbol	Definition	Units
A_1	Upstream flow channel area	cm^2
A_2	Nozzle or orifice projected open area	cm^2
A_3	Ratio A_2/A_1; Eq. (6)	
b	Parameter in Eq. (38), dependent on particle size	
B	Permeability	cm^2
B_1	Permeability of cloth when yarns impermeable	cm^2
B_o	Permeability of yarns in cloth	cm^2
c	Concentration	
c_1	Concentration	g/cm^3
c_t	Concentration at commencement of cake filtration, Eq. (37)	
C_D	Discharge coefficient	
d	Differential operator	
$\bar{d}$	Mean particle diameter	cm
d_p	Diameter of pore or capillary, or width of square aperture	cm
d_f	Fiber diameter	cm
d_y	Yarn diameter; multifilament or plain-weave monofilament	cm
$\bar{d}_y$	Mean yarn diameter; twill monofilament	cm
d_1	Weft yarn diameter	cm
d_2	Warp yarn diameter	cm
D	Effective diameter of pore; monofilament fabric	cm
(ec)	Warp yarns per centimeter	cm^{-1}
$\underline{f}$	Drag force per unit length	g/sec^2
g	Acceleration of gravity	cm/sec^2
h	Separation of axes of cylinders or wires	cm
I	Constant in Eq. (41b)	
k	Constant in Eq. (13)	cm^{-1}
k_1	Constant in Eq. (45)	
k_2	Constant in Eq. (46)	
k_f	Filtration constant, Eq. (34)	sec/g^1
K	Parameter in Eq. (31)	
K_1	Cake constant in Eq. (32)	sec/cm^3
K_2	Cloth constant in Eq. (33)	sec/cm^3

K_3	Constant in Eq. (44)	cm
K_a	Constant in Eq. (41a)	
K_o	Kozeny constant	
ℓ_e	Effective pore length in Eq. (35)	cm
L	Thickness of filter cake or medium	cm
m	Exponent in Eq. (46)	
m'	Proportionality constant in Eq. (37)	
m_1	Constant in Eq. (42)	
n	Exponent in Eq. (31)	
n'	Exponent in Eq. (42)	
N	Number of pores per unit area of filter medium	cm^{-2}
O	Parameter in Eq. (36)	
P	Bubble pressure	$dynes/cm^2$
ΔP	Pressure difference over filter cake or medium	$dynes/cm^2$
(pc)	Weft yarns per centimeter	cm^{-1}
Q	Volumetric flow rate per pore or orifice	cm^3/sec^1
r	Specific cake resistance	cm/gm
r_a	Specific cake resistance at zero velocity	cm/gm
r_{bp}	Bubble pressure radius	cm
r_c	Interyarn pore radius from cloth construction	cm
r_f	Interfiber pore radius from permeability	cm
r_o	Effective initial pore radius in standard blocking	cm
r_p	Pore radius calculated from permeability	cm
r_t	Specific cake resistance at concentration c_t	cm/gm
r_y	Interyarn pore radius from permeability	cm
r_∞	Specific cake resistance at infinite concentration	cm/gm
R	Effective resistance of filter medium	cm^{-1}
R_b	Resistance reduction due to bleeding	cm^{-1}
R_c	Resistance reduction due to bridging	cm^{-1}
Re	Reynolds number	
R_o	Clean medium resistance	cm^{-1}
R_T	Resistance of medium and initial layer of cake	cm^{-1}
S	Exponent in Eq. (36)	

S_o	Specific surface	cm^{-1}
t	Time	sec
u	Superficial or approach velocity	cm/sec
u_o	Initial filtration velocity	cm/sec
u_p	Mean velocity in capillary or pore	cm/sec
v_s	Particle sedimentation velocity	cm/sec^1
V	Volume of filtrate at time t	cm^3
w	Weight of solids per unit area of filter	g/cm^2
W	Wetted perimeter of pore	cm
W_c	Cloth weight	gm/cm^2
W_p	Wetted perimeter of plain pore	cm
W_T	Wetted perimeter of twill pore	cm
x	Parameter in Eq. (29); $x = d_y h^{-1}$	
y	Exponent in Eq. (38)	
α_o	Effective fraction open area of filter cloth	
α_p	Effective fraction open area of plain pore	
α_T	Effective fraction open area of twill pore	
β	Ratio: B/B_1	
ε	Porosity	
ε_p	Packing porosity of solids	
$\bar{E}(k,\pi/2)$	Elliptical integral of the second kind with variable k where $\bar{K} = \left[1 + \left(\frac{d_1}{\frac{1}{ec} - d_2}\right)^2\right]^{-1/2}$	
λ	Constant in Eqs. (18)-(20)	
μ	Fluid viscosity	$g/cm^1 sec^1$
ρ	Fluid density	g/cm^3
ρ_s	Solid density	g/cm^3
π	Ratio used in Eq. (40)	
σ	Surface tension	$dynes/cm^1$
ϕ	Cloth parameter in Eqs. 18, 20; $= [(ec)(d_1 + d_2)]^{-1}$	
ψ	Constant in Eq. 42; $\psi = \left[1 + \frac{m_1}{(\bar{d}/d_p + d_y)^{n'}}\right]$	

REFERENCES

1. G. D. Dickey, *Filtration*, Reinhold, New York (1961).
2. R. C. French, *Chem. Eng.* (1963).
3. J. P. Kovacs, *Chem. Eng.* (1960).
4. W. Shoemaker, *Filtrat. Separat.*, *12*, 61 (1975).
5. W. B. Dobie, *Liquid-Solid Separation*, H.M.S.O. London, 1966, p. 191.
6. D. B. Purchas, *Industrial Filtration of Liquids,* Leonard Hill, London, 1967.
7. P. M. Heertjes, P. J. Bakker, and J. L. Hibou, *Chem. Eng. Sci.*, *10*, 139 (1959).
8. British Standards Handbook No. 11, London, 1963, p. 313.
9. A. Rushton and D. J. Green, *Filtrat. Separat.*, *5*, 516 (1968).
10. A. E. Scheidegger, *The Physics of Flow Through Porous Media*, Toronto University, Toronto,
11. J. F. Zeivers, *Chem. Eng. Progr.*, *53*, 493 (1951).
12. H. P. Grace, *A.I.Ch.E. J.*, *2*, 316 (1956).
13. F. M. Tiller, *Ind. Eng. Chem.*, *53* (7), 529 (1961).
14. A. Rushton, *The Chemical Engineer*, London, 1970, p. 88.
15. P. M. Heertjes, *Chem. Eng. Sci.*, *6*, 269 (1957).
16. S. Backer, *Text. Res. J.*, *20*, 650 (1948).
17. A. F. Robertson, *Text. Res. J.*, *20*, 838 (1950).
18. A. F. Robertson, *Text. Res. J.*, *20*, 844 (1950).
19. H. P. Grace, *A.I.Ch.E. J.*, *2*, 307 (1956).
20. C. N. Davies, *Proc. Inst. Mech. Eng.*, B, *1*, 185 (1952).
21. N. Kimura and G. Ionya, *Chem. Eng.* (Tokyo), *23*, 792 (1959).
22. S. Aiba, *J. Gen. Appl. Microbiol.* (Tokyo), *8*, 169 (1969).
23. R. M. Werner and L. A. Clarenburg, *Ind. Eng. Chem., Ind. Intl. Ed.*, *4*, 288 (1965).
24. W. L. Ingmanson, B. D. Andrews, and R. C. Johnson, *TAPPI*, *42*, 840 (1959).
25. A. Rushton, D. J. Green, and H. E. Khoo, *Filtrat. Separat.*, *5*, 213 (1968).
26. R. McGregor, *J. Soc. Dyers Colour.*, *81*, 429 (1965).
27. R. R. Sullivan and K. L. Hertel, *J. Appl. Phys.*, *11*, 761, 1940.
28. O. Emersleben, *Z. Phys.*, *26*, 601 (1925).

29. R. R. Sullivan, *J. Appl. Phys.*, *13*, 725 (1942).

30. E. M. Sparrow and A. L. Loeffler, *A.I.Ch.E. J.*, *5*, 125 (1959).

31. J. Happel and H. Brenner, *Low Reynolds Number Hydrodynamics*, Prentice-Hall, Engelwood Cliffs (1955).

32. H. C. Brinkman, *Appl. Sci. Res.*, *A1*, 27 (1947).

33. H. C. Brinkman, *Appl. Sci. Res.*, *A1*, 81 (1948).

34. M. J. Denton, *Shirley Inst. Mem.* (Manchester, England), *35*, 41 (1962).

35. R. Ben Aim, P. Le Goff, and P. Le Lec, *Powder Technol.*, *5*, 51 (1971/72).

36. A. C. Wrotnowski, *Chem. Eng. Progr.*, *58*, 63 (1962).

37. A. C. Wrotnowski, *Chem. Eng. Progr.*, *53*, 313 (1957).

38. A. C. Wrotnowski, *Filtration Characteristics of Felted Structures* (Paper presented at the 42nd National Meeting A.I.Ch.E.), Atlanta, Ga., February, 1960.

39. C. E. Cunningham, G. Broughton, and R. R. Kraybill, *Ind. Eng. Chem.*, *46*, 1196 (1954).

40. P. V. R. Griffiths, *M.Sc. Thesis*, Manchester University, Manchester, 1969.

41. C. P. Kyan, D. T. Wasan, and R. C. Kintner, *Ind. Eng. Chem., Fundam.*, *9*, 596 (1970).

42. R. R. Sullivan, *J. Appl. Phys.*, *12*, 503 (1941).

43. E. Lord, *J. Text. Inst.*, *46*, T191 (1955).

44. S. Backer, *Text. Res. J.*, *21*, 703 (1951).

45. S. E. Penner and A. F. Robertson, *Text. Res. J.*, *21*, 775 (1951).

46. G. C. Pedersen, *Fluid Flow through Monofilament Fabrics* (Paper presented at the 64th National Meeting A.I.Ch.E.), New Orleans, La., March, 1969.

47. A. Rushton and P. V. R. Griffiths, *Trans. Inst. Chem. Eng.*, *49*, 49 (1971).

48. A. S. Iberall, *J. Res. Nat. Bureau Stand.*, *45*, 398 (1950).

49. H. C. Brinkman, *Research* (London), *2*, 190 (1949).

50. R. A. Mott, *Some Aspects of Fluid Flow*, Edward Arnold, London, 1971.

51. V. C. L. Hutson, *Chem. Eng.* (London), no. 232, 362 (1969).

52. H. Gonsalves, *Rec. Tran. Chim.*, *69*, 873 (1950).

53. P. M. Heertjes and H. van der Hans *Rec. Tran. Chim.*, *68*, 361 (1949).

54. E. E. Baumann and C. S. Oulman, *Filtrat. Separat.*, *17* (6), 682 (1970).

55. K. J. Ives, *Proc. Inst. Civ. Eng.*, *16*, 189-93 (1960).

56. V. Mackrle et al., *Int. Atomic Energy Agency Contact Rep. no. 98*, Czech Academy of Sciences, Institute of Hydrodynamics, Prague, 1965.

57. P. N. Heertjes and C. F. Lerk, *Trans. Inst. Chem. Eng.*, *45*, 7738 (1967).

58. A. K. Deb, *J. Sam. Eng. Div.*, Proceedings of the American Society of Civil Engineers, *95* (SA3), 399 (1969).

59. E. G. Smith, *Chem. Eng. Progr.*, *47*, 545 (1951).

60. E. Hatschek, *J. Soc. Chem. Inc.*, *27*, 528 (1908).

61. P. H. Hermans and H. L. Bredee, *J. Soc. Chem. Ind.*, *55* IT (1936).

62. P. M. Heertjes, *Chem. Eng. Sci.*, *6*, 190 (1957).

63. P. M. Heertjes, *Chem. Eng. Sci.*, *7*, 15 (1958).

64. T. F. Dmitrieva and A. B. Pakshver, *Khim. Prom.*, *11*, 20 (1951).

65. E. Kehat, A. Lin, and A. Kaplan, *Ind. Eng. Chem., Process Des. Develop.*, *6*, 48 (1967).

66. J. E. English, *Filtrat. Separat.*, 195 (1974).

67. A. Rushton and P. V. R. Griffiths, *Filtrat. Separat.*, *1* (1972).

68. A. Rushton and M. Spear, *Filtrat. Separat.* (1970).

69. A. Rushton, N. Turner, and R. J. Wakeman, Paper no. 29, *1st World Congress on Filtration*, Paris, April, 1974.

70. A. Rushton and A. Rushton, *Filtrat. Separat.* (1973).

71. A. Hixson, L. Work, and J. Odell, *Trans. Amer. Inst. Min. Met. Eng.*, *73*, 225 (1926).

72. A. Rushton and A. Rushton, *Filtrat. Separat.* (1971).

Chapter 4

INDUSTRIAL GAS FILTRATION

Koichi Iinoya

Department of Chemical Engineering
Kyoto University
Kyoto, Japan

I.	Introduction	310
II.	Bag Filters and Baghouses	310
	A. Mechanical Shaking and Reverse Flow	313
	B. Reverse Jets	318
	C. Sonic Cleaning Method	320
III.	Fibrous Mat Filters	324
	A. Continuous Types	324
	B. Panel Types	238
	C. Paper Filters	238
	D. High-Performance Filters	239
IV.	Mist Collectors	335
V.	Aggregate and Porous Filters	337
VI.	Performance Criteria	337
	A. Pressure Loss	339
	B. Collection Efficiency	343
	C. Life Expectancy	346
	D. Maximum Dust Load on a Fibrous Filter	346
VII.	Gas Cooling	347
	A. Air Mixing	347
	B. Heat Exchangers	348
	C. Spray Towers	348
VIII.	Typical Applications and Comparative Cost Data	351

Notation 354
References 356

I. INTRODUCTION

Filtration is the oldest and generally the most reliable method by which dusts, fumes, and mists may be removed from gases. Filters usually give very satisfactory collection efficiencies with moderate power consumption. They are mostly used dry, i.e., without water washoff [1-4].

There are three basic kinds of industrial gas filters: a fabric filter usually in the form of a bag or sleeve, a panel filter usually involving a packed fibrous mat, and an aggregate packed column. Description of the performance of filters involves specification of the pressure loss, collection efficiency, and service life factors for both the filter media and the other components of the system [5].

II. BAG FILTERS AND BAGHOUSES

Each fabric filter element consists of a woven or felted textile material in the shape of a tube or flat supported envelope. Many individual filter elements are contained in a single housing having gas inlet and outlet connections, a dust storage hopper, and cleaning mechanisms. Fabric filters may be used for control of dust concentrations in the range of 1 mg/m^3 to 100 g/m^3 with particles down to submicron fumes. Special fabrics permit operation at high temperatures (250°C) and afford resistance to corrosive chemical constituents of the filtered gas or collected materials [6].

A comparison of the collection efficiencies of bag filter cloths is not basically important, because almost all commercial filters remove as much as 99% of the dust or fume from gas streams. Greatest variety in design is found, however, in the mechanism by which the filter bags are cleaned. This is an important factor because it not only affects the size of the baghouse, but also its cost, maintenance requirements, and dependability [7].

The basic parameter of baghouse size vs. capacity is the so-called air-to-cloth ratio, i.e., the total flow rate (m^3/min) of gas to be filtered divided by the total area (m^2) of cloth comprising the filter. Most of the factors that affect the air-to-cloth ratio are linked to application, such as gas temperature, dust loading, filter cloth material, and dust characteristics.

The failure of even a small proportion of the filter bags seriously compromises the overall efficiency of a unit. A well-designed filter installation must therefore allow quick and easy access for the removal and replacement of individual bags. Figure 1 shows several types of bags, and Fig. 2 gives the relation between fiber diameter and denier with fiber density as a parameter.

Electrostatic charges of up to several thousand volts may be developed within the deposited layer on a filter bag. When such charges are present, it is very difficult to clean the cloth by any cleaning method. Antistatic fabrics, woven with several percentages of electroconductive fibers or stitched with metal wires, have been developed to minimize the effects of deposit charges. Table 1 gives examples of their electric resistances.

TABLE 1 Characteristics of Polyester Fabric Filters[a] with Metal Fibers

Filter cloth	Weight (g/m^2)	Weight (oz/yd^2)	Electrical resistance (ohm cm)
Polyester fiber only	330	9.1	3.3×10^{12}
Polyester with 3% metal thread	337	9.8	2.8×10^2
Polyester with 5% metal thread	359	10.5	1.4×10^2
Polyester with 10% metal thread	380	11.0	7×10^1

[a]Air permeability at a pressure loss of 12.7 mm H_2O of about 10 cm/sec (20 ft/min).

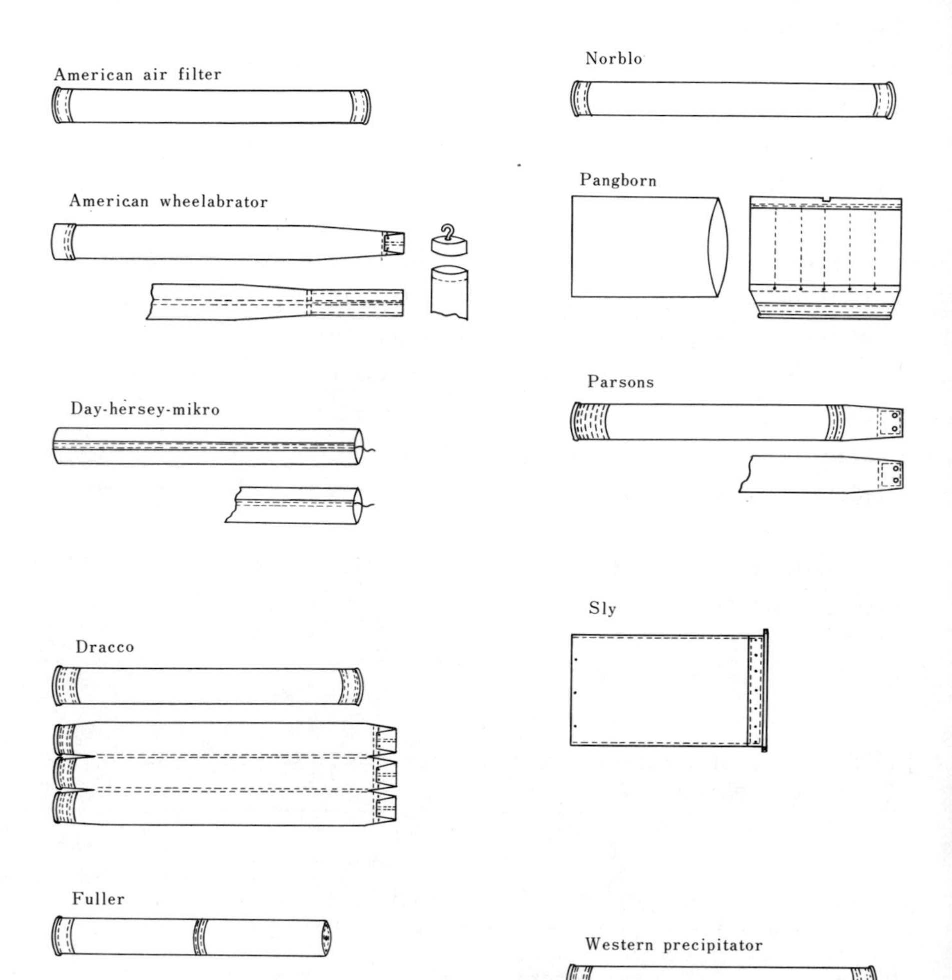

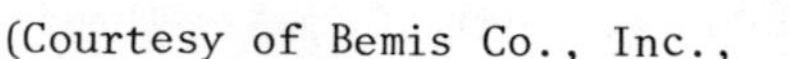

FIG. 1. Several bag shapes. (Courtesy of Bemis Co., Inc., St. Louis, Missouri.)

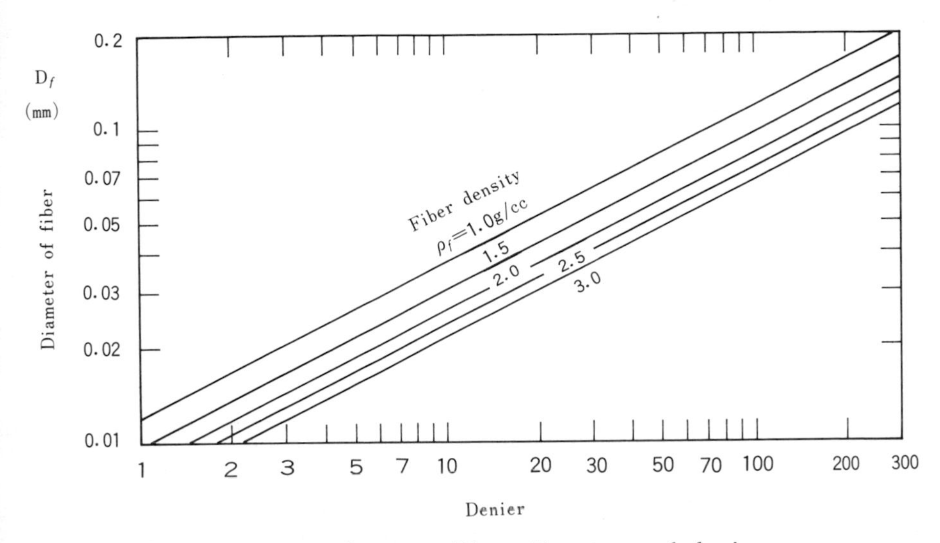

FIG. 2. Relation between fiber diameter and denier.

A. Mechanical Shaking and Reverse Flow

The principle of removing collected dust from a filter cloth by shaking or a back-flow of gas, accomplished either manually or automatically, is probably as old as gas filtration itself. In some cases, both shaking and back-flow are used simultaneously. The section of the unit containing the bags being cleaned is necessarily out of service during this portion of the cycle. It is difficult to say which cleaning method reduces bag flexing the most and hence leads to the greatest bag life. The maximum dust load (holding capacity) may be of the order of 0.1-0.5 kg of dust per m^2 of cloth surface area, although it depends on the characteristics of the dust and the filter medium. Air-to-cloth ratios during back-flow range from as low as 0.5 (m/min) to as high as 2 (m/min). Figure 3 shows a typical bag filter with mechanical shaking, and Figs. 4a and 4b show examples of the reverse-flow type. Figure 5 gives the relationship between shaking amplitude and duration in order to obtain the minimum residual drag of a bag cloth [8]. Figure 6 presents typical variations in pressure drop and gas flow rate for a conventional bag filter with mechanical shaking [9].

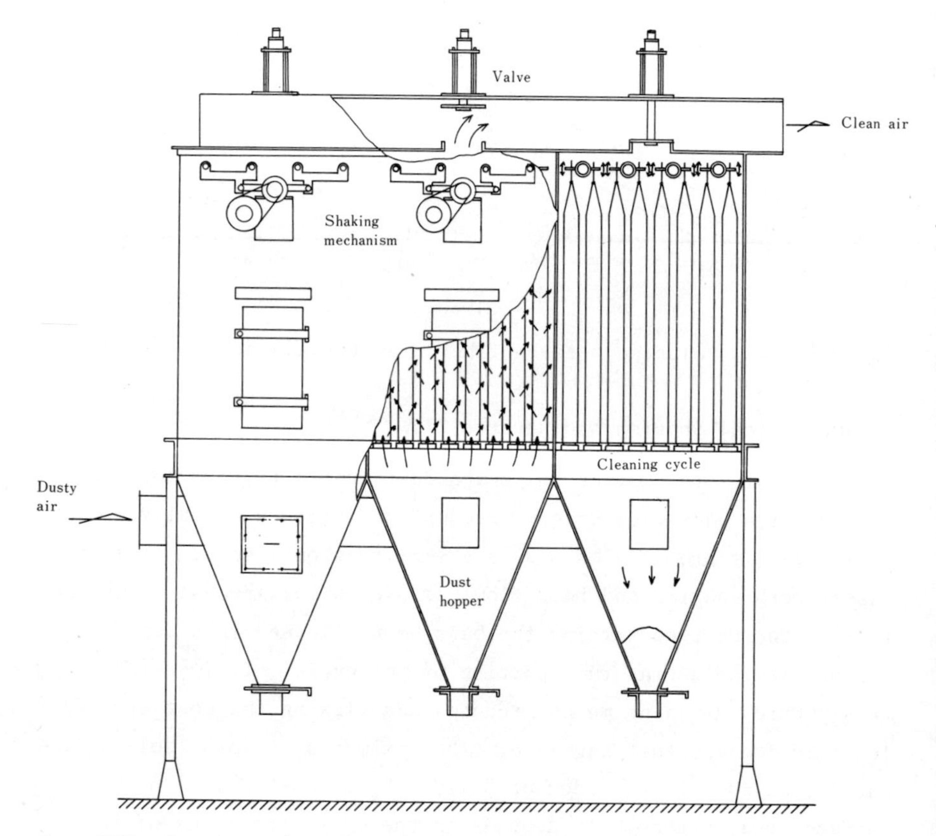

FIG. 3. A typical bag filter with mechanical shaking (Shinto Dust Collector).

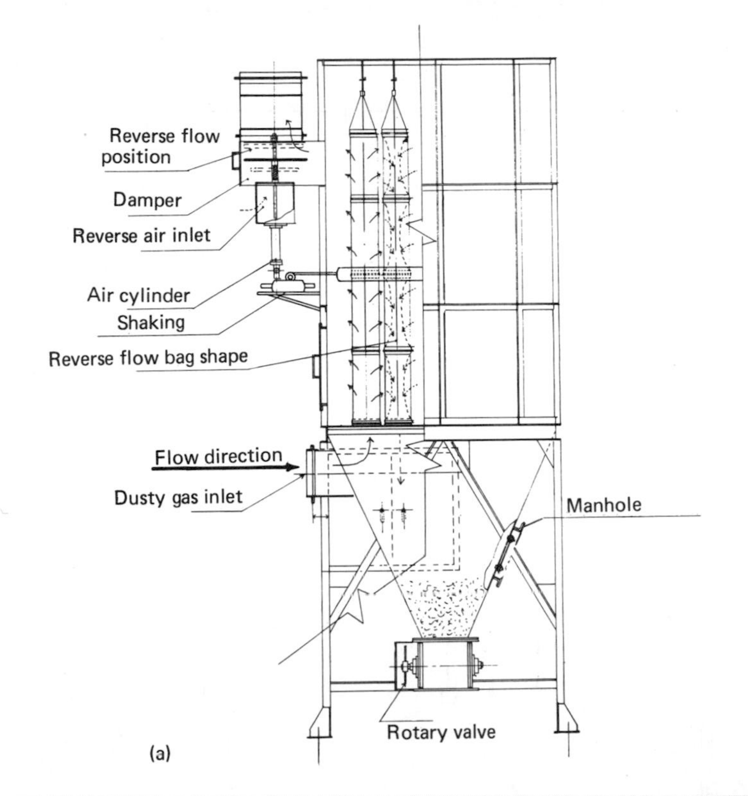

FIG. 4. (a) An example of bag filter with reverse flow cleaning mechanism (Nihon Spindle). (b) Reverse-flow type of bag filter (1,000 m^3/min for painting works). (Lugar-Sanko.)

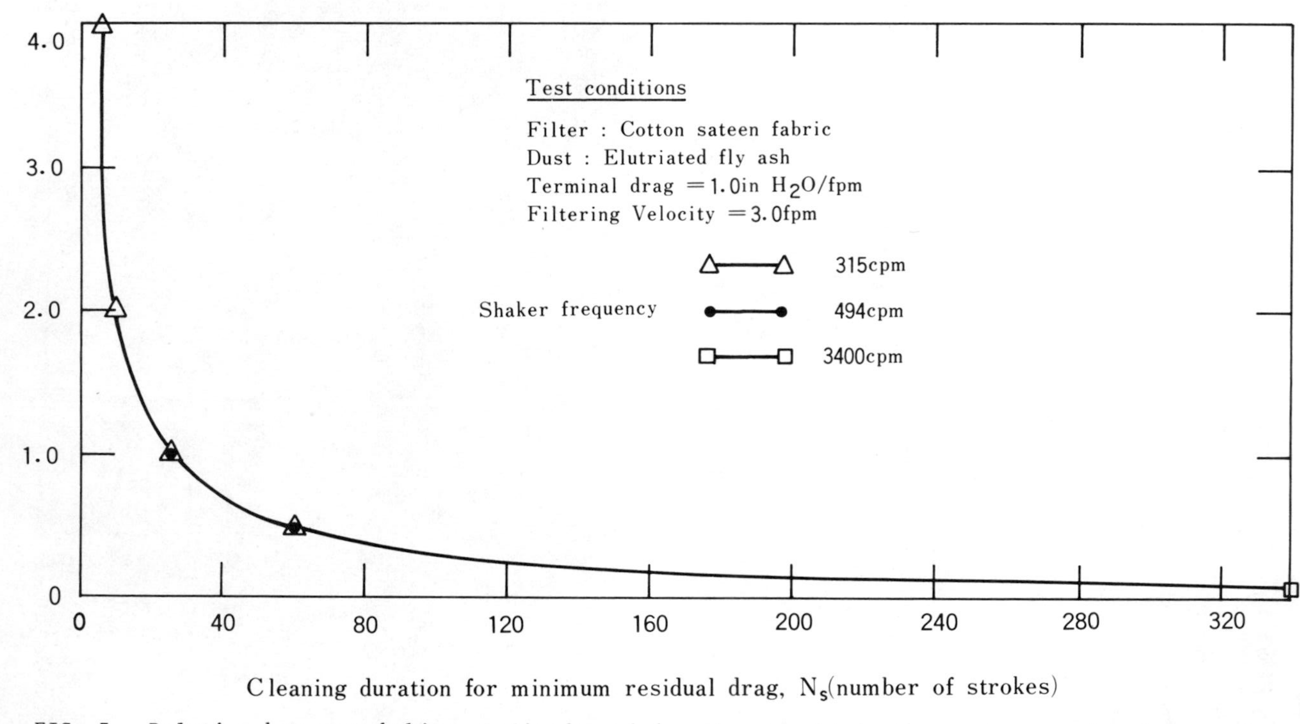

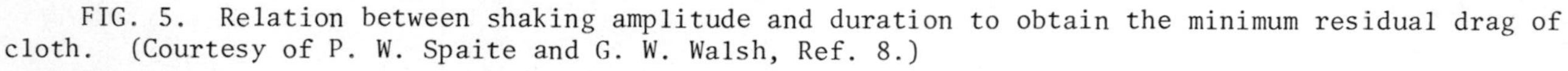
FIG. 5. Relation between shaking amplitude and duration to obtain the minimum residual drag of cloth. (Courtesy of P. W. Spaite and G. W. Walsh, Ref. 8.)

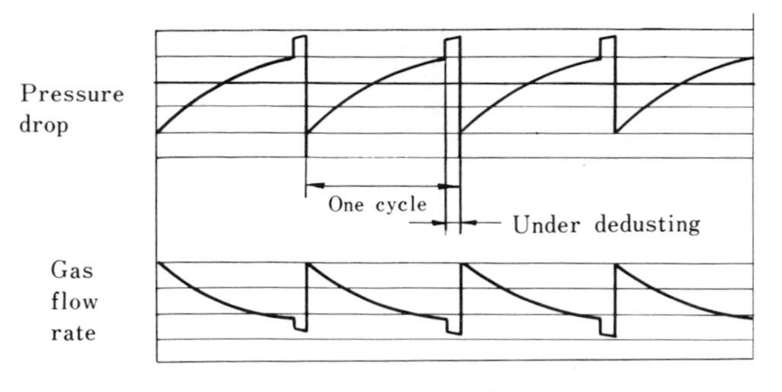

FIG. 6. Variations of pressure drop and gas flow rate for a conventional bag filter with mechanical shaking.

A new type of reverse flow cleaning is shown in Fig. 7. This filter incorporates a dedusting system making use of reversed and pulsating air flow. The bag frame, utilizing leaf-type filters, is fixed to a plate provided in a cylindrical structure. The air pulsation generating unit for bag cleaning is mounted on top of the tank.

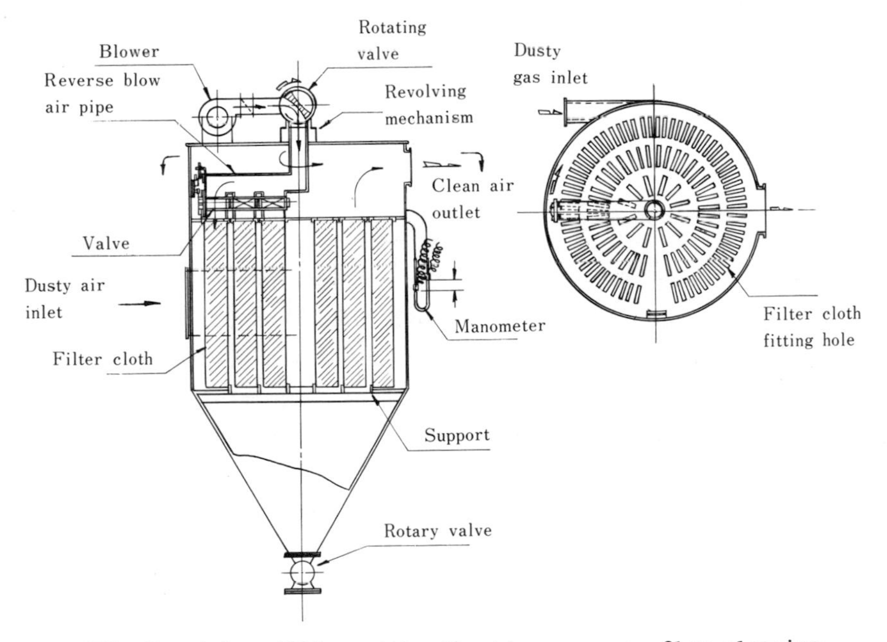

FIG. 7. A bag filter with vibrating reverse-flow cleaning. (Kurimoto.)

A blow pipe to convey air pulses from the upper bag frame is provided in the clean gas chamber, and a slide valve revolving on the plate to which the filters are fixed is attached to the end of the blow pipe. Most of the cloth elements continue the filtration, but one filter at a time overlaps the slide-valve arrangement and is cleaned. Thus, the unit as a whole carries out filtration and cleaning continuously and simultaneously. Amplitude of the cloth is less than 1 mm; the vibration frequency is 1,000-1,500 c/min; the reverse air pressure is 400-500 mm H_2O; and the vibration period is about 0.5 sec. Dust adhering on the cloth is released by the vibrating action, even if the dust contains moisture, oil, or mist. Industrial filters contain from 15 to 200 individual filters. As a result, the variations in pressure drop and gas flow rate are much less than in many other conventional reverse-flow filters. When inert gas is used for cleaning, even explosive dusts can be handled safely.

B. Reverse Jets

In recent years, cleaning techniques have been developed that allow significant increases in the air-to-cloth ratio without affecting the cloth life and maintenance requirements. Ejector cleaning, for example, is capable of operating at air-to-cloth ratio up to 5 (m/min).

In this type of filter system, unlike the conventional shaker or back-flow type, gas flows from the outside to the inside. The filter elements are supported with wire retainers to maintain their shape. As dust accumulates on the filter elements, periodic cleaning is accomplished by the introduction of a momentary jet of high-pressure air. Figure 8 shows the construction of this collector as well as the flow of air. The three filter elements on the left are filtering dust-laden air, while the one on the right is being cleaned. This is the so-called pulse-jet type of filter.

A solenoid valve on a compressed-air line is energized periodically with a timer for about 0.1 sec to blast air inside each filter

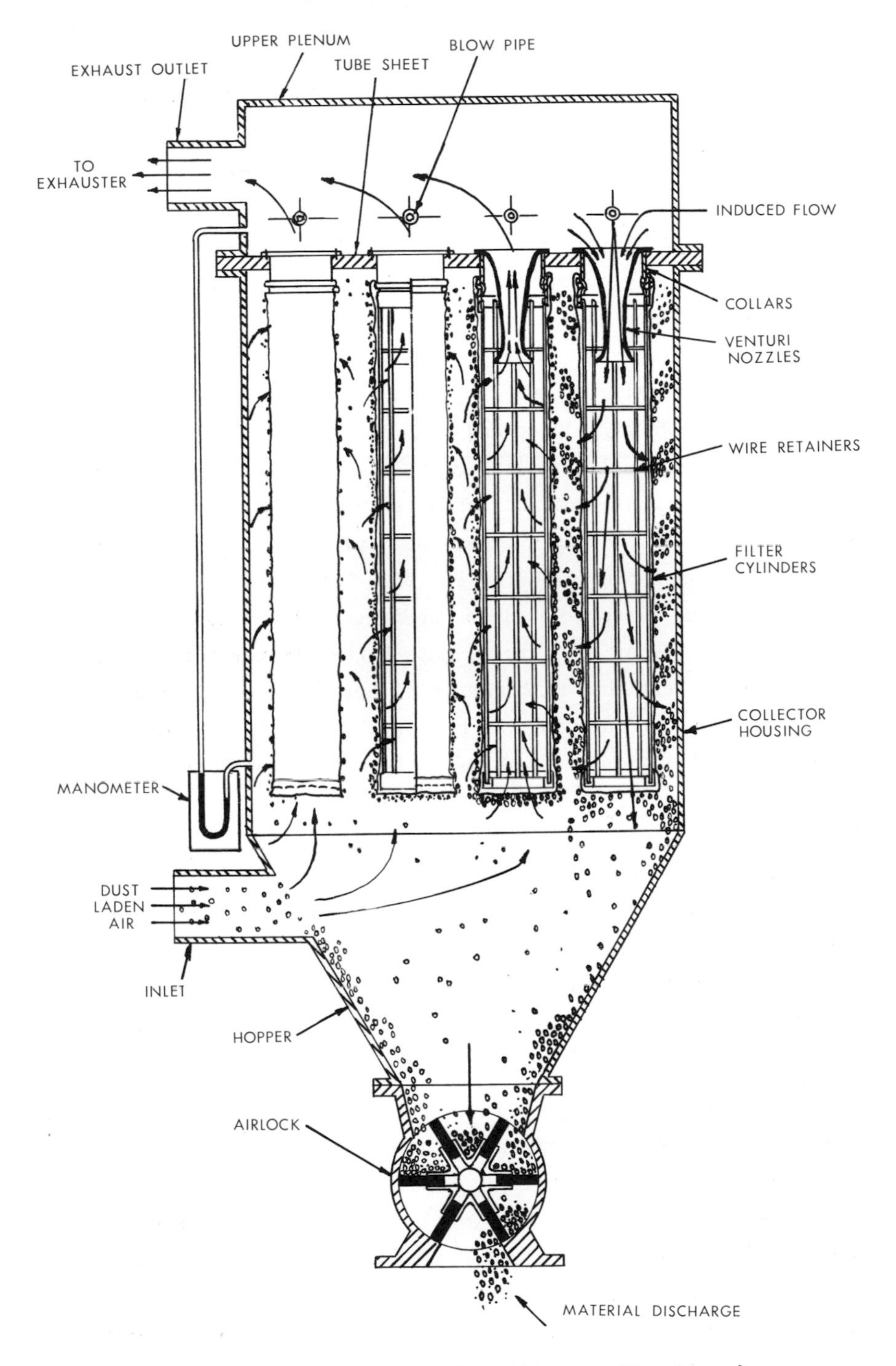

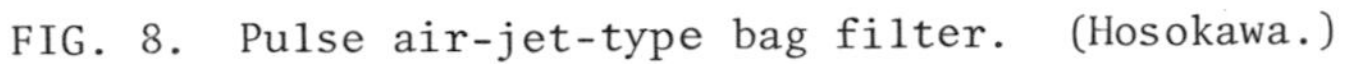
FIG. 8. Pulse air-jet-type bag filter. (Hosokawa.)

element. This primary high-pressure air amounts to a temporary 5 to 7 times increase in volumetric flow rate and thus produces a momentary expansion of the filter elements. Deposited dusts are thus swept away from the filter. Pulses occur usually every 6 to 10 sec. The disadvantage of this type of filter is that it is difficult to achieve effective cleaning action with long bags. Table 2 presents typical performance data.

An alternative type of filter is called the reverse jet, which utilizes air emerging from a slit in a ring that encircles each bag. The ring moves vertically up and down the outside of the entire bag length. The high-velocity air jet passing through the cloth removes the accumulated dust from the inside surface of the bag. Figure 9a shows the principle of this type, and a picture of the jet-ring mechanism [10] is presented in Fig. 9b. This type is mechanically complicated and is restricted to specific applications [11].

With both types of filters, much higher filter ratings can be achieved with felt than with fabric filter media. Filter cleaning times are of short duration, so pressure fluctuations are almost negligible. Therefore, it can often be safely considered that all the bag surfaces are in filtering service all the time. However, this is sometimes inadequate as the jet stream may be ineffectual in cleaning. Proper design then requires that each bag be sealed off as it is blown free of deposit, thus producing a positive back-flow action [12]. This development greatly extends the range of applications and the reliability of the jet-type filter while retaining its other advantages [13].

C. Sonic Cleaning Method

Occasionally, sonic generators are used to provide additional fabric vibrations for cleaning action. Sonic generators thus shake the dust loose from the collecting fabric. They are sometimes used to supplement repressuring and reverse-flow cleaning. Some carbon black, zinc oxide, or cement installations are using repressuring and sonic horns to clean fiber glass bags [7] which deteriorate

TABLE 2 Typical Performance Data for Pulse-air Filters

Filtered material	Bag material	Filtering rate [(m^3/min)/m^2]	Temp. (°C)	Dust loading (gm/m^3)	Pressure differential (mm H_2O)
Acrawax	Dacron	2.7			127
Acrawax	Wool	2.5			150
Alumina	Orlon	1.8	82-93	6.9	89
Alumina	Wool	3.0	66	275	89
Aluminum hydrate	Wool	1.2		275	76-89
Aluminum hydrate	Wool	2.0	49-54	275	76-89
Aluminum silicate	Orlon	3.2			
Ammonia perchlorate	Wool	3.3	21-27	115-344	180-200
Asbestos	Wool	2.7	49	170	24-27
Asbestos	Wool	3.7	21-32	6.9-9.2	76-127
Aspirin	Dacron	3.05	75	1.6	75
Aspirin	Orlon	3.0	49	6.9	
Attaclay	Orlon	2.3	121	18.5	
Bentonite	Orlon	2.2	93	2.3	
Benzazimide	Orlon	2.5	60	41	118
Bone charcoal	Wool	3.7	Amb.	11.5	102
Borax	Wool	4.0	43	4.6-9.2	
Carbon black	Dacron	2.2	60-127	2.3-18	15
Carbon black	Orlon	1.3	82	23	178
Carbon black	Wool	1.5		4.6-46	63
Carbon black		3.0	30	16	254
Cement	Wool	1.7		458	
Cement	Wool	2.0		458	
Feed, animal	Wool	2.1		755	
Flour	Wool	2.6	66	480	
Flour	Wool	3.0		149	
Fly ash	Dacron	2.0	121	9.2-92	
Graphite	Orlon	0.6	27-32	1.15	127
Gypsum	Orlon	3.0	72	4.6	89
Iron oxide	Wool	2.2	Amb.	4.6	50-89
Magnesium oxide	Dacron	1.74	32	66	76-102
Metal powder	Wool	2.2	66-93	103	38-51
P.V.C.	Orlon	2.26	66	160	76
Rubber chemicals	Orlon	3.2	32	46-69	
Sugar-flour mix.	Wool	2.13	4	182	25-102
Sulfur	Orlon	1.8	121	137	
Uranium compound	Wool	3.6	30	21.5	152

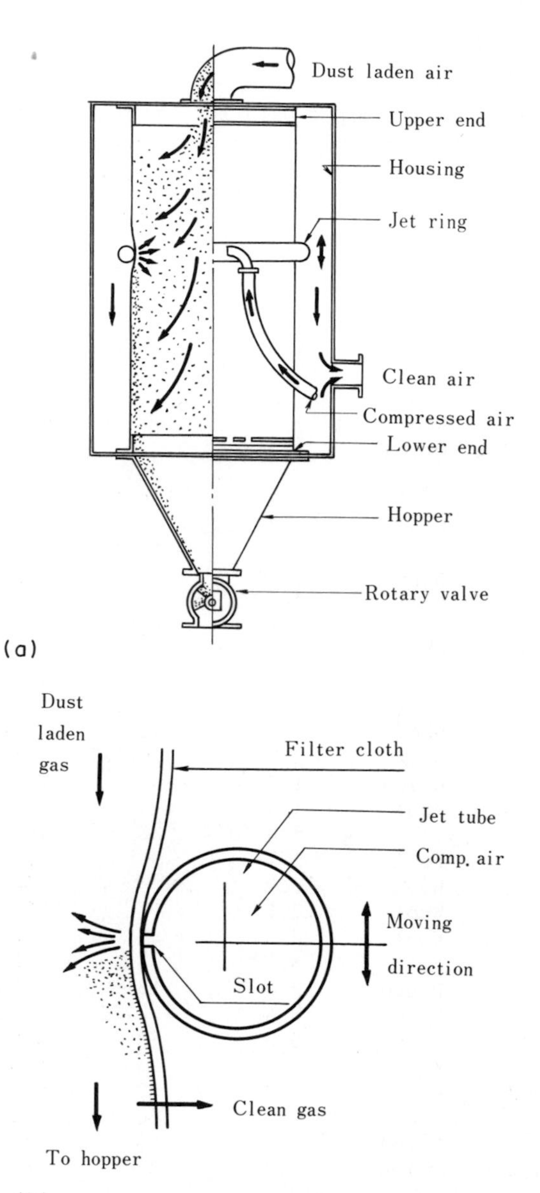

FIG. 9. (a) Principle of the reverse-jet-type bag filter. (b) Inside of the reverse-jet-type bag filter. (Hercy-Sanko.)

rapidly when subjected to repeated mechanical forces, especially at high temperatures. Figure 10 gives an example of an installation incorporating low-frequency sonic cleaning. The sound is usually produced by plate vibration, siren, or resonance whistle, the latter two using compressed air. Effective sound intensity needs to be more than 155 db; however, noise levels should be barely discernable outside the filter baghouse.

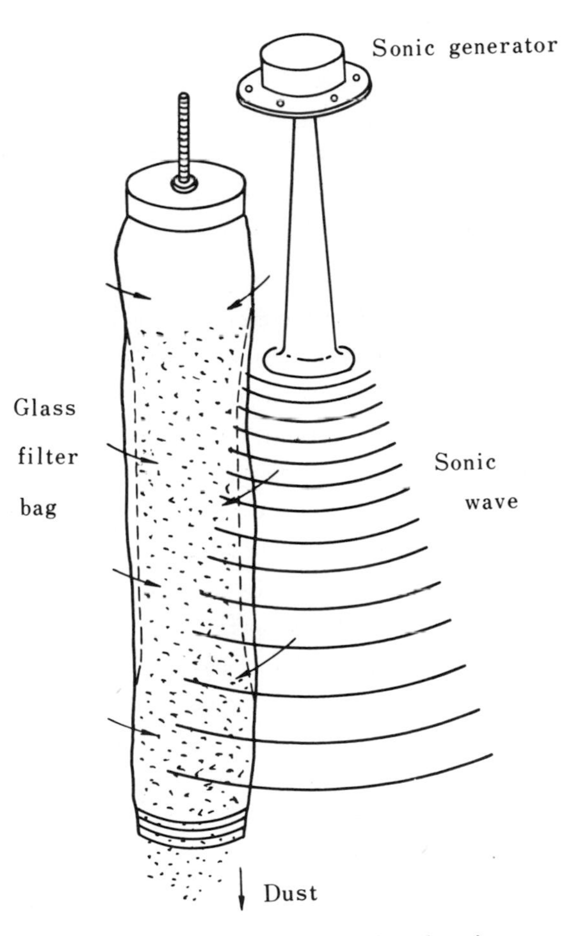

FIG. 10. Principle of sonic cleaning.

III. FIBROUS MAT FILTERS

Fibrous air filters are usually produced in rectangular forms or as continuous rolls. Application of mat filters is always restricted to low dust concentrations, e.g., air ventilation systems, because these filters are intended to be replaced once the dust deposit becomes excessive. Superficial gas velocity with mat filters usually range from 0.1-3 m/sec, and their highest pressure drops from 20-100 mm H_2O, depending on the intended performance. The media of mat filters should be fireproof; while most are discarded after a period of service, some mat filters can be washed with water and reused. In the automatic roll filter, the filter medium seeing service is intermittently replaced by the action of a pressure switch or a timer.

A. Continuous Types

Figure 11 shows an example of the continuous-roll type of fibrous mat filters [14]. The maximum dust load (holding capacity) on such a filter is usually restricted by an allowable maximum pressure loss, which is set by the blower performance characteristics. As the filter medium moves across the air stream, the build-up of the dust layer causes a local reduction in the air velocity through the medium for a constant pressure loss. Wide fluctuation in the total gas flow rate can be minimized in this type of filter by an appropriate schedule of shifting the medium.

Electrostatic fibrous air filters have both the merits of mechanical filters, i.e., easy care and low cost, and those of two-stage electrostatic precipitators, i.e., high collection efficiency for fine particles. Such filters are composed of charging zones and collecting zones as shown in Figs. 12a and 12b. Dust particles in the air are charged in a positive corona-discharge field within the charging zone and collected by the dielectric fibrous filter in a high electric field. Figure 13 shows some experimental data on the performance of a dielectric glass-fiber mat filter without corona discharge [15]. Alternating electric current is also effective, and a higher dielectric constant for the fiber is favorable.

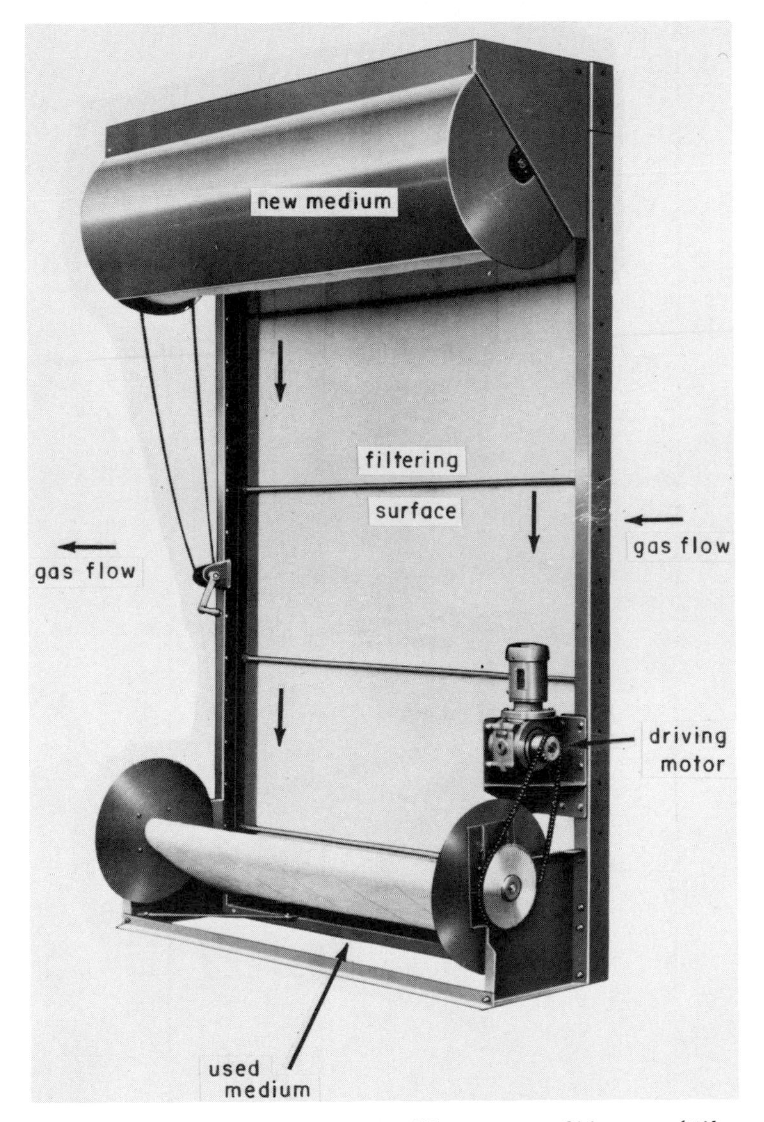

FIG. 11. Continuous-roll-type fibrous mat filter. (Nihon Spindle.)

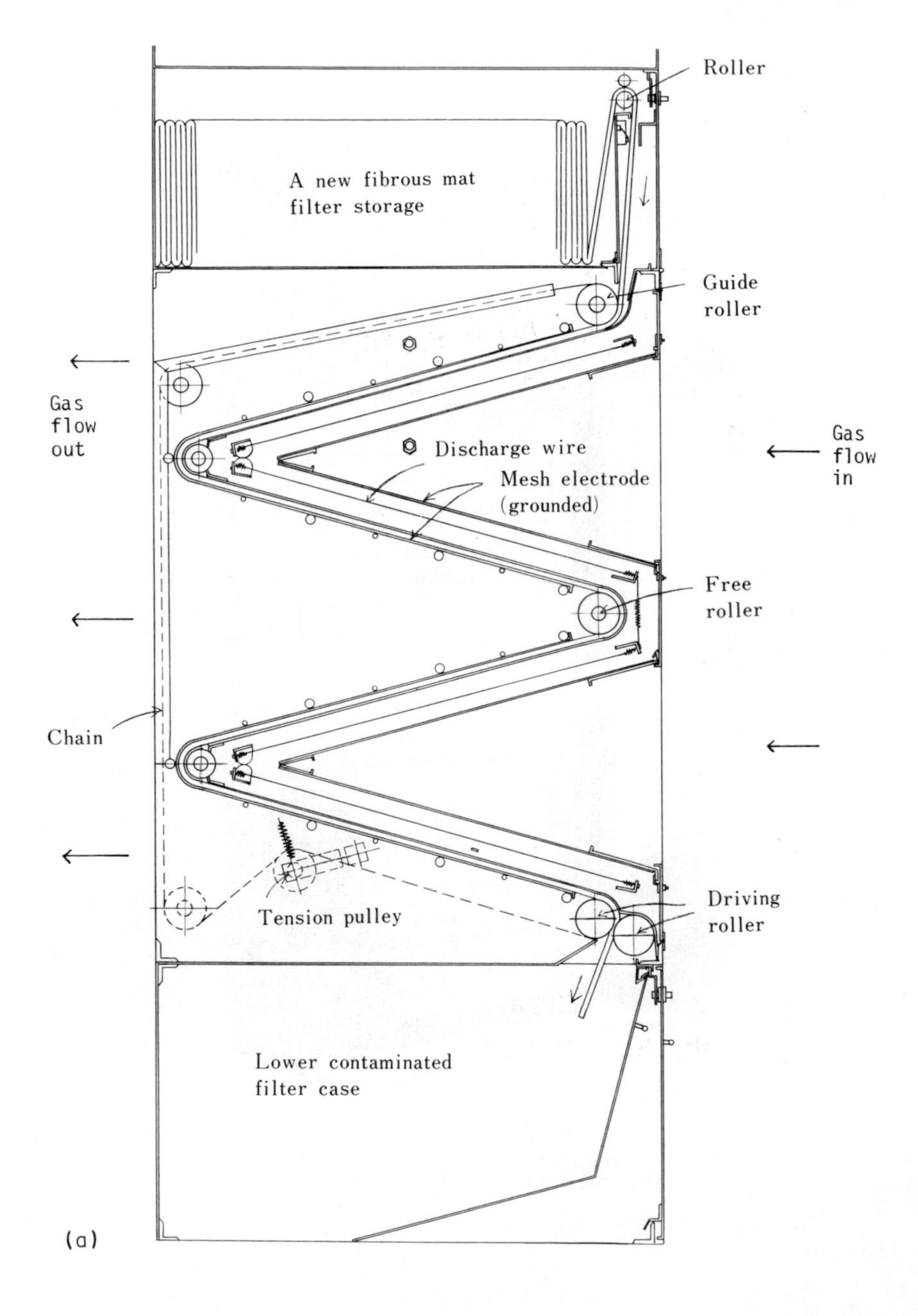

Roller
A new fibrous mat filter storage
Guide roller
Gas flow out
Gas flow in
Discharge wire
Mesh electrode (grounded)
Free roller
Chain
Tension pulley
Driving roller
Lower contaminated filter case
(a)

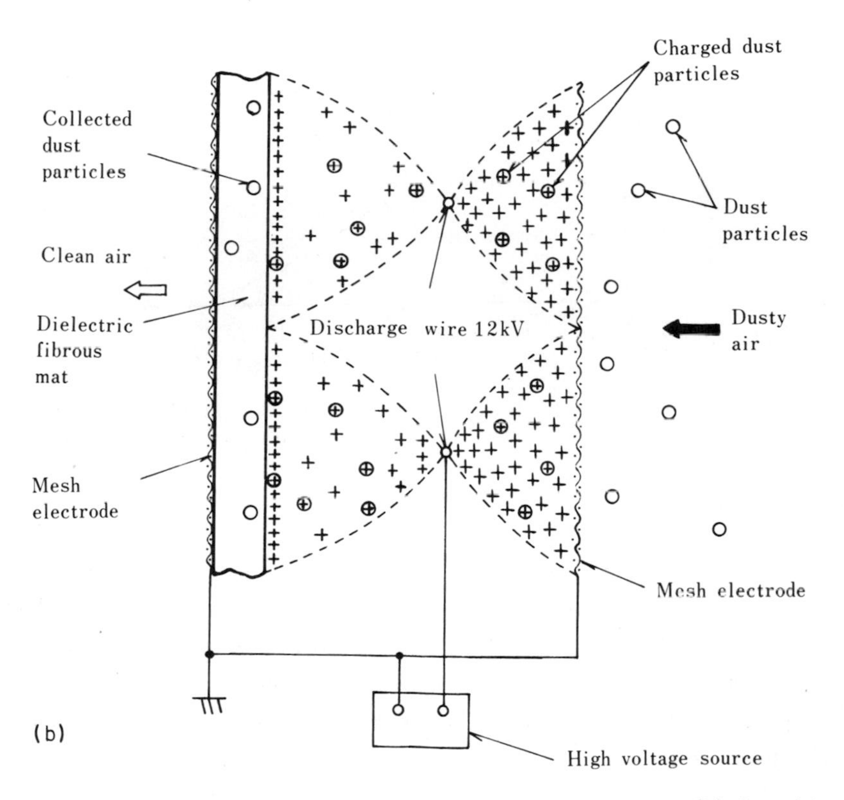

FIG. 12. (a) Electrostatic fibrous mat filter. (b) Details of electrostatic fibrous air filter. (Mitsubishi Electric.)

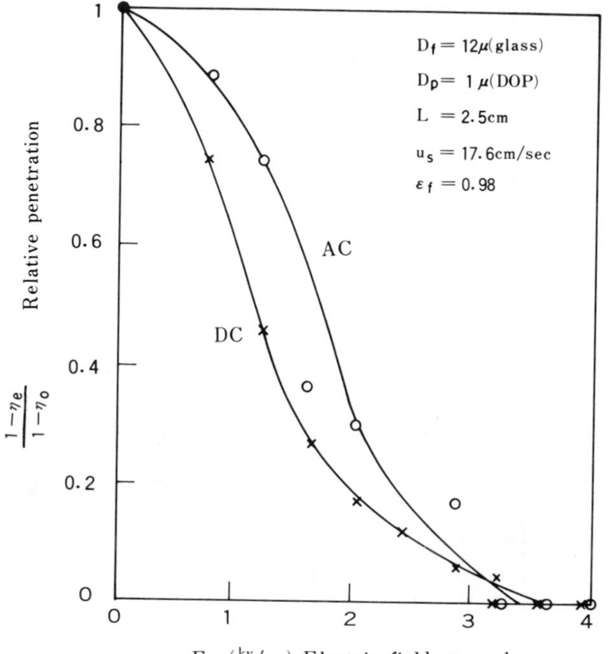

FIG. 13. Effect of electric voltage on the performance of dielectric glass-fiber mat filter (by light-scattering penetrometer).

B. Panel Types

Panel filters composed of meshed wire or synthetic and glass fibers are included in the panel filter category. These may be cleaned by washing with water or by blowing off the dust accumulation with a blast of air. Examples of renewable panel-type air filters in which the media are supported by a framework are given as Figs. 14a and 14b. Figures 15 and 16 show alternate types of panel filters.

C. Paper Filters

Two kinds of paper filters are used for gas filtration; one utilizes conventional filter paper similar to that for chemical analy-

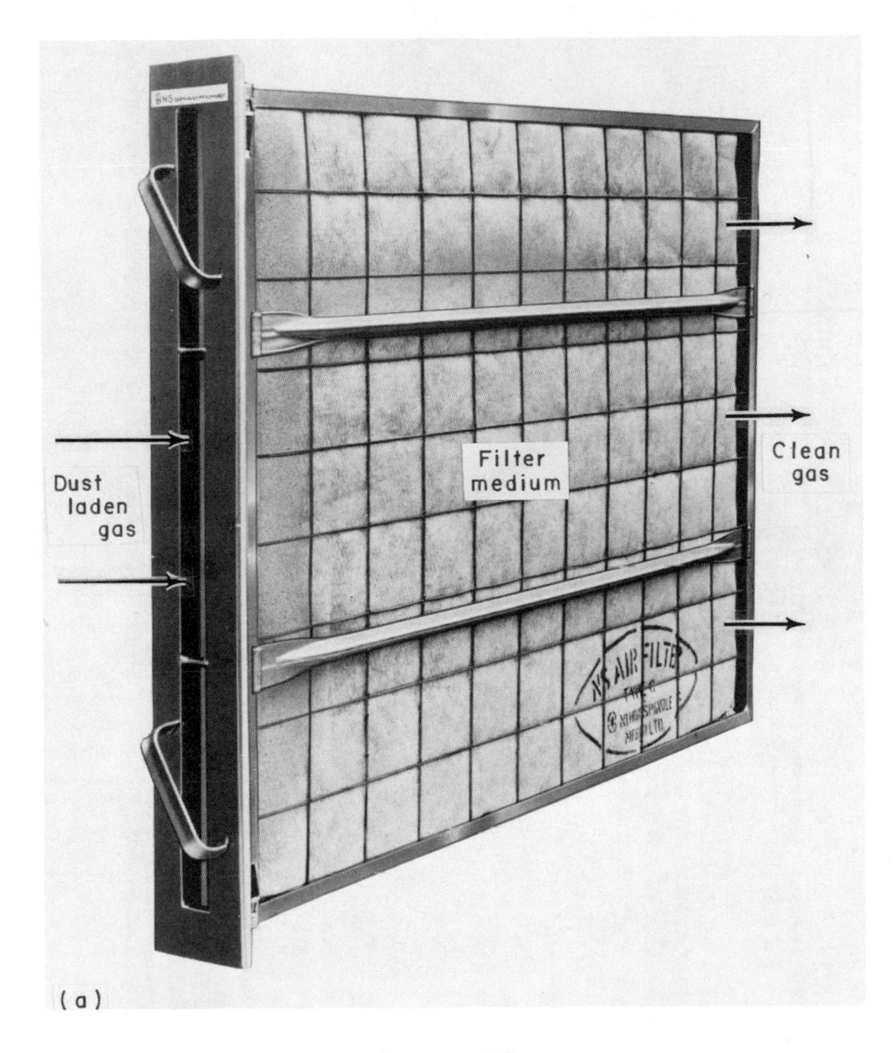

Fig. 14a. See legend on p. 330.

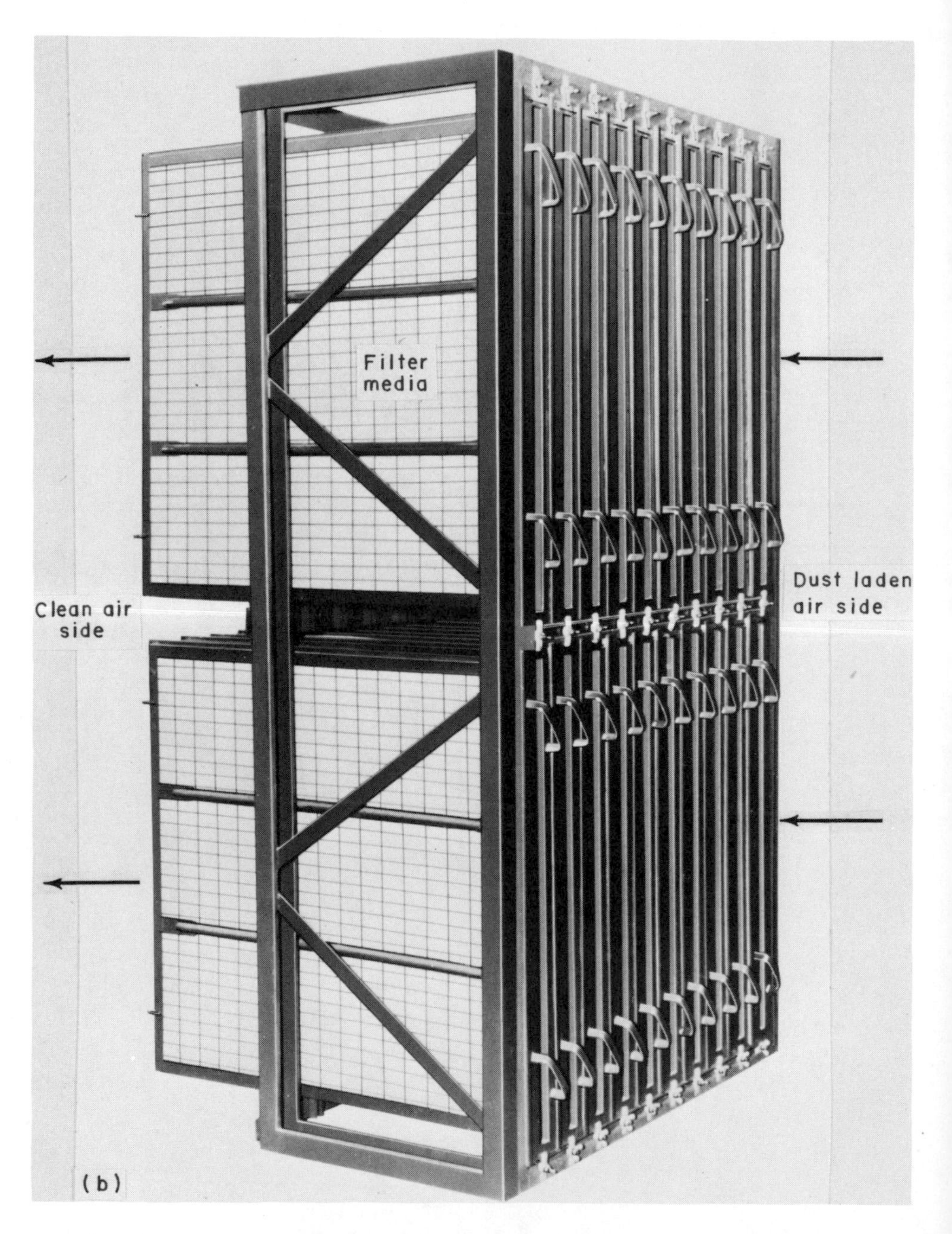

FIG. 14a and b. Recession (V)-type panel filters. (Nihon Spindle.)

FIG. 15. Renewable panel-type fibrous filter. (Nihon Spindle.)

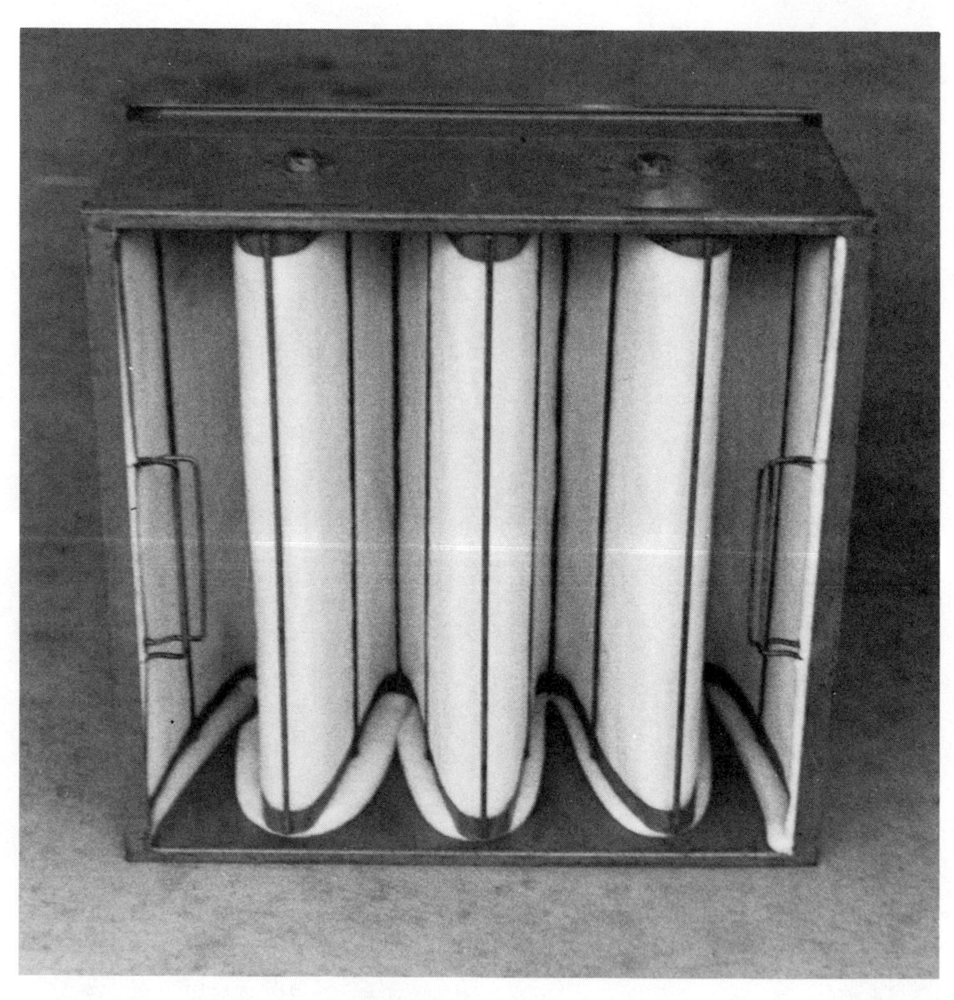

FIG. 16. Panel-type fibrous filter. (Nihon Spindle.)

sis having less than 90% collection efficiency for 0.3 μm particles [16], and the other is a high-performance filter paper which has about 99.9% collection efficiency for the same particle size. The former is employed in engine air cleaners, household vacuum cleaners, and the like. Pressure drop with them is about 10 mm H_2O at 1 cm/sec of superficial filtering velocity at no dust load and is proportional to the velocity either with or without dust load (deposit). Correlations between pressure loss and deposit dust load for various kinds of paper filter [17] are presented in Figs. 17 and 18.

D. High-Performance Filters

High-performance paper filters are used for emission control in atomic reactor installations [18], in clean rooms, and other situations requiring tight control. Their performance characteristics are shown in Figs. 17 and 18 along with those of conventional paper filters, the pressure-drop behavior of both being similar. The relation between the collection efficiency of a single fiber (η_ε) and pressure drop (Δp) with dust load as a parameter is generally taken as follows:

$$\frac{\eta_{\varepsilon d}/\eta_\varepsilon - 1}{\Delta p/\Delta p_o - 1} = (5.2 \times 10^{-3})m^{-2/3} \tag{1}$$

where $\eta_{\varepsilon d}$ is the collection efficiency of a single fiber with dust load and Δp_o the pressure drop of the filter medium.

The so-called High-Efficiency Particulate Air Filter (HEPA) has a minimum collection efficiency of 99.97% for 0.3 μm diameter dioctylphthalate (DPO) smoke particles and a maximum pressure drop of 25 mm H_2O when clean and operated at its rated air flow capacity [19]. Such a highly efficient filter can remove effectively airborne particles in sizes as small as 0.1 μm diameter [20].

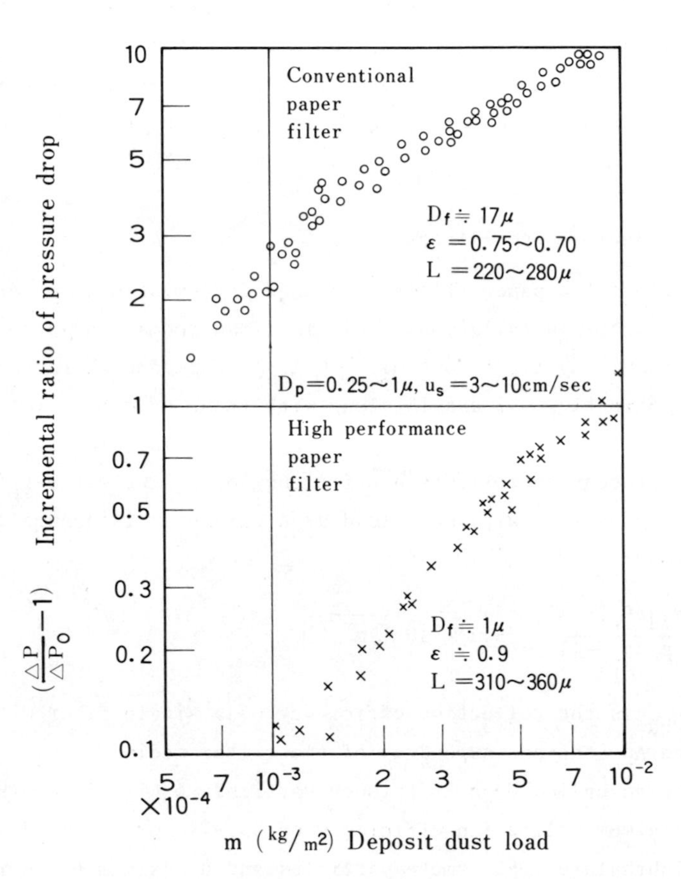

FIG. 17. Relation between incremental ratio of pressure drop across a filter and dust load.

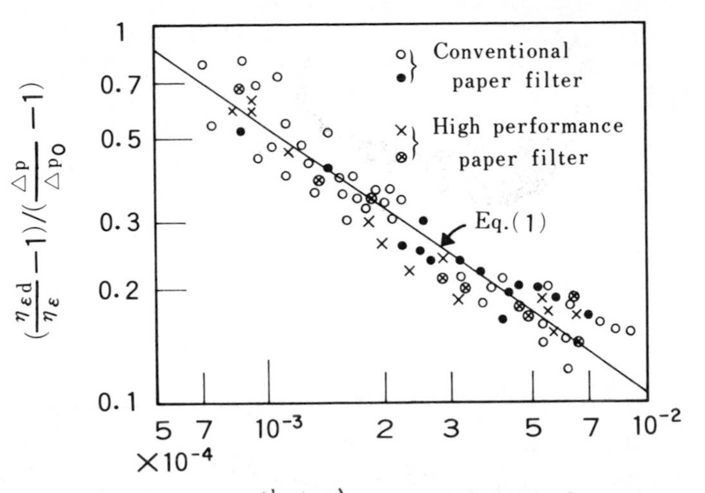

FIG. 18. Typical performances of two kinds of paper filters. Note: $1\mu < D_f < 18\mu$; $0.22mm < L < 0.31mm$; $0.71 < \varepsilon_f < 0.91$; $0.25\mu < D_p < 1\mu$; $3cm/sec < u_s < 10cm/sec$.

IV. MIST COLLECTORS

There are mainly two types of mist collectors. The one for fine mists is like a packed wire-mesh mat filter [21] as shown in Fig. 19. The other is a kind of inertia separator, e.g., a louver or zigzag separator, for coarse mist having particles larger than 10 μm. Figure 20 shows typical particle-size distributions for sulfuric acid mists [22]. There have been relatively few investigations of mist collection, even though the problem is widely encountered not only in chemical processing but in industry as a whole.

Fluidized beds are often employed as mist collectors. For a given bed, the removal efficiency of acid mists, composed of 2 to 14 μm diameter droplets, is improved with increasing bed weight and with increasing superficial gas velocity. The efficiency is substantially constant during the life of the bed and is independent of the entering mist concentration. However, beds of nonporous materials show an impractically short life [23]. The development of composite meshes has increased the separation efficiency for the

FIG. 19. A typical mist collector called "Demister®." (Courtesy of Otto H. York Co. Inc., Ref. 21.)

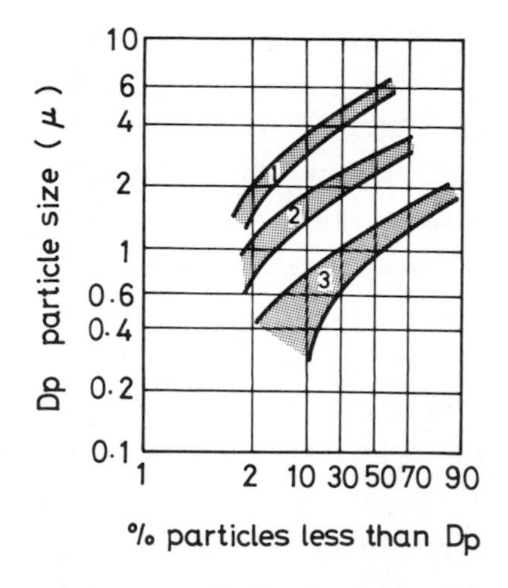

FIG. 20. Typical ranges of particle size of acid mist in absorber tower tail gas from sulfuric acid plants. (O. H. York Co. Inc.)

removal of fine-size mists [24]. Fiber mist eliminators for superficial velocities of 10-30 m/min are smaller and less expensive than conventional designs [25].

V. AGGREGATE AND POROUS FILTERS

Aggregate filters are exemplified by coke beds for acid mist recovery, gravel beds for exhaust gas cooling and collection [26], and beds of plastic or ceramic shapes (spheres, Raschig rings, etc.). Applications of this type of filter are restricted to special conditions, such as high temperatures or corrosive-gas treatment. They are operated both statically and dynamically. Dynamic operation achieves regeneration of the filter media by means of fluidization or continual, slow removal and addition of aggregate. A static bed must be cleaned by replacing, vibrating, agitating, or washing the filter media. Figure 21 shows a typical aggregate bed filter with a glass sphere packing and a water spray for regeneration.

Two kinds of porous filters—a soft (foam rubber) type and a rigid (ceramic or sintered metal) type—are employed for small-scale industrial gas filtration. The foam rubber filter is often employed for airline cleaning, e.g., for process control air, and the rigid filter for process gas sampling. Figure 22 shows a ceramic filter for gas sampling.

VI. PERFORMANCE CRITERIA

Gas flow rate, pressure drop, collection efficiency, and durability are the four performance characteristics of primary concern in industrial gas filtration. Each is judged from the economic standpoint.

The gas flow rate is established from the process and the environmental requirements. A complete filter unit should be designed with an upper limit to the flow rate because, in general, lesser rates result in better performance, i.e., higher efficiency and lower pressure loss. The service life of the filter media is usually the most important factor economically.

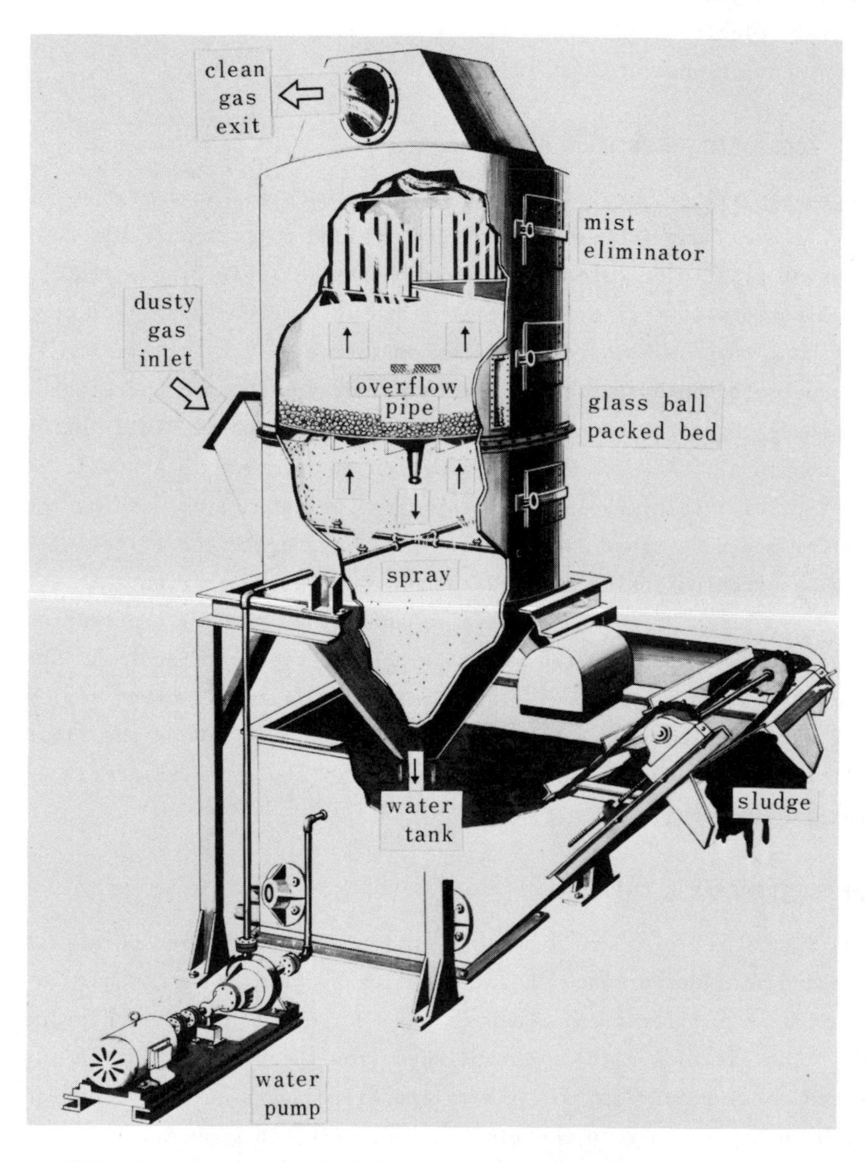

FIG. 21. Aggregate bed filter with ball packing and water spray. (Shinto Dust.)

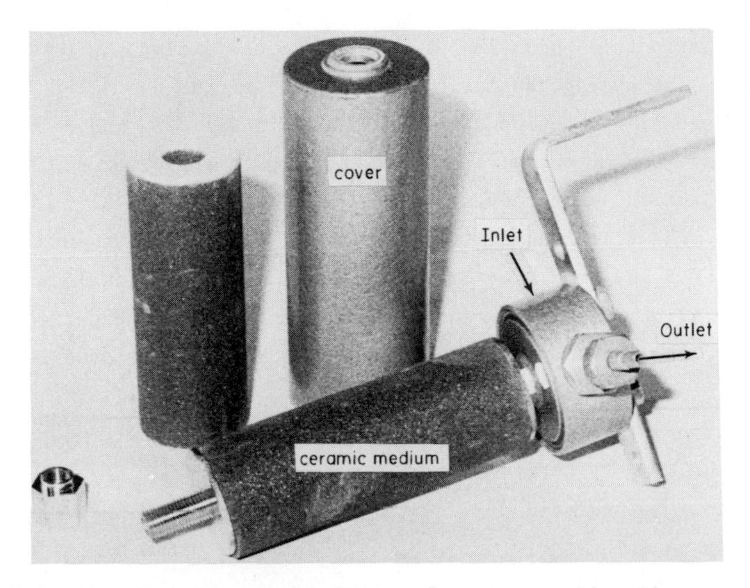

FIG. 22. Porous ceramic filter for gas sampling line. (Shimadzu.)

A. Pressure Loss

Pressure loss (or drop) Δp affects the main power consumption of most filter systems as follows: Power requirement HP $\propto Q\Delta p$, where Q is the gas flow rate by volume.

Power cost, on the average, is more than 10% of the aggregate annual cost. The pressure loss is the sum of losses due to the filter medium and the collected particle layer. It is usually proportional to the superficial filtering velocity u_s and gas viscosity μ because laminar flow is maintained in most filters. The following equation is appropriate for a fabric filter:

$$\Delta p = \Delta p_o + \Delta p_d = (\zeta_o + \zeta_d) \frac{\mu u_s}{g_c} \tag{2}$$

$$\zeta_d = \bar{\alpha}\, m \tag{3}$$

The loss Δp_o of the filter medium is negligibly small in comparison to that of the dust layer Δp_d on a bag filter. However, the pressure loss coefficient ζ_o for a clean filter, which depends upon the fabric structure, has some effect on the coefficient ζ_d of the dust layer. The average specific resistance $\bar{\alpha}$ of the deposited particle layer usually is not constant, as shown in Fig. 23, and depends upon the dust load m, the particle size, the volumetric voids of the layer, and the characteristics of the filter medium [27]. For clean fabric, the filtering air velocity u_s at which the pressure loss Δp_o equals 0.5 in. of H_2O (= 12.7 mm H_2O) is called the permeability. The relation between the above coefficient ζ_o and the permeability is given as shown in Fig. 24. Figure 25 gives a performance curve in terms of the pressure loss ratio for a typical filter cloth. The following values are common at normal operating conditions for conventional bag filters:

Δp = 50-250 mm H_2O m = 0.02-0.3 kg/m^2

Δp_o = 5-20 mm H_2O u_s = 0.5-5 m/min

$\bar{\alpha} = 10^9 - 10^{11}$ m/kg

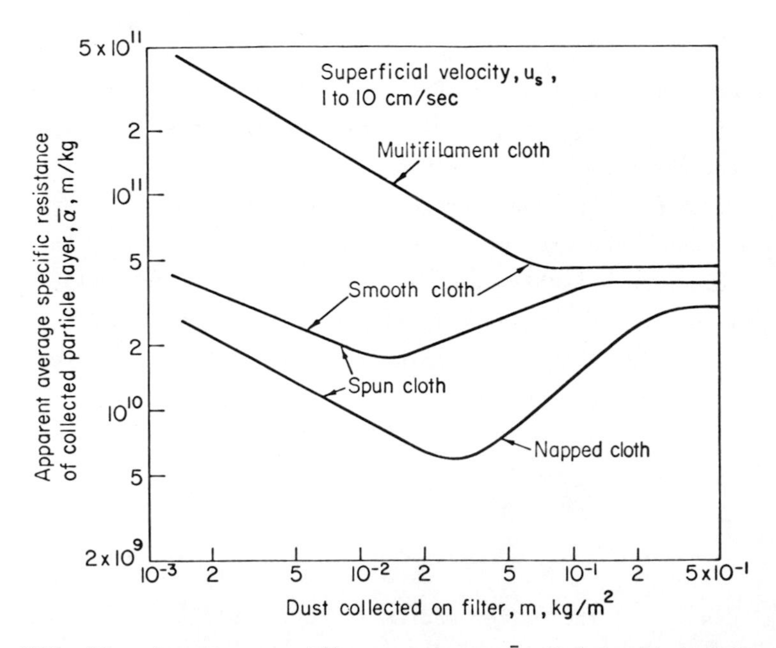

FIG. 23. Average specific resistance $\bar{\alpha}$ of deposit dust layer on filter cloth.

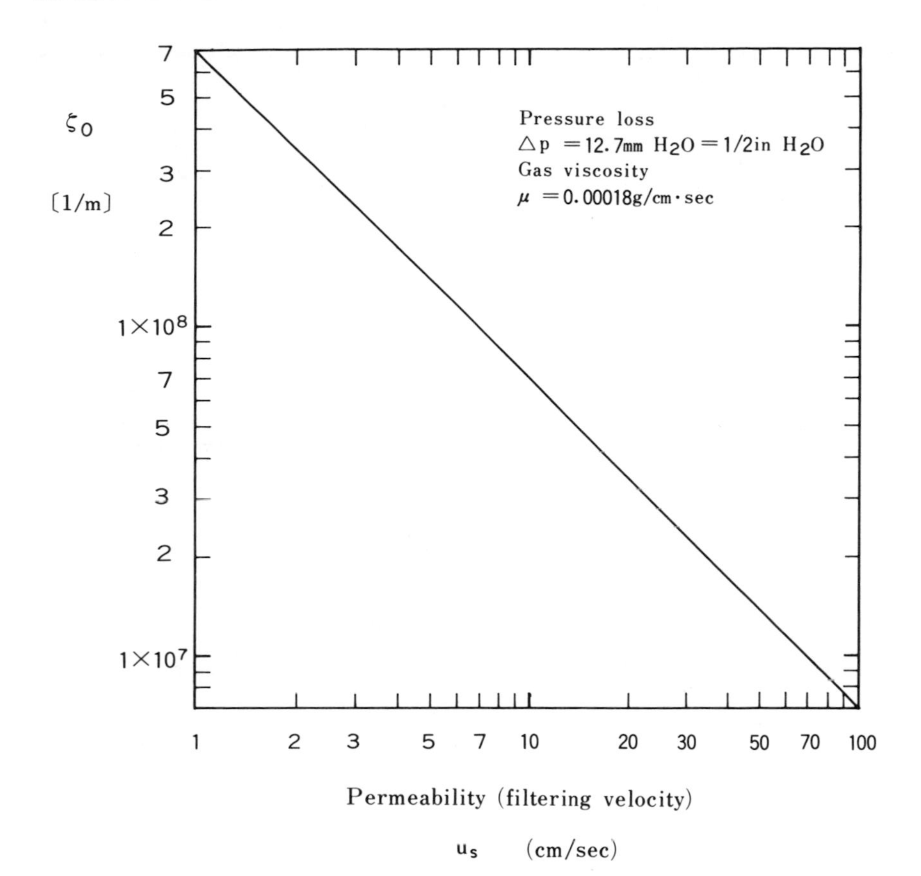

FIG. 24. The relation between pressure loss coefficient ζ_0 and permeability of filter cloth at atmospheric pressure.

For a fibrous filter of uniform fiber diameter, the pressure loss of a clean filter may be estimated by use of the following equations [28] and the curve of Fig. 26:

$$\Delta p = C_D \frac{2\rho L(1 - \varepsilon)u_s^2}{\pi g_c D_f \varepsilon^2} \tag{4}$$

$$C_D = 0.6 + \frac{4.7}{\sqrt{Re}} + \frac{11}{Re} \tag{5}$$

where C_D is the pressure drag coefficient, Re (= $D_f u_s \rho/\mu\varepsilon$) the

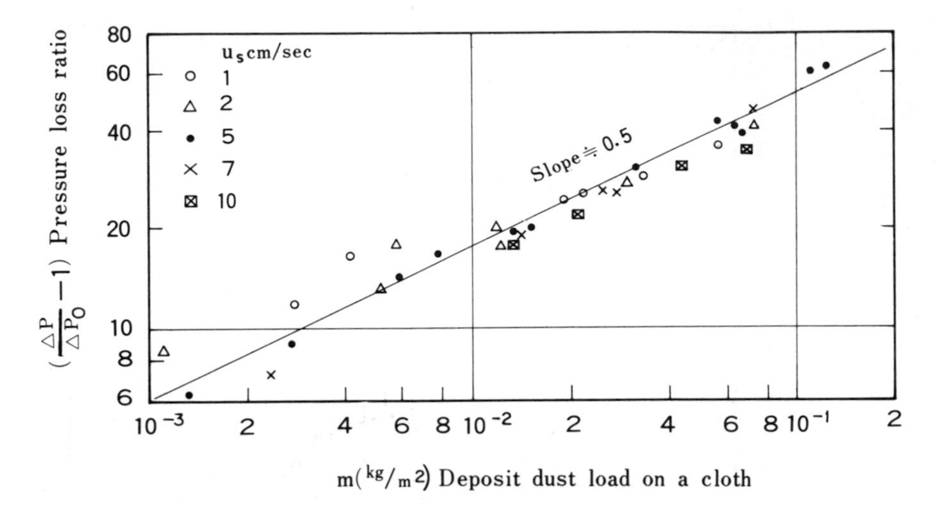

FIG. 25. Pressure loss ratio of filter cloth with dust load.

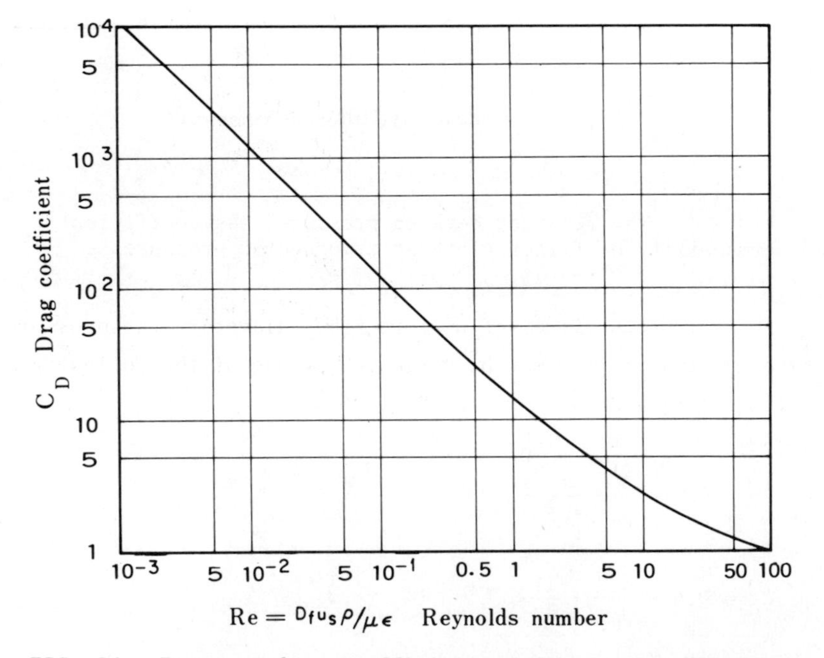

FIG. 26. Pressure drag coefficient of fibrous mat filter.

Reynolds number based on the fiber diameter and is usually less than 10 because of laminar flow, ρ the density, L the thickness of the filter, ε the volumetric void of the filter (0.98-0.98), g_c the gravitational conversion factor, D_f the fiber diameter, and u_s the superficial filtering velocity (0.1-3 m/sec).

The pressure loss of a fibrous filter with dust load is not estimatable in advance by any analytical or experimental method. The maximum value of loaded pressure loss depends on the process and many other factors. It will usually be less than 50 mm H_2O for ventilating or air cleaning purposes.

Open-pore foam filters show little pressure increase during dust loading, unlike conventional fibrous filters [29].

B. Collection Efficiency

Collection efficiency represents the main purpose of a filter and should be as high as possible. In general, lower filtering velocity gives higher collection efficiency with lower pressure loss. However, it requires a larger filtering surface area, i.e., a bigger installation.

There are at least two kinds of collection efficiency based on the weight of collected material. One is the instantaneous efficiency, and the other is the average, or cumulative, efficiency for a certain time period. They are different because collection ability generally increases with the dust load (deposit) on the filter. Therefore, the initial efficiency of a clean filter is usually the lowest under normal conditions.

Fabric filters nearly always collect over 90% of the entering particulate matter, and many collectors operate at 99%, or higher, during their continuous cycles. In fact, the efficiency is usually so high that it has received insufficient attention. Under usual operations, the collected particle layer on a cloth acts as the filter medium, even just after the cloth has been cleaned by shaking or exposed to a reverse flow. Failure to achieve high collection efficiency is almost always due to excessive cleaning, torn bags, bypass leakage, or an excessive gas flow rate which produces pinholes in the deposit particle layer. Figure 27 shows typical test results of the instantaneous collection efficiency.

On the other hand, the efficiencies of air cleaning or ventilation filters range from 20 to 99%. Figure 28 gives a typical collection efficiency curve for an electrostatic fibrous-mat filter as measured by the Dill dust spot tester. Figure 29 presents calculated collection efficiencies for an isolated fiber with monosize particles.

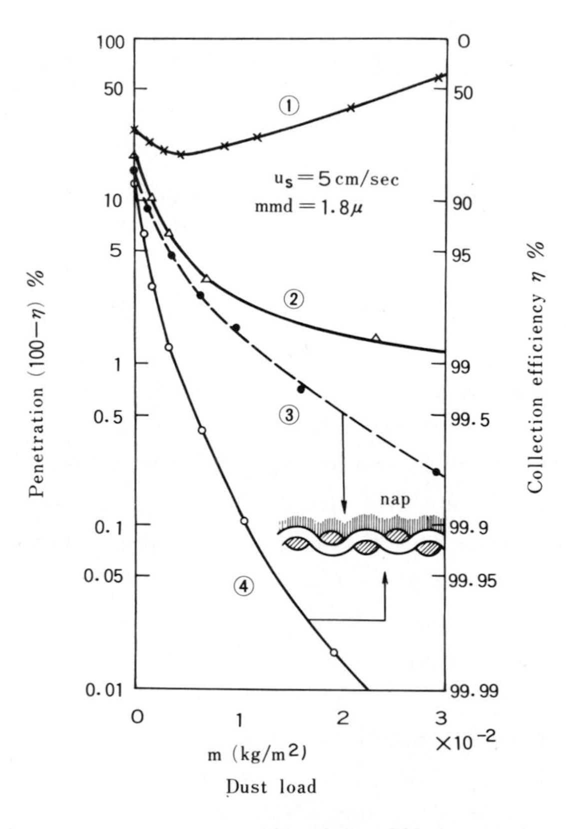

FIG. 27. Instantaneous collection efficiency of a fabric filter with dust load. (1) Blow off through pinholes of dust layer. (2) Little napped, flow in from napped surface (side). (3) Flow in from napped surface (side). (4) Flow in from unnapped surface (side) of a napped cloth.

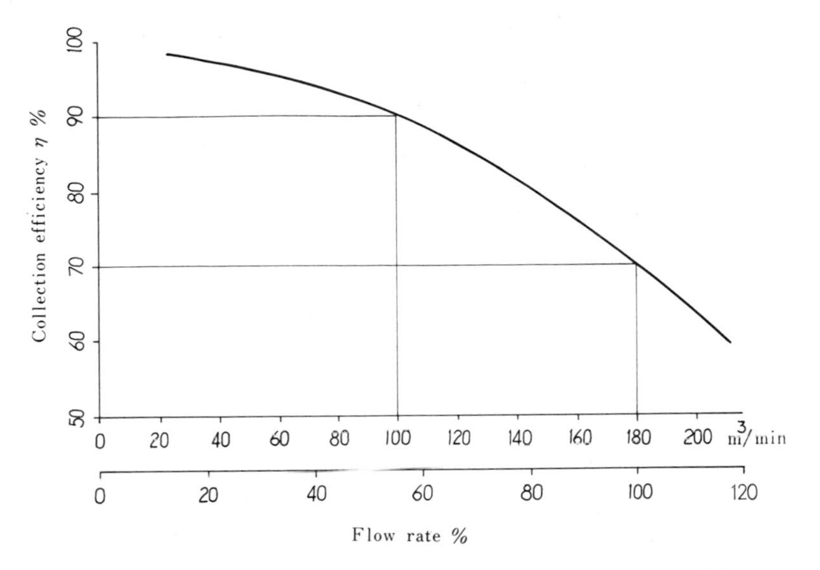

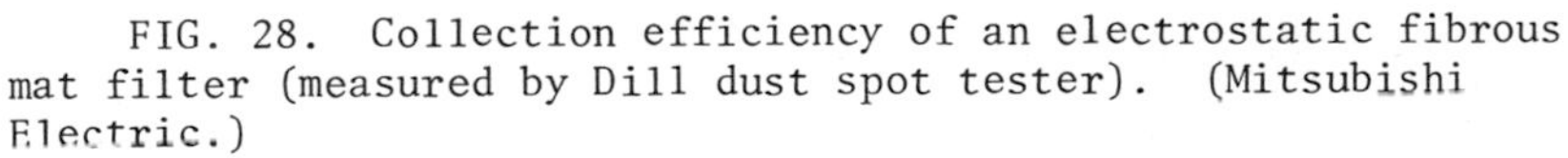

FIG. 28. Collection efficiency of an electrostatic fibrous mat filter (measured by Dill dust spot tester). (Mitsubishi Electric.)

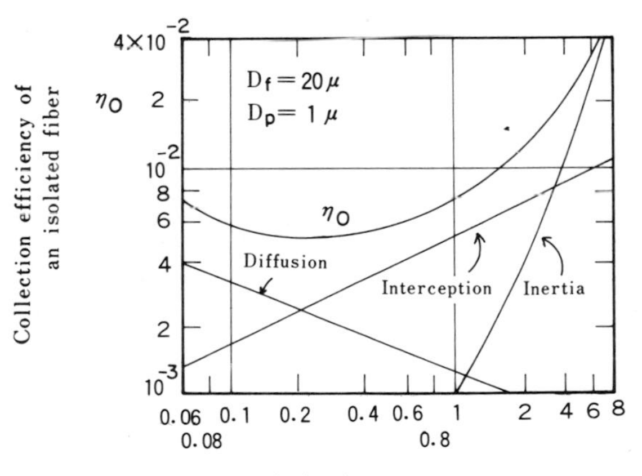

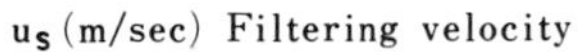

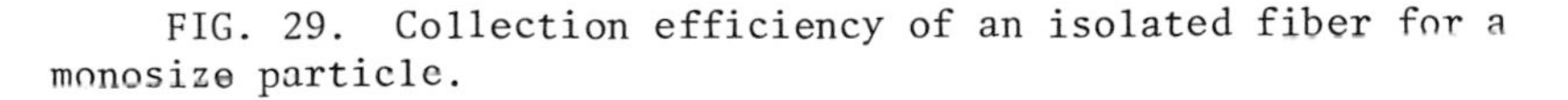

FIG. 29. Collection efficiency of an isolated fiber for a monosize particle.

C. Life Expectancy

Filter life generally is greatly improved if operating temperatures are held below the limits indicated in Table 2. Other factors upon which fabric life depends are the cleaning mechanism, the characteristics of the dust, the nature of the gas, and the care with which the installation is designed for the particular application. Because these vary widely, fabric life is difficult to assess. However, generally one to two years of service is considered acceptable [30].

Frequent or partial replacement of fabric bags in a baghouse is troublesome and expensive. Finishing techniques—such as heat setting, scouring, resin treating, calendering, or napping—will improve bag life, dimensional stability, permeability, and ease of cleaning in some instances.

D. Maximum Dust Load on a Fibrous Filter

Considerable improvement in both economy and dust-holding capacity can be achieved with a roll-mat air filter by continuously renewing the filter medium instead of periodically replacing the entire surface when the pressure loss reaches a maximum value. For any specified maximum pressure drop and the same air flow rate, the dust-holding capacity with continuous replacement is about twice that when the entire medium is replaced periodically, as shown in Table 3. Therefore, it is desirable from the economic standpoint to keep the filter area replacement as small as possible, if continuous replacement is not practical. The total length of filter roll or strip should be seven or more times as long as the length of the filtering surface, because first and last parts of a filtering medium will not be used effectively from the standpoint of dust-holding capacity.

TABLE 3 Maximum Dust Load as a Function of Roll-Type Filter Replacement Schedule[a] (Japan Vilene Company)

Replacement area (%)	Dust holding capacity per filtering area (gm/m^2)	Relative dust entrapment
100 (intermittent)	550	1.00
50	710	1.29
20	840	1.53
10	1020	1.86
5 (nearly continuous)	1100	2.00

[a]Conditions: Maximum pressure loss = 16 mm H_2O. Average air velocity = 2.5 m/sec. Test dust = AFI test dust (fine + carbon black + cotton linter). Dust concentration = 22 mg/m^3. Filter media = FR 285 (trade name). Collection efficiency = 86% by weight.

VII. GAS COOLING

Furnace and other process gases must often be cooled before filtering in order to protect the filter media and to ensure economic filter life. One of three methods—addition of tempering air; cooling by radiation, convection, or waste-heat boilers; and water-spray cooling—or a combination of these methods, is usually employed. It is generally most satisfactory when two or more of the above methods are arranged in a single installation to supplement each other, thereby making possible better temperature control.

A. Air Mixing

Admission of outdoor air permits precise and prompt regulation of temperature when full modulating bleed-in dampers are used. However, because the filter is required to handle a greater volume of gas owing to the tempering air, the filter must necessarily be larger and more expensive, and the operating cost of the blower becomes greater. Therefore, the tempering air method is usually the most expensive one and is generally employed for emergency use or auxiliary cooling in conjunction with automatic temperature control in large installations. Following air admission, the volume flow rate Q changes approximately as follows (see Fig. 30):

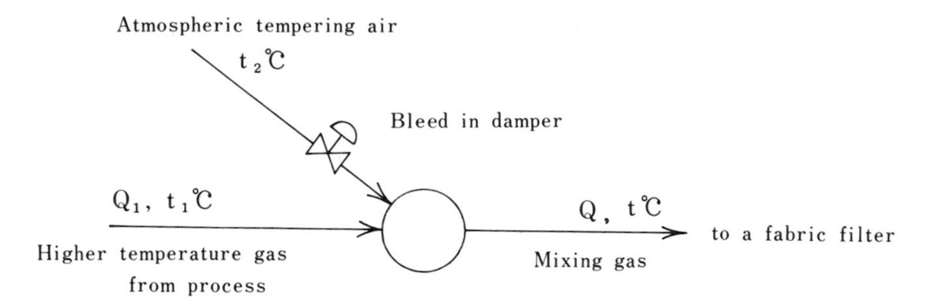

FIG. 30. Air mixing method for gas cooling.

$$\frac{Q}{Q_1} = \frac{(t + 273)(t_1 - t_2)}{(t_1 + 273)(t - t_2)} \tag{6}$$

For example, when the gas is cooled from $t_1 = 250°C$ to $t = 120°C$ with the ambient air at $t_2 = 30°C$, the final volumetric flow rate Q is increased by 75%. Therefore, a filter operating at 250°C requires only 57% (= 100/175) of the cloth area at 120°C.

B. Heat Exchangers

Cooling by radiation, convection, or heat exchanger involves a considerable investment in equipment, requires a relatively large amount of space, and lacks flexibility with respect to the temperatures that can be attained. However, operating cost for the blower is smaller, because the gas volume handled by the filter is a minimum. Air-cooled tubes and water-cooled vertical tubes are used as the heat exchanger, as shown in Fig. 31.

C. Spray Towers

Direct water-spray cooling is feasible if the gas is dry. Direct sprays are often utilized with smelting furnaces in the main exhaust duct where the gas temperature may be greater than 1,000°C. When all the water is not evaporated, but some wets the inner walls of following ducts and the baghouse, dust adhesion and metal corrosion may become troublesome. However, in this method, the final gas

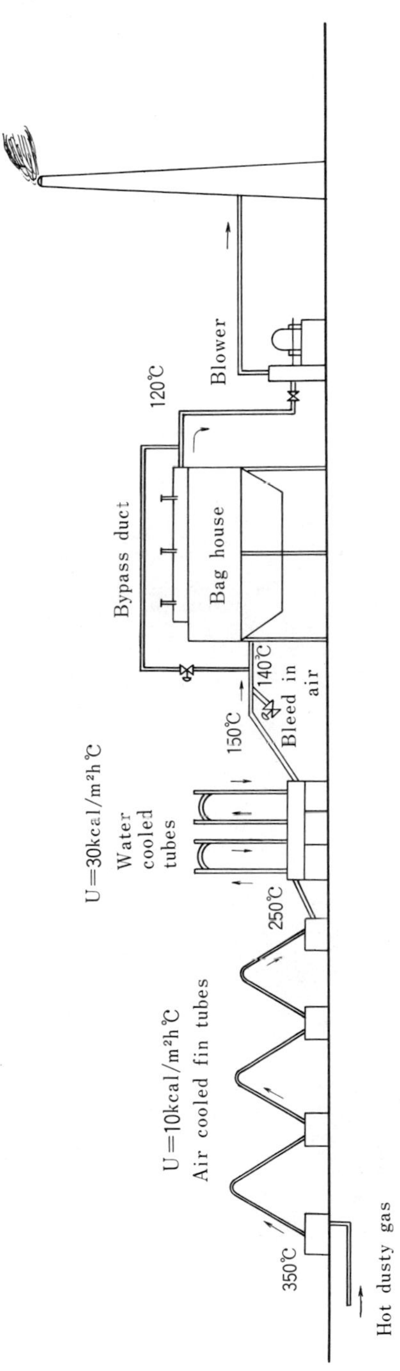

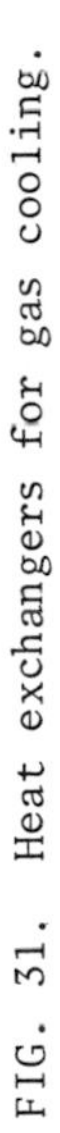

FIG. 31. Heat exchangers for gas cooling.

volume is lower than that of the tempering air method, and installation costs may be lower than those of the heat exchanger method. Figure 32 shows a typical arrangement of a spray tower.

FIG. 32. Vibrating reverse-flow-type bag filter with spray cooling tower in lead smelting. (Kurimoto.)

VIII. TYPICAL APPLICATIONS AND COMPARATIVE COST DATA

The following examples of filter collector application are presented to show how they may be applied to control particulate matter from various sources.

Compared to other particulate collectors, the filter usually exhibits a higher maintenance cost, especially if the medium is of the throw-away type. In addition, the cost of the blower, motor, structural steel, duct work, and installation, taken together, may exceed that of the basic filter unit. If the filter can be installed for pressure operation, i.e., with the blower located ahead of the filter, the initial cost is less than if a suction installation is required. This is so because the outside housing does not need to be airtight. An example is shown in Fig. 33. Dust adhesion and erosion of the blower blades tend to cause trouble in the case of pressure operation.

FIG. 33. Pressure-type installation of baghouse. (6,200 m^2 polyester cloth for electric furnace.) (Shinto Dust.)

One of the largest baghouses in existence is operated in conjunction with an electric-furnace steelmaking plant; it handles about 15,000 m^3/min using 3,120 silicone-treated glass-fiber bags, 30 cm diameter by 9 m long [31,32]. Following charging or tapping of a furnace, the fume not captured directly by hoods over the furnace accumulates in the building until the draft created by the baghouse fans gradually expels it through the filters.

The initial current cost of a pulse-air filter is approximately \$40.00 per m^3/min of capacity, which is comparable to that of an electrostatic precipitator. A fabric filter provided with mechanical cleaning costs between \$20.00 and \$30.00 per m^3/min, which is approximately twice that of cyclone collectors. High-temperature capability and the consequent need for a lower filtering velocity will, of course, make the initial cost greater.

Polypropylene is the cheapest filter cloth, while Nomex (glass), once one of the most expensive, is now moderate in cost. The initial cost of fibrous mat filters is generally about one-tenth that of cloth filters, but their maintenance, including replacement charges, can easily reverse the situation.

Annual maintenance costs, exclusive of power costs, are usually around 10% to 30% of the fabric filters themselves and may reach the initial installed cost for throw-away filters. From the standpoint of construction and operating cost, the optimum superficial velocity for a cloth filter is usually between 1 and 5 m/min. This is so because lower velocities require larger filtering areas and greater initial cost, while higher velocities are accompanied by shorter fabric life which gives rise to higher maintenance costs are shown in Fig. 34. The actual superficial velocity is often set lower than the above figure in order to obtain a lower pressure loss, greater collection efficiency, and longer cloth life.

Figure 35 presents an example of purchase cost, installed cost, and annualized cost of operation of fabric filters [33]. These cost curves represent the following types of filter installations: Curve A represents a fabric filter installation with high-temperature

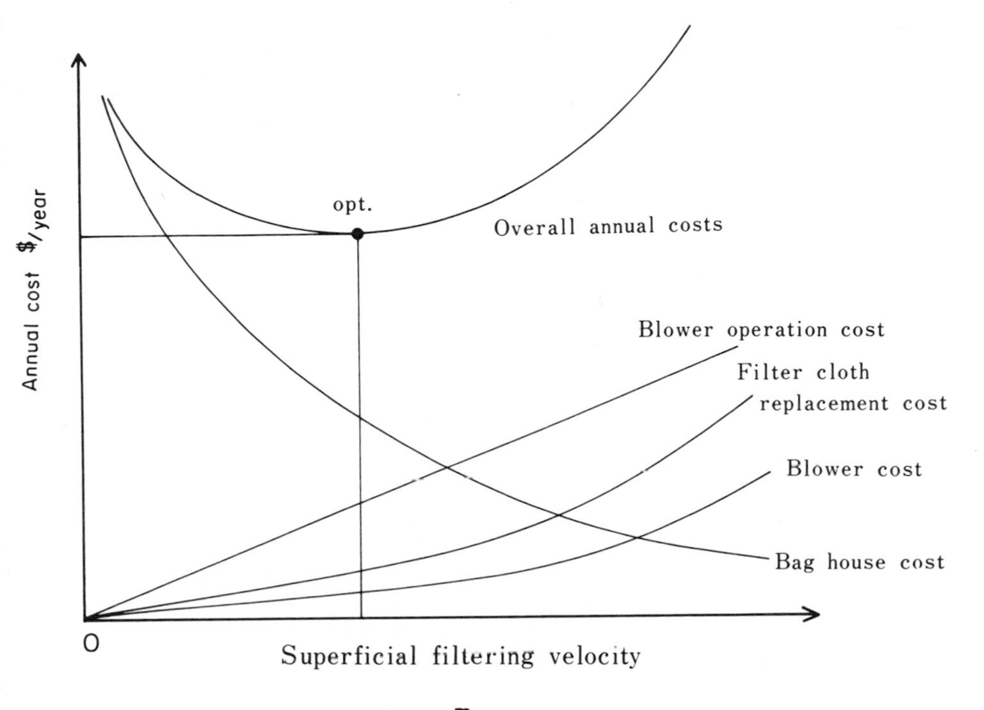

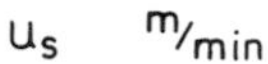

FIG. 34. Relations between annual costs and filtering velocity.

synthetic woven fibers (including fiber glass) and felted fibers, cleaned continuously and automatically. Curve B represents an installation using medium-temperature synthetic woven and felted fibers, cleaned continuously and automatically. Curve C is the least expensive installation; woven cotton fibers are used in a single compartment. The filters are intermittently cleaned.

The above curves do not include the cost for hoods, ductwork, and coolers; and the following conditions are assumed: annual operating time = 8,760 hr, pressure drop = 100 mm H_2O, power cost = \$0.011 per kw hr, and maintenance cost = \$1.8 per m^3/yr.

A filter baghouse is less expensive than an electrostatic precipitator on the basis of first costs. However, baghouses operating under average conditions are subject to high maintenance costs,

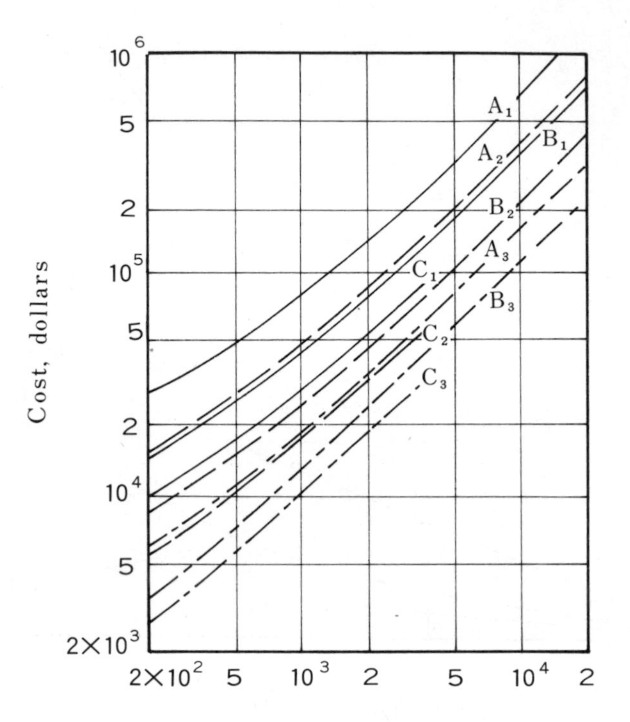

FIG. 35. Fabric filter costs. (1) ——— Installed cost. (2) — — — Purchase cost. (3) — - — - — Annualized cost of operation.

since each baghouse contains a great number of bags, each of which must be replaced as it fails. Therefore, when the required collection efficiency is more than 99% on the basis of particle mass and the gas flow to be treated is less than 2,000 m^3/min [34,35], a filter installation would be more economical than an electrostatic precipitator or a venturi scrubber of comparable efficiency as shown in Figs. 36a, 36b, and 36c.

NOTATION

C_D	Pressure drag coefficient of fibrous mat filter	
D	Diameter	m or μm

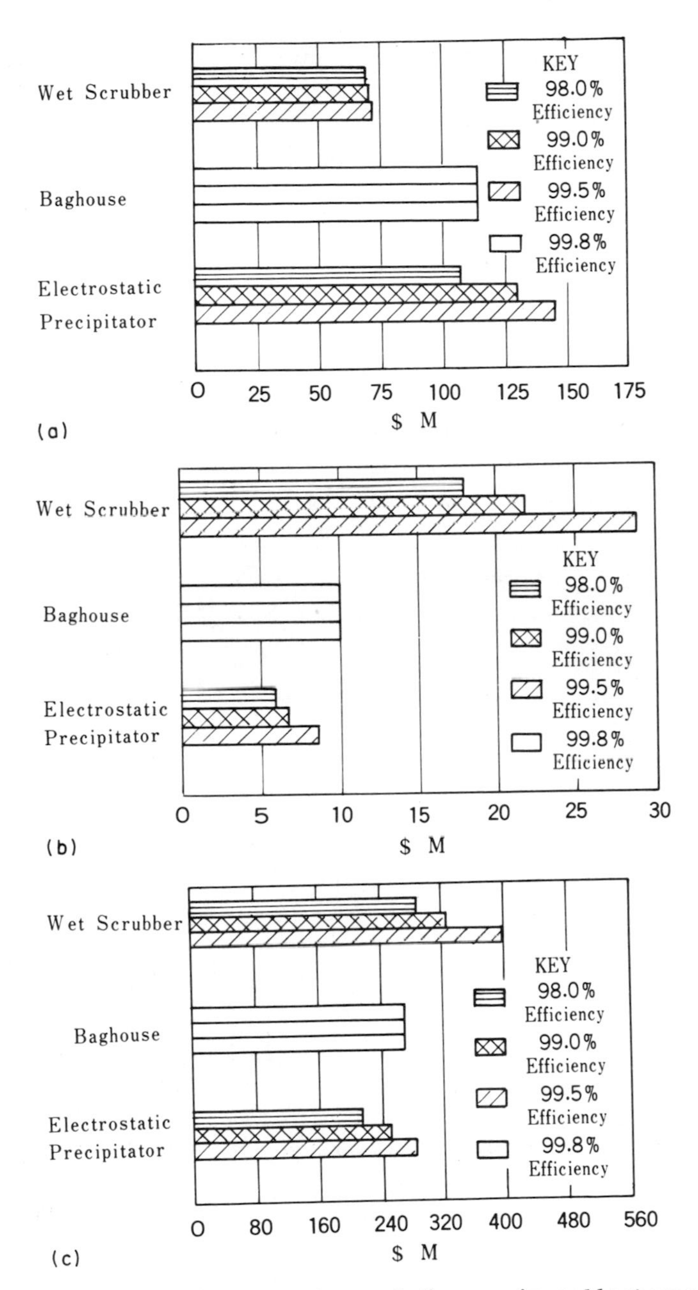

FIG. 36. Economic comparison of three main collectors: (a) installed cost; (b) annual operating cost; (c) total 10-yr. cost (straight-line depreciation). (Courtesy of J. L. Smith and H. A. Snell, and A.I.Ch.E., Ref. 34.)

g_c	Gravitational conversion factor	9.8 kg.m/Kg/sec^2
L	Thickness of a filter	m
m	Deposit dust load on filter media	kg/m^2
HP	power consumption	
Δp	Pressure loss	mm H_2O
Q	Gas flow rate by volume	m^3/min
Re	Reynolds number	
t	Temperature	°C
u_s	Superficial filtering velocity	m/min
$\bar{\alpha}$	Average specific resistance	m/kg
ε	Volumetric void of filter	(always < 1)
μ	Gas viscosity	g/(cm sec)
ρ	Density	g/cm^3
η	Collection efficiency	
ζ	Pressure loss coefficient	

Subscripts

d	(Deposit) dust
o	Filter media only
f	Fiber
p	Particle
ε	Single fiber with interference effect
εd	Single fiber with dust load

REFERENCES

1. P. A. F. White and S. E. Smith, *High-Efficiency Air Filtration*, Butterworths, London, 1964, p. 194 (bag filter and p. 133 (air filter).
2. W. Strauss, *Industrial Gas Cleaning*, Pergamon, London, 1966, p. 244.
3. A. C. Stern, *Air Pollution*, 2nd ed., vol. III, Chap. 44, Academic, New York, 1968, p. 409.
4. C. Pazar, *Air and Gas Cleanup Equipment*, Pollution Control Handbook No. 1, Noyes Data Corporation, 1970, p. 99.

5. L. S. Dollinger, How to Specify Filters, *Hydrocarbon Processing, 48* (10), 88 (1969).

6. C. A. Snyder and R. T. Pring, Design Considerations in Filtration of Hot Gases, *Ind. Eng. Chem., Process Des. Develop, 47* (5), 960 (1955).

7. F. R. Culhane, Air Pollution Control Production Baghouses, *Chem. Eng. Prog. 64* (1), 65 (1968).

(a) C. E. Billings and J. E. Wilder, *Handbook of Fabric Filter Technology, Vols. I and II*, PB-200-648 and 649, National Technical Information Service, U. S. Department of Commerce, December, 1970.

(b) *The User and Fabric Filtration Equipment*, Niagara Frontier Section, Air Pollution Control Association, October, 1973, p. 164.

(c) B. E. Kester, *Design, Operation, and Maintenance of High-Efficiency Particulate Control Equipment*, Greater St. Louis Section, Air Pollution Control Association, March, 1973, p. 147.

(d) Special Issue of Fabric Filters, *J. Air Pollut. Control Ass., 24* (12) (1974).

8. G. W. Walsh and P. W. Spaite, *Preprint of ASME, Annual meeting*, December, 1960, and *Preprint of APCA Annual meeting*, June, 1961.

9. D. G. Stephan and G. W. Walsh, Residual Dust Profiles in Air Filtration, *Ind. Eng. Chem., 52* (12), 999 (1960).

10. K. J. Caplan, A Self-Cleaning Air Filter, *Chem. Eng. Progr., 50* (8), 410 (1954).

11. W. B. Harris and M. G. Mason, Operating Economics of Air-Cleaning Equipment Utilizing the Reverse Jet Principle, *Ind. Eng. Chem., 47* (12), 2423 (1955).

12. K. J. Caplan, The CS Filter—A New High-Performance Cloth Arrestor, *A.I.H.A. J., 28* (6), 567 (1967).

13. J. C. Walling, Ins and Outs of Gas Filter Bags, *Chem. Eng., 77* (23), 162 (1970).

14. H. J. Strauss, Vorschlag für ein Verfahren zur Prüfung von Rollbandfiltermedien, *Staub, 29* (10), 404 (1969).

15. K. Makino and K. Iinoya, Experiments on Collection Efficiency of Dielectric Fiber Mat Filter, *Kagaku Kogaku, 32* (1), 99 (1968).

16. D. Rimberg, Penetration of IPC 1478, Whatman 41, and Type 5G Filter Paper as a Function of Particle Size and Velocity, *A.I.H.A. J., 30* (4), 394 (1969).

17. K. Iinoya, K. Makino, O. Inoue, and T. Imamura, Dust Collection Performance of Paper Filter, *Kagaku Kogaku, 34* (6), 632 (1970).

18. C. A. Burchsted and A. B. Fuller, Design, Construction, and Testing of High-Efficiency Air Filtration Systems for Nuclear Application, OR NL-NSIC-65, UC-80, Oak Ridge National Laboratory, January 1970.

19. H. J. Ettinger, J. D. Dofield, D. A. Bevis, and R. N. Mitchell, HEPA Filter Efficiencies using Thermal and Air Jet Generated Dioctyl Phthalate, *A.I.H.A. J., 30* (1), 20 (1969).

20. E. Stafford and W. J. Smith, Dry Fibrous Air Filter Media, *Ind. Eng. Chem., 43* (6), 1346 (1951).

21. O. H. York, Performance of Wire Mesh Demisters, *Chem. Eng. Progr., 50* (8), 421 (1954).

22. O. H. York and E. W. Poppele, Two Stage Mist Eliminators for Sulfuric Acid Plants, *Chem. Eng. Progr., 66* (11), 67 (1970).

23. H. P. Meissner and H. S. Mickley, Removal of Mists and Dusts from Air by Beds of Fluidized Solids, *Ind. Eng. Chem., 46* (6), 1238 (1949).

24. O. H. York and E. W. Poppele, Wire Mesh Mist Eliminators, *Chem. Eng. Progr., 59* (6), 45 (1963).

25. J. A. Brink, W. F. Burggrabe, and J. A. Rauscher, Fiber Mist Eliminators for Higher Velocities, *Chem. Eng. Progr., 60* (11), 68 (1964).

26. H. L. Engelbrecht, The Gravel Bed Filter, *J. Air Pollut. Control Ass., 15* (2), 43 (1965).

27. N. Kimura and K. Iinoya, Pressure Drop Characteristics of Filter Cloth for Dust Collection, Kagaku Kogaku, *29* (3), 166 (1965); also Kagaku Kogaku (abridged edition in English), *3* (2), 193 (1965).

28. N. Kimura and K. Iinoya, Experimental Studies on the Pressure Drop Characteristics of Fiber Mats, Kagaku Kogaku, *23* (12), 792 (1959).

29. N. Kimura, F. Hayashi, and K. Iinoya, Dust Collection Characteristics of Porous Filter Medium (Urethane Foam), Kagaku Kogaku, *29* (8), 662 (1965).

30. J. E. Wilder and C. E. Billings, Performance of Fabric Filtration Systems, *Preprint of 2nd International Clean Air Congress,* Na EN-16 B, Washington, D. C., December, 1970.

31. J. L. Venturi, Operating Experience with a Large Baghouse in an Electric Arc Furnace Steelmaking Shop, *J. Air Pollut. Control Ass., 20* (12), 808 (1970).

32. B. J. Squires, Electric Arc Furnace Fume Control and Gas Cleaning, *Filtrat. Separat.*, 7, 447 (1970).

33. National Air Pollution Control Administration, *Control Techniques for Particulate Air Pollutants*, Part II, PB-190253 (p-3167-2), January, 1969, p. 102.

34. J. L. Smith and H. A. Snell, Selecting Dust Collectors, *Chem. Eng. Progr.*, *64* (1), 60 (1968).

35. J. S. Munson, Dry Mechanical Collectors, *Chem. Eng.*, *75* (22), 147 (1968).

Chapter 5

FILTRATION IN THE CHEMICAL PROCESS INDUSTRY

Frank M. Tiller

Department of Chemical Engineering
University of Houston
Houston, Texas

Antoine Alciatore

Dicalite Division, GREFCO
New Orleans, Louisiana

Mompei Shirato

Department of Chemical Engineering
Nagoya University
Chikusa, Nagoya, Japan

I. Introduction 362

II. Analysis of Filtration Operations 365

III. Product Specification 366

IV. Slurry and Cake Properties 368

V. Initial Choice of Equipment Class 369

VI. Modification of Slurry Properties 370

VII. Washing and Deliquoring 375

VIII. Description of Equipment 377
- A. Batch Filters 379
- B. Semicontinuous Filters 385
- C. Continuous Filters 386
- D. Miscellaneous Filters 390
- E. Filter Applications 391

IX. Flow Equations for Porous Media 394
- A. Frictional Drag on Particles 395
- B. Porosity and Flow Resistance 396
- C. Average Filtration Resistance 402
- D. Pumping Mechanisms 403
- E. Material Balance 405

X. Batch Cake Filtration 406
- A. Constant Pressure 407

B. Constant Rate 411
C. Centrifugal Pump 412

XI. Cake Porosity 416
A. Porosity as a Function of Distance 417
B. Average Porosity as a Function of Applied Pressure 419
C. Hydraulic Deliquoring 423
D. Expression 427

XII. Cake Washing 437
A. Deliquoring by Suction 440
B. Air Rate 442

XIII. Experimental Testing Procedures 442
A. Practical Filtration Testing 443
B. Addition of Filter Aid 447

XIV. Cycle Analysis 452

XV. Continuous Filtration 459

Notation 469

References 472

I. INTRODUCTION

In 1923, the Chemical Catalog Company published as part of The Modern Library of Chemical Engineering a text by Arthur Wright [1] entitled Industrial Filtration which included Part I, "Theory of Filtration"; Part II, "Mechanics of Filtration"; and Part III, "Filter Practice." It was clearly evident that a considerable amount of machine development had been accomplished in solving practical problems, while theory simply served as an adjunct to interpretation of basic tests or operating problems. The basic situation has changed very little in the intervening half century. Under "Mechanics of Filtration" very detailed descriptions were given of the following process filters: bag, plate and frame, suction leaf, pressure leaf-Kelley, Sweetland, Vallez, rotary vacuum-Oliver, American disc, Feinc, Merrill, Oliver table, and a few specialized units. Thus, it is evident that on the way to solving the filtering problems posed by the young chemical, sugar, and metallurgical industries, most of the basic types of filters were developed. Before categorizing this equipment, it is useful to look at the entire field of solid-liquid separation. For purposes of analysis, it can be divided as shown in Fig. 1.

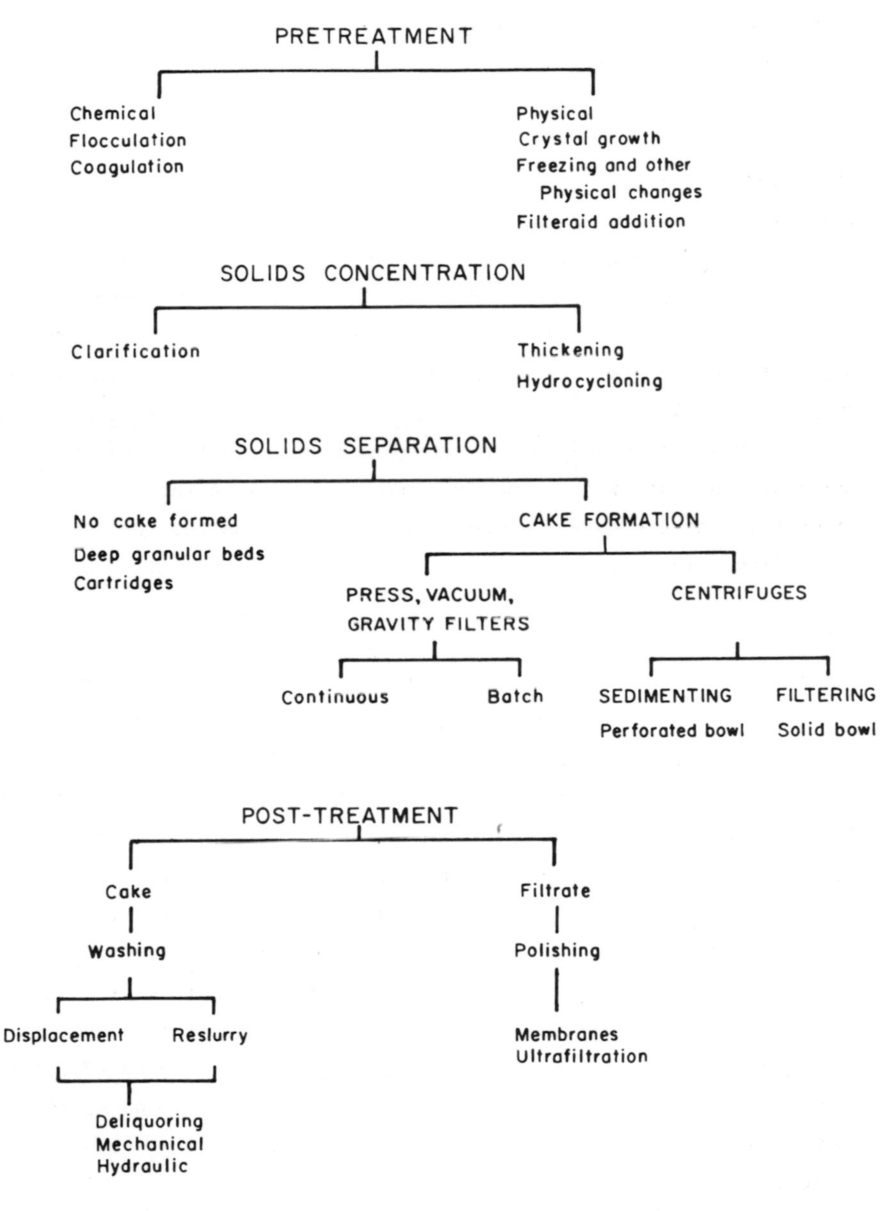

FIG. 1. Stages of solid-liquid separation.

Any design of a separation system must treat all stages of pretreatment, solids concentration, solids separation, and post-treatment. Pretreatment is utilized to alter the properties of the slurry so that it is easier to filter. For example, particle size may be increased by chemical treatment, producing flocculation or coagulation. In solids concentration, part of the liquid may be removed by thickening or hydrocycloning so that the load on the filter is decreased. It is also common practice in the chemical industry to add filter aids such as diatomaceous earth or expanded perlite to the slurry and thereby increase the permeability of the cake.

Filter operations can be divided into two broad categories of cake and depth filtration. Cake filtration can be further subdivided into centrifugal and pressure, vacuum, and gravity (PVG) operations. In cake filtration, the particles from a slurry are stopped at the surface of a supporting porous medium while the fluid passes through. In depth filtration, the particles are captured in the interstices of the solid, and no cake is formed on the surface of the medium. In many processes, a stage of depth filtration precedes the formation of a cake. The first particles may enter the medium, and with very dilute slurries there may be a time lag before a cake begins to form. Smaller particles enter the medium while larger particles bridge the openings and start the build-up of a surface layer. In general, depth filtration is used for taking out small quantities of contaminants. Cake filtration is primarily employed for more concentrated slurries. In a typical chemical plant both cake and depth filtration operations may be encountered. Influent plant water as obtained from surface sources is treated and filtered in deep-bed sand filters. Effluent water may be handled in a similar manner or by PVG filtration depending on the concentration. In the actual processing operations, PVG filtration and centrifugation are both encountered. Many separations can be effected equally well with either PVG or centrifugal equipment; and cost, operating, and maintenance considerations determine the actual choice.

II. ANALYSIS OF FILTRATION OPERATIONS

In analyzing an operation involving solid-liquid separation, it is essential to preserve the proper perspective of the entire process. Too often engineers do not have sufficiently adequate information for design. A group that would refuse to design a fractionating column without the necessary thermodynamic data will blithely call for a filter without knowing particle size distribution, settling rates, rate of cake build-up, moisture content of the cake as a function of pressure, or other important information. It is not possible to find experimental data in the literature or handbooks for the design of solid-liquid separation processes as can be done with operations like heat transfer and distillation. Consequently, it is necessary either to depend upon past experience or systematically to investigate the properties of the slurry to be separated into liquid and solid. A blueprint for consideration of major steps is laid out below.

As a general approach to analysis of a problem, the following is a guide to major decisions that must be made in the design of a filtration system involving cake formation:

1. Product specification
   a. Filtrate clarity: permissible particle size and amount in filtrate
   b. Solids: solubles in cake, liquor content of cake
   c. Contamination of cake or filtrate
2. Properties of slurry and cake
   a. Particle size distribution: semiquantitative estimates in absence of accurate data
   b. Mass and volume fraction of solids in slurry and cake
   c. Rate of cake growth as a function of pressure and rate
   d. Physical and chemical properties: vapor pressure, densities, corrosiveness, and interaction with filter aids
3. Modification of slurry properties
   a. Chemical pretreatment: coagulation, flocculation

   b. Particle fragility and possible degradation of original or flocculated particles by pumping, agitation, or mechanical treatment
   c. Thickening
      (1) Gravity settling
      (2) hydrocyclones
      (3) Delayed cake formation
   d. Addition of filter aid
4. Washing
   a. Displacement
   b. Repulping in countercurrent or cocurrent operation
5. Deliquoring
   a. Effect of filtration pressure on average cake porosity
   b. Effect of mechanical pressure on porosity, expression
   c. Change of direction of flow (right-angled in presses) to effect porosity decrease
6. Selection of equipment
   a. Production rates
   b. Limitation of physical or chemical properties
   c. Filter
   d. Pump
   e. Media or media plus precoat
   f. Methods of discharge
7. Cycle analysis
   a. Capacity as a function of external variables: cycle length, rotational speed of drums, or velocity of belt
   b. Cost as a function of variables affecting capacity and cycle length
8. Design and optimization

Certain data are critical at various stages and must be obtained experimentally.

III. PRODUCT SPECIFICATION

Strangely, many engineers requesting quotations on solid-liquid separation equipment do not even know what specifications are required for the final product. It is essential to know how clean a filtrate is needed. Can 0.1, 1.0, or 5.0 μm particles be permitted in the filtrate and in what quantity? Without that information, it is impossible to specify properly the correct filter media or precoat material and to know whether or not a polishing step with a membrane may be needed on the final filtrate. If there is no experience upon which to base specifications, then educated estimates must be made. Examples of typical specifications used in the chemical process industries are:

Filtrate will have a Tyndall or Jackson (Fluorazin) Turbidity Unit reading less than a specified value

Filtrate will be visually clear

Filtrate will have trace of solids on a 1.2-μm membrane disc

Filtrate will have undissolved solids not in excess of a fixed concentration in mg/liter or parts per million (ppm)

Filtrate will be visually clear at ambient temperature at a specified time after filtration

Normally, specifications on retained solids refer to freedom from soluble materials and average liquor content. It is not necessarily possible to specify a low liquor content (or average porosity) without considering the economics of different filters. In general, low liquor content as attained by using high filtration pressure, squeezing mechanically, blowing with air, or aiding drainage on a rotary vacuum filter by means of steam is restricted to particles greater than approximately 10 μm in diameter. Capillary pressures for 1.0-μm particles are so high that it is difficult to blow air or steam through a cake with such particles. If a rotary vacuum filter were chosen, there would be a limitation on porosity reduction for cakes with small particles unless an additional unit for mechanical expression were available. In general, it is necessary to balance the additional cost of producing a dry cake against thermal requirements of driers.

IV. SLURRY AND CAKE PROPERTIES

Properties of the slurry are absolutely fundamental to design. It is desirable to know the particle size range as the method of separation is directly related to the particle characteristics. For example, a slurry containing 0.05% of clay in the 0.1-1.0 μm range mixed with 2% fibers varying from 10-50 μm would receive totally different treatment from a material with particles distributed uniformly over 5-75 μm. The former might be subjected to a two-stage process in which the fibers were separated first and the clay removed with filter aids or subjected to a deep-bed granular filter with backwash.

If an accurate particle analysis is not available, it is frequently possible to obtain a qualitative notion by sedimentation tests. Velocities are directly proportional to the difference in density of the solid and liquid, and only rough estimates can be made without density data. If a slurry retains a cloudy appearance for hours or days, it contains submicrometer particles. Some experienced individuals can estimate sizes simply by feeling the solids.

Generally, the mass fraction of solids in the slurry is known. However, frequently no attempt is made to determine the average porosity or liquid content of the cake. A rough estimate can be made in a sedimentation test. When particles settle out of a slurry, the porosity of the cake formed at the bottom of the vessel will approximate the loosest arrangement obtained at a very low filtration pressure. A more precise estimate can be obtained from a cake formed under pressure. When filtration tests are performed, the average liquor content of the cake should be obtained as a routine step.

Perhaps the single most important item of information is the rate of cake build-up. If the process conditions represent a typical filtration encountered elsewhere, initial data may be obtained from an equipment or filter aid supplier. If the application is unique or available data must be checked or augmented, representative samples must be taken for laboratory filtration tests to determine the rate of cake build-up along with other data.

V. INITIAL CHOICE OF EQUIPMENT CLASS

In the first phase of an investigation, an initial choice of the general class of equipment should be tentatively made on the basis of rate of cake growth. Selection of equipment adapted, on one hand, to filtering materials rapidly and, on the other, to resistant cakes can be made by considering the average rate of build-up. As a broad general rule continuous or semicontinuous filters operating under vacuum are well suited for rapid cake deposition, whereas pressurized units serve best for solids having low permeability. Flood et al. [2] classified equipment in accord with its suitability for different rates of cake build-up. A simplified and modified summary of these data are shown on Table 1.

Materials that build up at the rate of roughly 1 cm in 1-10 sec settle rapidly and require only a small head (gravity or vacuum) to produce adequate rates. At lower rates of growth, continuous filters of the rotary-drum, disc, or horizontal-belt type are suitable choices. With still lower rates of cake growth, it becomes necessary to use enclosed pressure filters generally of a batch type to increase flow rates.

As materials become even more difficult to filter, other strategies must be employed to maintain satisfactory flow rates. If a relatively dilute slurry with less than approximately 0.1% by volume of solids is to be filtered, a granular bed or other depth-type filter such as a cartridge can be used. Also a rotary drum with a

TABLE 1 Initial Equipment Selection

	Slurry	Cake build-up	Type of filtering equipment
1.	Rapid filtering	cm/sec	Vacuum or gravity: pans, screens, drum, belt
2.	Medium filtering	cm/min	Vacuum: drum, disc, belt, pans
3.	Slow filtering	cm/hr	Pressure filters: presses, leaf, tray, plate
4.	Clarification	Negligible cake	Cartridges and granular beds

thick precoat can advantageously be employed for highly resistant, large-volume, dilute slurries. Solids forming low-permeability cakes are removed partially as a thin cake and partially in the surface layers of the precoat. Part of the precoat is removed in each revolution, so that the surface is continuously renewed.

VI. MODIFICATION OF SLURRY PROPERTIES

Chemical pretreatment can be considered as a method for improving filtration characteristics. In Chap. 3, the use of flocculants is discussed as a means for increasing effective particle size. The classification of a slurry according to the rate of cake build-up is often dramatically improved by chemical or physical pretreatment. For example, a difficult-to-filter material can be flocculated into particles which settle rapidly. Thickening in the form of gravity sedimentation can then be used to produce a more concentrated slurry which leads to a more rapid rate of cake growth. The supernatant liquor may be treated separately in a clarification operation.

Admixes of filter aids can also be used to lower the average resistance of filter cakes. With the addition to slurries of materials like expanded perlite, diatomaceous earth, and carbon, the classification of the slurry with respect to the most suitable equipment for its handling may change when the rate of cake growth is increased.

In the chemical process industries the use of inert filter aids is so widespread that some engineers refer to the area as filter aid filtration. One of the most important filter aids, diatomite, has been in use for over half a century. Diatomite is a sedimentary rock composed of skeletons of single-celled, aquatic plant life called diatoms. The skeletons are principally composed of silica and occur in myriad forms. Extensive experience has led to close coordination of filter design, auxiliary equipment, and filter media with the employment of filter aids, resulting in a simplification of many of the operations.

Diatomite is mined in open quarries. Since diatomite as mined may contain 60-70% water, it is usually stockpiled for sun drying and curing. Because of the high water content and low density of the crude, transportation costs force processing plants to be located near the mine. Crude diatomaceous earth is brought to a processing plant where finished filter aids are produced by milling, further drying, heat treatment, and air classification. A "natural" filter aid may be produced by simply milling, drying, and classification. This type will produce the highest product clarity but will do so at a low flow rate. To accelerate the rate of flow the diatomite must be calcined to smooth the diatom surface and reduce frictional fluid drag. This results in a slight loss of product clarity but a major gain in flow rate. The color of the diatomite varies from light tan to shades of gray and gray pink for the naturals. The calcined product tends toward a soft pink. During calcination diatomite may be fluxed with soda ash. This produces a white flux-calcined group of filter aids comprising the medium to high flow-rate group.

Permeabilities of filter aids cover roughly a hundred-fold range. The more resistant materials retain particles as small as 0.1 μm whereas the less-resistant filter aids may permit 1.0 μm particles to pass through a bed. These materials are recommended at use rates of 10 lb per 100 ft^2 of filter surface but the average use is closer to 13 lb. A coarse grade may be employed to assist the separation of the cake from the medium, or a fine grade may be chosen for increased clarity. As with deep beds and media, the particle size captured by various filter aids may vary because of such factors as fluid viscosity, cake or bed depth, and flow rate.

When used as admix the filter aid beneficiates the flow characteristics of the cake. Within normally operating filter aid systems, the objective is to add sufficient filter aid to produce a less resistant cake while using a grade that will render a clarity of filtrate to meet the requirement of the filtration. The specific resistance α_{av} (different scales) of calcium carbonate, titanium dioxide, and clay as a function of concentration of Dicalite (perlite) 426 is shown in Fig. 2. A good filter aid will greatly reduce the resistance as indicated by the curve for titanium dioxide and clay. An addition of 10% by weight of perlite 426 to TiO_2 results

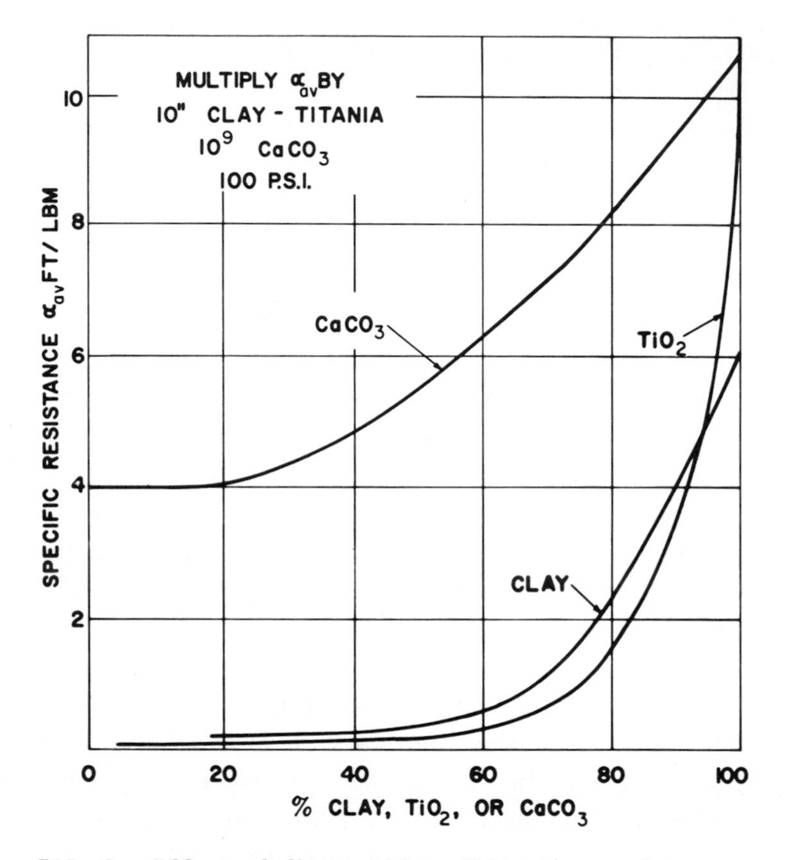

FIG. 2. Effect of filter aid on filtration resistance.

in a 60% decrease in the resistance, whereas 20% of filter aid lowers α_{av} by 85%. With $CaCO_3$ the drop is less dramatic, and other filter aids would be investigated if needed. Actually, the $CaCO_3$ has a relatively low resistance and could be filtered easily without assistance.

As a filter aid is added to a system, the average flow rate through the resultant cake is increased. However, the added mass of solids acts adversely on rate and fills the cake space sooner on batch pressure filters. There is, therefore, usually a region where incremental additions of filter aid are neither helpful nor economical.

In Eq. (29) as developed subsequently, it can be seen that the pressure drop across the cake is given by the product $w\alpha_{av}$ of the total mass of dry solids (w) in the cake per unit area and the average filtration resistance (α_{av}). The total mass is obtained as $w/(1 - x)$ with x equal to the mass fraction of filter aid. Thus, as the mass fraction of filter aid increases, $(1 - x)$ decreases and $w/(1 - x)$ increases. Comparisons of different quantities of filter aid should be made on the basis of $\alpha_{av}/(1 - x)$.

In Table 2, a comparison is made of the relative effect of perlite 426 on $CaCO_3$, TiO_2, and clay. It is clear that no improvement would be realized in the case of $CaCO_3$ whereas as much as 50% by weight of filter aid might be usefully employed with the TiO_2 or clay. To obtain a complete picture of the most economical addition of filter aid, it would be necessary to compare the cost of filter aid with the decrease in equipment and operating costs.

Perlite is the second most important mineral filter aid and is a relative newcomer, having been developed after World War II. Perlite is a name first used by Beudant in 1822 to describe a particular variety of naturally occurring glassy volcanic rock, characterized by perlite or onion-like, splintery breakage planes. The rock in the natural strata may vary in color from gray, grayish black, to hues of blue, green, red, or brown. Diatomite crude is light and

TABLE 2 Effect of Filter Aid Concentration on Relative Resistance at 100 psi

x, mass fract. perlite	$CaCO_3$		TiO_2		Clay	
	α_{av}	$\frac{\alpha_{av}}{(1 - x)}$	α_{av}	$\frac{\alpha_{av}}{(1 - x)}$	α_{av}	$\frac{\alpha_{av}}{(1 - x)}$
0	10.5[a]	10.5[a]	9.0[b]	9.0[b]	6.0[b]	6.0[b]
0.2	8.2	10.2	1.6	2.0	2.6	3.3
0.4	6.2	10.4	0.3	0.5	0.6	1.0
0.6	4.9	12.2	0.12	0.3	0.32	0.8
0.8	4.5	22.5	0.08	0.4	0.2	1.0

[a]Multiply by 10^9.
[b]Multiply by 10^{11}.

often wet, whereas perlite crude is dense and relatively dry. Whereas producing plants for diatomite are normally near the quarry, perlite is dense and may be shipped economically long distances for processing, thus making a wide range of delivery systems feasible. When delivery systems, dosage rates, and the basic price per ton of the filter aid are included, perlite is often a lower cost performer. After drying, grinding, and screening, the perlite ore is blended into various grades of filter aid. With proper crushing and heating, it will expand in an explosive fashion to about ten times its original volume and whiten.

Since diatomite was well established as a filter aid before the advent of perlite, grade levels for water permeabilities were selected for perlites corresponding to the established diatomite grades. There are differences between the two materials which require recognition in any comparative evaluation. For example, perlite normally produces less-dense cake than diatomite, but the range varies with producer. This density (Table 3) must be considered in operating studies as it may affect dosage and does affect cake volume.

For any given permeability, a packed bed of perlite will not produce the same degree of clarity as diatomite. In any evaluation of the two filter aids, required clarity should first be established; other parameters may then be investigated.

TABLE 3 Relative Flow Rates of Diatomite and Perlite

Relative flow rate	Diatomite (lbm/ft^3)	Perlite (lbm/ft^3)		
		Light	Standard	Heavy
1	19-25		24	30
2	21-22		18	23
3	19-22	13	15	19
7	18-22	11	13	19
10	18-22	11	13	19
25	18-24	6.5	12	19

One of the noticeable characteristics of an expanded perlite filter aid is the nonwetting "floating" particles that are present in the medium-to-high flow grades. The amount of "float" as measured in water in the low-permeability grades is from a trace to about 0.5% by weight; for mid-permeability, it ranges from 0.5 to 2.0%. High permeability, low-density materials contain as high as 5 to 10% floats which behave like an admix, and if they enter into the cake they affect flow rate, clarity, and dosage analysis. If they float free of the cake, they are useless.

There are a number of other materials used as filter aids, including carbon, asbestos, and cellulose. Both perlite and diatomaceous earth are primarily silicates which may not be chemically compatible with some systems. Then it may be necessary to use carbon as an aid. Carbon usually has some degree of surface activity.

Asbestos is used in either the acid-washed or natural form. It may be employed alone; however, it is usually incorporated with diatomite in an amount of about 7-1/2% by weight. It forms a very tight mat which enhances effluent clarity. Additional functions include cake stabilization (by improving bridging and diminishing cracks) and closing off mechanical leaks in screens and seals.

Cellulose is widely used in the filtration of products that cannot tolerate silica. It is a fibrous filter aid which stabilizes cakes and may be used alone as a precoat or as an admix with mineral materials. Should a mineral precoat be desired, it is added on top of the cellulose.

Typical uses for filter aids in the chemical process industries are given in Table 4.

VII. WASHING AND DELIQUORING

Post-treatment consists primarily of washing and deliquoring the cake and, possibly, further processing of the filtrate to remove fine particles. Washing can be effected by simply displacing the slurry by clean wash liquid at the end of filtration. Removal of the solute from the mother liquor is partially by direct displacement and

TABLE 4 Typical Filter Aid Uses

Diatomite	Perlite	Cellulose	Asbestos mix
	INORGANICS		
Alum	Alum	Cell brines	Nitrates
Metal salts	Potassium chlorate	Plating solutions	
Potassium chlorate	Cell brines	Waste water	
Sulfur	Sulfur	Alumina	
Phosphoric acid	Nickel oxides		
Ammoniated phosphoric acid			
Plating solutions			
	ORGANIC		
Additives	Melamine	Amines	Urea
Resins	Polyethylene	Waste water	Waste water
Nylon intermediates	Nylon intermediates		
Amines	Waste water		
Waste water			

partially by diffusion from stagnant areas into the body of wash liquid. While displacement washing is quite efficient in the early stages, efficiency drops as diffusion becomes the controlling process. Repulping of the cake and carrying on a counter- or cocurrent washing operation is less efficient at first but improves in comparison with the displacement washing as solute concentration decreases. Simplicity of operation favors simple displacement operation when it is satisfactory.

Deliquoring of cakes has become increasingly important as environmental standards have become more rigid. In some cases, simply increasing filtration pressure at the end of the filter cycle is sufficient. Where water must be subsequently removed by drying, hydraulic squeezing of the wet cakes at pressures ranging into thousands of pounds per square inch can be very effective.

In some presses and in at least one commercial leaf filter, deliquoring is accomplished by continuing the flow after the space for the cake is completely full. Once the two cake surfaces have merged, flow takes place at right angles to the original patterns. That causes compaction of the surface layers which normally are quite "soupy."

VIII. DESCRIPTION OF EQUIPMENT

Since many filter designs are used in the chemical process industries, an overall understanding of the characteristics of basic filters is essential. Equipment may be classified according to:

1. Driving force
   a. Gravity
   b. Vacuum
   c. Pressure: centrifugal and positive displacement pumps
2. Continuity of operations
   a. Continuous
   b. Semicontinuous
   c. Batch
3. Filtering surface
   a. Moving or stationary
   b. Geometrical shape: flat, cylindrical, leaf, plate
   c. Media for retaining cake or precoat
4. Vessel containing cake and slurry
   a. Closed tanks under pressure
   b. Open tanks: top feed on belts and drums, bottom feed for drums and discs, banks of vacuum leaves
   c. Frames
   d. Recessed plates
5. Post-treatment
   a. Washing
   b. Deliquoring by hydraulic pressure or mechanical expression
   c. Drying

6. Cake discharge
    a. Wet discharge
    b. Dry discharge
    c. Continuous filters

Other factors chiefly involved with mechanical features, flow channels, and automation are utilized for classifying equipment. In terms of continuity of operation, PVG filters may be grouped as follows:

A. Batch filters
    1. Press, plate and frame, recessed plate (Fig. 3)
    2. Leaf, horizontal, vertical (Fig. 4)
    3. Plate
    4. Tray (Fig. 5)
    5. Tube (Fig. 6) or candle
    6. Nutsche
    7. Vacuum leaf
    8. Special designs
B. Semicontinuous filters
    1. Table or tilting pan (Fig. 7)
    2. Intermittent belt
    3. Automatic press
C. Continuous filters
    1. Drum (Fig. 8)
    2. Disc (Fig. 9)
    3. Belt (Fig. 10)
    4. Staged thin-cake pressure filter

With new automatic unloading devices now being used for batch filters, it is difficult to assign a precise category to some filters.

For cake filtration the majority of pressure filters are batch while most vacuum filters are continuous or semicontinuous. Advantages and disadvantages are covered in subsequent discussions in a general manner, it being recognized that such considerations vary widely depending on the particular filtration problem.

A. Batch Filters

1. *The Press*

The filter press (Fig. 3) consists of a skeleton framework made up of two end supports connected by two horizontal parallel bars. On the bars a varying number of filter chambers are assembled which consist of (1) medium-covered recessed filter plates or (2) medium-covered plates alternated with frames which provide space for the cake. The chambers are closed and tightened by a screw or hydraulic ram which forces the plates and frames together, making a gasket of the filter cloth. The charge enters the filter press under pressure and fills each chamber approximately simultaneously. The liquid passes through the filter medium, which in turn retains the solids. The clear filtrate is removed at a discharge outlet. There are various ways of arranging feed inlet, filtrate outlet, and wash-liquor inlet port. The advantages of a filter press are adaptability to high pressures, production of a dry washed cake, durability, media and frames easily alterable, and flexibility; disadvantages include

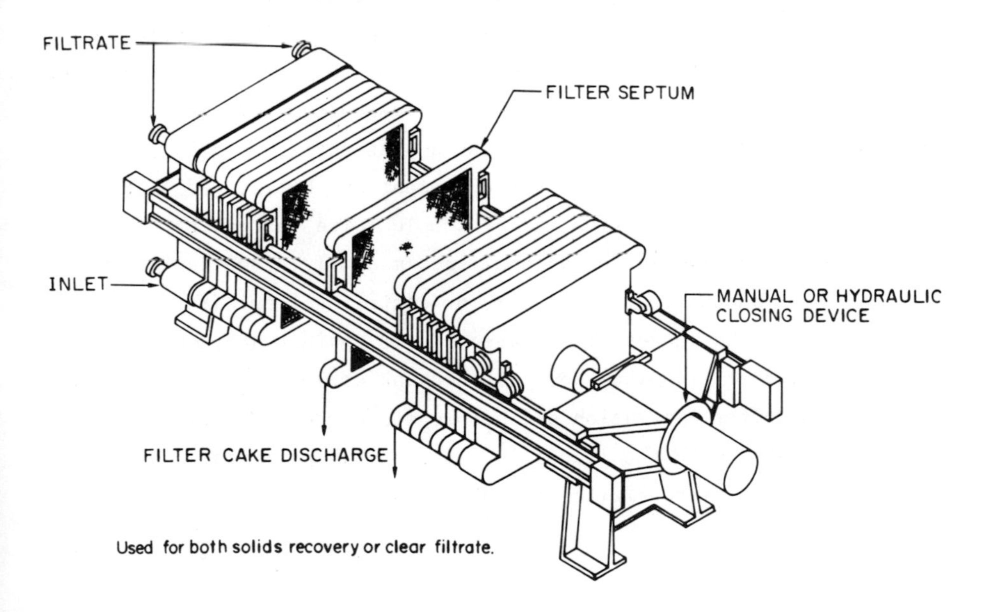

FIG. 3. Plate and frame filter.

high labor cost unless mechanized, leakage, and exposure of product to air during discharge. The provision of mechanized systems to move the filter plates and to open and close the filter press at the end and commencement of the filtration cycle can reduce operating manpower requirements. Filter-press materials of construction include cast iron, wood, stainless steel, glass-filled polyester, polypropylene without fillers, and aluminum; metal plates and frames may have a rubber or phenolic coating, or they may be entirely of plastic.

2. *Tank-Leaf*

The batch, pressure, tank-leaf filter (Fig. 4) has a variety of designs. This unit employs leaves as the filtering element. Common variations involve the arrangement of the tank and leaves into various vertical and horizontal combinations. Vertical tanks normally contain vertical leaves, while horizontal tanks may have either vertical or horizontal leaves. The leaves may be stationary or rotating and may be square, rectangular, trapezoidal, circular, a segment of a circle, or elliptical in shape. Drainage outlets may be located at the bottom, top, and center or in various combinations. Each leaf design lends itself to several discharge methods. For wet discharge, elements may be cleaned by a stationary, rotating, oscillating, or traveling sluice; or the element may be rotated past a fixed sluice. Sluice action may be coupled with sparge, or vibration. Dry discharge may be accomplished manually or mechanically using sudden shock, vibration, air, or centrifugal spinning. With some filters, the leaves are rotated slowly against a brush to assist in cake release. In other designs the tank or leaves are sometimes rotated 90° after filtration and before discharge. The filtration is accomplished with the leaves in a horizontal position, and the discharge is carried on with the leaves in a vertical position to take advantage of gravity.

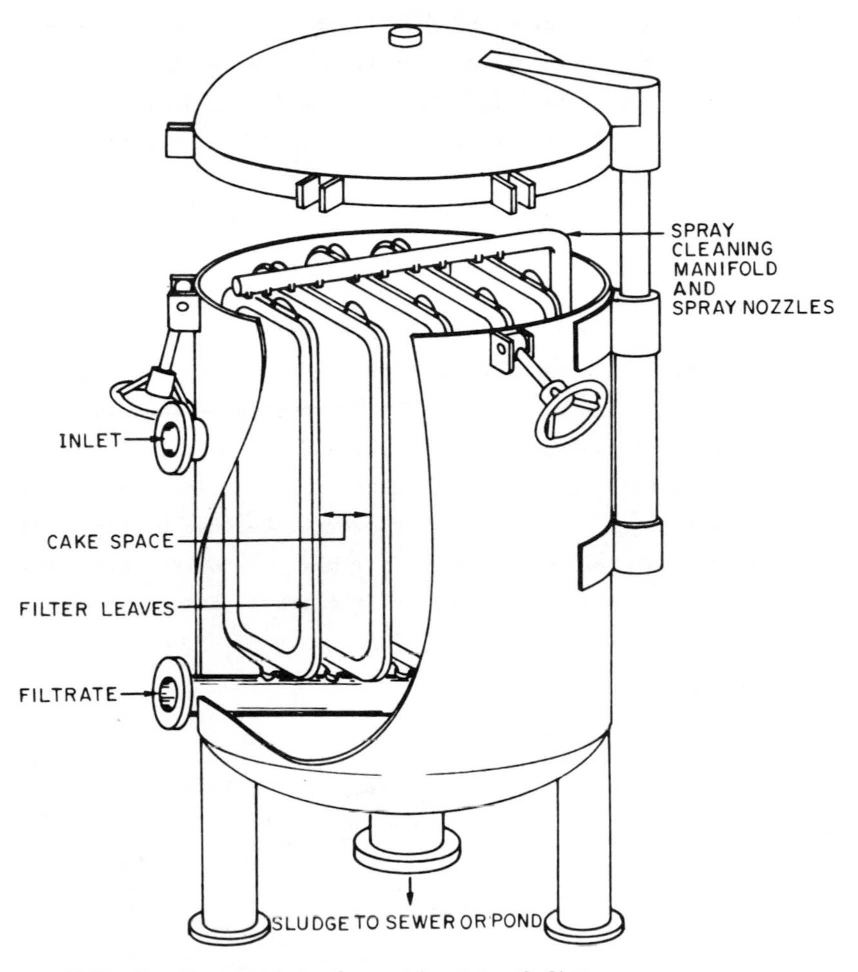

FIG. 4. Vertical tank-vertical leaf filter.

3. *Plate Press*

The plate filter is a vertical tank unit with a cartridge of horizontal plates the medium of which acts as both gasket and filter. Flow is downward through the medium on the upper side of the circular plate where the cake is formed. Flow with gravity gives excellent cake stability. The advantages of this type of filter are adaptability to use with paper media, excellent cake stability, good for small batches, and good for intermittent filtration; its disadvantages are high labor cost, leakage if plates become warped, and size limitations. Materials of construction include mild steel and 304 and 316 stainless steel. Plates normally are designed for 50 psi maximum pressure.

4. *Tray Filter*

The tray filter (Fig. 5) consists of a series of trays in a horizontal tank open at one end. Each tray is normally drained through an individual outlet to a manifold outside the tank. Piping is arranged so that liquid may be transferred from the bottom of the tank to the top tray, and to the other trays by overflow. The tray filter, having an enclosed tank, gives an efficient drying cycle. This allows almost complete solid and filtrate recovery. Because of the horizontally formed, tray-held cake, the filter may be run intermittently without the cake moving or falling off the tray. Batches may be run

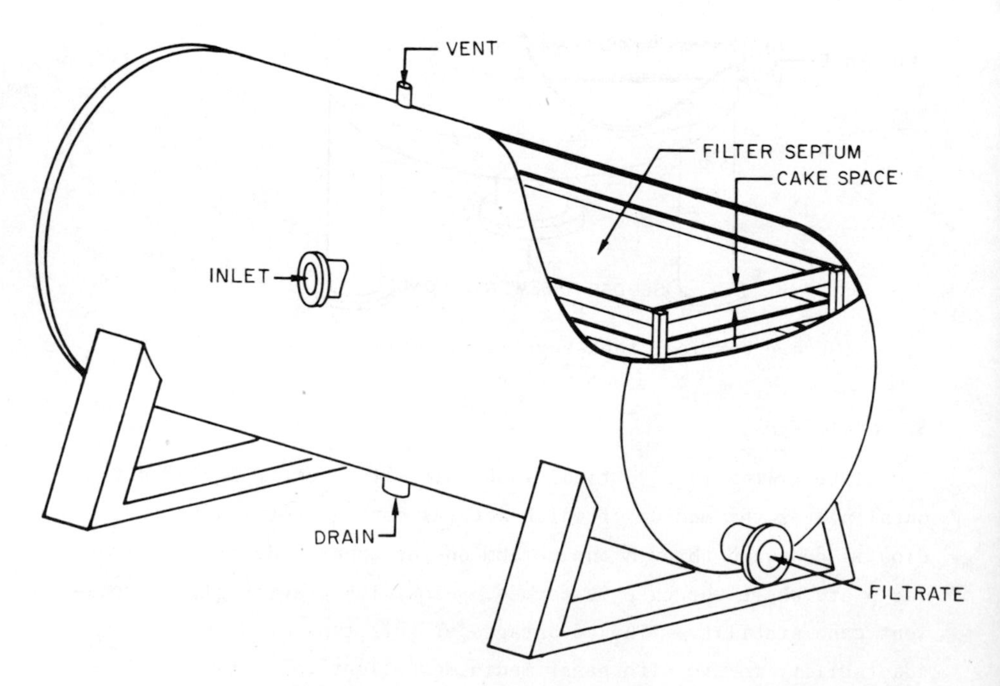

FIG. 5. Horizontal tray filter.

up to the cake-holding capacity of the tray. If required, solids may be dissolved or melted in the tray. The primary advantage of the tray filter is that complete recovery of solids and liquid is practical. Its disadvantages arise from the fact that filtration occurs on the upper surface of the tray only; it is size-limited; and operating costs are higher than in two-sided filtration units. Standard materials of construction are mild steel and stainless 304 and 316 steels. Metals such as monel, hastelloys, titanium, and nickel are available when required.

5. *Tank-Tube*

A tank-tube filter (Fig. 6) normally contains vertical tubes consisting of wire-wound elements, packs of rings, woven synthetic media, flexible wire meshes, porous ceramics, and porous carbon. The tubes may be mounted for bottom or top discharge. Cake removal typically depends on the backwash action with or without an air or gas pump. The primary attributes of this filter are low space requirement, comparatively low cost construction, and simple operation. Its disadvantages come about because its tubes must be uniformly cleaned by backwashing to prevent ensuing problems; dry discharge is not generally feasible except for some designs; and the construction of the tube must be able to withstand repeated, reverse-flow, backwash action. Tank-tube filters can be constructed of metals similar to other tank units.

6. *Nutsche*

Similar in operation to the well-known Buchner funnel, the Nutsche filter is usually constructed as a cylinder divided into two parts by a horizontal filter plate covered with a convenient medium. Slurry is placed in the upper chamber, and filtration is accomplished by vacuum or gravity. A Nutsche filter has advantages of simplicity of construction and operation, effective displacement washing, suitability for testing procedures and pilot plant studies, suitability for free-flowing crystalline materials, and suitability for corrosive materials. Its use demands a high labor cost for cake

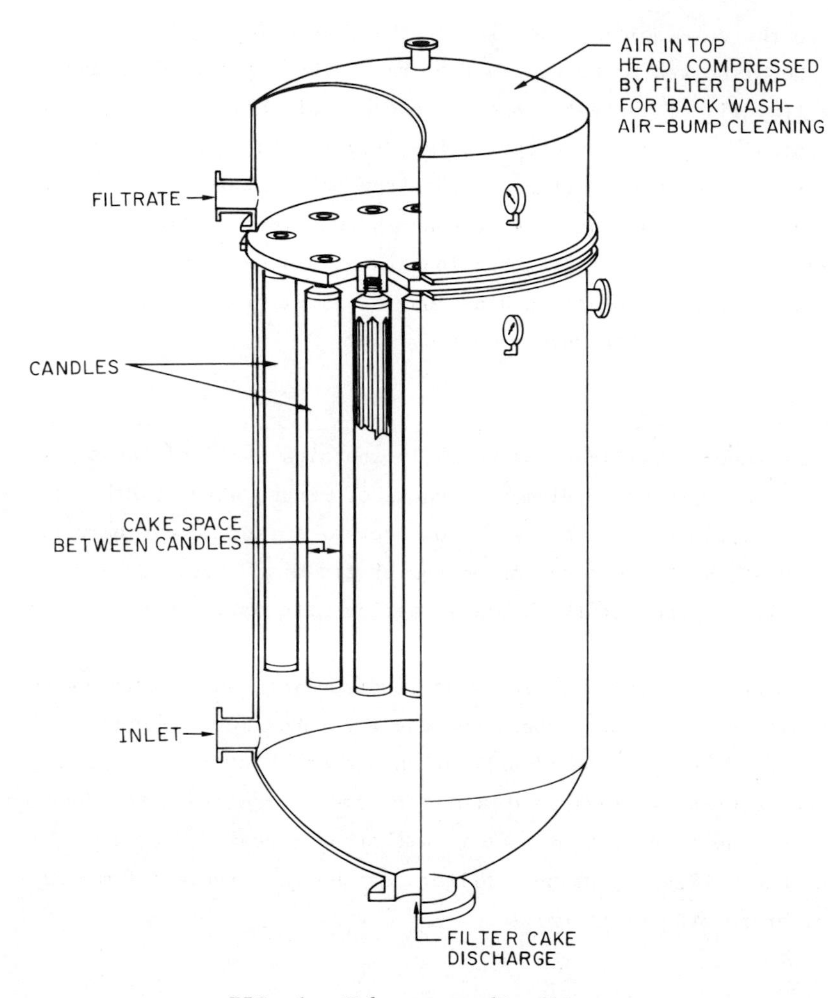

FIG. 6. Tube or candle filter.

removal, and it occupies a large floor space per unit filtering area. It may be constructed of a wide range of materials including reinforced plastic ceramics.

7. *Vacuum Leaf*

The vacuum leaf filter is an open leaf filter operating under suction. An early version, the Moore filter, consisted of an open-tank suction leaf unit for filtration with the capability for the leaves to be transferred under pressure differential to subsequent tanks for washing and cake discharge. If the slurry is agitated sufficiently to prevent cake formation, it can be used as a thickener. Elements can be made in tubular form for clarification filtration. Advantages of this filter include a wide range of available construction materials; low cost in large units; low labor cost for easily discharged cake; and easy inspection of elements, medium, precoating, and cake formation. Its disadvantages arise from a limitation on volatile liquids and temperature controls imposed by its open construction and a pressure differential limitation imposed by its use of a vacuum. Its tank can be constructed of a wide range of materials, including wood and concrete. Elements may be constructed of metal and plastic.

B. Semicontinuous Filters

1. *Horizontal Pan*

This filter (Fig. 7) consists of a series of open pans located in a horizontal plane rotating about a central, vertical axis. The pans form numerous wedge-shaped sections which slope toward and connect directly to a common filter valve underneath the center of the unit. The filter cake may be washed with spray weirs after original

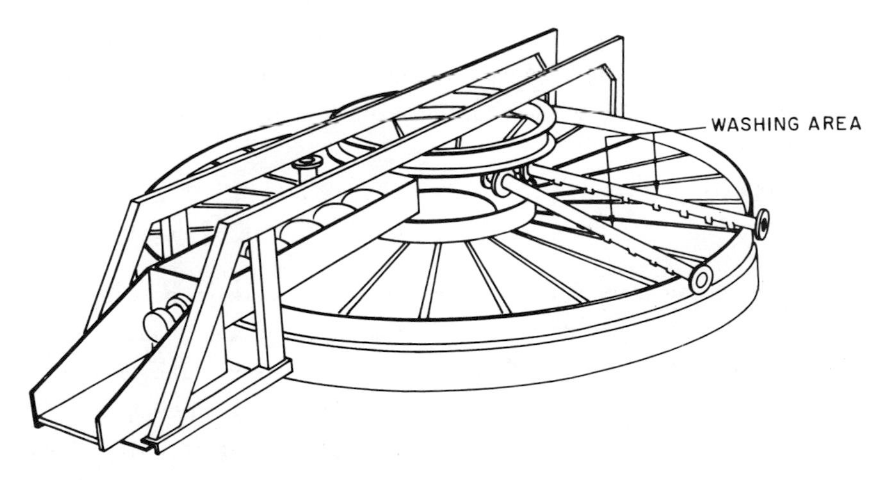

FIG. 7. Rotating tray, table, or tilting pan filter.

dewatering takes place. Countercurrent washing is possible. Cake discharge is accomplished by separately driven scrolls or paddles, or simply by tilting each sector at the completion of filtration (tilting-pan filter). The filter is suitable for granular product dewatering with or without washing and for very high solid loads and high hydraulic requirements. As disadvantages, it utilizes one-sided element filtration; it requires a large space; it is not suitable where media may be easily blinded; and its cost is relatively high.

2. *Automatic Press*

It is debatable whether the automatic press should be listed as a semicontinuous unit or as a batch unit. Basically the unit consists of a series of horizontal frames with a removable filter medium mounted on an endless belt which weaves back and forth among the frames. At the end of the cycle, the frames are separated and the endless filter belt put into motion. Cake is discharged as the belt leaves the filter chambers. Fresh medium is introduced into the filter, and the used sections are washed and cleaned while the next cycle is in progress. A flexible membrane is utilized for pressing and deliquoring the cake by hydraulic pressure. Such a filter produces a cake with a low residual moisture content; the blinding of cloth is minimized, and it is flexible in its usage. High cost is its primary disadvantage. As with nonautomatic press filters it can be furnished with rubber and plastic coated frames.

C. Continuous Filters

1. *Drum Filter*

In this device, a drum (Fig. 8), or cylinder, having a porous wall and covered on the outside with filter media, rotates about a horizontal axis with a portion submerged in the slurry. The filter operates continuously through stages of cake formation, washing, drying, and discharge. Vacuum is the normal driving force, although enclosed pressurized units are sometimes built. The drum is usually

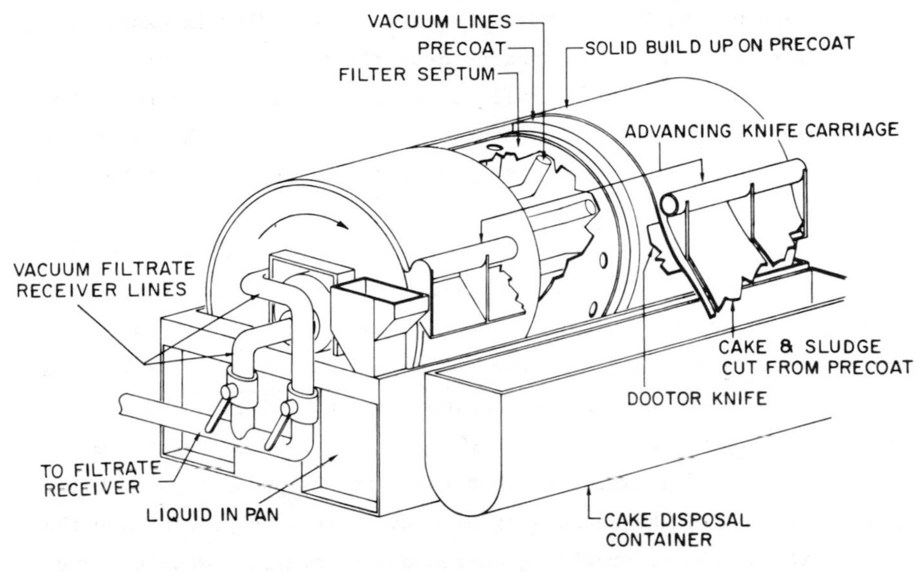

FIG. 8. Rotary drum filter.

subdivided into a number of separate compartments so that the various stages of filtering, washing, drying with air or steam, and discharging with an air blow can be carried out. Some drum filters, however, are furnished with a single internal compartment. Rotational speeds usually vary from about 0.25 to 3.0 rpm.

While rotary drum filters are normally employed for materials of moderate to low resistance, they can be used as precoat filters for highly resistant slurries. A thick coat of several inches of filter aid is deposited on the drum prior to the start of filtration. The precoat is then gradually cut off as the solids are deposited, thereby retaining a low overall flow resistance.

Classification of drum filters is based upon the method of handling the media and discharging the cake. In addition to simple scraper discharge, string or coil, belt, and roller discharge methods may be employed. In the string discharge drum filter, endless strings are wound over the filter medium to lift off and remove the filter cake. The strings normally pass from the drum over a discharge roll where, because of the small radius of the roll, they make a sharp turn and dislodge the cake. The medium is passed as an endless belt in belt discharge drum filters over the drum for filtering and then over a small roller to dislodge the cake. The medium may be cleaned as it returns. With roller discharge, a roller is rotated close to the surface of the drum. The cake passes to the roller and is then scraped off.

The string discharge removes some cakes very well. However, thin cakes cannot be discharged with it, and some noncohesive materials may not be lifted off. The belt discharge, like the automatic press, allows for good cleaning of the media. It is relatively expensive and requires care in keeping the media aligned. The roller discharge depends upon adhesion to the roller instead of the media. It is simple and inexpensive when the cake has properties such that it can be used. The advantages of drum filters in general are that they are capable of providing continuous filtration, washing, and partial drying. With relatively large particles, it is possible to produce nearly dry cakes by using steam; and they permit wide latitude in operating conditions through control of submergence, vacuum, and rotational speed. As disadvantages, they have a relatively high cost, especially in small sizes; they are not applicable with cakes that build up slowly; they are not applicable to batch-type processes; precoat type may necessitate high cost of precoat filter aids; and pressure units are normally limited in size by cost and, in addition, depend upon the efficacy of cake removal from the tank in dry form.

2. *Disc Filter*

Disc filters (Fig. 9) consist of a series of thin, slotted discs revolving on a common shaft and partially submerged in the slurry. Each disc is composed of multiple pie-shaped sectors and each is covered by the filter medium. The filtrate passes through the medium and into separate outlets for each section. The outlet piping for each sector passes through the center shaft and to the main filter valve which controls the vacuum and flow. Disc filters are noted for providing the least cost per square foot of effective

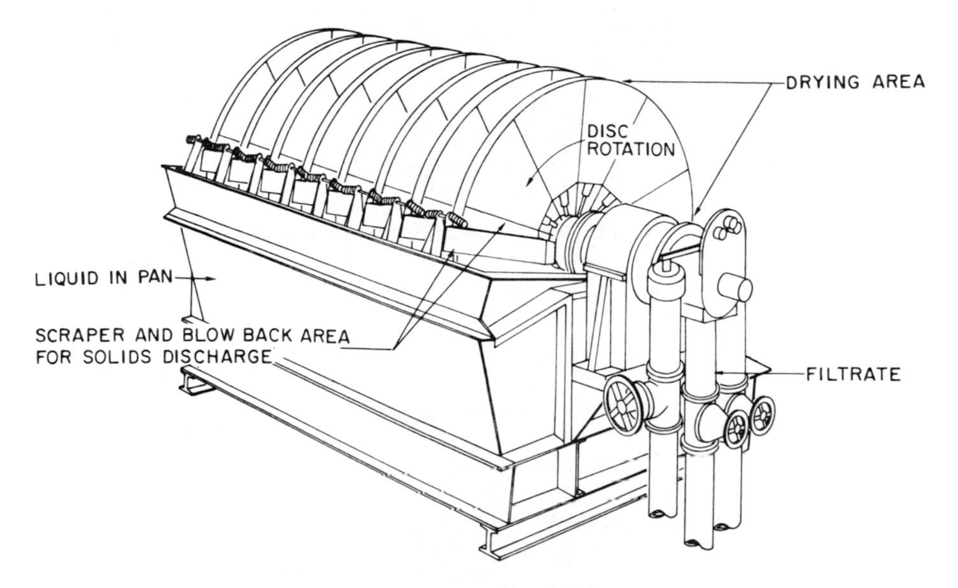

FIG. 9. Vacuum disc filter.

filtering area and for having the greatest filtration area in a minimum of space. Their disadvantages are that they require readily suspended solids and effective cake formation with a minimum of cloth blinding. They are unsatisfactory for washing or precoat operation and have poor dewatering characteristics. Disc filters can be constructed of a variety of materials, including wood and plastic.

3. *Band or Horizontal Belt*

An endless horizontal belt (Fig. 10) passing over suction boxes under vacuum identifies a band or horizontal belt filter. The belt may be washed on its return beneath the filter. Advantages of this arrangement are that the relative areas devoted to filtration, washing, and drying can be more readily adjusted than with rotary drums; settling does not disturb the filtration; and countercurrent washing of cake is possible. Disadvantages arise from high first cost and problems associated with belt construction and sliding surfaces.

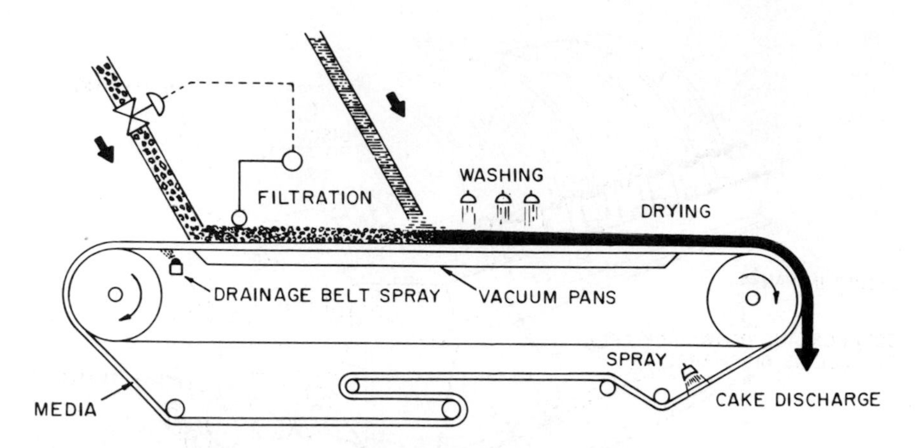

FIG. 10. Horizontal belt filter. Used in metals, fertilizer and dewatering heavy sludges, etc. where solids recovery is prime product and filtrate secondary.

4. *Staged Thin-Cake Pressure*

The Artisan thin-cake filter consists of fixed plates alternating with moving vanes which wipe away a large portion of the cake. Operating at pressures up to 350 psi, the filter maintains high rates. At the outlet of the filter, a concentrated suspension is periodically discharged. The rheological properties of the suspension are important in determining the degree to which concentration is possible.

D. Miscellaneous Filters

Despite the many filters of widely different types on the market, new innovations are continuously being announced. Among unique types are the Guva Tower press (Fig. 32), the English China Clay (ECC) tube press, and the VC filter.

The Guva Tower press consists of two endless media belts arranged vertically. The belts are slightly farther apart at the top than at the bottom. Slurry is introduced at the top. Filtration takes place under the influence of gravity. At the bottom of the filter where the cake is discharged, the belts are brought together and the cake is squeezed to reduce its liquor content.

The ECC tube press consists of a series of filter tubes inside concentric outer walls. Cake is formed on the outside of the filter tube in the annular region. After filtration is completed, the cake is squeezed hydraulically with a membrane under pressures as great as 4,500 psi. Very dry cakes can be produced.

The VC filter is similar to the ECC tube press in that there are two concentric cylinders. However, in the VC filter, the inner cylinder rotates, thereby reducing resistance during filtration (assuming no particle degradation). At the end of filtration, an inflatable membrane is used to press the cake.

E. Filter Applications

Sometimes preliminary data will easily indicate the type of filter to be used. For example, small intermittent batches requiring filter aids suggest a vertical, horizontal-plate, tank filter. However, it is more often the case that selection depends on small variations among similar designs; for example, a decision may be based on the merits of a center-discharge sluice horizontal-tank filter versus a bottom discharge one. As an aid in making this choice, typical applications are listed in Table 5.

TABLE 5 Typical Filter Applications

Press:

- Lube oil additives
- Alginates
- Pharmaceuticals
- Resins
- Cellulose xanthate
- Sugar clarifier muds
- Metallurgical
- Catalyst recovery
- Edible oils
- Sewer sludge

Leaf:

- Kelly design
    - Alumina production
    - Sulfur

TABLE 5 (continued)

Leaf:

- Sweetland
  - Cane and beet sugar
  - Lube oil additives
- Horizontal-tank center-discharge sluice
  - Effluent water
  - Sugar liquors
  - Beer—primary and polish
  - Caustic chlorine brines
  - Process water
  - Nickel sulfate
  - Polyethylene
- Horizontal-tank bottom discharge—sluice or dry cake
  - Edible oils
  - Beer—primary and polish
  - Sugar liquors
- Vertical tank, vertical leaf
  - Aluminum sulfate
  - Plating solutions
  - Nylon salts
  - Sodium chlorate

Plate:

- Batch pharmaceuticals
- Tall oil
- Resins
- Vodka
- Varnish
- Rum

Tray:

- Precious metal recovery
- Alkyd resins

Tube:

- Sodium hydroxide
- Condensate
- Plating solutions
- Effluent water
- Brines

Nutsche:

- Laboratory tests

Vacuum leaf:

- Titanium dioxide production
- Water

TABLE 5 (continued)

Special designs:

Vertical tank, horizontal leaf, centrifugal cake removal
- Melamine
- Edible oils

Table or tilting pan:

- Iron ore concentrates
- Aluminum trihydrate
- Phosphoric acid
- Fluid bed cracking
- Catalyst

Drum:

Precoat
- Corn processing, glucose, oils
- Pharmaceuticals, antibiotics
- Enzyme production
- Plant waste
- Lube oil additives
- Lube oil
- Slop oil
- Wine production

Scraper discharge
- Cane sugar clarifier muds
- Metallurgical slimes
- Chemical salts
- Lime
- Dewaxing

Belt discharge, string, coil
- Sewage sludge
- Cobalt sulfate
- Calcium sulfate
- Chalk slurries

Disc:

- Metallurgical dewatering
- Cement slurry
- Dewatering coal flotation concentrates

Belt:

- Paper manufacturing (Fourdrinier)
- Pigments
- Gypsum crystals
- Pharmaceuticals

IX. FLOW EQUATIONS FOR POROUS MEDIA

Basic laws (readers primarily interested in application can skip directly to Eq. 29) governing the flow of liquids through uniform, incompressible beds serve as a basis in developing formulas for more complex, nonuniform, compressible cakes. Substantial quantities of data are available for the flow of air and water through widely different kinds of solids. Those data serve as a basis for various types of mathematical formulation. Darcy's law can be expressed in the form (see Shirato et al., Ref. 3, for a more sophisticated equation)

$$\frac{dP_L}{dx} = \frac{\mu}{K} q \tag{1}$$

where dp_L/dx is the hydraulic pressure gradient, K the permeability, and q the superficial flow rate expressed as volume/(unit area)(time). In filtration, it is customary to use the mass of dry solids per unit area w instead of the distance from the media x. The mass dw is given by

$$dw = \rho_s(1 - \varepsilon)\, dx \tag{2}$$

where ρ_s is the true density of solids and ε is the porosity of fraction voids. Substituting Eq. (2) in Eq. (1) yields

$$\frac{dp_L}{dw} = \frac{\mu q}{K\rho_s(1 - \varepsilon)} = \mu\alpha q \tag{3}$$

where α is the local filtration or flow resistance. The values of permeability and flow resistance are related by

$$K = \frac{1}{\rho_s\alpha(1 - \varepsilon)} \tag{4}$$

While the value of $\rho_s(1 - \varepsilon)$ varies widely (10-100 lbm/ft^3), a value of 50 can be conveniently used to compare magnitudes of the permeability and flow resistance, as is done in Table 6. It is apparent that flow resistance varies widely in going from granular sand beds to gelatinous cakes with submicrometer particles.

TABLE 6 Comparison of Flow Resistance and Permeability

Material	Flow resistance (ft/lbm)	Permeability (ft^2)
Two-inch pipe in laminar flow	--	1×10^{-3}
Sand bed	10^6	2×10^{-8}
Filter aid	10^9	2×10^{-11}
Clay	10^{11}	2×10^{-13}
Gelatinous cake	10^{14}	2×10^{-16}

A. Frictional Drag on Particles

When suspended solids are deposited during cake filtration, liquid flows through the interstices of the compressible bed in the direction of decreasing hydraulic pressure. The solids are retained by a screen, cloth, porous metal, or other structure known as the septum or filter medium. The solids forming the cake are compact and relatively dry at the medium, whereas the surface layer is in a wet and soupy condition. The porosity is a minimum at the point of contact between the cake and medium where x = 0 (Fig. 11) and a maximum at the surface (x = L, the cake thickness) where the liquid enters. The drag on each particle is communicated to the next particle; and,

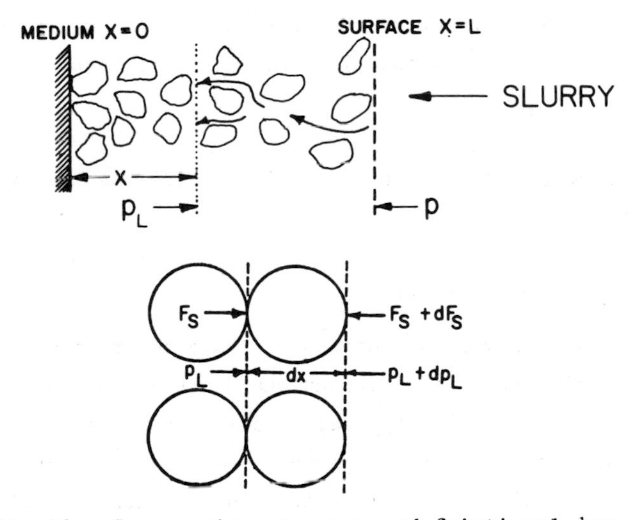

FIG. 11. Compressive pressure and frictional drag.

consequently, the net solid compressive pressure increases as the medium is approached, thereby accounting for the decreasing porosity.

Assuming that inertial forces are negligible, a force balance over the portion of the cake from x to L yields

$$F_s + Ap_L = Ap \tag{5}$$

where the applied pressure p may be a function of time but is independent of distance x. The term F_s represents the accumulated drag on the particles and is communicated through the points of contact. If the particles are in point rather than area contact, the hydraulic pressure p_L may be assumed to be effective over the entire cross-sectional area A of the cake. Dividing Eq. (5) by A and defining the compressive drag pressure by $p_s = F_s/A$ yields

$$p_L + p_s = p \tag{6}$$

The drag on the particles is a combination of skin and form drag produced by friction developed at the surface of the particles. The drag is transmitted through the points of particle contact. The cross-sectional area does not equal the surface area of the particles or the contact area. Thus p_s is a fictitious or pseudopressure which is introduced for convenience. Taking a differential with respect to x in the interior of the cake (assuming the pressure gradient to be a series of quasistatic states), there results

$$dp_s + dp_L = 0 \tag{7}$$

which simply states that drag pressure increases as the hydraulic pressure decreases.

B. Porosity and Flow Resistance

It is generally assumed in compressible cake theory that the local porosity and flow resistance are unique functions of the drag pressure. In Figs. 12 and 13, typical logarithmic plots of α and ε vs. p_s are shown. The pressure is given in kilo-newtons/meter^2, ($6.89\ \text{kN/m}^2 \equiv 1$ psi). Thus, 100 kN/m^2 approximately equals 1 atm.

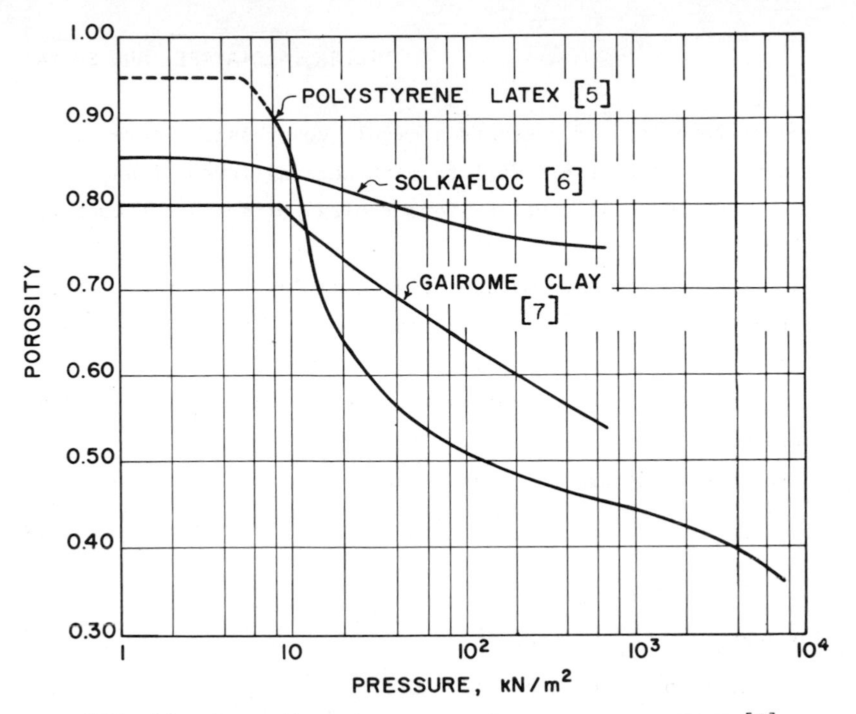

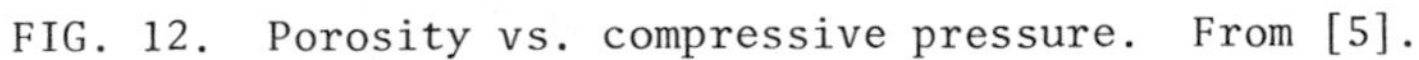
FIG. 12. Porosity vs. compressive pressure. From [5].

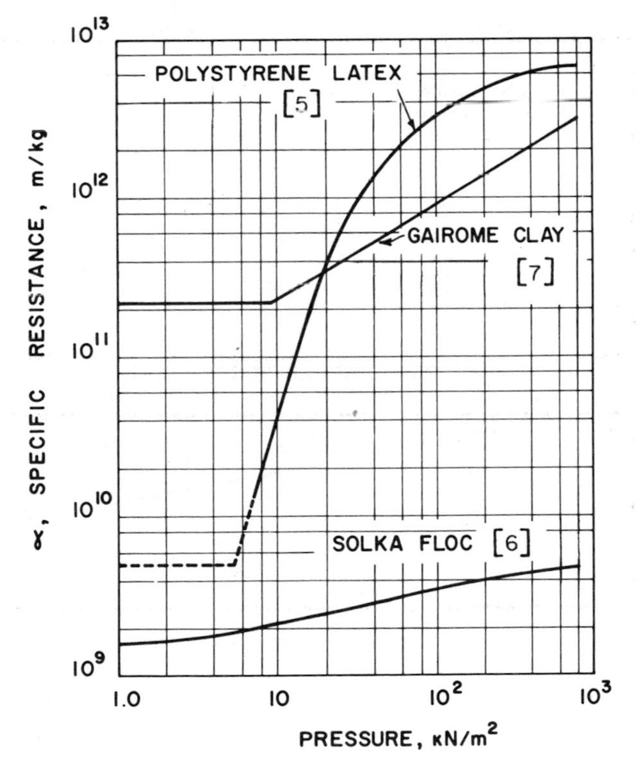

FIG. 13. Specific resistance vs. compressive pressure. From [4].

The polystyrene latex represents a highly compressible material which does not lend itself well to mathematical formulation. The Gairome clay and Solka Floc are more typical of solids encountered in filtration.

It is useful to have mathematical formulas to represent the data of Fig. 12 and 13. A satisfactory method for analyzing the data is illustrated in Fig. 14. It is assumed that the flow resistance and porosity take on constant values, α_i and ε_i, at the same low pressure p_i which is generally in the range of 1.0 psi. Above p_i, power functions are utilized to represent the data, thus

$$\varepsilon = Ep_s^{-\lambda} \qquad p_s \geq p_i \tag{8}$$

$$\varepsilon = \varepsilon_i = Ep_i^{\lambda} \qquad p_s \leq p_i \tag{9}$$

$$\alpha = ap_s^n \qquad p_s \geq p_i \tag{10}$$

$$\alpha = \alpha_i = ap_i^n \qquad p_s \leq p_i \tag{11}$$

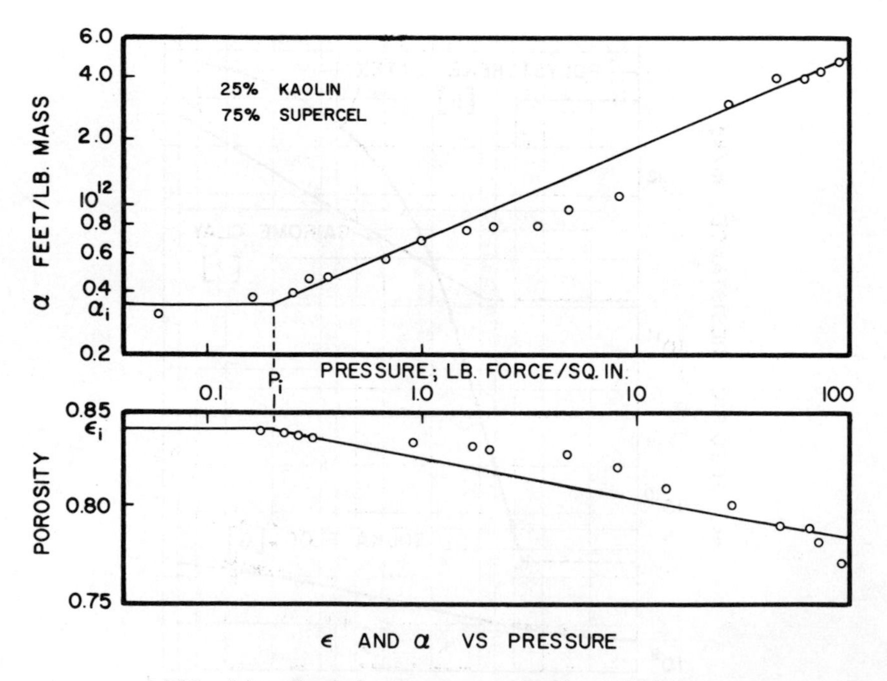

FIG. 14. Empirical representation of data.

It is useful to relate the permeability K or alternatively $(1 - \varepsilon)$ as a similar function of p_s. The form of Fig. 14 is identical to that of Fig. 15, a logarithmic plot of $1 - \varepsilon$ versus p_s. It is then possible to write

$$1 - \varepsilon = Bp_s^{\beta} \qquad p_s \geq p_i \tag{12}$$

$$1 - \varepsilon = 1 - \varepsilon_i = Bp_i^{\beta} \qquad p_s \leq p_i \tag{13}$$

The permeability can then be represented by

$$K = \frac{1}{\rho_s(1 - \varepsilon)\alpha} = \frac{p_s^{-n-\beta}}{\rho_s aB} \tag{14}$$

In Table 7, typical values for the constants in Eqs. (8)-(13) are shown.

Two integrals appear frequently in design formulas and will be derived at this point. First consider the integral

$$\int_0^{p_s} \frac{dp_s}{\alpha} = \int_0^{p_1} \frac{dp_s}{\alpha_i} + \int_{p_1}^{p_s} \frac{dp_s}{ap_s^n} \tag{15}$$

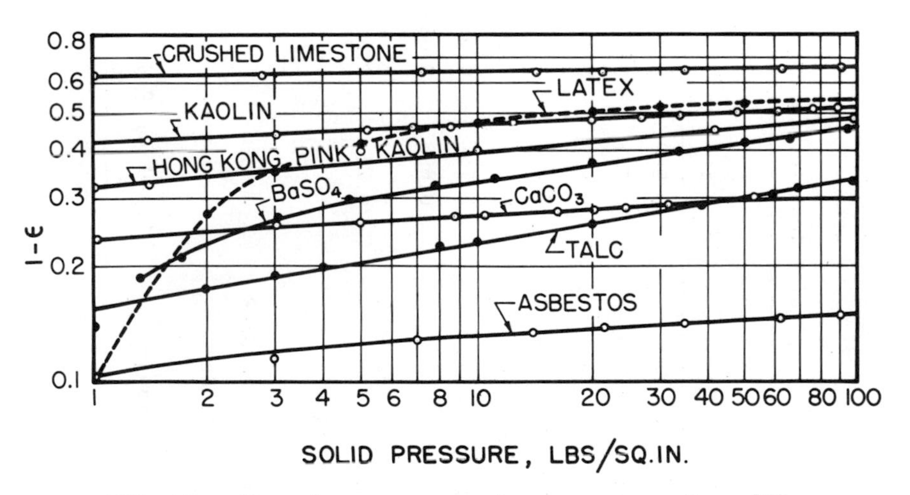

FIG. 15. $(1 - \varepsilon)$ vs. compressive pressure. From [8].

TABLE 7 Typical Values for the Constants in Equations (8) through (13)

Substance	ε_i	p_i	a ($\times 10^{10}$)	n	E	λ	B	β	$\frac{1-n-\beta}{\lambda}$
Asbestos	0.902	0.9	--	--	0.90	0.017	0.115	0.057	--
Calcium carbonate	0.771	0.9	--	0.19	0.77	0.034	0.235	0.063	22
Celite	0.872	0.9	--	0.14	0.90	0.017	--	--	--
Crushed limestone	0.375	1.0	--	--	0.375	0.015	--	--	--
Gairome clay	0.800	1.3	282	0.60	0.815	0.091	0.26	0.13	3.0
Ignition plug clay	0.78	0.6	--	0.56	0.75	0.07	0.27	0.128	4.5
Kaolin	0.698	0.03	--	--	0.59	0.045	0.42	0.054	--
Kaolin, Hong Kong pink	0.72	0.5	101	0.33	0.70	0.059	--	0.005	11.3
Solka Floc	--	--	0.0024	1.01	--	--	--	--	--
Talc	--	--	8.66	0.51	0.86	0.054	0.155	0.203	5.3
TiO_2, flocculated	--	--	32	0.27	0.80	0.038	--	--	--
Zinc sulfide	--	--	14	0.69	--	--	--	--	--
General range	0.4-0.95	0.01-1.5	10^8-10^{14}	0-1.2	0.4-0.95	0-0.1	0.1-0.5	0-0.25	--

It is divided into two parts. The first portion represents the region in which α is considered constant, and the second integral corresponds to the region in which α is represented by a power function. Integrating Eq. (15) leads to

$$\int_0^{p_s} \frac{dp_s}{\alpha} = \frac{p_i^{1-n}}{a} + \frac{p_s^{1-n} - p_i^{1-n}}{a(1 - n)} = \frac{p_s^{1-n} - np_i^{1-n}}{a(1 - n)} \qquad p \geq p_i \tag{16}$$

If n is 0.5 or less, the term in p_i can be neglected for pressures above 10 psi. However, if n is as large as 0.7, the p_i term must be included. When n becomes large, the power function approximation is less accurate, and numerical methods should be employed for the integration. When p_i is neglected, Eq. (16) becomes

$$\int_0^{p_s} \frac{dp_s}{\alpha} = \frac{p_s^{1-n}}{a(1 - n)} \tag{17}$$

In cake filtration, p_s would be replaced by $p - p_1$, the pressure drop across the cake. The integral $\int dp_s/\alpha(1 - \varepsilon)$ is also of importance and can be found in an analogous manner, thus

$$I = \int_0^{p_s} \frac{dp_s}{\alpha(1 - \varepsilon)} = \int_0^{p_i} \frac{dp_s}{\alpha_i(1 - \varepsilon_i)} + \int_{p_i}^{p_s} \frac{dp_s}{aBp_s^{n+\beta}} \tag{18}$$

where α_i and ε_i are considered constants. Integrating and substituting limits gives

$$I = \frac{p_i}{\alpha_i(1 - \varepsilon_i)} + \frac{1}{aB}\frac{p_s^{1-n-\beta} - p_i^{1-n-\beta}}{1 - n - \beta} \tag{19}$$

Substituting $\alpha_i = ap_i^n$ and $(1 - \varepsilon_i) = Bp_i^\beta$ and rearranging, Eq. (19) results in the expression

$$I = \frac{1}{aB}\frac{p_s^{1-n-\beta} - (n + \beta)p_i^{1-n-\beta}}{1 - n - \beta} \qquad p \geq p_i \tag{20}$$

For pressures above 5-10 psi, and $n + \beta$ less than 0.7, the p_i term can be neglected, yielding

$$I = \frac{p_s^{1-n-\beta}}{aB(1 - n - \beta)} \tag{21}$$

C. Average Filtration Resistance

The basic flow equation can be put in the form

$$\frac{dp_L}{dw} = - \frac{dp_s}{dw} = \mu\alpha q \tag{22}$$

where dp_L is replaced by $-dp_s$. As α is a function of p_s, it is preferable to replace p_L. Equation (22) can be rearranged into the form

$$-\mu q \; dw = \frac{dp_s}{\alpha} \tag{23}$$

Integration of Eq. (23) is carried out between the limits (see Fig. 11) of $w = 0$ at the medium and $w = w$ at the cake surface and of $p_s = p - p_1$ at the medium and $p_s = 0$ at the cake surface. The pressure p_1 at the exit of the filtrate from the cake is related to the medium resistance R_m by

$$p_1 = \mu q R_m \tag{24}$$

Integrating Eq. (23) and substituting limits leads to

$$\mu q w = \int_0^{p-p_1} \frac{dp_s}{\alpha} \tag{25}$$

The average value of any function y is given by

$$y_{av} = \frac{1}{X} \int_0^X y \; dx \tag{26}$$

Utilizing this definition and replacing y by $1/\alpha$ leads to

$$\frac{1}{\alpha_{av}} = \frac{1}{p - p_1} \int_0^{p-p_1} \frac{dp_s}{\alpha} \tag{27}$$

Substituting for the integral in Eq. (25) and eliminating p_1 by use of Eq. (24) produces

$$\mu q w = \frac{p - \mu q R_m}{\alpha_{av}} \tag{28}$$

Solving for q, there is obtained

$$\frac{dv}{dt} = q = \frac{p}{\mu(\alpha_{av} w + R_m)} \tag{29}$$

Many analyses of filtration start with Eq. (29). The derivation is not rigorous in that it has been assumed that q and the area are constant throughout the bed. Shirato et al. [3] investigated cakes in which a more complex equation resulted because of the variation of q with distance. Whenever area varies, as in radial flow, Eq. (29) must be modified.

Basically Eq. (29) states that

$$\text{Rate} = \frac{\text{pressure drop}}{\mu(\text{cake resistance + medium resistance})} \tag{30}$$

The total cake resistance changes as the mass of cake w grows with time. While it is assumed that R_m is constant, it probably changes with some media during filtration because of the migration of fine particles with subsequent deposition in the media. Unless the medium can be cleaned perfectly between runs, the resistance will gradually increase until finally the medium must be discarded or reworked.

D. Pumping Mechanisms [9]

For purposes of mathematical treatment, filtration processes are classified according to the variation of the pressure and flow rate with time. Generally, the pumping mechanism determines the flow characteristics and serves as a basis for division into the following categories: (1) constant pressure filtration (the actuating mechanism is compressed gas maintained at a constant pressure, or a vacuum pump), (2) constant rate filtration (positive displacement pumps of various types are employed), (3) variable-pressure, variable-rate filtration (the use of a centrifugal pump results in the rate varying with the back pressure on the pump), and (4) stepped pressure (for experimental purposes, it is possible to increase the pressure manually during filtration and simulate various pumping conditions).

Flow rate versus pressure characteristics for the four types of filtration are illustrated in Fig. 16. Arrows drawn on the curves point in the direction of increasing time. The constant pressure curve is represented by a vertical line, the downward arrows indicating that the rate decreases with time. Drawn horizontally, the constant-rate filtration curve has arrows pointing to increasing pressure with time. The rate for a filter actuated by a centrifugal pump will follow the downward trend of the variable-pressure, variable-rate curve. Depending upon the characteristics of the centrifugal pump, widely differing curves may be encountered. If the first portion of the curve is nearly flat, the pump will produce a filtration that is almost at constant rate. The dotted curve is approximately equivalent to a filtration carried out first at constant rate and then at constant pressure.

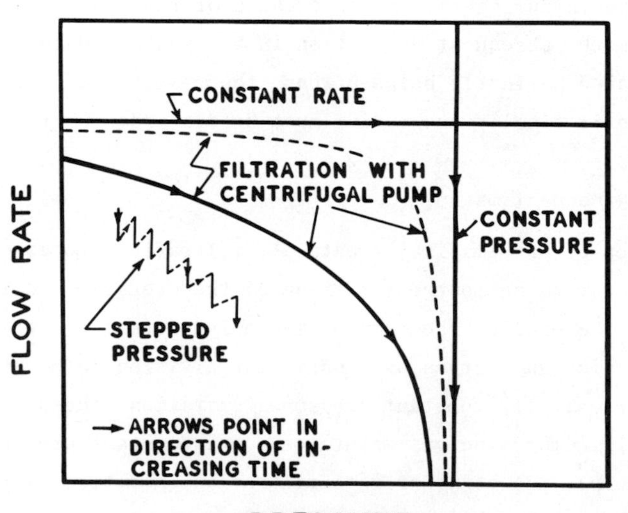

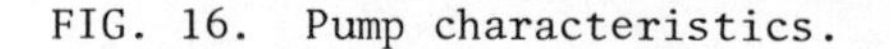

FIG. 16. Pump characteristics.

E. Material Balance

From an overall viewpoint, a material balance can be written on a unit area basis in the form:

Mass of slurry = mass of cake + mass of filtrate

$$\frac{w}{s} = \frac{w}{s_c} + \rho v \tag{31}$$

where w is the mass of dry solids per unit area, v the filtrate volume per unit area, s the average mass fraction of solids in the slurry, s_c the average mass fraction of solids in the cake, and ρ the density of the filtrate. Frequently s_c is replaced by 1/m, where m is the mass of wet cake per unit mass of dry cake. The quantity $s/s_c = ms$. Solving for v in Eq. (31) yields

$$v = \frac{1 - s/s_c}{\rho s} w \tag{32}$$

which is frequently written as

$$w = cv \tag{33}$$

where c equals the mass of solids per unit volume of filtrate. Differentiating v with respect to time yields the flow rate q of filtrate

$$q = \frac{dv}{dt} = \frac{1 - s/s_c}{\rho s} \frac{dw}{dt} + \frac{w}{\rho} \frac{1}{s_c^2} \frac{ds_c}{dt} \tag{34}$$

For the most part, s_c is considered constant, and ds_c/dt is placed equal to zero. In general, s_c varies in any filtration in which the pressure continuously rises. In constant-pressure filtration of talc, latex, and calcium carbonate, it was shown by Tiller and Cooper [8] that s_c or m reached a constant value in less than 0.5 min. For long batch filtrations, a variation over such a short period of time would be of little significance and could be neglected. However, in rotary drum filtration where the filtration time for a 120° submergence at 0.5 rpm is 40 sec, it is not possible to assume s_c or m constant. In general, sophisticated numerical methods would be required for continuous rotary filtration if accurate calculations were needed.

The average mass fraction of solids in the cake s_c frequently must be related to the average porosity ε_{av} of the cake. A simple calculation of s_c yields

$$s_c = \frac{\rho_s(1 - \varepsilon_{av})}{\rho_s(1 - \varepsilon_{av}) + \rho\varepsilon_{av}} \tag{35}$$

It is important to relate the cake thickness to both w and v. For the entire cake, Eq. (2) in combination with Eq. (33) yields

$$w = cv = \rho_s(1 - \varepsilon_{av})L \tag{36}$$

Solving for v and eliminating c produces

$$v = \sigma\left(\frac{1}{s} - \frac{1}{s_c}\right)\left(1 - \varepsilon_{av}\right)L \tag{37}$$

where $\sigma = \rho_s/\rho$. Either s_c or ε_{av} can be eliminated from Eq. (37). Eliminating s_c yields

$$v = \left\{\sigma\left(\frac{1 - s}{s}\right) - \varepsilon_{av}\left[\sigma\left(\frac{1 - s}{s}\right) + 1\right]\right\}L \tag{38}$$

or eliminating ε_{av} gives

$$v = \frac{\sigma(s_c - s)}{s[\sigma(1 - s_c) + s_c]}L \tag{39}$$

As most filters are designed on the basis of cake thickness, Eqs. (38) and (39) are important in converting v to L in formulas relating v to p and t.

X. BATCH CAKE FILTRATION

Equations (25), (29), and (36) form the basis for developing design equations for cake filtration. In Eq. (29) the variables include the following: filtrate flow rate/area, $q = dv/dt$; pressure, p; mass of dry solids/area, w; and cake resistance, α_{av}. Pressure and rate are related by pump characteristics. The mass of solids w is usually eliminated in favor of v by means of Eq. (36).

Eliminating w from Eq. (25) produces

$$\mu c v q = \mu c v \frac{dv}{dt} = \int_0^{p-p_1} \frac{dp_s}{\alpha} = \frac{p - p_1}{\alpha_{av}} \tag{40}$$

where p_1 is given by Eq. (24). Equation (40) is then the basis for most design procedures. Its use depends upon the assumptions that q is constant throughout the cake and that c does not vary with time. Frequently p_1 is neglected, and α_{av} is assumed constant in a constant pressure filtration.

A. Constant Pressure

An exact solution of Eq. (40) requires numerical techniques. However, if it is assumed that α_{av} is constant and a function of p, a simple formula relating v to t can be obtained. As α_{av} depends upon $p - p_1$ rather than p, the assumption of constant α_{av} is equivalent to assuming that p_1 is small.

It is simplest to use Eq. (29) directly for constant pressure filtration. Rearranging the equation leads to

$$\mu c \alpha_{av} v \frac{dv}{dt} + \mu R_m \frac{dv}{dt} = p \tag{41}$$

and integration of Eq. (41) yields

$$\mu c \alpha_{av} \frac{v^2}{2} + \mu R_m v = pt \tag{42}$$

This latter equation yields a parabolic relation (see Ref. 10 for exception to parabolic relation) between v and t. Frequently the term involving R_m is neglected leading to

$$\mu c \alpha_{av} \frac{v^2}{2} = pt \tag{43}$$

Experience is required to determine conditions under which Eq. (43) can replace Eq. (42). Generally, filtration lasting a relatively long time (perhaps 10-30 min) is amenable to analysis by means of

Eq. (43). Basically, the pressure drop across the medium must be a small fraction of the total pressure drop for the $R_m v$ term to be neglected.

An expression for α_{av} can be obtained by combining Eqs. (17) and (27) to give

$$\alpha_{av} = a(1 - n)(p - p_1)^n = (1 - n)\alpha \tag{44}$$

If p_1 is neglected and Eq. (44) is substituted into Eq. (43), there results

$$\frac{a(1 - n)\mu c v^2}{2} = p^{1-n} t \tag{45}$$

This equation is useful for estimating the effect of different pressures on the filtration time. It is primarily of importance in filtrations that last long enough for R_m to be neglected.

Constant pressure filtration has long been the favorite method for obtaining experimental data in the laboratory because of its simplicity. Interpretation of constant-pressure test data is generally based upon Eq. (41) rearranged in the form

$$\frac{dt}{dv} = \frac{2}{K} v + b \tag{46}$$

where

$$K = 2p/\mu c \alpha_{av} \tag{47}$$

$$b = \mu R_m/p \tag{48}$$

While it might appear easier to use Eq. (42) in the form

$$\frac{t}{v} = \frac{v}{K} + b \tag{49}$$

Eq. (46) is, in fact, better because it does not require identification of the precise time at which t = 0. Further, as the v-t relationship is parabolic, the value of dt/dv can be taken rigorously as the average flow rate at the average volume of any interval regardless of the size of Δv. Knowing the value of K, it is possible to obtain α_{av} and then construct a logarithmic graph of α_{av} vs. p. The medium resistance can be calculated from the intercept b.

Example 1

It is desired to test an aqueous slurry for its rate of cake build-up. A small 0.01-ft^2 filter working at a vacuum of 11.1 in. of mercury is to be used. The slurry consists of 1.0% by weight of TiO_2. Enough perlite filter aid, Dicalite 435, is added to bring the total concentration to 3.0% solids. The following data are available:

Temperature = 80°F

Viscosity = 0.861 Cp

Density of solids = 164.5 lbm/ft^3

Density of water = 62.2 lbm/ft^3

Moisture content of cake = 76.0%

Filtrate volume (ml)	110	170	225	290	340
Time (sec)	8.1	17.4	30.1	46.3	66.6

Equation (46) will be used to analyze the data. The volumes will be converted to volume per unit area, v, in ft^3/ft^2. There are 28,316 ml per ft^3, and thus the conversion factor is 100/28,316 = 0.00353. The average rate between two intervals (however large for a parabola) represents the tangent to the curve at the midpoint of the interval. Thus $\Delta v/\Delta t$ can be considered as the rate at a volume given by $0.5(v_1 + v_2)$. Table 8 has been constructed on the basis of the given filtration data. The data are plotted in Fig. 17. It is impossible to obtain a reliable value of R_m from the straight line since slight

TABLE 8 Analysis of Data

t	Δt	v	Δv	$v + \Delta v/2$	$\Delta t/\Delta v$
8.1	9.3	0.388	0.212	0.494	43.9
17.4	12.7	0.600	0.194	0.697	65.5
30.1	16.2	0.794	0.230	0.883	70.4
46.3	20.3	1.024	0.178	1.086	114.0
66.0		1.201			

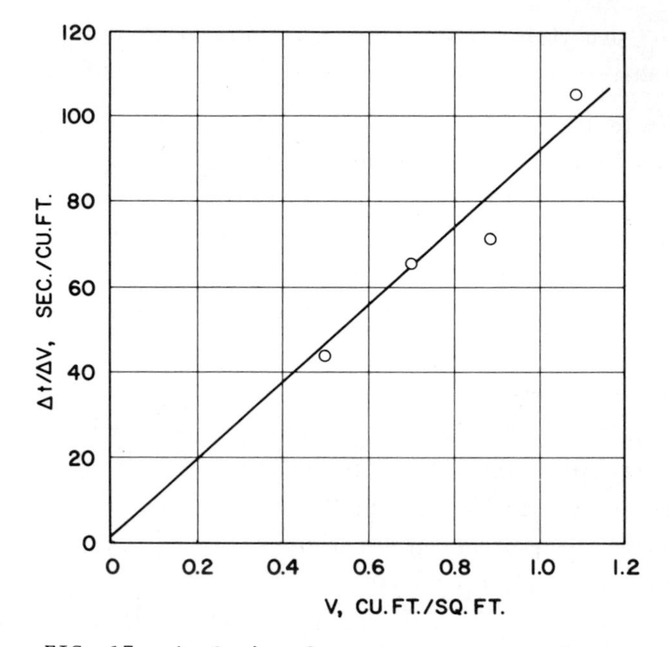

FIG. 17. Analysis of constant pressure data.

modification of its position would give changes of 100% in the intercept. The slope of the line is 89.5, and K = 2/89.5 = 0.0223. For further calculations, c is needed from Eqs. (33) and (34):

$$c = \frac{s}{(1 - s/s_c)} = \frac{62.2 \times 0.03}{1 - 0.03/0.24} = 2.13 \text{ lbm/ft}^3 \tag{50}$$

where $s_c = 1 - 0.76 = 0.24$. It is now possible to obtain α_{av} from Eq. (47). First, it is necessary to calculate p and μ in consistent units:

$$p = \left(\frac{11.1}{29.92}\right) 14.7 \times 144 \times 32.2 = 25{,}287 \text{ lbl/ft}^2 \tag{51}$$

$$\mu = 0.000672 \times 0.861 = 0.000579 \text{ lbm/ft sec} \tag{52}$$

Then

$$\alpha_{av} = \frac{2p}{\mu c K} = \frac{2 \times 25.287}{5.79 \times 10^{-4} \times 2.13 \times 0.0223} = 1.84 \times 10^9 \text{ ft/lbm} \tag{53}$$

This is a low value for filtration resistance. As no direct measurement was made of cake thickness, Eq. (39) must be used to calculate L. The value of σ is 164.5/62.2 = 2.64. Then

$$L = \frac{s[\sigma(1 - s_c) + s_c)]}{(s_c - s)} v = \frac{0.03(2.64 \times 0.76 + 0.24)}{2.64(0.24 - 0.03)} v$$

$$= 0.121v \tag{54}$$

With v = 1.201, the final cake thickness in 66.6 sec is 1.75 in. With this fast rate of build-up, a continuous filter of the rotary drum, horizontal belt, or disc types would represent reasonable choices.

B. Constant Rate [11]

The volume is related to the time by the simple relation

$$v = qt \tag{55}$$

Variation of pressure with time can be obtained from Eq. (40) by eliminating v to give

$$p = \mu c \alpha_{av} q^2 t + \mu q R_m \tag{56}$$

If α_{av} is replaced by Eq. (44), there is obtained

$$(p - p_1)^{1-n} = a(1 - n)\mu c q^2 t \tag{57}$$

This equation is quite similar to Eq. (45) for constant pressure filtration if p_1 is neglected. The average rate of filtration for constant pressure filtration is simply $q_{av} = v/t$. Eliminating v from Eq. (45) produces

$$p^{1-n} = \frac{\mu c}{2a(1 - n)} q_{av}^2 t \tag{58}$$

Equation (58) indicates that, for the same final pressure and time, the average rate with constant pressure will be $\sqrt{2}$ times the rate during constant-rate operation. A better comparison comes from comparing the times for two filtrations which terminate with the same final p and v. Eliminating q from Eq. (57) in favor of v yields

$$\frac{\mu c}{a(1 - n)} v^2 = (p - p_1)^{1-n} t \tag{59}$$

and comparing it with Eq. (45) shows that the time required for constant rate is twice that of constant pressure for the same final volume and pressure. Example 7 gives a treatment of constant-rate operation.

C. Centrifugal Pump Filtration [12]

If both pressure and rate vary, it is necessary to impose the pump characteristics on the equations. No simple formulas can be obtained to relate p to t, and a relatively simple numerical integration is required. Equation (40) can be rearranged into the form

$$v = \frac{1}{\mu c q} \int_0^{p-p_1} \frac{dp_s}{\alpha} = \frac{1}{\mu c q} \frac{p - p_1}{\alpha_{av}} \tag{60}$$

The rate q is a function of p, and α_{av} is a function of $p - p_1$. Consequently, it is possible to construct a relationship of v to p. Substituting for α_{av} yields

$$v = \frac{(p - p_1)^{1-n}}{\mu c a(1 - n) q} \tag{61}$$

Once v has been obtained as a function of p and q, t can be obtained by integration; thus

$$t = \int_0^t \frac{dv}{q} \tag{62}$$

If data obtained from a series of constant pressure tests yield values of α_{av} and s_c (or c), it is possible to utilize Eqs. (60) and (61) to give a volume vs. rate relationship which can then be employed for finding the time.

Example 2

A 2% by weight mixture consisting of two-thirds TiO_2 and one-third expanded perlite is to be filtered. The average density of the solids is 202.6 lbm/ft^3. A 250-ft^2 leaf filter is available, and it is desired to know what quantities can be filtered in the range of 20°-27°C with a centrifugal pump having the following characteristics:

p (psi)	0	10	20	30	40	50	60
q (gal/min)	500	469	418	345	251	136	0
(ft^3/sec)	1.114	1.046	0.931	0.769	0.560	0.304	0

For preliminary calculations, the medium resistance will be neglected.

A series of constant pressure tests was made on a small laboratory filter with the following results:

p (psi)	0.5	1.1	9.0	27.0	34.5	58.5	69.0	91.5
α_{av} ($\times 10^{-9}$)	2.1	2.28	5.73	11.1	16.2	17.1	18.6	21.1
ε_{av}	0.915	0.903	0.880	0.852	0.841	0.832	0.878	--
s_c	0.232	0.259	0.307	0.361	0.381	0.396	0.403	--

Values of α_{av} and $(1 - \varepsilon_{av})$ are plotted in Fig. 18. It is necessary to calculate c from Eqs. (32) and (33):

$$c = \frac{\rho_s}{1 - s/s_c} = \frac{62.3 \times 0.02}{1 - 0.02/s_c} \tag{63}$$

The thickness can be related to the volume filtered by Eq. (36), thus

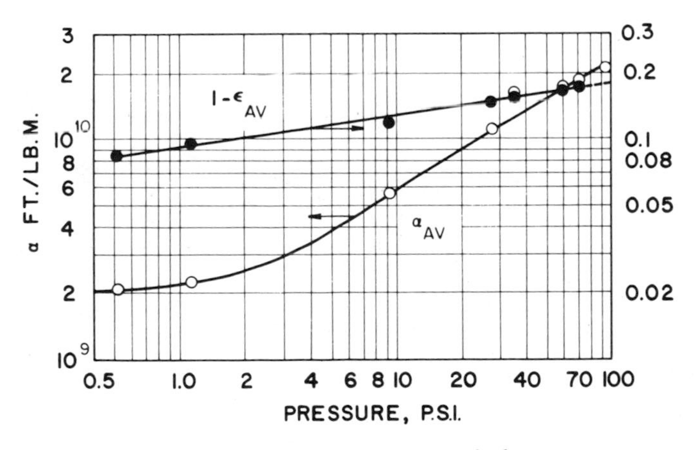

FIG. 18. $(1 - \varepsilon)$ and α_{av} vs. applied pressure.

$$L = \frac{c}{\rho_s(1 - \varepsilon_{av})} v = C_L v \tag{64}$$

The value of c and the multiplier C_L of v in Eq. (64) vary with p. Calculations indicate that c varies from 1.147 at 0.5 psi to 1.188 at 69 psi while C_L varies from 0.0666 to 0.0341 in the same range. Thus, while an average value of c could be used, the variation in C_L is too large for it to be considered constant.

Equation (60) gives v as a function of p

$$v = \frac{1}{\mu c \alpha_{av}} \frac{p}{q} \tag{65}$$

where c, α_{av}, and q are functions of p. The viscosity will be taken at 20°C as 0.000672 lbm/(ft sec). In terms of psi and q_g = (gal/min)/ft^2, Eq. (65) becomes

$$v = \frac{60 \times 7.48 \times 144 \times 32.17}{0.000672\ c\alpha_{av}} \frac{(\text{psi})}{q_g} = \frac{3.094 \times 10^9\ (\text{psi})}{c\alpha_{av} q_g} \tag{66}$$

Quantities necessary to the calculations are shown in Table 9. With values from Table 9, a plot is made of L, p, and 1/q vs. v as shown in Fig. 19. The time is found from the area under the curve of 1/q vs. v. The final results, as given in Table 10, are plotted in Fig 20. The cake builds up rather rapidly, and the cycle length would

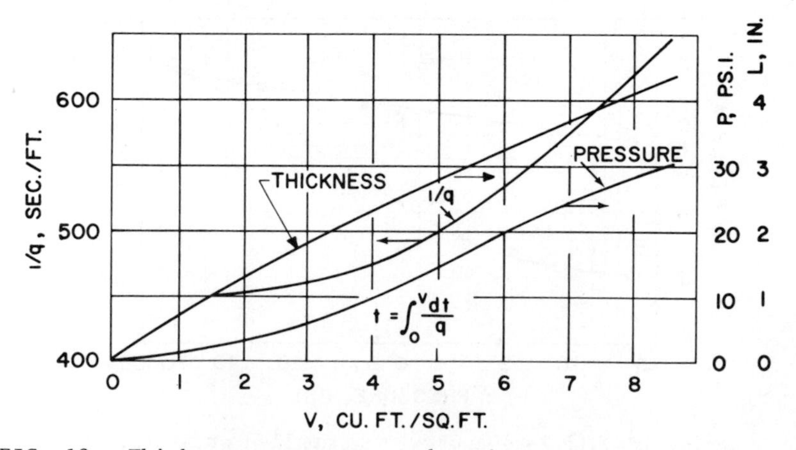

FIG. 19. Thickness, pressure, and reciprocal rate vs. volume.

TABLE 9 Data for Constructing Fig. 19

p	q (gal/min)	q_g	q $\frac{ft^3}{(sec)(ft^2)}$	α_{av} (ft/lbm) ($\times 10^{-9}$)	ε_{av}	s_c	c	C_L	v (ft^3/ft^2)	L (in.)
0.5	499	0.998	0.00224	2.10	0.915	0.232	1.364	0.0709	0.541	0.46
1	498	0.996	0.00222	2.22	0.905	0.254	1.352	0.0556	1.032	0.69
5	487	0.974	0.00217	4.00	0.880	0.307	1.332	0.0548	2.98	1.96
10	469	0.938	0.00209	6.04	0.869	0.329	1.327	0.0468	4.12	2.31
15	446	0.892	0.00199	7.72	0.860	0.346	1.322	0.0466	5.10	2.85
20	418	0.836	0.00186	9.21	0.853	0.359	1.319	0.0443	6.09	3.24
30	345	0.690	0.00154	11.8	0.846	0.371	1.317	0.0422	8.65	4.38

TABLE 10 Data for Constructing Fig. 20

v (ft /ft)	Incremental area (sec)	t (sec)	t (min)	p (psi)	L (in.)
1	448	445	7.4	0.9	0.7
2	450	895	14.9	2.6	1.3
3	460	1355	22.6	5.6	1.8
4	468	1823	30.4	9.7	2.3
5	486	2309	38.5	14.7	2.8
6	516	2825	47.1	20	3.2
7	555	3380	56.3	24.7	3.7
8	597	3977	66.3	28.4	4.1

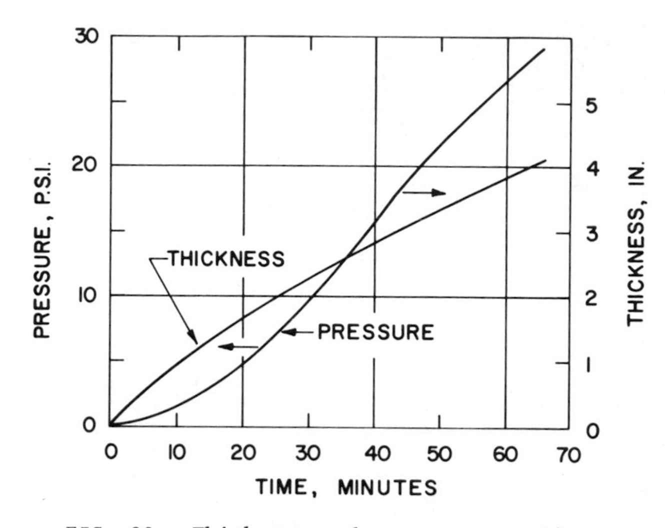

FIG. 20. Thickness and pressure vs. time.

depend upon the spacing of the leaves. For a cake thickness of 2 in., approximately half-hour cycles would be required.

The pump would be somewhat large for the slurry and press. If the cake thickness were restricted to 2 in., the rate would never drop below 90% of the maximum pump rating and the pressure would never rise above 10 psi.

XI. CAKE POROSITY

The final liquor content or average cake porosity is of importance in filter design. Increasing filtration pressure increases the dryness of the final cake. When satisfactory limits can be reached by simply supplying a sufficiently high final pressure, no other technique need be considered. However, if the cake is still too wet, hydraulic deliquoring or mechanical expression can be utilized.

In recent years, deliquoring of filter cakes has received increased emphasis. In disposal of sewage sludge, it is important to both disposal by incineration and landfill. In the chemical process industries, drying of wet cakes increases cost. Hence it is desirable to eliminate as much liquor as possible by nonthermal methods.

A. Porosity as a Function of Distance

Few authors have treated the question of porosity variation in compressible filter cakes, although it is fundamental to both qualitative and quantitative problems of design and operation of solid-liquid separation equipment. Hutto [13] and Shirato et al. [14,15] have both published data giving porosity as a function of distance through the bed. Tiller and Cooper [8] developed analytical formulas for porosity and hydraulic pressure as a function of fractional distance based on Eqs. (8)-(13). They provided criteria by which it was possible to classify the widely divergent porosity behavior of different kinds of materials.

Equation (1) can be rearranged in the form

$$\mu q \, dx = -K dp_s = -\frac{dp_s}{\rho_s \alpha (1 - \varepsilon)} \tag{67}$$

where p_s is chosen for the independent variable, and K, α, and ε are assumed to be functions of the solid compressive pressure. Equation (67) can be integrated over a portion of the cake and then the entire cake. The limits of integration are given by

	x	p_s	p_L
At medium	0	p	0
Inside cake	x	p_s	p_L
At surface	L	0	p

where it is assumed that the medium resistance is negligible. Integrating from x to L and 0 to L yields [16]

$$\mu q (L - x) = \frac{1}{\rho_s} \int_0^{p_s} \frac{dp_s}{\alpha (1 - \varepsilon)} \tag{68}$$

$$\mu q L = \frac{1}{\rho_s} \int_0^{p} \frac{dp_s}{\alpha (1 - \varepsilon)} \tag{69}$$

Equation (69) provides a relationship between the flow rate, thickness L, and the pressure drop p across the cake.

If Eq. (68) is divided by Eq. (69), there is obtained

$$1 - \frac{x}{L} = \frac{\int_0^{p_s} \frac{dp_s}{\alpha(1 - \varepsilon)}}{\int_0^{p} \frac{dp_s}{\alpha(1 - \varepsilon)}} \tag{70}$$

This equation gives the fractional distance through the cake as a function of the upper limit p_s of integration. It indicates that x/L vs. p_s curves are independent of flow and total thickness. In turn, p_L (equal to $p - p_s$) and ε can be related to x/L through their functional relationships to p_s.

If the conditions underlying Eq. (21) hold, the integrals in Eq. (70) may be replaced by approximate formulas [see Eq. (21)] to yield ($R_m = 0$)

$$1 - \frac{x}{L} = \left(\frac{p_s}{p}\right)^{1-n-\beta} = \left(\frac{p - p_L}{p}\right)^{1-n-\beta} \tag{71}$$

This equation relates the hydraulic pressure drop to the distance through the cake. In Fig. 21, the fractional hydraulic-pressure drop is shown as a function of the fractional distance through the cake for different exponents in Eq. (71). As the exponent (and compressibility) increase, the hydraulic pressure drop becomes smaller over a substantial portion of the cake. Consequently, the drag pressure is small throughout a large fraction of the cake which remains unconsolidated. Tiller and Green [4] demonstrated that a dense "skin" can develop next to the media with highly compressible materials.

The porosity can be introduced by utilizing Eq. (8), thus

$$1 - \frac{x}{L} = \left(\frac{p_s}{p}\right)^{1-n-\beta} = \left(\frac{\varepsilon_1}{\varepsilon}\right)^{(1-n-\beta)/\lambda} \tag{72}$$

where ε_1 is the porosity at the medium where $p_s = p$. Equation (72) is valid for $\varepsilon \leq \varepsilon_i$. Values of the exponent in Eq. (72) are shown for a few substances in Table 7.

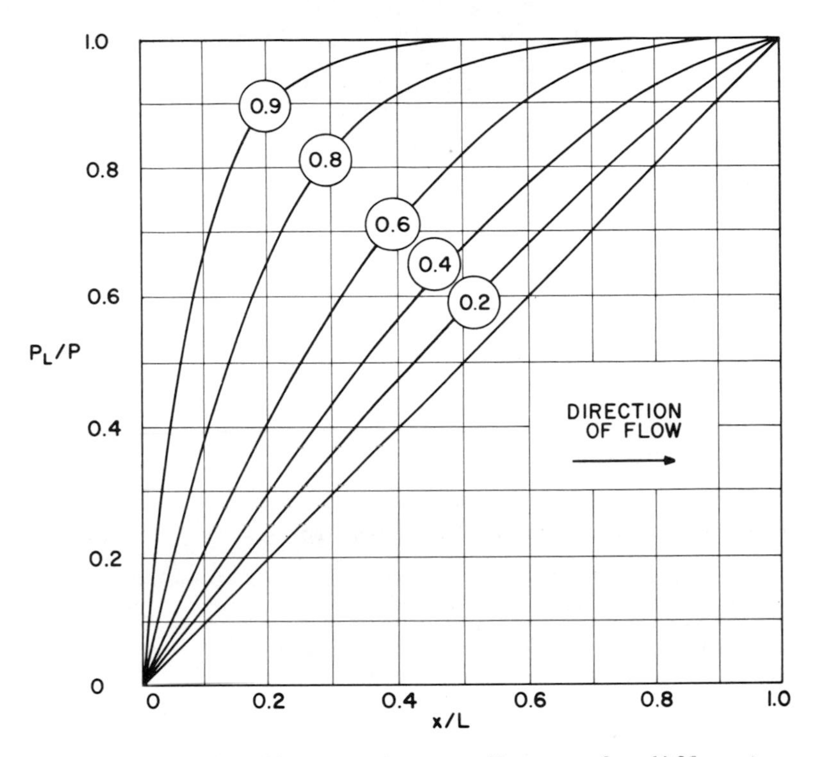

FIG. 21. Hydraulic pressure vs. distance for different compressibility coefficients.

B. Average Porosity as a Function of Applied Pressure

While it is useful to know how the porosity varies throughout the solid, a designer is primarily concerned with the average porosity. The average porosity as defined by

$$\varepsilon_{av} = \frac{1}{L}\int_{o}^{L} \varepsilon \, dx \tag{73}$$

On changing the variables of integration to p_s, there is obtained

$$\varepsilon_{av} = \frac{1}{L}\int_{p}^{o} \varepsilon \frac{dx}{dp_s} \, dp_s \tag{74}$$

The term dp_s/dx can be obtained from Eq. (3) in the form

$$\frac{dp_s}{dx} = - \mu\rho_s\alpha(1 - \varepsilon)q \tag{75}$$

Substituting it into Eq. (74) produces

$$\varepsilon_{av} = \frac{1}{\mu\rho_s qL}\int_0^p \frac{dp_s}{\alpha(1 - \varepsilon)} \tag{76}$$

Substituting for the thickness as obtained from Eq. (69) leads to

$$\varepsilon_{av} = \frac{\int_0^p \varepsilon\ dp_s/\alpha(1 - \varepsilon)}{\int_0^p dp_s/\alpha(1 - \varepsilon)} \tag{77}$$

This equation can be rearranged into the form

$$\varepsilon_{av} = 1 - \frac{\int_0^p dp_s/\alpha}{\int_0^p dp_s/\alpha(1 - \varepsilon)} \tag{78}$$

Substituting Eqs. (17) and (21) in place of the integrals yields

$$1 - \varepsilon_{av} = B\,\frac{1 - n - \beta}{1 - n}\,p^{\beta} \tag{79}$$

For practical purposes, Eq. (79) can be improved if an experimental value of ε_{av} is used to evaluate the multiplier of the pressure term. Assuming the porosity to be equal to ε_{avo} at p_o yields

$$\frac{1 - \varepsilon_{av}}{1 - \varepsilon_{avo}} = \left(\frac{p}{p_o}\right)^{\beta} \tag{80}$$

Basically, Eq. (80) shows that a logarithmic plot of the average volume fraction of solids $(1 - \varepsilon_{av})$ versus the applied pressure should yield a straight line. The approximation is generally valid for pressures above 10 psi and materials that are not too compressible.

In Fig. 22, the porosity of a number of substances is plotted against the fractional distance through the cake. The curve for the highly compressible polystyrene latex was obtained by numerical

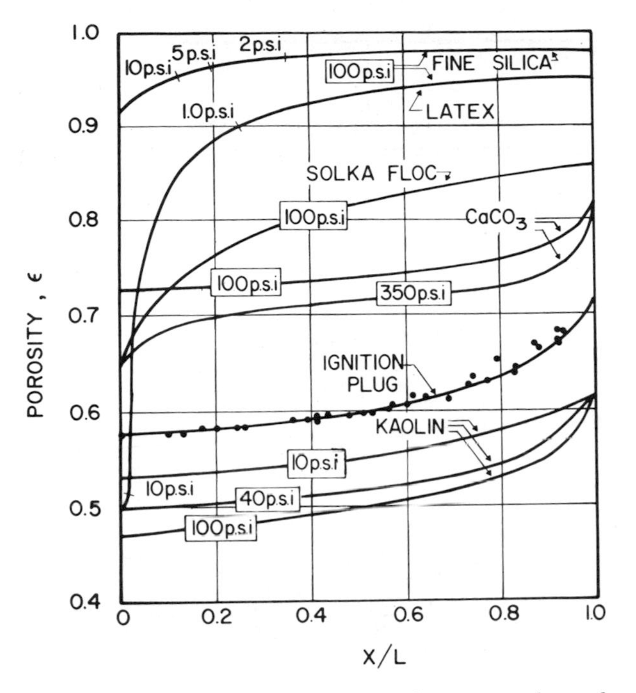

FIG. 22. Porosity vs. fractional distance through cake.

integration using Eq. (70). The others lend themselves equally well to numerical integration or to application of Eq. (72). In Fig. 23, the average porosity is plotted against the total pressure for the same substances. The behavior of the polystyrene latex is in contrast to the other materials in that increasing pressure does not reduce the average porosity after it reaches a fixed value. Such behavior is expected from materials that have large changes in porosity and flow resistance in the low compressive-pressure range.

Example 3

Samples of a cake were taken from a filter press operating at 20 and 60 psi. The solid has a specific gravity of 2.1, and the liquid is water. It is desired to estimate the moisture content at 100 psi. A small part of the surface of each cake was removed in order to obtain estimates of ε_i (which is needed for Example 4). With a known value of ε_i, s_c at $p_s = 0$ can be calculated. Values follow:

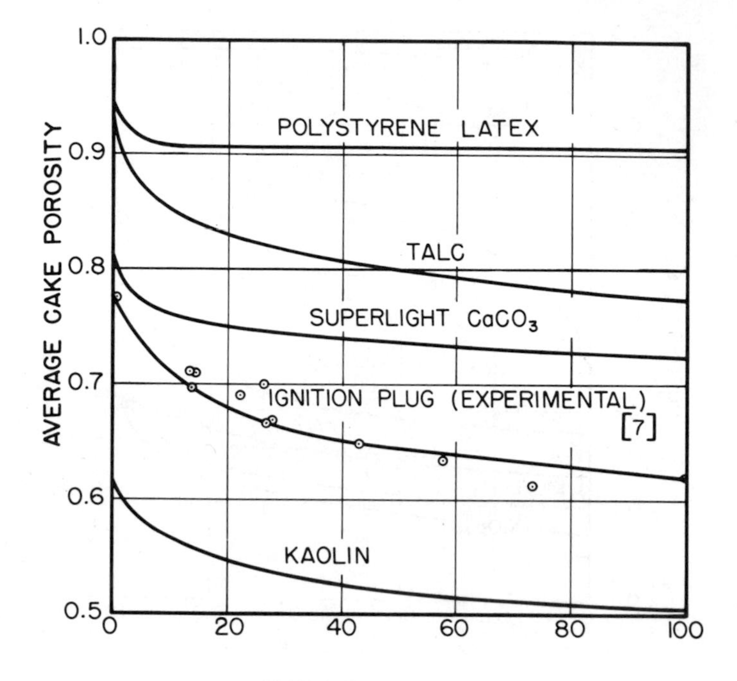

FIG. 23. Average cake porosity vs. applied pressure.

p_s (psi)	0	20	60
s_c	0.172	0.285	0.370

To solve the problem, it is convenient to use Eq. (80). The average porosity can be obtained by rearranging Eq. (35) to give

$$\varepsilon_{av} = \frac{\sigma(1 - s_c)}{\sigma(1 - s_c) + s_c} \tag{81}$$

which then yields the following values:

p_s (psi)	0	20	60
ε_{av}	0.910	0.840	0.781

At $p_s = 0$, $\varepsilon_{av} = \varepsilon_i$. Solving for β in Eq. (80) gives

$$\beta = \frac{\log(1 - \varepsilon_{av})/(1 - \varepsilon_{avo})}{\log p/p_o} \tag{82}$$

$$= \frac{\log 0.219/0.16}{\log 60/20} = 0.285 \quad (83)$$

It is now possible to estimate the average porosity at 100 psi, thus

$$1 - \varepsilon_{av} = 0.16(100/20)^{0.285} \quad (84)$$

This yields $\varepsilon_{av} = 0.747$. The average moisture content is calculated with Eq. (35), giving $s_c = 0.415$. Increasing the pressure to 100 psi would thus remove 17.2% of the liquid present at 60 psi.

It would obviously have been better to have more than two values to obtain β, as there is always uncertainty in extrapolation.

C. Hydraulic Deliquoring

If the porosity cannot be decreased sufficiently by a direct increase of pressure, it is possible to utilize hydraulic deliquoring in certain types of filter units such as plate and frame or recessed-plate filters and specially designed leaf filters in which a blank surface is inserted between leaves. The porosity depends upon the drag pressure $p_s = p - p_L$ which remains low over a large fraction of the cake as illustrated in Fig. 21. By changing the flow direction, it is possible to produce high drag compressive pressures where previously p_s had been low. As the compressive process is essentially irreversible, the portion of the cake previously consolidated does not change; and the soupier layers of the cake undergo compression and deliquoring.

A simple way to effect hydraulic pressure changes is to flow at right angles to the original pattern. If the entrance and exit ports of a press are appropriately designed, flow can continue after the press has been normally filled and the filtration presumably terminated. In Fig. 24, the condition at the end of an ideal filtration in a press is illustrated. The two surfaces have just closed, and the press is "full." The porosity curves at the end of the process are shown along with arrows denoting the flow direction.

Cutting off the normal exit for the filtrate and forcing the flow to pass at right angles through the cake produces the situation

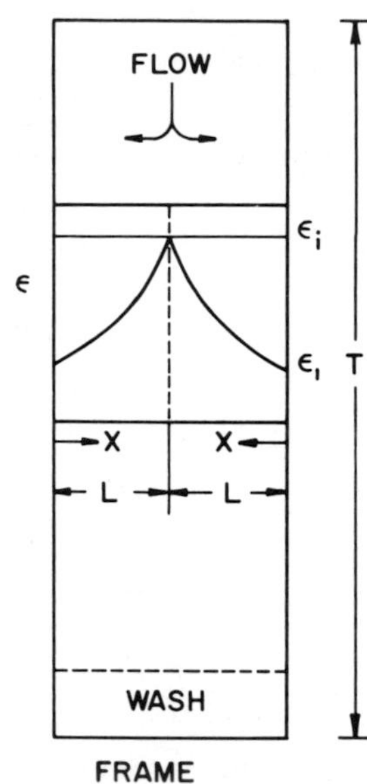

FIG. 24. Porosity distribution at end of cycle in plate and frame press.

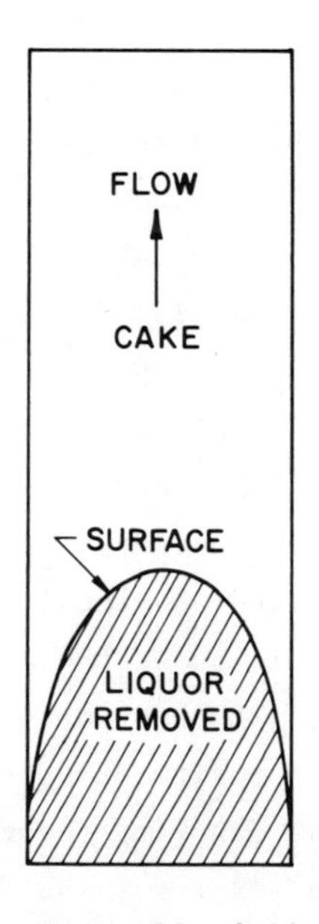

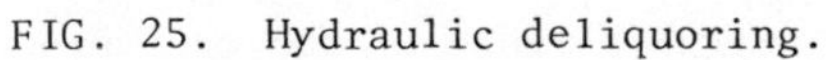

FIG. 25. Hydraulic deliquoring.

depicted in Fig. 25. Even if flow continues through the normal filtrate channels (most probable in current designs), deliquoring will still be accomplished although on a somewhat less-effective basis. A new set of porosity curves are established as shown in Fig. 26 and as approximated by straight lines in Fig. 27. Theoretical calculation of the curve in Fig. 26 is beyond the scope of this chapter. Nevertheless, the straight-line approximation in Fig. 27 can be used to give a rough estimate of deliquoring possible with this scheme.

It is necessary to estimate the decrease in the T dimension of the cake in order to obtain a formula for the deliquoring operation. At the centerline of the press, the porosity equals ε_i at the end of filtration while the average for the entire cake is ε_{av}. After the

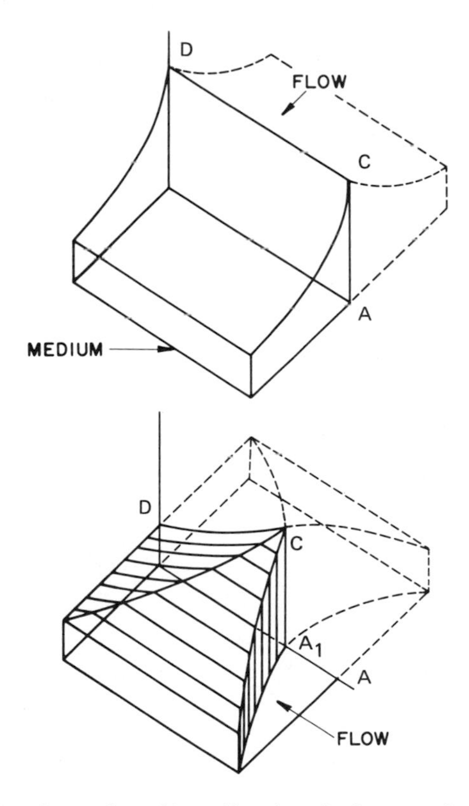

FIG. 26. Porosity distribution before and after deliquoring.

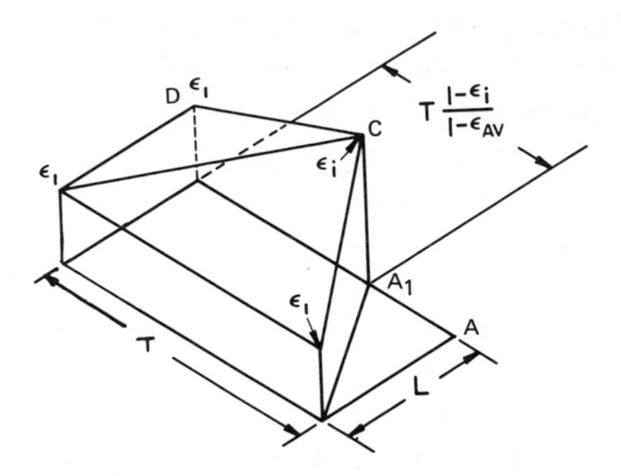

FIG. 27. Simplified view of porosity relationships.

new equilibrium is established, point A moves to point A_1, and curve CD is equivalent to the old porosity distribution with ε_{av} as an average. A balance over the solid along the centerline yields

$$T_2(1 - \varepsilon_{av}) = T(1 - \varepsilon_i) \tag{85}$$

where T_2 is the decreased cake dimension during deliquoring. This relationship assumes that pure liquid is used for the hydraulic deliquoring process. As a practical matter, slurry would be employed, and the formulas would have to be modified accordingly. Because of the approximate nature of this development, no corrections have been made.

The original volume of liquid is given by $LT\varepsilon_{av}$. The final volume after deliquoring is

$$\text{vol.} = \left(\frac{1 + \dfrac{1 - \varepsilon_i}{1 - \varepsilon_{av}}}{2}\right)\varepsilon_1 LT + \left(\frac{1 + \dfrac{1 - \varepsilon_i}{1 - \varepsilon_{av}}}{2}\right)\frac{(\varepsilon_i - \varepsilon_1)}{3} LT \tag{86}$$

Dividing by the original volume gives 1 - F, where F is the fraction of liquid removed. Thus

$$1 - F = \left[\frac{1 + (1 - \varepsilon_i)/(1 - \varepsilon_{av})}{2}\right]\left[\frac{\varepsilon_i + 2\varepsilon_1}{3\varepsilon_{av}}\right] \tag{87}$$

TABLE 11 Hydraulic Deliquoring

Substance	ε_i	ε_1	ε_{av}	1 - F	F
$CaCO_3$ (100 psi)	0.82	0.73	0.75	0.87	0.13
(350 psi)	0.82	0.65	0.72	0.83	0.17
Kaolin (100 psi)	0.63	0.47	0.55	0.86	0.14
Talc (100 psi)	0.90	0.67	0.75	0.70	0.30

A summary of values of F for several materials is given in Table 11.

D. Expression [14-20]

The term "expression" is the separation of a liquid from a solid-liquid mixture by compression. In filtration, the original mixture is sufficiently fluid to be pumpable; in expression this is not always true, and the material may be either semisolid or a thick slurry.

1. *Equipment [16,17,18]*

Various types of equipment for expression are available. Hydraulic presses of the batch type have been used for centuries. Principal batch types are the box, pot, platen, curb, and cage. For continuous operation, the screw press, the disc press (Fig. 28), and various types of roller mills are in common use. In recent years, new equipment has been introduced which either reduces the moisture content or shortens the deliquoring time. The common character of new expression equipment is the compression of filter cakes inside the filter by elastic membranes.

Figure 29 illustrates a belt filter with compression rollers; Fig. 30 shows a filter chamber of a fully automatic press with membrane; Fig. 31 is a horizontal automatic pressure filter with compression device; and Fig. 32 is a tower belt filter which continuously employs gravity filtration followed by roll expression regulated by a hydraulic system.

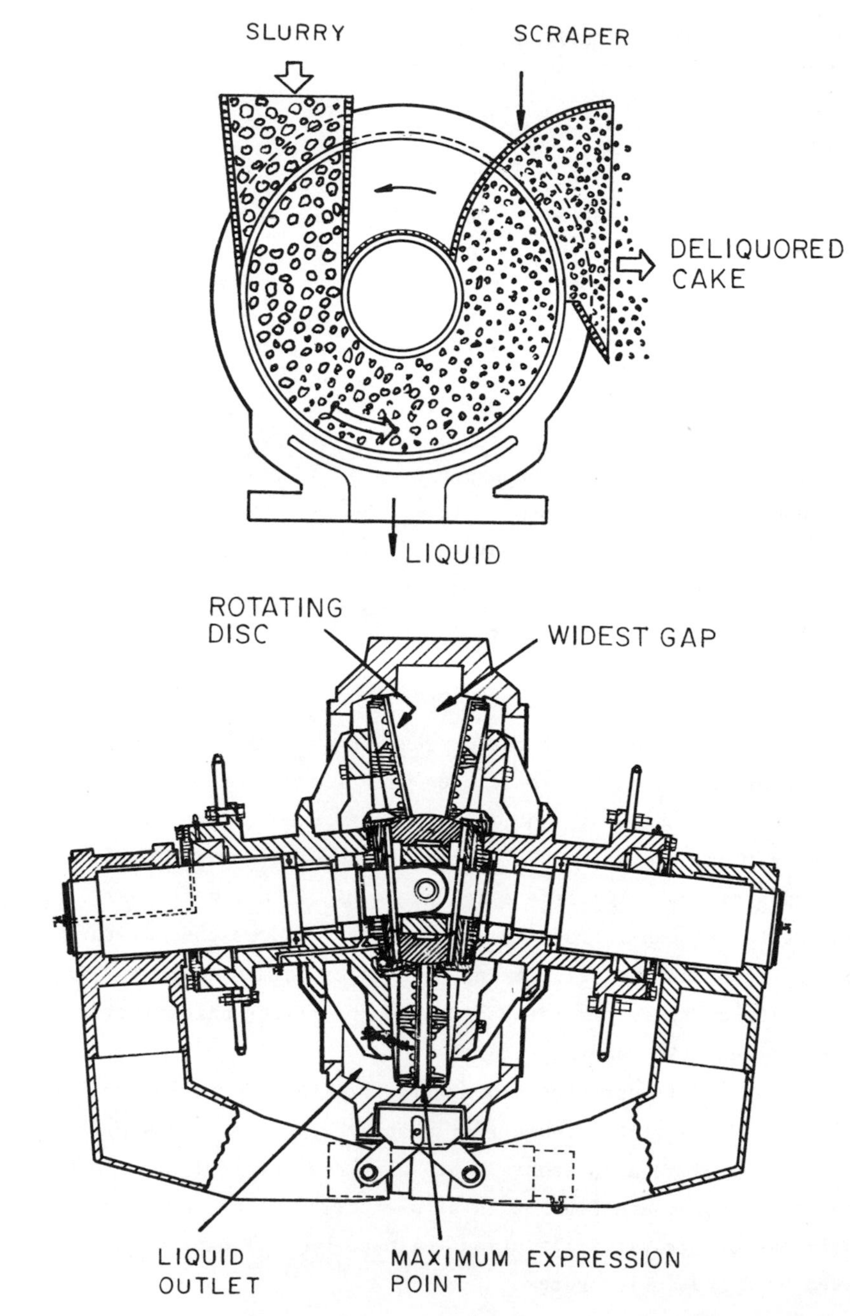

FIG. 28. V-Disc press. (Asahi Koki Co. and EIMCO Corp.)

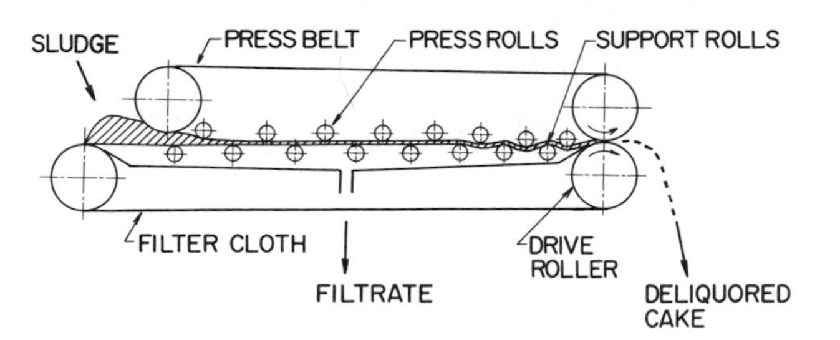

FIG. 29. Belt filter with compression rollers.

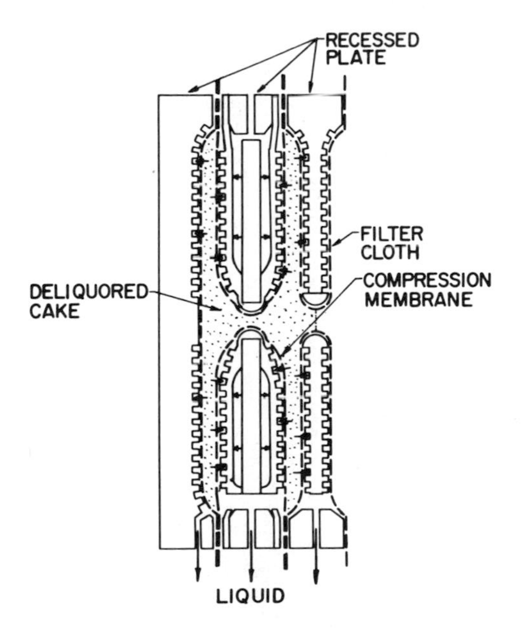

FIG. 30. Filter plate with compression membrane. (MF Automatic Press Filter, Kurita Machinery Mfg. Co.)

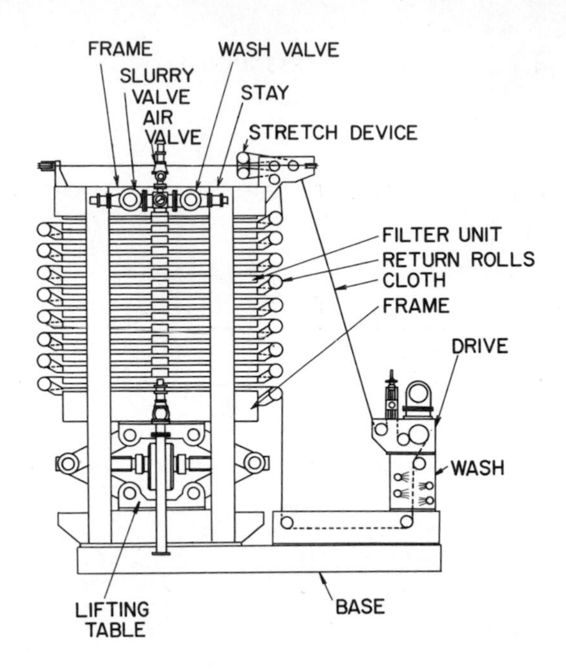

FIG. 31. $\phi\pi$ AKM(FPAKM) Horizontal automatic filter. (Tsukishima Kikai Co.)

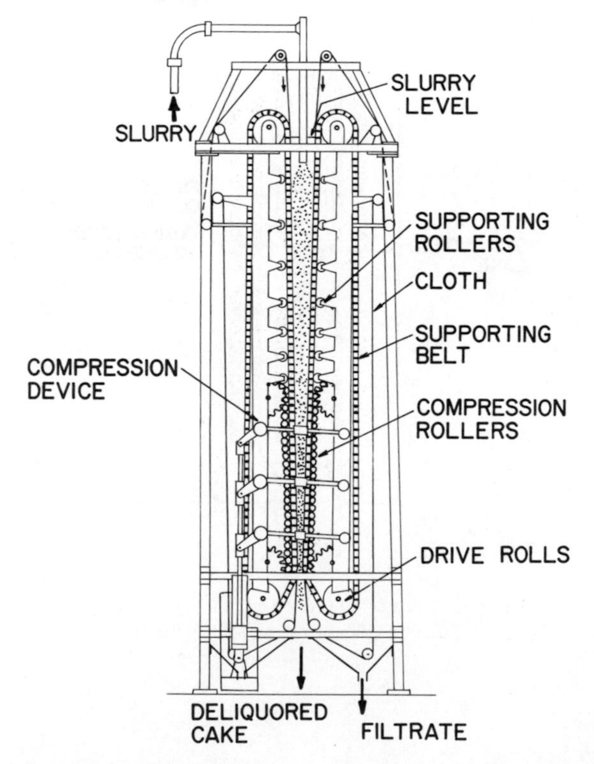

FIG. 32. Tower belt filter. (Prefiltec AG and Miura Chem. E. Co.)

2. *Theory*

The mathematics of expression is divided into two parts. In the first the mechanism is actually filtration, and in the second the mechanism is consolidation as in soil mechanics. In general, the original mixture to be expressed will be thick and may appear to be either a semisolid or a slurry. Whereas the particles of a cake may carry a structural load, a slurry acts like a liquid. When the original mixture is a slurry, applying constant mechanical pressure causes a sudden increase of hydraulic pressure uniformly throughout the slurry. Expression then proceeds as it would during filtration at constant pressure. Therefore, the expression process can be analyzed by filtration theory. When the thickness L of the slurry mixture decreases to a definite thickness L_1, the filtration period ends, and further expression proceeds on the principle of consolidation.

The basic flow equation, Eq. (22), will be employed to derive the equation for the consolidation period. This equation may be rewritten

$$u_w = \frac{1}{\mu\alpha}\left(\frac{\partial p_L}{\partial w}\right)_t \tag{88}$$

where u_w is the apparent velocity of liquid relative to the w-plane. In the conventional Terzaghi [24] theory of consolidation, the x-coordinate distance is exclusively used as shown in Fig. 33. However, as consolidation proceeds not only liquid but also solid particles in a layer of infinitesimal thickness dx move toward the drainage surface. Consequently, in deriving the consolidation equations it is more convenient and more accurate to use solid-particle distribution w instead of the x-coordinate, where w is a moving plane which contains w g of solids per unit cross-sectional area between the plane and the septum.

The coefficient of volume change m_v of the infinitesimal layer dw under compression is defined by the following equation in soil mechanics:

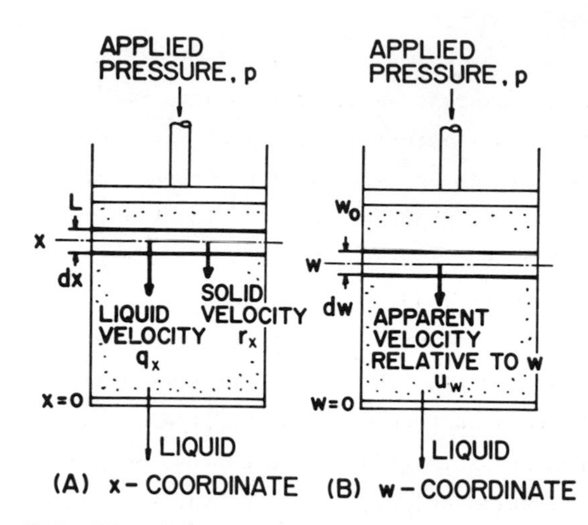

FIG. 33. Cake under consolidation (i = 1).

$$m_v = \frac{1}{1+e}\frac{de}{dp_s} \tag{89}$$

The mass balance of liquid in the layer leads to the continuity equation in the form

$$\frac{1}{\rho_s}\frac{\partial e}{\partial t_c} = \frac{\partial u_w}{\partial w} \tag{90}$$

where e is the local value of void ratio defined by $e = \varepsilon/(1 - \varepsilon)$ and t_c the expression time.

Differentiating Eq. (22) with respect to w yields

$$\frac{\partial u_w}{\partial w} = -\frac{1}{\mu}\frac{\partial}{\partial w}\left(\frac{1}{\alpha}\frac{\partial p_s}{\partial w}\right) \tag{91}$$

and substituting Eq. (89) into Eq. (91) gives

$$\frac{\partial u_w}{\partial w} = \frac{\partial}{\partial w}\left(\frac{C_e}{\rho_s}\frac{\partial e}{\partial w}\right) \tag{92}$$

Combining Eqs. (92) and (90) and rearranging yields

$$\frac{\partial e}{\partial t_c} = \frac{\partial}{\partial w}\left(C_e\frac{\partial e}{\partial w}\right) \tag{93}$$

Substituting Eqs. (89) and (91) into Eq. (69) leads to

$$\frac{\partial p_s}{\partial t_c} = C_e\left[\frac{\partial^2 p_s}{\partial w^2} - \frac{1}{\alpha}\frac{d\alpha}{dp_s}\left(\frac{\partial p_s}{\partial w}\right)^2\right] \tag{94}$$

where C_e is a local value of the modified coefficient of consolidation represented by

$$C_e \equiv \frac{\rho_s}{\mu\alpha m_v(1 + e)} \tag{95}$$

If C_e can be assumed to be constant, Eq. (93) leads to the well-known form of the diffusion equation

$$\frac{\partial e}{\partial t_c} = C_e \frac{\partial^2 e}{\partial w^2} \tag{96}$$

Since C_e may change with w and t_c, Eq. (96) may be used as an approximation with a proper mean value of C_e considered a constant.

To obtain mathematical solutions of the consolidation equations, it is essential to know the initial condition, specifically the hydraulic pressure distribution in the filter cake. Hydraulic pressure distributions in filter cakes of moderate compressibility are shown in Fig. 34; they approximate a sinusoidal curve.

The mathematical solution for Eq. (96) is given by

$$U_c = 1 - \exp\left(-\frac{\pi^2 T_c}{4}\right) \tag{97}$$

assuming for the initial condition a sinusoidal hydraulic-pressure distribution. U_c indicates average degree of consolidation over the total thickness and is called the average consolidation ratio, defined by

$$U_c \equiv \frac{L_1 - L}{L_1 - L_2}$$

where L_1 denotes the initial thickness and L_2 the theoretical final thickness of compressed sample. T_c is the consolidation time factor

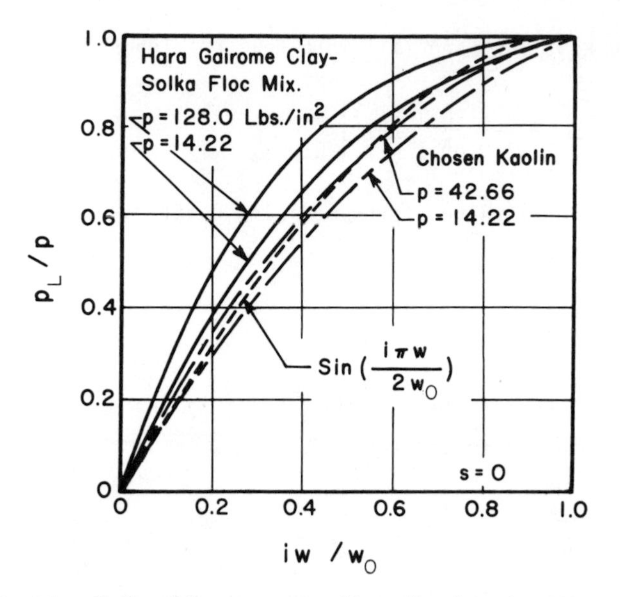

FIG. 34. Hydraulic pressure distribution in filter cake.

defined by

$$T_c \equiv \frac{i^2 C_e t_c}{w_o^2} \tag{99}$$

where i is the number of drainage surfaces and w_o the total mass of dry solids per unit cross-sectional area. Consolidation following the filtration of a Hara Gairome clay-Solka Floc mixture is illustrated in Fig. 35.

Equations (97) and (98) indicate that the time required for attaining a specified value of U_c is proportional to the square of w_o if other parameters are held constant. The approximate solution (Eq. 97) can be used for a semisolid mixture having an initial concentration higher than a certain limiting value.

Compression-permeability data for α and e can be represented in the following forms:

$$\alpha = \alpha_o + \alpha_1 p_s^n \tag{100}$$

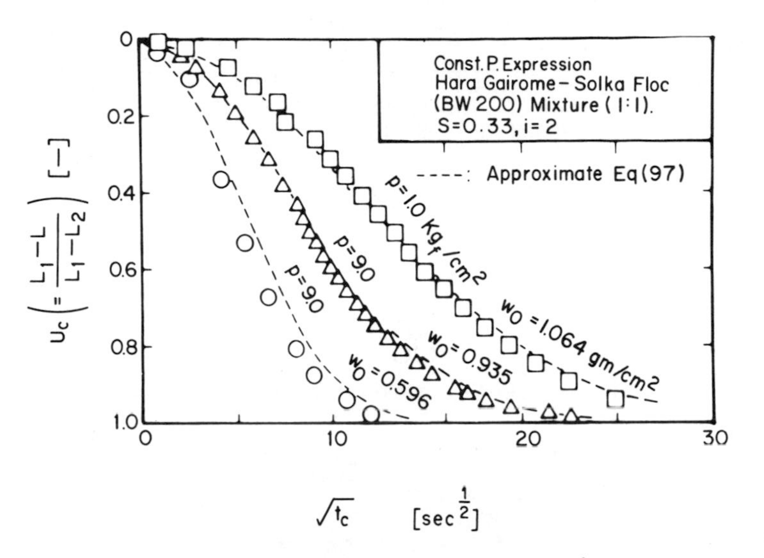

FIG. 35. Constant pressure expression.

$$e = e_o - c_c \ell_n p_s \qquad \text{(Terzaghi and Peck's form [27])} \tag{101}$$

Eliminating m_v in Eq. (95) by means of Eq. (89), and using the empirical Eqs. (100) and (101), the following equation for the modified average consolidation coefficient C_e can be derived:

$$C_e = \frac{\rho_s p_s}{\mu c_c(\alpha_o + \alpha_1 p_s^n)} \tag{102}$$

Equation (102) gives a prediction for C_e, thus the consolidation process can be predicted from calculations based upon data of compression-permeability measurements. Figure 36 illustrates calculated and experimentally determined C_e values.

According to the same arguments as employed in developing the constant pressure expression, constant-rate and variable-rate operations can also be theoretically analyzed. Recently, more rigorous solutions for expression operations have been developed on the basis of other models.

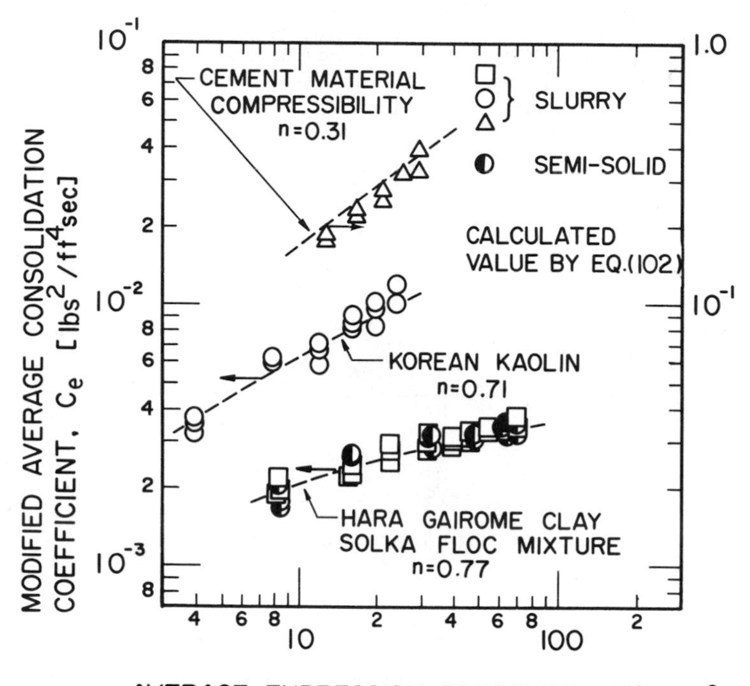

FIG. 36. Values of C_e.

Example 4.

It is desired to see how much additional liquid might be removed by expression and hydraulic deliquoring with right-angled flow patterns. In order to use Eq. (87), it is necessary to have an estimate of ε_1 which is the local porosity at 100 psi and not the average calculated in Example 3. As the value β is available, Eq. (12) could be used to calculate ε provided B were known. It will be supposed that a sample of cake from close to the media has been obtained at 60 psi and yields a porosity of 0.68. Then from Eq. (12) B is given by

$$B = (1 - \varepsilon)p_s^{-\beta} = 0.32 \times 60^{-0.285} = 0.1025 \tag{103}$$

Solving for ε at 100 psi gives

$$\varepsilon = 1 - 0.1025 \times 100^{0.285} = 0.619 \qquad (104)$$

This value equals the ε_1 of Eq. (87). It is now possible to substitute in Eq. (87) to find the approximate fraction of liquid which may be expressed hydraulically. Substituting appropriate values yields

$$1 - F = \left(\frac{1 + 0.09/0.253}{2}\right)\left(\frac{0.91 + 3 \times 0.619}{3 \times 0.747}\right) = 0.84 \qquad (105)$$

The original volume of liquid can thus be reduced about 16% over and above that remaining at 100 psi by hydraulic deliquoring. The percent solids would increase from 41.5% to 45.8%. If the operation were carried out under vacuum, the final moisture content would be in the neighborhood of 25%. The successive fraction of liquid removed over and above vacuum operation would be as follows:

Operation	Pressure (psi)	s_c	Additional fraction removed over and above vacuum
Vacuum pressure	10-14	0.25	0
	20	0.286	0.15
	60	0.370	0.44
	100	0.415	0.53
Right-angled deliquoring	100	0.458	0.60
Mechanical expression	100	0.573	0.72

Process needs would dictate which of the procedures should be chosen. In practice, the moisture contents shown for operation at low pressures might be considerably improved by blowing or sucking air through the cake at the end of filtration (see Example 8).

XII. CAKE WASHING

When wash liquid first enters a filter cake, it displaces liquid having the same concentration of solubles as the original slurry. As shown in Fig. 37 wash liquid flows viscously and eventually breaks through the cake. If there were no internal mixing or diffusion, the liquid leaving each capillary in the cake would consist

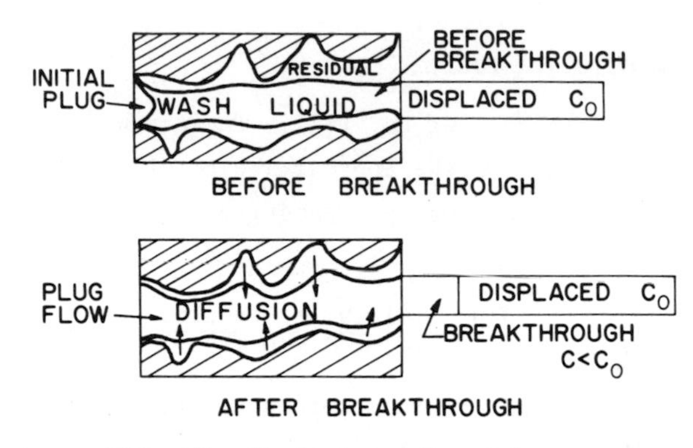

FIG. 37. Mechanism of washing.

of displaced liquor plus pure breakthrough wash liquid. The liquid in the cake would be either pure wash or residual liquor with concentration C_o. However, there is low-velocity diffusion of solute from the "dead" areas into the wash liquid. Consequently, the concentration of the wash is higher than its initial value. Inside the cake, there are two separate streams: (1) the residual liquid from which diffusion takes place and (2) the wash liquid. The proportion of wash to displaced liquor leaving the cake is different from the ratio of wash in plug flow inside the cake to the existing residual liquid. Consequently, the instantaneous concentration of filtrate leaving the cake differs from the average concentration in the cake. This is indicated in Fig. 37. The exit liquid has concentration C_o, while the average concentration in the cake is less than C_o.

The displacement phase of washing is highly efficient. However, as displacement loses its effectiveness and diffusion becomes important, the efficiency drops off rapidly. There comes a point at which it is more economical to repulp, filter, and wash in additional stages either with countercurrent or cocurrent flow.

Experimental wash curves represented as fraction of solute remaining $(C_o - C)/(C_o - C_w)$ vs. the wash ratio (ratio of wash to void volume of cake) are conveniently plotted semilogarithmically as shown in Fig. 38. No experimental point should fall on the left

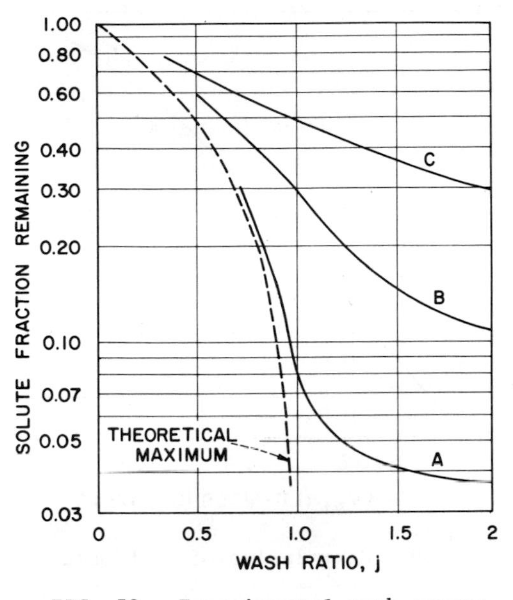

FIG. 38. Experimental wash curves.

of the maximum theoretical curve which represents perfect displacement.

For cakes with low porosity and relatively high resistance, wash rates are low but efficiency as represented by curve A in Fig. 38 is good. Curve B is typical of the majority of cases. It is difficult to remove more than 90% of the original solute.

The amount of wash liquid employed varies considerably between batch and continuous filtration. In the latter, the maximum wash ratio is restricted to values in the range of 1.5 to 2.0 on rotary drum and disc filters. In batch pressure operation, washing can be continued for longer periods, although the efficiency drops off quite rapidly.

Cake wash time is the most difficult variable to correlate. Filtration theory suggests three correlations:

1. Wash time vs. wv_w
2. Wash time vs. jw, where j = wash ratio
3. Wash time/form time vs. wash volume/form volume

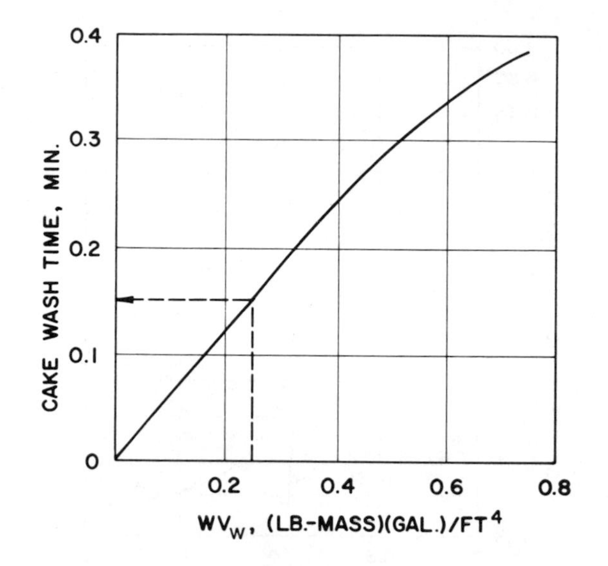

FIG. 39. Correlation of wash data.

Fortunately, the first correlation usually gives satisfactory results. This curve starts as a straight line and then frequently falls as the volume of wash water increases (Fig. 39). If for some reason this correlation is not satisfactory, the others may be tried. More complete information about washing can be obtained by consulting Michaels et al. [25], Han and Bixler [26], and Wakeman [27].

A. Deliquoring By Suction

In vacuum operations after filtration is completed, the vacuum may be left on in order to remove some of the liquid. Similarly, in pressure filtration, compressed gas may be used to blow out a portion of the liquor. Lloyd [28] points out that removal of liquor by suction or blowing depends upon the capillary suction or back pressure. In general, particles should be larger than 10 μm in diameter if any substantial amount of liquor is to be removed.

Nelson and Dahlstrom [29] and Dahlstrom [30] have shown that the following factor in consistent units is useful for correlating cake moisture content in continuous filtration:

$$\text{Correlating factor} = \frac{\text{ft}^3/\text{min}}{\text{ft}^2} \frac{\Delta P}{w} \frac{t_d}{\mu} \tag{106}$$

where $(\text{ft}^3/\text{min})/\text{ft}^2$ is the air flow rate through the filter cake and t_d the drying time. Figure 40 shows the general shape of the curve. The correlating factor chosen for design should be somewhere past the knee of the curve. Values to the left of the knee approach an unstable operating range wherein a small change in operating conditions can result in a relatively large change in cake moisture content. If tests are made at constant temperature and vacuum, the pressure drop and viscosity terms can be eliminated from the expression. Often, air rate data are not available. Fortunately, correlations can be obtained without air rates, particularly if the cakes are relatively nonporous. The correlating factor is then reduced to the simplified term t_d/w, involving only drying time and cake weight per unit area per revolution. A substantial degree of data scatter is normally encountered in a moisture-content correlation. Any point selected on the correlation will represent an average operating condition. To assure that the moisture content of the cake will never exceed a particular value,

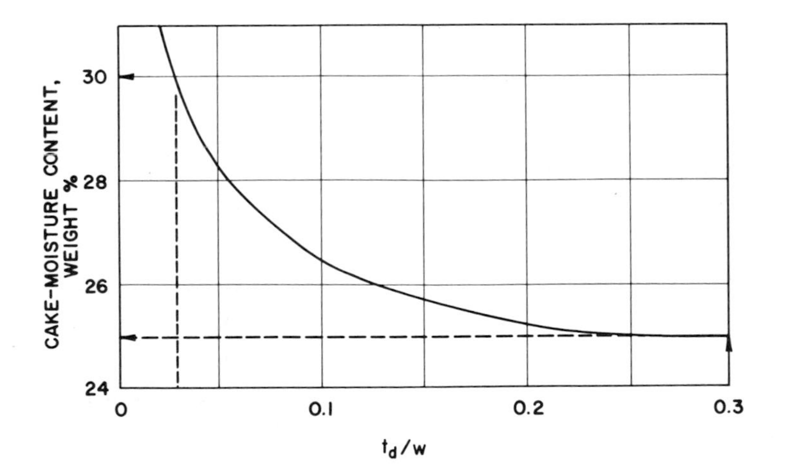

FIG. 40. Correlation of cake moisture content during suction.

the correlating factor at the desired minimum should be multiplied by 1.2 before calculating the required drying time.

B. Air Rate

Air flow through the cake and thus vacuum pump capacity can be determined from measurements of flow rate as a function of time. If for a material the following data were obtained:

Time (min)	0.05	0.1	0.2	0.3	0.4	0.6	0.8	1.0	1.5
Flow rate [(ft^3/min)/ft^2]	2.5	4.2	5.9	6.8	7.45	8.2	8.6	8.75	9.2

integration of these data over the times involved in the first and second stages of drying in continuous filters would indicate the required vacuum pump capacity.

XIII. EXPERIMENTAL TESTING PROCEDURES

It is impossible to find experimental data in the technical literature permitting design of cake filtration systems; thus obtaining experimental data is essential. It is also impossible to present a completely general approach to experimentation; nevertheless, a logical methodology based upon the material presented above can be devised.

It is presumed that at least a tentative decision has been reached on filtrate clarity, permissible soluble content of the cake, and the average porosity or liquor content of the cake. Basic decisions must yet be made with respect to the following: (1) use of flocculants and filter aids, (2) thickening by sedimentation, hydrocycloning, or by prevention of cake formation with prefilters, (3) choice of filter based upon rate of cake build-up, (4) filter media, (5) washing, and (6) deliquoring hydraulically or by mechanical expression. The use of flocculants and thickening operations will not be covered in the experimental methodology of this chapter except to say that they are a fundamental part of any analysis. It will be assumed that both have been investigated as a routine part of the analysis. Basically, this chapter concentrates on the rate of cake build-up after appropriate pretreatment has taken place.

Unless data are available from previous experience, the solution to a filtration problem must involve practical testing. There are two broad approaches; one relying on the equipment specialist and vendor, and the other placing the basic reliance on the personnel of the plant involved. The latter approach has advantages as it tends to familiarize the personnel with filtration and, more specifically, the variations in the particular problem under study. The former has the advantage of placing experienced personnel with equipment on the project, but has the universal high risk that the tests could incorporate a major sampling error.

A. Practical Filtration Testing

One simple test method is to employ a Buchner funnel or Nutsche filter and obtain filtration rates as a function of pressure drop (vacuum). Such tests rapidly give general information on the rate of cake build-up and the difficulty of filtration. They give an indication as to whether or not pretreatment should be seriously investigated and, if carefully made, can yield design data for vacuum filters. If initial tests indicate that pressure filtration should be considered, it is advisable to test further using some kind of constant pressure apparatus. Constant pressure is not extensively used in practice, but for testing it has the advantage of involving pressure, temperature, area, medium, filter aid and slurry as constants, any one of which may be varied. One such unit in use both for filter aid production control and filter aid selection is the constant-pressure bomb filter, designs of which vary around that shown in Fig. 41. Silverblatt et al. [25] discuss bench-scale testing with vacuum leaves and bomb filters.

If there is no need for adding filter aid, the bomb filter can be used for prediction and pressure, volume, and time relationships. If filter aid additions are required, the bomb can be used to determine the optimum quantity and type.

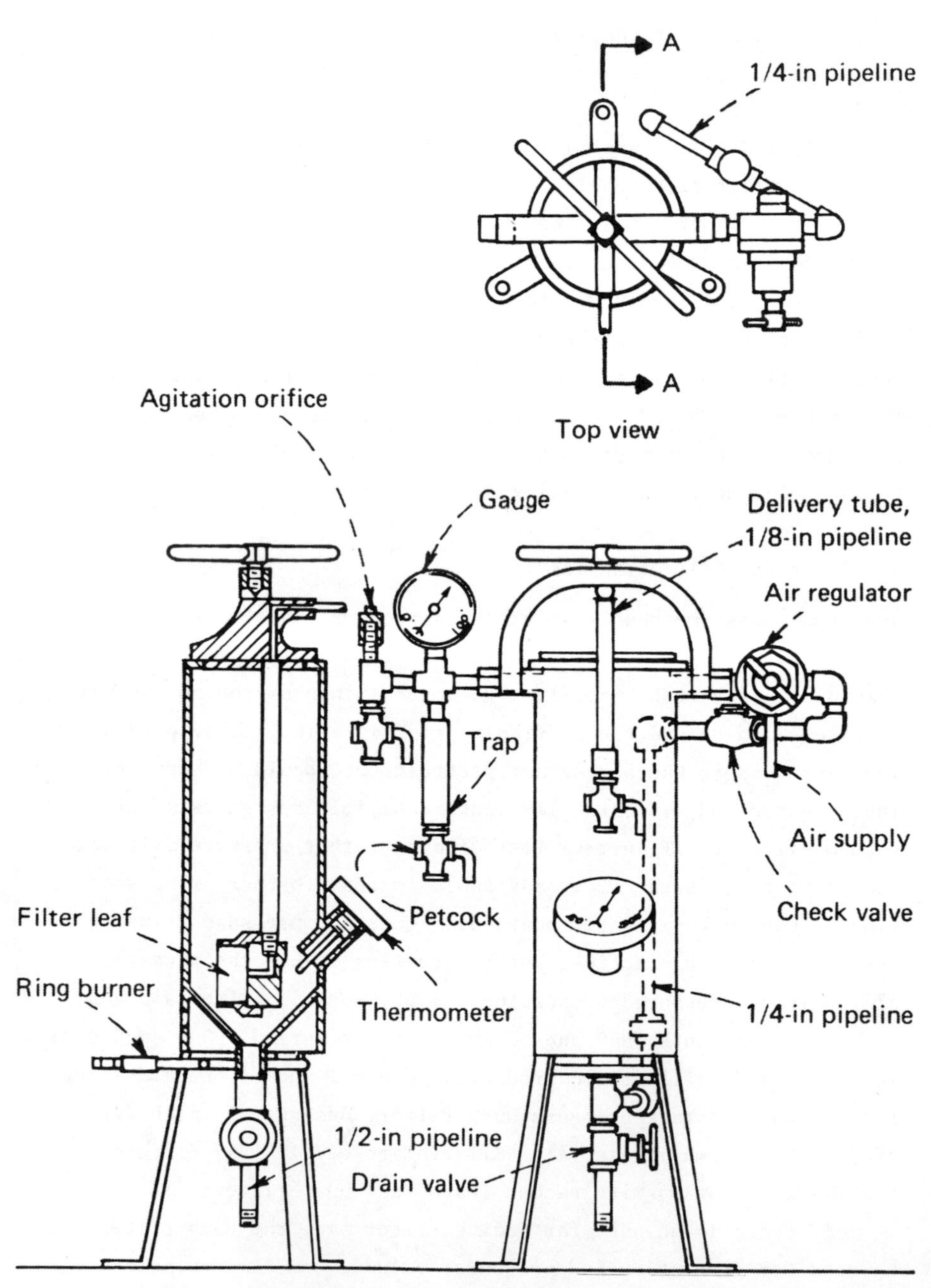

FIG. 41. Bomb filter. (Dicalite/GREFCO, Inc., Subsid. General Refractories Co.)

Different companies use varying factors for scale-up of area. No standards are available. Correction factors in use range from a 25% to a 100% increase in the area determined on small-scale equipment. It is likely that a cake formed on a small area will differ markedly from that for the same material filtered on plant-scale equipment. The method of deposition, wall effects, and nonuniformity influence the result.

Although theory points to the use of Eqs. (46) and (49) as best for analyzing experimental data, another method [32] utilizes logarithmic plots of v vs. t. Equation (49) can be rearranged to

$$t = \frac{v^2}{k}\left(1 - \frac{bk}{v}\right) \tag{107}$$

Taking logarithms yields

$$\log v = 0.5 \log t - 0.5 \log k - 0.5 \log(1 + bk/v) \tag{108}$$

As v increases, the last term approaches zero, and the slope of a logarithmic plot of v vs. t approaches 0.5 as shown by curve A in Fig. 42. Frequently in small laboratory tests, the slurry is charged into a closed container with provision for agitation by bubbling gas through the slurry. When the experiment is prolonged, there may be settling with a consequent increase in the slurry concentration. During the first part of the test, an interface between the supernatant liquid and settling solids could form. Then as the slurry

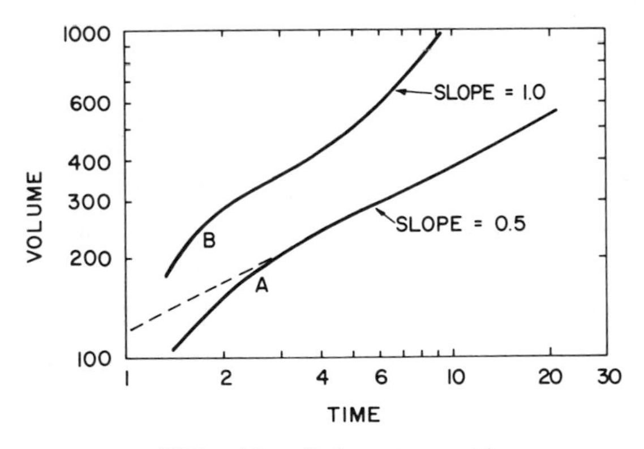

FIG. 42. Volume vs. time.

level decreases, the surface where the cake is being formed will first experience an increasing concentration, then a decreasing concentration, and finally a period in which clear liquid flows through the cake. When the slurry become devoid of solids with s = 0, Eq. (42) assumes the form

$$t = bv \tag{109}$$

or

$$\log t = \log v + \log b \tag{110}$$

The slope of this line will be unity. Thus, if a rapidly settling slurry were filtered in a bomb filter until all the liquid was removed, the curve might take the form indicated by B in Fig. 42. The first portion of B represents the region where the medium resistance is of importance, and the final part corresponds to the passage of clear liquid.

If a bomb filter is used to study clarification (settling is less of a problem) where negligible cake is formed, the basic laws governing flow are changed. For standard blocking [33] where the pores are assumed to fill gradually, the volume vs. time relationship is given by

$$v = \frac{t}{\frac{1}{2}k_1 t + 1/q_o} \tag{111}$$

When t is small, Eq. (111) reduces to $v = q_o t$. As t becomes large and the media becomes completely blocked, v approaches $2/k_1$. Thus a logarithmic plot will give an initial slope of unity and a final slope of zero. Other blocking laws give similar results.

The major point for an experimentor to remember is the possibility, and sometimes probability, that the basic conditions involved in developing theoretical formulas may not be fulfilled in the actual experimentation. Caution should always be observed in interpretation.

B. Addition of Filter Aid

After initial tests have been completed, it may be decided to employ a filter aid. Testing would then incorporate procedures for selection of grade and amount of precoat and admix. Such tests could be utilized during early design or later review.

Processes utilizing aids can be classified as follows:

1. Slurry forms a resistant cake. Filter aid is added to reduce the average resistance.
2. A dilute slurry forms no cake but requires clarification.
    a. A fixed precoat is laid down to serve as a depth filter.
    b. A thick precoat is laid down on a rotary drum filter and then gradually scraped off.
    c. An admix is added to the slurry to provide a cake which serves as the equivalent of a depth filter.
3. Filter aid serves as a base (precoat) over a screen to hold the cake and prevent passage of small particles.

Experimental procedures require a representative sample of the slurry in sufficient quantity to run a complete series of tests. These are divided into preliminary tests, basic series, and checks when necessary. A typical bomb unit of the type shown in Fig. 41 is operated with 3500 ml of slurry, and preliminary tests involving a single filter aid are directed toward determining whether filter aid is required and, if so, approximately how much. The basic series then follows in which different filter aids are compared. If a filter aid gives the clarity specification desired, then a move to a finer filter aid will only reduce throughput. The objective is to select a filter aid that will provide the required clarity with the maximum throughput at minimum cost. While information from the bomb-type unit inherently depends upon the quality of the representative sample, there are errors due to wall effects, and, for very viscous liquors, the unit must be modified to provide larger filtering areas. In the main, however, it is a reliable method for initial filter aid selection. Constant pressure filter tests are an important phase of the majority of manufacturers test programs.

Interpretation of data varies somewhat, depending upon whether clarification or cake filtration is involved. Data are plotted in the form of log v vs. log t, and conclusions depend upon the theory as outlined in Eqs. (107)-(112). Generally the medium consists of cotton twill with a small resistance, and log v vs. log t data yield a line of slope 0.5 when a cake is formed. With clarification tests, the slope for the raw slurry is usually less than 0.5. As it is generally desirable to have a true cake formed, experiments are usually started with zero filter aid, and then increments of filter aid of a middle grade are added until approximately a 0.5 rate-slope is obtained. The pressure may also be varied to provide a range of basic test values. Then additional tests are made in a similar fashion except that slurries are prepared with different filter aids.

Filtration is continued for a period of 21 min, a time that has empirically proved satisfactory. Longer time can be employed where desirable. The volume vs. time data generally plot as a straight line on logarithmic paper, and extrapolations may be made to longer periods such as 2, 8, and 16 hr.

Example 5.

A request was made to a filter aid manufacturer to determine the proper filter aid dosage and the grade for clarifying 2000 gpm of copper sulphate solution. The solution, referred to as "liquor" in this example, is obtained by leaching copper ore with H_2SO_4 and contains up to 100 ppm of suspended solids. The desired maximum pressure differential in a constant rate filtration of the liquor is 40 psi and the temperature is 20°C.

In order to approximate results for constant rate filtration, a constant pressure differential of 20 psi was employed in the test filter.

Initially, Dicalite 476 was used for determination of the proper dosage. The test leaf area was chosen to be 0.005 ft^2. The data are shown in Fig. 43a. As a starting point, the filter aid dosage was chosen to be equal to the solids content of the liquor. As the dos-

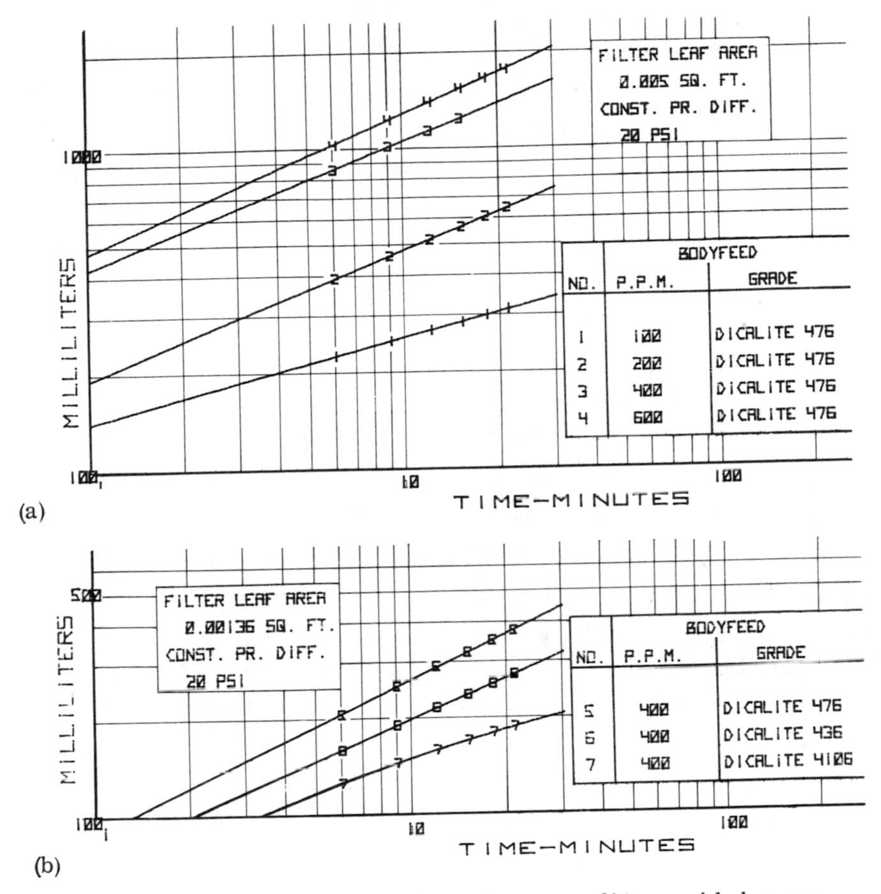

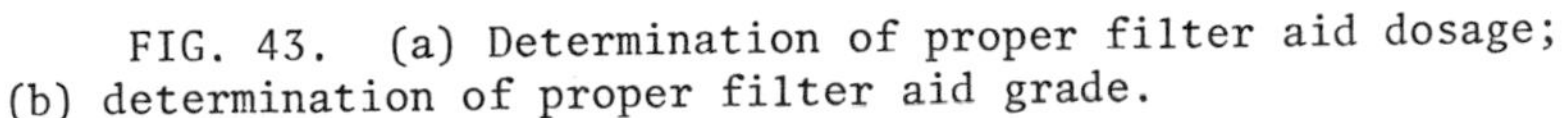

FIG. 43. (a) Determination of proper filter aid dosage; (b) determination of proper filter aid grade.

age was increased from 100 to 600 ppm, the throughput increased. But, beyond 400 ppm the filter aid addition did not substantially increase the throughput. Therefore, 400 ppm is the desired dosage. The required number of tests can be minimized by varying the dosage by a factor of 2 in successive runs.

It can be seen from Eq. (108) that a logarithmic plot of total volume vs. time, instead of V vs. time, would also be a straight line of slope 0.5 at extended times because the leaf area term is a constant and can be combined with the middle term on the right hand side of Eq. (108). Before attempting to draw a straight line through the logarithmic plot, the data for the first 5 min is customarily neglected in order to let the medium resistance term in Eq. (108) become insignificant. By extrapolating the straight line, the throughput obtainable can be estimated up to several hr. However, extrapolation beyond 16 hr is unreliable. The slope of the straight line is almost never equal to 0.5 but is less than 0.5 due to deviations from the assumptions made in deriving Eq. (108).

In the second part of the test, the dosage was held constant at 400 ppm and two other Dicalite perlite grades were investigated. A test leaf of 0.001363 ft^2 area was employed due to a shortage of the liquor. Fig. 43b displays the data obtained with Dicalite 4106 and Dicalite 436 which are, respectively, coarser and finer than Dicalite 476.

Experience dictates that the optimum dosage is independent of the filter aid grade. Dicalite 476 yielded more throughput than the other two and is the proper perlite grade. Dicalite 4106 is too coarse for the solids to be filtered and they penetrate the cake. As a result, the volume vs. time plot, on log-log paper, is not a straight line and is known as a "plugging curve."

In addition to the throughput, cake thicknesses were measured at the end of each run. The data obtained on cake thicknesses and extrapolated filtrate volumes appear in Table 12.

It is convenient to convert the volume, obtained by extrapolation of the volume vs. time plot, into volume per unit area. A factor that converts the total volume in ml into gal/ft^2 is known as the conversion factor for the leaf. The conversion factors for 0.005 ft^2 and 0.00136 ft^2 leaves are 0.0528 and 0.194, respectively. The throughput converted into gal/ft^2, is divided by the cycle length to obtain the average flowrate in gph/ft^2. The results are shown in Table 12.

TABLE 12 Data on Cake Thicknesses and Extrapolated Filtrate Volumes

Test No.	Leaf area ft^2	Filter aid	Dosage (ppm)	Cake thickness at end of test (in.)	Volume at end of test (ml)	Extrapolated volume (ml) at cycles of 4 hr	8 hr	16 hr
1	0.005	Dicalite 476	100	0.02	313	590	710	850
2	0.005	Dicalite 476	200	0.07	646	1700	2250	2960
3	0.005	Dicalite 476	400	0.12	1231	3690	4860	6400
4	0.005	Dicalite 476	600	0.40	1757	5050	6820	9180
5	0.00136	Dicalite 476	400	0.25	373	1200	1670	2300
6	0.00136	Dicalite 436	400	0.15	272	772	1050	1400
7	0.00136	Dicalite 4106	400	0.14	186	Plugging curve		

The leaf area employed in test Nos. 1 through 4 was 0.005 ft^2 while the leaf area in test Nos. 5 through 7 was 0.00136 ft^2. Due to the extra wall friction in the small leaf, the cake compaction was less than that with the larger leaf. Consequently, the flowrate per unit area through the 0.00136 ft^2 leaf is more than that through the 0.005 ft^2 leaf under otherwise identical conditions, as can be seen by comparing test Nos. 3 and 5. Therefore, the data collected on the 0.00136 ft^2 leaf are useful only for comparison. Before utilizing the test filter data to predict the performance of commercial size filters, it is advisable to try a leaf of the next larger size and see if the flowrate per unit area remains unchanged.

For a given liquor and a body feed, the cake thickness is approximately proportional to the volume throughput per unit area. Thus, cake thicknesses at various cycles of several hours are calculated by extrapolation of the data in Fig. 43a; the results are presented in Table 13.

The results of test No. 3 shown in Table 12 indicate that at a constant flowrate of 0.535 gpm/ft^2, a cycle length of 8 hr can be obtained. The corresponding cake thickness would be 0.47 in. In order to calculate the filtration capacity, a cake discharging and precoating time of 2 hr is customarily added to the filtration cycle time. In that case, the effective flowrate is 0.428 gpm/ft^2. As the desired total flowrate is 2000 gpm, a filter area of 4672 ft^2 would be calculated. Because such calculations are only approximations, a 20% excess area is recommended.

TABLE 13 Calculational Results for Example 5

Test No.	Bodyfeed grade	Dosage ppm	Line avg slope	Flowrate in gph/ft^2 at cycles of			Cake thickness in inches at cycles of		
				4 hr	8 hr	16 hr	4 hr	8 hr	16 hr
1	Dicalite 476	100	0.262	7.8	4.7	2.8	0.04	0.05	0.05
2	Dicalite 476	200	0.399	22.5	14.9	9.8	0.19	0.24	0.32
3	Dicalite 476	400	0.396	48.8	32.1	21.1	0.36	0.47	0.62
4	Dicalite 476	600	0.432	66.7	45.0	30.3	1.20	1.50	2.10
5	Dicalite 476	400	0.476	58.0	40.3	28.0	0.80	1.10	1.50
6	Dicalite 436	400	0.429	37.5	25.3	17.0	0.42	0.56	0.77
7	Dicalite 4106	400	Plugging curve						

XIV. CYCLE ANALYSIS

In batch filtration, the time of the filtration portion of the cycle markedly affects the overall capacity. In general, the following operations form a part of a filtration cycle: precoating, filtration, washing, hydraulic deliquoring, mechanical deliquoring (expression), displacement of liquid (so-called drying) by blowing with air or suction under vacuum, emptying unfiltered liquor through a scavenger plate, discharge of cake, and cleaning and reassembly. Average capacity of the filter is given by the mass of dry solids or filtrate produced in the filtration step (and in the emptying of unfiltered liquor, if included) divided by the time for all of the operations. Only a few simple cases have been analyzed in the technical literature.

Constant pressure filtration is the easiest to analyze. Suppose medium resistance is negligible and the cycle consists of filtration with time t_F, washing t_w, and discharge t_D. The discharge and reassembly is considered to be independent of the filtration time. Washing rates are presumed to be proportional to the final rate of filtration. Volume is related to time by Eqs. (45) or (47) in the form

$$v^2 = Kt \tag{112}$$

The final rate of filtration is obtained by differentiation, giving

$$\frac{dv}{dt} = \frac{K}{2v} = \frac{1}{2}\sqrt{\frac{K}{t_F}} \tag{113}$$

where t_F is the total filtration time. The rate of washing is given by

$$qw = \frac{\delta}{2}\sqrt{\frac{K}{t_F}} \tag{114}$$

where δ depends on the filtration operation. In a simplistic approach $\delta = 1$ for leaf filters and 0.25 for plate and frame units. In a leaf filter, the wash liquid follows directly behind the slurry and encounters identical conditions if the pressure and liquid viscosity are the same. In a plate and frame filter, the wash is introduced in such a way that it passes through both thicknesses of cake. Thus, with a doubling of the thickness (see Fig. 24) and halving of the area, the rate is reduced to about one-fourth of the final filtration rate. The ratio j of the wash liquid to the void volume in the cake is given by

$$j = \frac{\varepsilon_{av} L}{v_w} \tag{115}$$

where v_w is the volume of wash liquid per unit area. Substituting for L from Eq. (36) yields

$$j = \frac{c\varepsilon_{av}}{\rho_s(1 - \varepsilon_{av})}\frac{v}{v_w} \tag{116}$$

Then combining terms

$$v_w = Jv = J\sqrt{Kt_F} \tag{117}$$

where $J = c\varepsilon_{av}/j\rho_s(1 - \varepsilon_{av})$, which indicates that the wash liquid is proportional to the filtrate. Wash time is given by

$$t_w = \frac{v_w}{q_w} = \frac{J\sqrt{Kt_F}}{(\delta/2)\sqrt{K/t_F}} = \frac{2J}{\delta} t_F \tag{118}$$

Thus the washing time is proportional to the filtration time.

The time per unit volume of filtrate is given by

$$\frac{t_F + t_w + t_D}{v} = \frac{t_F(1 + 2J/\delta) + t_D}{\sqrt{Kt_F}} \tag{119}$$

Differentiating Eq. (119) and setting it equal to zero yields

$$0 = \frac{d}{dt_F} f(t_F) = \frac{1}{2} \frac{1}{\sqrt{Kt_F}} (1 + 2J/\delta) - \frac{1}{2} \frac{t_D}{\sqrt{K}\, t_F^{3/2}} \tag{120}$$

and solving for t_F yields

$$t_F(1 + w/\delta) = t_F + t_w = t_D \tag{121}$$

The total cycle time is simply twice the dumping time, and the maximum average rate is simply half the average rate during filtration ($t_w = 0$).

The question to be asked in constant rate or centrifugal pump filtration depends upon the possibility of changing flow rate through adjustments in pump speed. Assuming that a maximum pressure is to be attained in constant rate operation, Eq. (57) can be used. Neglecting p_1, the product q^2t will be constant if the operation is carried out to some fixed p. Then

$$q^2 t_F = K_R \tag{122}$$

The wash rate is q, and wash time is given by

$$t_w = \frac{Jv}{\delta q} = \frac{J}{\delta} t_F \tag{123}$$

as $v = qt_F$. To find the minimum time per unit volume of filtrate, consider

$$\frac{t_F + t_w + t_D}{v} = \frac{t_F(1 + J/\delta) + t_D}{qt_F} \tag{124}$$

Eliminating q with Eq. (122) gives

$$f(t) = \frac{t_F(1 + J/\delta) + t_D}{\sqrt{Kt_F}} \tag{125}$$

This equation is identical in form to Eq. (119) and yields a result equivalent to Eq. (121). The rate should be adjusted to a value such that

$$q^2(t_D - t_w) = K_R \tag{126}$$

Example 6

Cycle analysis for an aqueous slurry is to be investigated by constant pressure filtration. A constant pressure apparatus is being utilized. The slurry contains 1.0% by weight solids having a specific gravity of 2.67. The temperature of the slurry is 68°F. Tests are made at pressures of 10, 20, 35, and 50 psi. The area of the medium is 0.01 ft^2. Average moisture content s_c of the cake is determined after each test. Data for 10 psi are given in Table 14 along with calculated values of v and t/v. Analysis of the constant pressure data is made according to Eq. (49) resulting in a plot of the form of t/v vs. v as shown in Fig. 44. Values of R_m and α are calculated from the slopes and intercept of the lines. A summary of experimental and calculated values appears in Table 15. Analytical equations to represent α and $(1 - \varepsilon_{av})$ are given by

$$\alpha_{av} = 3.50 \times 10^{11}(\text{psi})^{0.3} \tag{127}$$

$$(1 - \varepsilon_{av}) = 0.15(\text{psi})^{0.13} \tag{128}$$

TABLE 14 Experimental v vs. t Data[a]

t (min)	vol (ml)	v (ft^3/ft^2)	t/v (sec/ft) (× 10^3)
1	18	0.06356	0.94
3	42.5	0.150	1.20
5	59.0	0.208	1.44
10	96.0	0.338	1.78
15	120.0	0.424	2.12
20	143.	0.505	2.37
25	165.	0.583	2.58
30	181.	0.639	2.82

[a]For p = 10 psi and A = 0.01 ft^2.

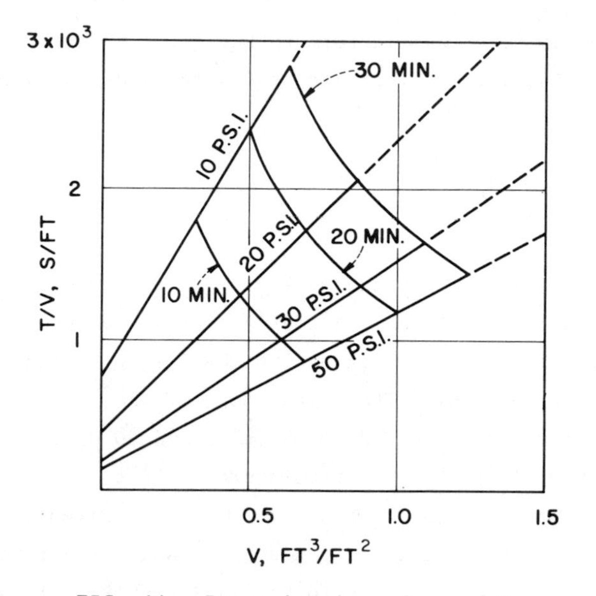

FIG. 44. Determination of α and R_m.

Suppose that 720 gal/hr of a 10% by weight slurry with the same solids as just tested are now to be filtered. It is useful to analyze filter area and cake thickness as a function of cycle time. The average throughput equals the filtrate volume divided by the sum of the actual filtration time plus the time required for additional operations such as washing, drying, discharging, cleaning, reassembling, and precoating. An analysis of the data is also shown in Fig. 45. The average rates over the entire cycle are shown for downtimes of 5 to 60 min. While spectacularly larger

TABLE 15 Calculated Values for Constant Pressure Filtration

p (psi)	s_c	c	ε_{av}	L (in.)	A slope ($\times 10^3$)	B inter-cept	α (ft/lbm) ($\times 10^{11}$)	R_m (ft^{-1}) ($\times 10^{10}$)
10	0.403	0.640	0.798	0.144	3.252	0.72	6.99[a]	4.97[b]
20	0.431	0.639	0.779	0.180	1.946	0.370	8.6	5.10
35	0.455	0.638	0.762	0.2102	1.314	0.200	10.167	4.82
50	0.470	0.638	0.750	0.231	1.025	0.140	11.32	4.83

[a]Multiply by 10^7.

[b]Multiply by 10^{10}.

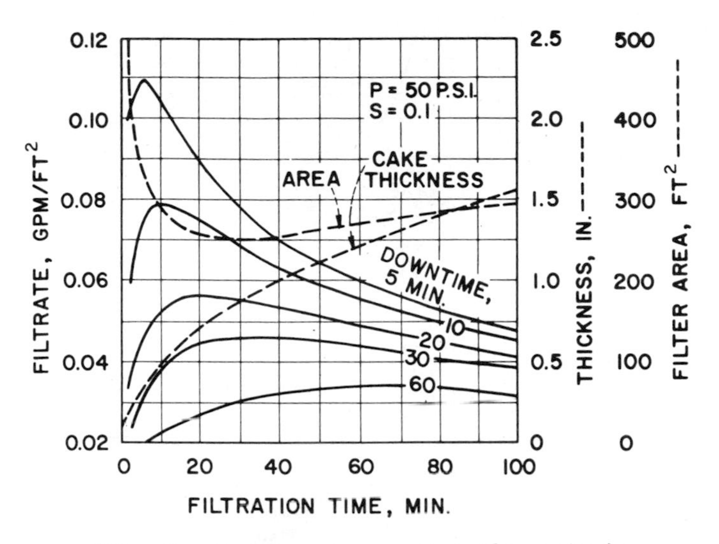

FIG. 45. Constant-pressure cycle analysis.

rates can be obtained with very short downtimes, most equipment requires longer running times. It is apparent that the maximum average rate occurs at a point where the filtration time nearly equals the downtime.

The total filter area shown in Fig. 46 is for a downtime of 30 min. The minimum, corresponding to a cake thickness of 0.85 inch at 30 min, is about 250 ft^2. Longer runs would not have a large effect on the area because of the relatively low slope of the area vs. filtration time curve. Both 10 and 100 min filtration times (add 30 min to obtain cycle time) give the same average rate of 0.0225 gal/(min ft^2) and require the same area of about 290 ft^2 with cake thickness about 0.5 and 1.75 in. If leaf or plate and frame presses were used, the same number of leaves or plate would be required for both 10 and 100 min runs. However, the spacing between leaves or the thickness of the frames could vary substantially. Larger equipment cost would be involved with the thicker cakes. However, because of the longer cycle, the labor costs would probably be less.

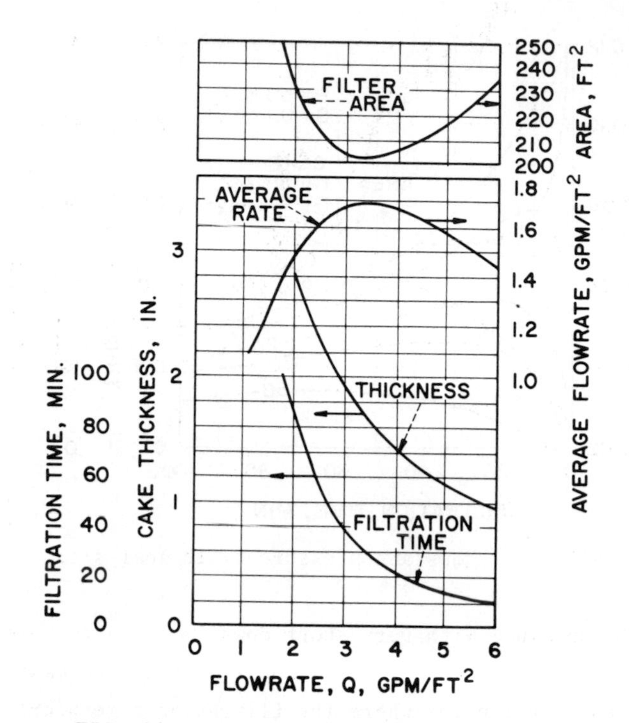

FIG. 46. Analysis of constant rate data.

It is unlikely that a constant pressure driving force would be used in actual practice. With either a centrifugal or constant rate pump, the average rate is smaller than with constant pressure filtration because the pressure drop increases gradually. Thus the actual area would be larger than that shown in Fig. 46, for a constant rate filtration the area required being exactly twice the area of a constant pressure run. A centrifugal pump would give results somewhere in between.

Example 7

A cycle analysis of constant-rate slurry filtration is made using a constant rate pump. Tests are made at 2, 3, and 4 gal/(min ft^2), and the results are plotted logarithmically as shown in Fig. 47 as pressure vs. time in minutes. In accordance with Eq. (57), the data are replotted as $p - p_1$ vs. q^2t, which yields a

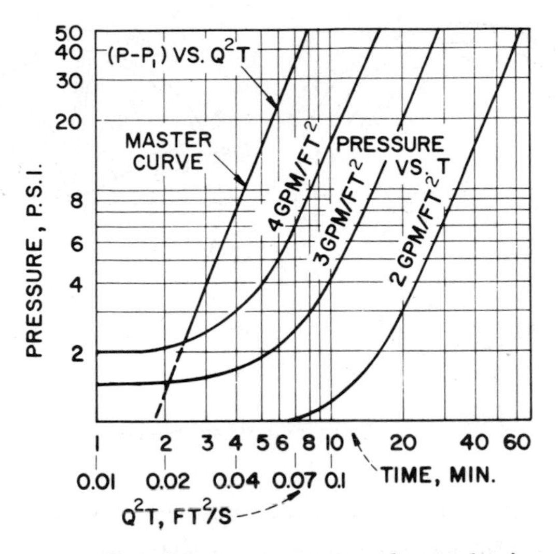

FIG. 47. Constant-rate cycle analysis.

straight line for $p - p_1$ above 2 to 3 psi. Deviation from the theoretical line is due to the inaccuracy of the approximation for α at low pressures. The pressure drop across the medium is given by $p_i\,(\text{psi}) = 0.5q[\text{gal}/(\text{min ft}^2)]$.

With this master curve, it is possible to predict volume vs. time curves for any rate. If the power law for α as given by Eqs. (10) and (11) is not valid, the master q^2t vs. $p - p_1$ curve will not be a straight line on a logarithmic plot.

For a slurry feed rate of 340 gal/min, a cycle analysis is made assuming that the maximum pressure is 50 psi and the downtime is 30 min. Results of these calculations are also shown in Fig. 46 where average filtrate rate for the entire cycle, area cake thickness, and filtration time are plotted against rate.

XV. CONTINUOUS FILTRATION

The salient features of rotary drum filtration are illustrated in Fig. 48 where a cylindrical drum having a permeable surface is revolving counterclockwise partially submerged in a slurry. A pressure differential is usually maintained between the outer and inner

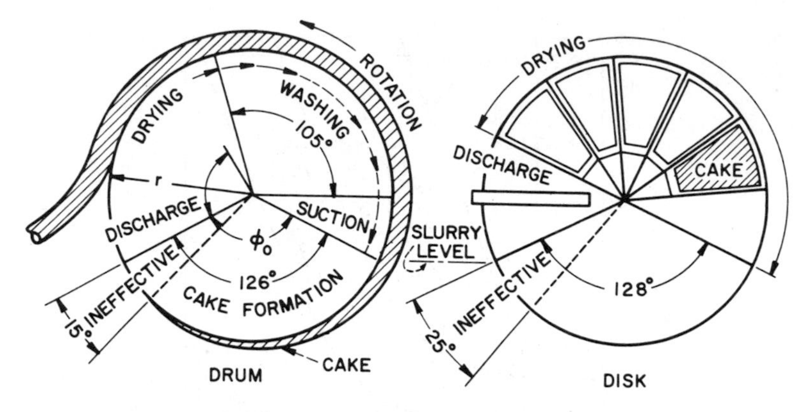

FIG. 48. Drum and disc filter.

surfaces by means of a vacuum pump. However, the drum might be enclosed and operated under pressure. In addition to the vacuum or pressure, each point on the periphery of the drum is subjected to a hydrostatic head z of slurry.

The drum of radius r rotates at an angular velocity of ω radians/sec (N rpm). The portion of the drum submerged in the slurry is subtended by an arc ϕ_o. The remaining parts of the drum are utilized for washing, drying, and discharge. Filtration is assumed to begin at the instant the drum enters the slurry, although practically there is a time lag in establishing the full vacuum because of the necessity for maintaining a vacuum seal as each compartment enters the slurry. Continuous multicompartment drum filters are normally used on materials that are relatively concentrated and easy to filter. Rates of cake build-up are in the range of 0.05 in./min to inches per second. Submergence normally runs from 25-75% (40% being quite common) with rotational speeds of from 0.1 to 3 rpm. With those conditions, filtration times could range from 5 sec to 7.5 min, the great majority of industrial filtrations falling within these limits. Drum diameters typically range from 6 to 12 ft, although larger diameters are occasionally encountered. The cake build-up is also shown in Fig. 48.

With 40% submergence and a 12-ft diameter drum, the hydrostatic head may be up to 5.7 ft, which is a significant fraction of the driving force in vacuum filtration. As the slurry will have a density greater than water, the effective head may be as great as 6 to 7 ft. The relationship among the pressure driving forces is illustrated in Fig. 49a. The total pressure p is given by the sum of the pressures due to vacuum (or pressure) p_v and the hydrostatic head p_H. As the drum enters the slurry, all of the pressure is taken up by the drop p_1 across the medium. The effective-pressure driving force across the cake is given by

$$\Delta p_c = p_v + p_H - p_1 = p - p_1 \tag{129}$$

If the hydrostatic head amounts to a large fraction of the vacuum and the medium pressure falls off quickly, Δp_c will rise to a maximum and then fall as demonstrated in Fig. 49a.

From the figure, it can be seen that the total pressure is given by

$$p = p_v + \rho_o gr(\cos \phi - \cos \phi_o/2) \tag{130}$$

Finding the average pressure leads to

$$p_{av} = p_v + \frac{\rho_o gr}{\phi_o}(2 \sin \phi_o/2 - \phi_o \cos \phi_o/2) \tag{131}$$

In Figs. 49b and c, the hydraulic distribution in the cake is illustrated for both incompressible and compressible cakes. The dotted lines give the total pressure drop across the cake. In Fig.

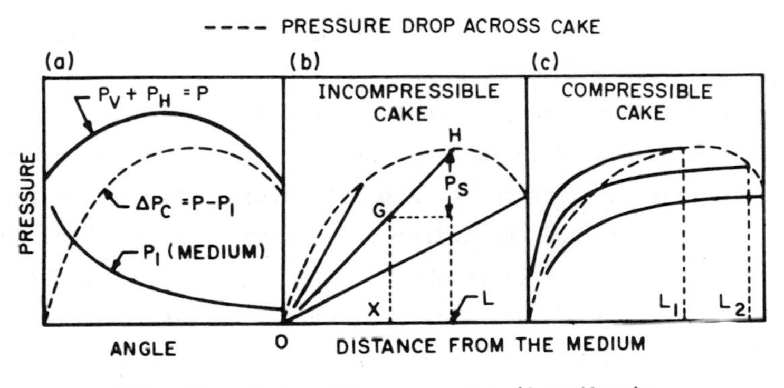

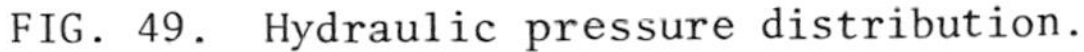
FIG. 49. Hydraulic pressure distribution.

49a, Δp_c is shown as a function of the cake thickness L. The solid lines in Fig. 49b and 49c represent the hydraulic pressure drop $p_L - p_1$ in the cake as a function of the thickness x. When the cake thickness is L, the total pressure drop across the cake $\Delta p_c = p_V + p_H - p_1$ is given by point H as shown by Fig. 49b. At distance x when the thickness is L, the local pressure drop $p_L - p_1$ is given by point G.

For an incompressible cake, the hydraulic distribution, $p_L - p_1$ vs. x, is shown as a series of straight lines. For a highly compressible cake as shown in Fig. 49c, the $p_L - p_1$ curves indicate a steep slope close to the medium with most of the pressure drop occurring in layers adjacent to the devices supporting the cake.

The pressure $p_s = p - p_L$ causes the cake to compress. As p_s reaches its maximum value at the medium where $p_L = p_1$, compressible cakes have their minimum porosity at the medium where x = 0. In Fig. 49b, a graphical interpretation of p_s is shown for the cake represented by OGH. At point X in the cake, the value of $p_L - p_1$ is found from point G. The pressure drop across the cake, Δp_c, is located at H; and the difference between G and H equals p_s.

In view of the complexity of the pressure drop relationships, it is doubtful that the simplified theory developed by Ruth and Kempke [34] and presented by later authors [35] could be considered adequate for treating continuous rotary drum filtration. In addition to the pressure drop problem, it is doubtful that the flow resistance α reaches its equilibrium values as required by Eqs. (10)-(11). Thus rotary drum filtration theory must be carefully evaluated.

Relationships developed on the basis of considerable simplification have been used in much of the literature. It is customary to assume that Δp_c does not change, the average filtration resistance is constant, and the medium resistance is negligible. Under such conditions, the relationship of the volume filtered per unit time per unit area of the actual submerged filtration surface is given by Rushton and Hameed [35]

$$Y = \left(\frac{2pcf}{\mu\alpha t_R}\right)^{1/2} \tag{132}$$

where p is the assumed total constant pressure neglecting p_H and p_i, f the fraction of the surface devoted to cake formation, t_R the time for one revolution, μ the viscosity, and c the ratio of the mass of dry cake to the filtrate volume. While the use of Eq. (132) is dubious, the term in parentheses may be used to correlate data. Either the factor in parentheses or the dimensionless quantity $(2pcf/\mu\alpha t_R Y^2)$ can be profitably used for analyzing data.

Equation (43) can easily be adapted to continuous filtration on either drum (Fig. 48) or horizontal belt filters. The volume v of Eqs. (42) or (43) represents the volume of filtrate per square foot per cycle, which is $ft^3/(ft^2\ rev)$. To convert to Q (ft^3/sec), v must be multiplied by ωrh, for a drum filter where ω is the angular velocity, r radius, and h drum width. For a horizontal belt filter, uh would be used where u is the linear velocity. The rate is frequently treated as Q_D $ft^3/(ft^2$ of drum surface · sec), which is related to Q and v by

$$2\pi rhQ_D = Q = \quad rhv \tag{133}$$

$$Q(\text{belt}) = uhv \tag{134}$$

Substituting these expressions in Eq. (43) and replacing t by ϕ_o/ω (or z/v for a belt) yields

$$Q = 2\pi rhQ_o = rh\sqrt{\frac{2p_{av}\omega\phi_o}{\mu\alpha c}} \tag{135}$$

where ϕ_o is the angle of cake formation and z is belt length. The mass/time is given by W = cQ or

$$W = rh\sqrt{\frac{2p_{av}c\omega\phi_o}{\mu\alpha}} \tag{136}$$

where p_{av} is the sum of the vacuum and the hydrostatic head. Equations (135) and (136) are useful for predicting changes in operating variables on the overall rate.

For purposes of analyzing vacuum-leaf test data, it is preferable to deal directly with Eq. (42) and consider v to be $ft^3/(ft^2 \text{ rev})$. It is also preferable to use w = cv lbm/(ft rev) in the form

$$w = \sqrt{\frac{2cp_{av}t}{\mu\alpha}} \tag{137}$$

The rate is then given by w/t or

$$\frac{w}{t} = \sqrt{\frac{2cp_{av}}{\mu\alpha}}\, t^{-0.5} = Xt^{-0.5} \tag{138}$$

Taking logarithms yields

$$\log \frac{w}{t} = -0.5 \log t + \log X \tag{139}$$

Equations (138) and (139) will be used here for interpreting test data.

In Fig. 50a and 50b, data are plotted in accordance with Eqs. (138) and (139). A 0.5-in. cake is produced in a form time of 0.88 min at a rate of 245 lbm of dry cake/(ft^2 hr). If the submergence were 120°, then it would take (360/120)(0.88) = 2.64 min/rev to produce a half-inch thick cake. Cutting the thickness to 0.25 in. would double the rate and increase the drum speed four-fold.

Lines drawn with a slope of negative unity in Fig. 50b represent constant cake thickness. If L is held constant in Fig. 50b then the value of X in Eq. (138) must change. The dotted line in Fig. 50b parallel to the experimental curve represents what would occur if p were reduced 20%. As X is proportional to the square root of p, the new line is drawn at $\sqrt{0.8}$ = 0.895, or at a distance 10.6% below the old line. Whereas a 0.5-in. cake was produced in 0.88 min under the original conditions, 1.06 min would be required at the new reduced pressure.

The slope of curves plotted according to Fig. 50b and Eq. (139) will be -0.5 when the medium resistance is zero and no irregularities occur while taking data. However, medium resistance is not zero, and other theoretical problems enter into consideration. Consequently, the slope may not equal -0.5. The vast majority of

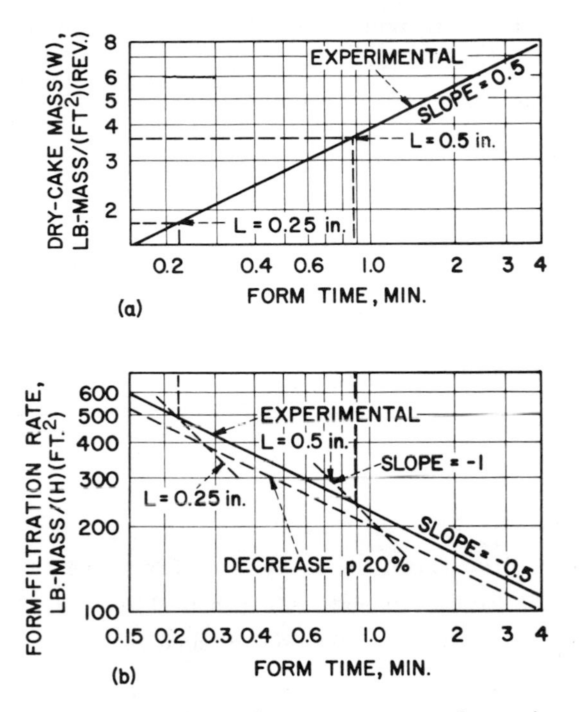

FIG. 50. Cake and mass rate vs. form time.

substances yield curves in the range of -0.5 to -0.65. Commonly the rate curve is nearly horizontal at very short form-times, with the slope gradually increasing to a value of -0.5 or somewhat steeper.

In continuous filtration on drums, the cycle is completed in one revolution. As shown in Fig. 48, the various stages of cake formation, washing, drying, and discharge must be balanced by drum geometry. One of the stages will control the cycle. For example, if washing requires the entire 105°, there may be too much cake produced during the form period, making it necessary to restrict cake thickness by adjustment of the bridge in the filter value to decrease filter time.

Discharge characteristics of different filters require certain minimum thickness as tabulated in Table 16.

TABLE 16 Minimum Cake Thickness for Discharge[a]

Filter type	Minimum design thickness
Drum	
Belt	1/8 to 3/16
Roll discharge	1/32
Standard scraper	1/4
Coil	1/8 to 3/16
String discharge	1/4
Horizontal belt	1/8 to 3/16
Horizontal table	3/4
Tilting pan	3/4 to 1
Disc	3/8 to 1/2

[a]From [31].

Example 8 [31]

A belt-type drum filter is to be used for filtering, washing, and drying a cake having properties given by Figs 38, 39, 40, and 50. The following conditions will be assumed.

Slurry concentration	40% solids
Solute concentration in liquid	2%
Final cake moisture	25%
Wash ratio	1.5
Cake thickness (Eq. 36)	w = 7.2L (in.)
Maximum apparent submergence	35% or 126°
Maximum effective submergence	30% or 108°
Maximum arc for washing	29% or 104°
Maximum arc available excluding discharge and resubmergence	75% or 270°

Based upon the data of Table 16, it will be assumed that a 0.25-in. thick cake can be used for initial design purposes. The cake mass is then given by $w = 7.2 \times 0.25 = 1.8$ lbm/ft^2. From Fig. 50a or b the form time equals 0.22 min and the filtration rate is 491 lb/(hr ft^2) of drum surface. Based upon an effective submergence of 30%, the minimum cycle based on cake formation is 0.22/0.3 = 0.73 min/rev.

It is next necessary to determine if washing or drying can be accomplished within the times available at the 0.73 min/rev. Looking at Fig. 48 it can be seen that there will be a minimum suction or drying time during passage through 27° (7.5%) of the perimeter. Then

Drying time $0.075 \times 0.73 = 0.06$ min

Correlating factor $t_d/w = 0.06/1.8 = 0.033$

Based on Fig. 47 the dewatered but unwashed (D/U) cake will have a moisture content of 30%. Then, with a wash ratio of 1.5,

Liquid in D/U cake $= (30/70)(1.8) = 0.77$ lbm/(ft^2 rev)

Quantity of wash $= 1.5 \times 0.77 = 1.155$ lbm/(ft^2 rev)

$= 0.14$ gal/(ft^2 rev)

To calculate wash time, $wv_w = 1.8 \times 0.14 = 0.25$. From Fig. 39, the wash time is 0.15 min. This corresponds to $0.15/0.77 = 0.20$. As up to 29% of the circumference can be used, washing offers no problems.

For a final moisture content of 25%, the simplified correlating factor (Fig. 40) is $0.3 = t_d/w$. With $w = 1.8$, $t_d = 0.54$ min or $0.54/0.73 = 0.739$, or nearly three-fourths of the circumference. Thus a lower speed will have to be used. As a first estimate, note that the maximum arc for wash plus final dry as given by Fig. 39 is the following:

75 - cake formation - suction = 75 - 30 - 7.5 = 37.5% of periphery

Using the originally calculated wash plus dry times of 0.54 + 0.15 = 0.69 min, then

$0.69/0.375 = 1.84$ min/rev

Washing arc $= 0.15/1.84 = 0.0815$ rad or 29°

Starting the wash at the horizontal centerline will require careful adjustment of wash sprays to prevent runback and will require at least two headers.

Minor adjustments can be made by recalculating each quantity with each change of conditions.

Initial dry	=	1.84 × 0.075 = 0.14 min
t_d/w	=	0.14/1.8 = 0.08
D/U moisture (Fig. 40)	=	27%
Liquor in D/U cake	=	(27/73)(1.8) = 0.67 lbm/(ft^2 rev)
wv_w	=	1.8 × 0.12 = 0.22
Wash time (Fig. 39)	=	0.14 min

The cycle time is now (0.14 + 0.54)/0.375 = 1.81 min/rev. The required effective submergence is (0.22/1.81)(100) = 12.2%. This is much less than the 30% available. The filter valve bridge must delay start of vacuum application, or the slurry level can be reduced. If the level is reduced, additional initial dry time will be available, reducing the angle required for washing. The design cycle at 0.55 rpm is

Operation	Time (min)
Form	0.22
Initial dry	0.14
Wash	0.14
Final dry	0.54
Discharge and resubmergence	0.77
	1.81

The filtration rate is calculated as 491 × 0.122 = 59.9 lbm/(hr ft^2). A scale-up factor of 0.8 based on experience is applied to give 0.8 × 59.9 = 47.9 lbm/(hr ft^2). The scale-up factor is not intended as a safety factor to allow for increased production; it corrects for deviation due to the size of the test equipment, media blinding, etc.

For solute recovery, it will be assumed that curve B of Fig. 38 applies. With j = 1.5, the fraction remaining is 0.145. To be on the safe side a value of 0.2 will be used. The following calculations are needed

Solute in feed (60/40)(0.02)	= 0.03 lbm/lbm solid
Solute in D/U cake (27/73)(0.02)	= 0.0074 lbm/lbm solid
Solute in washed cake 0.0074 × 0.2	= 0.0015

The fractional recovery equals (0.030 - 0.0015)/0.03 = 0.95. Using 0.145 instead of 0.2 would yield 0.964 for the wash efficiency.

The air rate can be calculated based upon the data previously presented in Sec. 5. XII. B. During the 0.14 min of initial drying the average rate is found to be 2.95 $(ft^3/min)(ft^2\ rev)$. The average rate during the final 0.54 min of drying is 5.85 $(ft^3/min)(ft^2\ rev)$. The total air rate is given by $0.14 \times 2.95 + 0.54 \times 5.85 = 3.57\ (ft^3/min)/(ft^2\ rev)$. As there are 1.81 min/rev, the air rate is $3.57/1.81 = 1.91\ (ft^3/min)/ft^2$.

Example 9

Determine the size and velocity of a horizontal belt filter required to produce 30,000 lbm/hr of dry solid with a 0.5-in. cake. The properties of the slurry are such that the following cycle results with $w = 3.6\ lbm/ft^2$ cycle:

	Time (min)
Form time	0.15
Initial dry	0.20
Wash	0.60
Final dry	0.50
	1.45

A scale-up factor of 0.8 will be used with a 9-ft-wide belt.

$$\text{Belt speed} = 30{,}000/60 \times 9 \times 3.6 = 15.4\ ft/min$$

Length of belt under vacuum is given by

$$Z = 15.4 \times 1.45/0.8 = 27.9\ ft$$

The design filtration rate is given by the product of w, the number of cycles (60/1.45) per hour, and the scale-up factor, or 119 $lbm/(hr\ ft^2)$.

NOTATION

Units are given in terms of pounds mass (lbm), poundals (lbl), feet, and seconds. For the S.I. system, the following substitutions can be made: kilograms, kg, for lbm; newtons, N, for lbl; meters, m, for ft. Multiplying pounds force by the conversion factor g_c = 32.16 yields poundals. Where units are meaningless or unimportant, they are omitted. The symbol [-] indicates that the quantity is dimensionless.

a	Empirical constant in Eq. (10)
A	Cross-sectional area, ft^2
b	Intercept in Eq. (46), $\mu R_m/p$, sec/ft
B	Empirical constant in Eq. (12)
c	Mass of solids/unit vol. of filtrate, Eq. (33), lbm/ft^3
c_c	Empirical constant in Eq. (101)
C_e	Modified coefficient of consolidation, Eq. (95), $lbm^2/(ft^4\ sec)$
C_L	Multiplier in Eq. (64)
C	Concentration
C_o	Concentration of exit liquid
C_w	Concentration of wash liquid
e	Void ratio, [-]
e_o	Empirical constant in Eq. (101)
E	Empirical constant in Eq. (8)
f	Fraction of drum surface in active filtration, [-]
F	Fraction of liquor removed in right-angled deliquoring, Eq. (87), [-]
F_s	Frictional drag on solids, lbl
g	Gravitational constant, 32.16 ft/sec^2
h	Drum or belt width, ft
I	Integral as defined in Eq. (21)
i	Number of drainage surfaces, Eq. (99), [-]
j	Wash ratio, volume of wash/void volume of cake, [-]
J	Constant defined by Eq. (117)
k_1	Empirical constant defined by Eq. (111)
K	Darcy constant, Eq. (1), ft^2
K	Constant in Eq. (46), sec/ft^2
K_R	Empirical constant defined by Eq. (122)
l_n	Empirical constant in Eq. (101)
L	Cake thickness, ft
L_1, L_2	Initial and equilibrium cake thicknesses during expression, Eq. (98), ft
m_v	Coefficient of volume change, Eq. (89), ft^2/lbl
n	Compressibility coefficient, Eq. (8)
p	Applied pressure, lbl/ft^2
p_o	Fixed pressure, Eq. (80), lbl/ft^2

p_{av}	Average pressure on rotary drum filter, Eq. (131), lbl/ft^2
p_i	Low pressure below which α and ε are considered constant, lbl/ft^2
p_L	Hydraulic pressure, lbl/ft^2
p_s	Accumulative drag pressure on solids, $p - p_L$, Eq. (6), lbl/ft^2
p_v	Pressure due to vacuum, lbl/ft^2
p_H	Hydrostatic head, lbl/ft^2
q	Superficial flow rate, $ft^3/(ft^2\ sec) = ft/sec$
q_o	Empirical rate constant in Eq. (111), ft/sec
q_g	Rate in $gal/(min\ ft^2)$
Q	Rate for continuous filter, ft^3/sec
Q_D	Rate for rotary drum filter, $ft^3/(ft^2$ of drum) $\cdot$ sec
r	Radius of rotary drum filter, ft
R_m	Medium resistance defined by Eq. (24), 1/ft
s	Mass fraction of solids in slurry, [-]
s_c	Average mass fraction of solids in cake, [-]
t	Time, sec
t_c	Time for consolidation, Eq. (90), sec
t_d	Drying time for continuous filters, min
t_w	Time for washing, sec
t_D	Dumping, cleaning, and reassembly time, sec
t_F	Filtration time, sec
t_R	Time for one revolution of rotary drum filter, min
T	Dimension of frame in filter press, Fig. 27, ft
T_2	Decreased cake dimension during deliquoring, ft
T_c	Consolidation time factor, Eq. (99), [-]
u	Linear belt velocity, ft/sec
u_w	Apparent liquid velocity relative to w-plane, Eq. (88), ft/sec
U_c	Average consolidation ratio, Eq. (98), [-]
v	Volume of filtrate per unit area, $ft^3/ft^2 = ft$
v_w	Volume of wash per unit area, gal/ft^2
w	Mass of dry cake per unit area, lbm/ft^2
w_o	Total mass of dry solids per unit area, lbm/ft^2
W	Production rate of dry material in continuous filtration, lbm/sec
x	Distance through cake measured from medium, ft

x	Mass fraction of filter aid in Table 6.2; variable of integration in Eq. (26)
X	Limit of integration in Eq. (26); group of parameters in Eq. (138)
y	Integrand in Eq. (26)
y_{av}	Average value of y
Y	Volume filtered/(time)(area) in rotary drum filtration, Eq. (132), $ft^3/ft^2 \cdot sec = ft/sec$
Z	Length of belt filter, ft
α	Specific filtration resistance, ft/lbm
α_o, α_1	Empirical constants in Eq. (100)
α_i	Constant value of α when $p_s \leq p_i$, ft/lbm
α_{av}	Average value of α, ft/lbm
β	Empirical exponent in Eq. (12)
δ	Ratio of wash rate to final filtration rate in batch filters, [-]
Δ	Difference
ε	Porosity, void fraction, [-]
ε_1	Value of porosity at medium, [-]
ε_i	Constant value of porosity when $p_s \leq p_i$, [-]
ε_{av}	Average value of ε, [-]
ε_{avo}	Value of ε_{av} at pressure P_o, [-]
λ	Exponent defined by Eq. (8)
μ	Viscosity, lbm/(ft sec)
ρ	Density of liquid, lbm/ft^3
ρ_o	Density of slurry, lbm/ft^3
ρ_s	True density of solids, lbm/ft^3
σ	Ratio ρ_s/ρ, [-]
ϕ	Variable angle, Fig. 48, radians
ϕ_o	Angle of submergence, Fig. 48, radians
ω	Angular velocity, rad/sec

REFERENCES

1. A. Wright, *Industrial Filtration*, Modern Library of Chemical Engineering, Chemical Catalog Co., 1923.
2. J. E. Flood, H. E. Porter, and F. W. Rennie, "Filtration Practice Today," *Chem. Eng.*, *73*, 163-181 (June 30, 1966).
3. M. Shirato, M. Sambuichi, H. Kato, and T. Aragaki, "Internal Flow Mechanism in Filter Cakes," *A.I.Ch.E. J.*, *15*, 405-409 (1969).

4. F. M. Tiller and T. C. Green, "The Role of Porosity in Filtration, Part 9, Skin Effect with Highly Compressible Materials," *A.I.Ch.E. J.*, *19*, 1266-1269 (1973).

5. H. P. Grace, "Resistance and Compressibility of Filter Cakes. I and II," *Chem. Eng. Prog.*, *49*, 303-318, 367-376 (1953).

6. Wei-Ming Lu, "Theoretical and Experimental Analysis of Variable Pressure Filtration and the Effect of Side Wall Friction in Compression-Permeability Cells, Ph.D. Dissertation, University of Houston, 68-110, 149, University Microfilms, Ann Arbor, Mich., 1968.

7. M. Shirato and S. Okamura, "Behavior of Gairome Clay Slurries under Constant Pressure Filtration," *Kagaku Kogaku*, *20*, 678-683 (1956).

8. F. M. Tiller and H. R. Cooper, "The Role of Porosity in Filtration, Part 5, Porosity Variation in Filter Cakes," *A.I.Ch.E. J.*, *8*, 445-449 (1962).

9. F. M. Tiller and C. J. Huang, "Theory of Filtration Equipment," *Ind. Eng. Chem.*, *53*, 529-537 (1961).

10. F. M. Tiller, "The Role of Porosity in Filtration, Part 1, Numerical Methods for Constant Rate and Constant Pressure Filtration Based on Kozeny's Law," *Chem. Eng. Progr.*, *49*, 467-479 (1953).

11. F. M. Tiller, "The Role of Porosity in Filtration, Part 2, Analytical Equations for Constant Rate Filtration," *Chem. Eng. Progr.*, *51*, 282-290 (1955).

12. F. M. Tiller, "The Role of Porosity in Filtration, Part 3, Variable-Pressure, Variable-Rate Filtration," *A.I.Ch.E. J.*, *2*, 171-174 (1958).

13. F. B. Hutto, "Distribution of Porosity in Filter Cake," *Chem. Eng. Progr.*, *53*, 328-332 (1957).

14. M. Shirato, H. Kato, T. Murase, and M. Shibata, "Studies on Expression-Fundamental Analysis for Mechanism of Dehydration by Expression," *J. Ferm. Tech.*, *43*, 255-265 (1965).

15. M. Shirato, T. Aragaki, K. Ichimura, and N. Ootsuji, "Porosity Variation in Filter Cake under Constant Pressure Filtration," *J. Chem. Eng.* (Japan), *4*, 172-177 (1971).

16. F. M. Tiller and H. Risbud, "Hydraulic Deliquoring of Filter Cakes," *First Pacific Chemical Engineering Conference, A.I.Ch.E.*-Soc. Chem. Eng. (Japan), vol. 2, 1972, pp. 80-87.

17. M. Shirato, T. Murase, H. Kato, and S. Fukaya, "Fundamental Analysis for Expression under Constant Pressure," *Filtrat. Separat.*, *7*, 277-282 (1970).

18. M. Shirato, T. Murase, M. Negawa, and T. Senda, "Fundamental Studies of Expression under Variable Pressure," *J. Chem. Eng.* (Japan), *3*, 105-112 (1970).

19. M. Shirato, T. Murase, M. Negawa, and H. Moridera, "Analysis of Expression Operation," *J. Chem. Eng.* (Japan), *4*, 263-268 (1971).

20. M. Shirato, T. Murase, H. Doi, and H. Shimaoka, "Analysis of Filtration Processes in Plate-and-Frame Press," *Kagaku Kogaku*, *36*, 781 (1972).

21. I. Kormendy, "New Apparatus to Study the Pressing Process, Experimental and Evaluation Methods, Correlation Between Pressure and Equilibrium Juice Yield in the Case of Apple Pulp," *Acta Alimentaria*, *1*, 315 (1972).

23. A. Hackel, H. Horrak, and H. Jelenz, "Zur Verringerung der Kuchenfeuchte durch Membranpressen," *Chem. Tech.*, *24*, 405 (1972).

24. K. Terzaghi and R. B. Peck, *Soil Mechanics in Engineering Practice*, John Wiley, N. Y., 1947.

25. A. S. Michaels, W. E. Baker, H. J. Bixler, and W. R. Vieth, "Permeability and Washing Characteristics of Flocculated Kaolinite Filter Cakes, Hydraulic Permeability and Particle Orientation," *Ind. Eng. Chem., Fundam.*, *6*, 29-33 (1967); "Washing Characteristics," *Ind. Eng. Chem., Fundam.*, *6*, 33-40 (1967).

26. C. D. Han and H. J. Bixler, "Washing of the Liquid Retained by Granular Solids," *A.I.Ch.E. J.*, *13*, 1058-1066 (1967).

27. R. J. Wakeman, "Prediction of the Washing Performance of Drained Filter Cakes," *Filtrat. Separat.*, *9*, 409-415 (1972)

28. P. J. Lloyd, "Particle Characterization," *Chem. Eng.*, *81*, 120-123 (April 24, 1974).

29. P. A. Nelson and D. A. Dahlstrom, "Moisture Content Correlation of Rotary Vacuum Filter Cakes," *Chem. Eng. Prog.*, *53*, 320-327 (1957).

30. D. A. Dahlstrom, "Scale-Up Methods for Continuous Filtration Equipment," International Symposium on the Scaling Up of Chemical Plant and Processes, London, May 28-29, 1967.

31. C. E. Silverblatt, F. M. Tiller, and H. Risbud, "Batch, Continuous Processes for Cake Filtration," *Chem. Eng.*, *81*, 127-136 (April 29, 1974).

32. A. F. Alciatore and E. Neu, "Correlation between Constant Pressure Bomb Filter and Full-Scale Plant Results," *Filtrat. Separat.*, *10*, 581-585 (1973).

33. H. P. Grace, "Structure and Performance of Filter Media," *A.I.Ch.E. J.*, *2*, 307-336 (1956).

34. B. F. Ruth and L. L. Kempke, "An Extension of the Testing Methods and Equations of Batch Filtration Practice to the Field of Continuous Filtration," *Trans. A.I.Ch.E.*, *34*, 33-83 (1937).

35. A. Rushton and M. S. Hameed, "The Effect of Concentration in Rotary Vacuum Filtration," *Filtrat. Separat.*, *6*, 136-139 (1969).

Chapter 6

ULTRAFILTRATION

*Richard P. de Filippi**

Abcor, Inc.
Cambridge, Massachusetts

I. Introduction 476

II. Distinguishing Characteristics 476

III. Membranes 480
- A. Structure 481
- B. Formation 482

IV. Mass Transfer 486
- A. Flux 486
- B. Rejection 499

V. Engineering and Design of Systems 504
- A. Tubular Membrane Configuration 505
- B. Thin-Channel Membrane Modules 507
- C. Hollow Fiber (Capillary Membrane) Systems 508

VI. Applications 509
- A. Comparison with Alternative Processes 509
- B. Concentration and Fractionation of Cheese Whey 510
- C. Recovery of Electrocoat Paint 512
- D. Concentration of Spent Emulsions of Metal Machining Oil 514

---

*Current affiliation: Arthur D. Little, Inc., Cambridge, Massachusetts.

E. Recovery of Polyvinyl Alcohol Sizing from Textile Mill Desizing Wastes 516
F. Other 516
Notation 516
References 517

I. INTRODUCTION

In many ways, ultrafiltration is equivalent to normal particle filtration simply extended to a finer particle size. However, some process characteristics are unique, and ultrafiltration can be considered a distinct departure from other size-related separations. Ultrafiltration is a membrane filtration process which separates "particles" of about 10 to 100 Å from their surrounding medium. In this size range, obviously, particles can be solutes in liquid solution, which indeed is a common class of mixtures treated by ultrafiltration. However, the process has also been successfully applied to colloidal suspensions which are difficult to separate practically by other techniques.

A feature of ultrafiltration, perhaps unique among filtration processes, is the ability to operate with steady filtration fluxes in the absence of external means for clearing the filter of accumulated solids. In ultrafiltration, the retained material always concentrates at the membrane-solution interface, but it is swept away by fluid-dynamic forces. There are exceptions to this feature, since in some cases membrane fouling can occur, causing a decline in ultrafiltration flux with time and necessitating cyclic operation with periodic clean-up of the membranes. Clean-up does not represent the method of recovering retained solids, however; rather, retentate is recovered as a fluid concentrate flowing continuously from the system.

II. DISTINGUISHING CHARACTERISTICS

The basic principle of operation of ultrafiltration is simple, as illustrated in Fig. 1. Flowing by the membrane is a solution containing two solutes: one of the molecular size too small to be

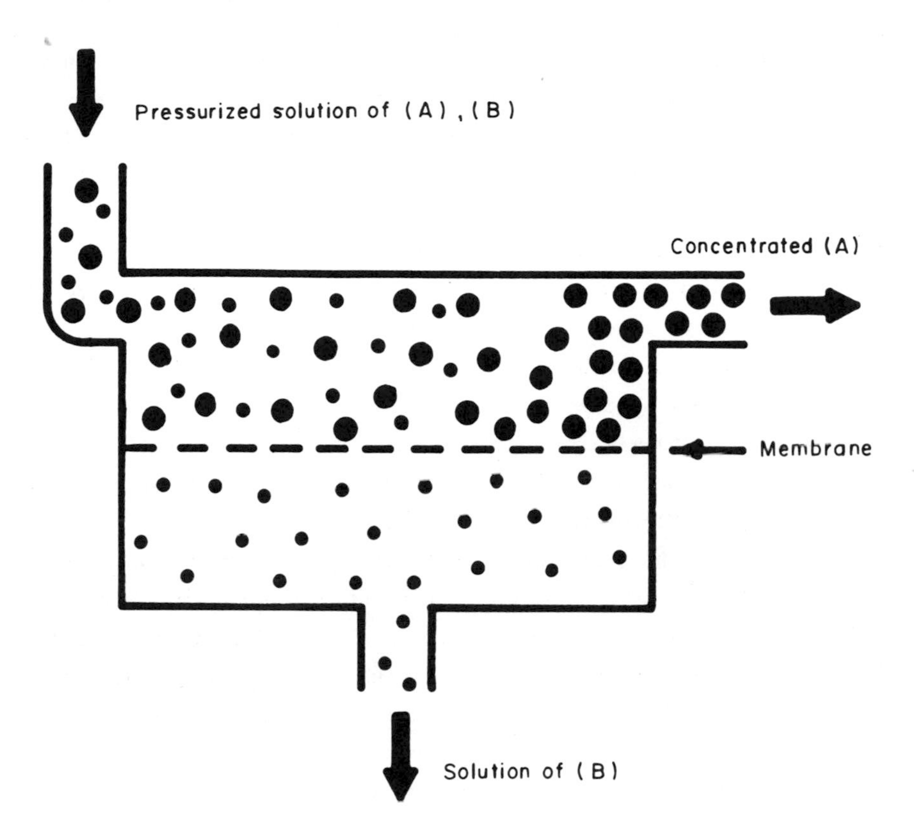

FIG. 1. Schematic diagram of membrane ultrafiltration process [12]. (From *CHEMTECH*, a publication of the American Chemical Society, reproduced with permission.)

retained by the membrane, and the other of larger size allowing 100% retention. A hydrostatic pressure is applied to the upstream side of the supported membrane, and solvent plus small-molecule solute pass through the membrane, while the large-molecule solute is retained (rejected) by the membrane. A fluid concentrated in the retained solute is collected from the upstream side of the membrane, and a solution of small-molecule solute is collected from the downstream side. Of course, where only a single solute is present and it is rejected by the membrane, the liquid collected downstream is pure solvent.

Retained particle size is one characteristic distinguishing ultrafiltration from other filtration processes. Viewed on a spectrum of membrane separation processes, ultrafiltration is only one of a series of membrane methods that can be used for molecular separations. For example, reverse osmosis, a membrane process capable of separating dissolved sodium chloride from water, falls further down on the same scale of separated particle size. In Fig. 2, that size spectrum is shown as a function of filtration flux. At the small-molecule/low-flux end of the spectrum lies the commercial cellulose acetate reverse-osmosis membrane, with the capability of retaining hydrated sodium and chloride ions. Next come ultrafiltration membranes with pores that span a size range of about 10^{-3} to 10^{-2} μm (10-100 Å) with filtration fluxes of about 0.5 to 10 gallons per square foot per day (gfd) per pound per square inch of pressure driving force. Microporous filters capable of virus and bacteria retention cover the size range about 0.01-1.0 μm, with fluxes of 10-1,000 gfd per psi ΔP. Finally, conventional industrial filters for normal particulate materials are capable of filtering particles of 1 μm or larger, with filtration fluxes above 1,000 gfd per psi of pressure loss. Differences in filtration flux among these processes are reflected as differences in processing cost per unit volume of filtrate. To the extent that membrane surface area controls process costs, very great differences in cost will exist between ultrafiltration and more conventional particle filtration. The practicality of ultrafiltration rests more with its ability to perform separations that are not technically feasible by other methods.

A second area of distinction between ultrafiltration and particle filtration lies in the nature of the filter medium, and the effects of the process environment on its properties. Ultrafiltration membranes can contain pores of a relatively narrow size distribution. They differ from microporous membranes, in that the pores active for selective filtration are not formed by compressed particle interstices which result in a more complex network. Rather, they may be closer in structure to Nuclepore filters, which are formed by etching cylindrical pores into a solid matrix. Because of the small pore size in ultrafiltration membranes, effects of external conditions on the solid membrane material can have an effect on the porous structure. For example, changes in temperature which

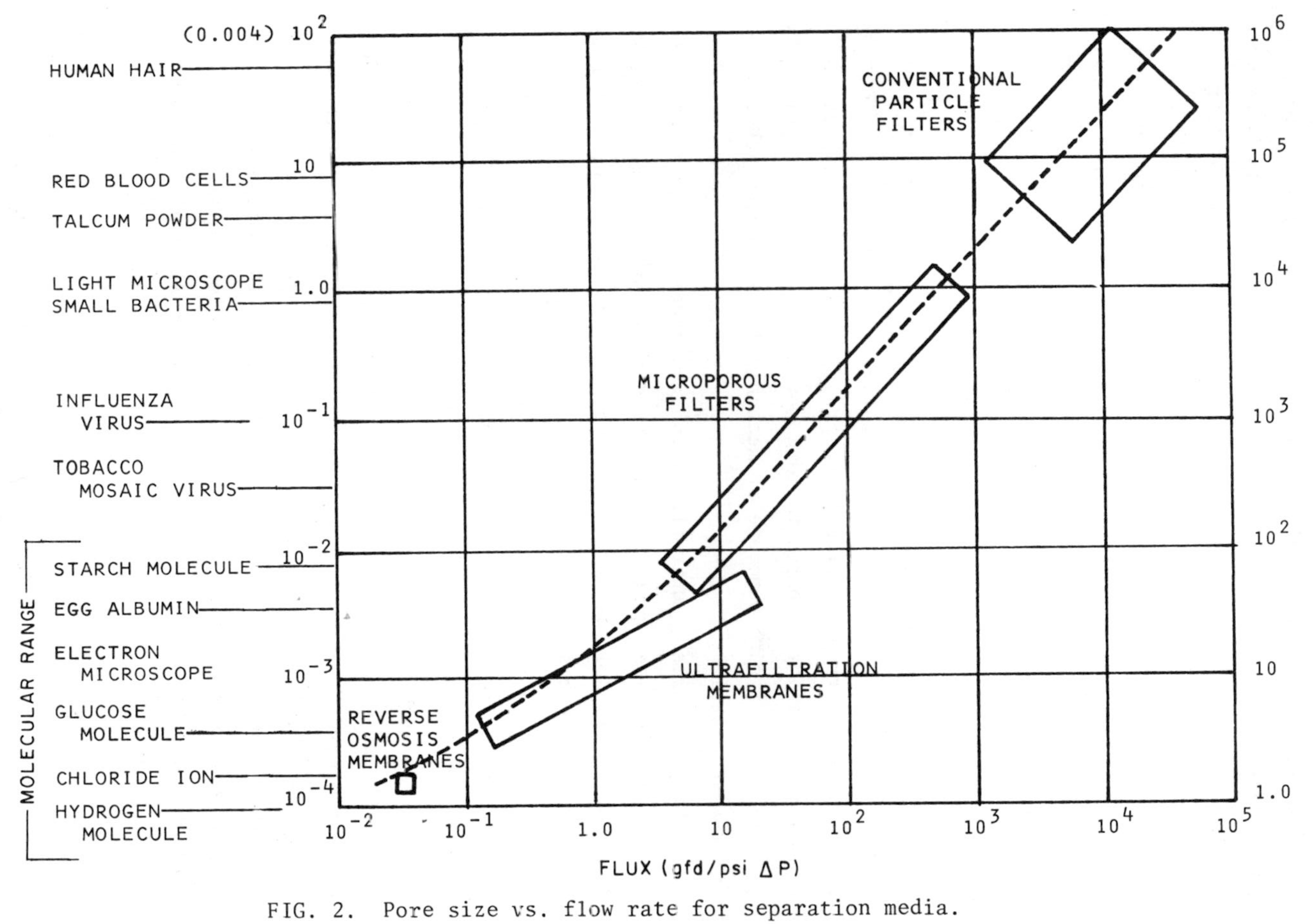

FIG. 2. Pore size vs. flow rate for separation media.

would change properties of the matrix (due to glass transitions, or alterations in swelling properties) may cause changes in the selective properties of the membrane. Similarly, in aqueous systems, extremes in pH which affect chemical composition of the membrane will also alter its ultrafiltration characteristics. However, they are relatively resistant to structural change in comparison to reverse-osmosis membranes. For example, cellulose acetate reverse-osmosis membranes are stable within a pH range of 4-8. Cellulose acetate ultrafiltration membranes maintain stable properties over the range of pH 3-9.

Relative to other filter media, ultrafiltration membranes are fragile and require supporting substrates, a need which is accentuated by the higher pressure differences used (10-100 psi) because of low permeability compared to more porous filter media. Membranes are also sensitive to puncture or mechanical abrasion, which can destroy selectivity.

The properties of ultrafiltration membranes lead to a range of process applications quite distinct from those of conventional filtration. Where solutes are being separated from solution, ultrafiltration can serve as a concentration or fractionation process for single-phase liquid streams. Thus, ultrafiltration competes with adsorptive and evaporative separation processes and has the potential for broader applicability than conventional filtration. Usually it will not perform the entire separation task, however, because it produces a concentrate rather than a pure product, and the concentrate requires further processing if a pure product is to be recovered. For waste streams, however, where adequate reduction in pollutants can be obtained by ultrafiltration, the single step may be all that is required. Disposal of the reduced volume of pollutant concentrate replaces the problem of disposal of the entire waste stream.

III. MEMBRANES

The capability of polymeric membranes to separate components of a liquid solution has been a familiar concept for many years. Membranes for retention of high molecular-weight solutes in aqueous solutions, using a pressure driving force for ultrafiltration, have been in practice in biological laboratories for at least 30 years. Conventional cellophane (regenerated cellulose) dialysis membranes have been employed for ultrafiltration of proteinaceous solutions, although flux rates were extremely low, and far from those required for economic industrial practice.

A. Structure

In the early 1960s, Loeb [1] perfected a technique for producing membranes with fluxes several orders of magnitude higher than those observed before. These membranes, now well established to possess an extremely thin selective layer (0.1-1.0 μm), have been the major factor responsible for the achievement of practical operations in ultrafiltration. While originally formulated for a far more stringent task, that of retaining salt in seawater to produce fresh water, modifications in fabrication techniques have led to membranes with desirable retentive characteristics for a number of industrial applications. Indeed, it has become possible to tailor membranes with a wide range of selective properties.

The structure of ultrafiltration membranes has not yet been completely elucidated; nevertheless, there is sufficient consistent experimental information to support the hypothesis of pores penetrating the active selective layer. The size distribution of these pores is surprisingly narrow in some cases. For example, it is possible to retain a solute of molecular weight of about 24,000 quantitatively (the enzyme trypsin) while passing lower molecular-weight proteins ranging from 11,000 to 19,000 in the ultrafiltrate [2]. These membranes are different from so-called solution-diffusion membranes, which have been the subject of study for a wide variety of gas- and liquid-phase molecular separations. The latter group possesses a permselective structure which is nonporous, and separation is effected on the basis of differences in solubility and molecular diffusivity within the actual polymer matrix. Alternatively, if solution-diffusion membranes were reviewed as porous, the pore size is of the order of magnitude of Angstroms, and these "pores" can discriminate between subtle differences in permeate molecular structure, such as in positional isomers. In porous ultrafiltration membranes, some degree of transport through the actual solid polymer matrix takes place, but its contribution to the overall flux is probably negligible.

A spectrum of permselective properties can be produced from the same membrane material, by variations in fabrication technique. Thus, an effective desalination membrane can be made from solution-cast cellulose acetate by employing a final annealing step, while a higher flux membrane incapable of rejecting salts can be made by simply eliminating annealing. Moreover, intermediate properties such as partial salt rejection can be obtained by varying the time-temperature history during annealing. Consequently, it appears that almost a continuous spectrum of structures can be formulated.

The ultrafiltration membrane in most common use is cellulose acetate. This is largely because the polymer (cellulose 2.5-acetate) was found to be optimal for water transport and selectivity in reverse-osmosis desalination systems, and much of the technology for the fabrication of membranes selective to a wide range of species, including macromolecules, came as a by-product. Presently, there is increasing use of other membrane materials, including polyelectrolyte complexes, aromatic polyamides, acrylics, and cellulose acetate-butyrate.

B. Formation

The development of techniques for production of high-flux asymmetric ultrafiltration membranes was the major factor leading to practical commercial applications. These membranes exhibit ultrafiltration fluxes 2 to 3 orders of magnitude higher than normal homogeneous membranes such as cellophane. These high fluxes are realized because the effective selective layer has been reduced in thickness to a dimension less than 1 μm, with the remaining portion of the structure acting as an open, highly porous, nonselective support for this thin layer. Thus, a membrane structure of a total thickness of 0.1 mm, and good mechanical strength, contains a selective membrane at one surface with a thickness less than 1% of the total.

Asymmetric ultrafiltration membranes are solution cast, using a casting dope often consisting of a mixture of good and poor solvent for the polymer. For example, one formulation [3] employs a mixture of 60 vol % of acetone (good solvent) and 40 vol % of formamide (poor solvent) as mixed solvent for cellulose acetate at a concentration of 200 g/liter of solvent. The cast film is dried briefly in air, which causes evaporation of solvent from a very thin layer at the surface, resulting in formation of the selective "skin" on the membrane. Once this consolidated skin is formed, further evaporation is slowed drastically, apparently due to a case-hardening effect because of low solvent diffusion rates in the dried polymer [4]. The membrane is then immersed in cold water, which causes gelation of the underlying cellulose acetate, forming a very open porous structure. The remaining solvent is leached out into the water. Thus, the membrane is formed in two steps, resulting in the formation of two layers with distinctly different properties. Flux and rejection properties can be changed by altering the composition and concentrations of solvent and the time-temperature history during membrane formation [1].

Recent work [5,6] has provided further quantitative information on the effects of casting-dope composition and casting conditions on membrane properties. In this system, the polymer was cellulose acetate, the solvent acetone, and the nonsolvent aqueous magnesium perchlorate. The ratio of nonsolvent to solvent in the dope had the most profound effect on flux (Fig. 3) and rejection (Fig. 4). In contrast, the polymer content, within the range of 14-19% cellulose acetate by weight, had little or no effect. Casting-solution temperature increases led to lower-flux, higher-rejection membranes; longer evaporation periods prior to immersion into the leaching water had the same effect. Both the higher casting-dope temperature and longer evaporation time permit a greater degree

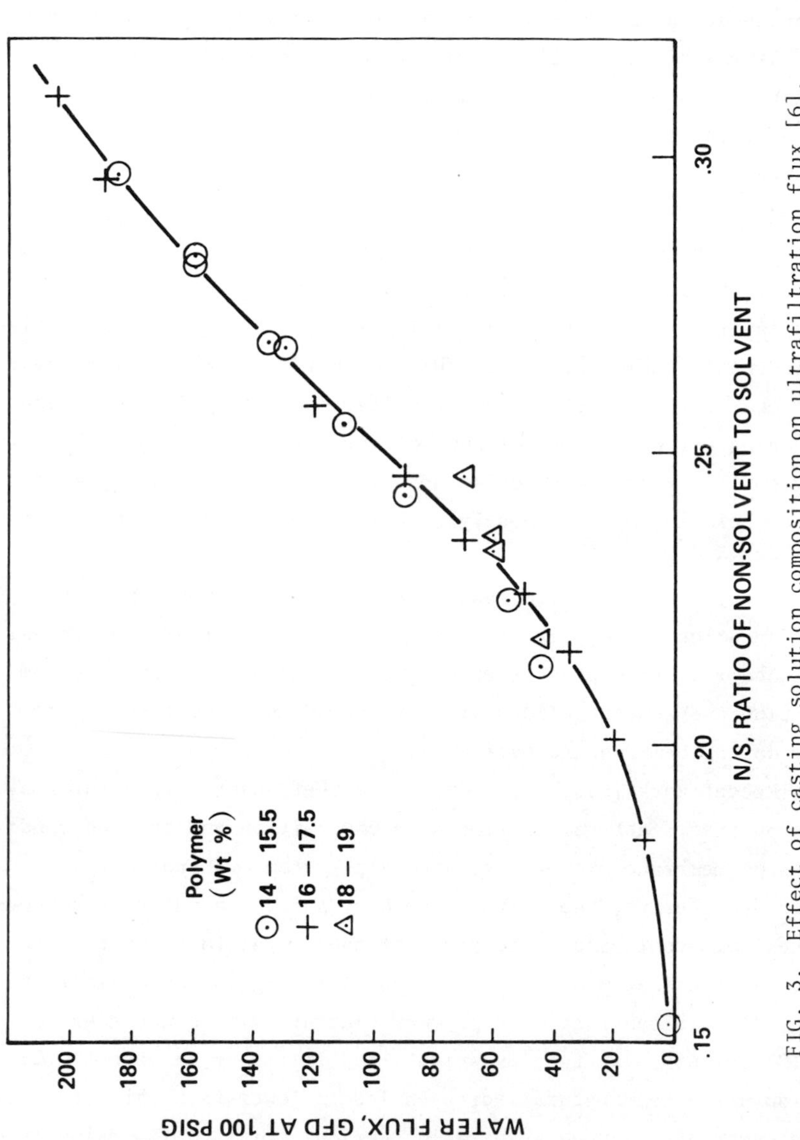

FIG. 3. Effect of casting solution composition on ultrafiltration flux [6].

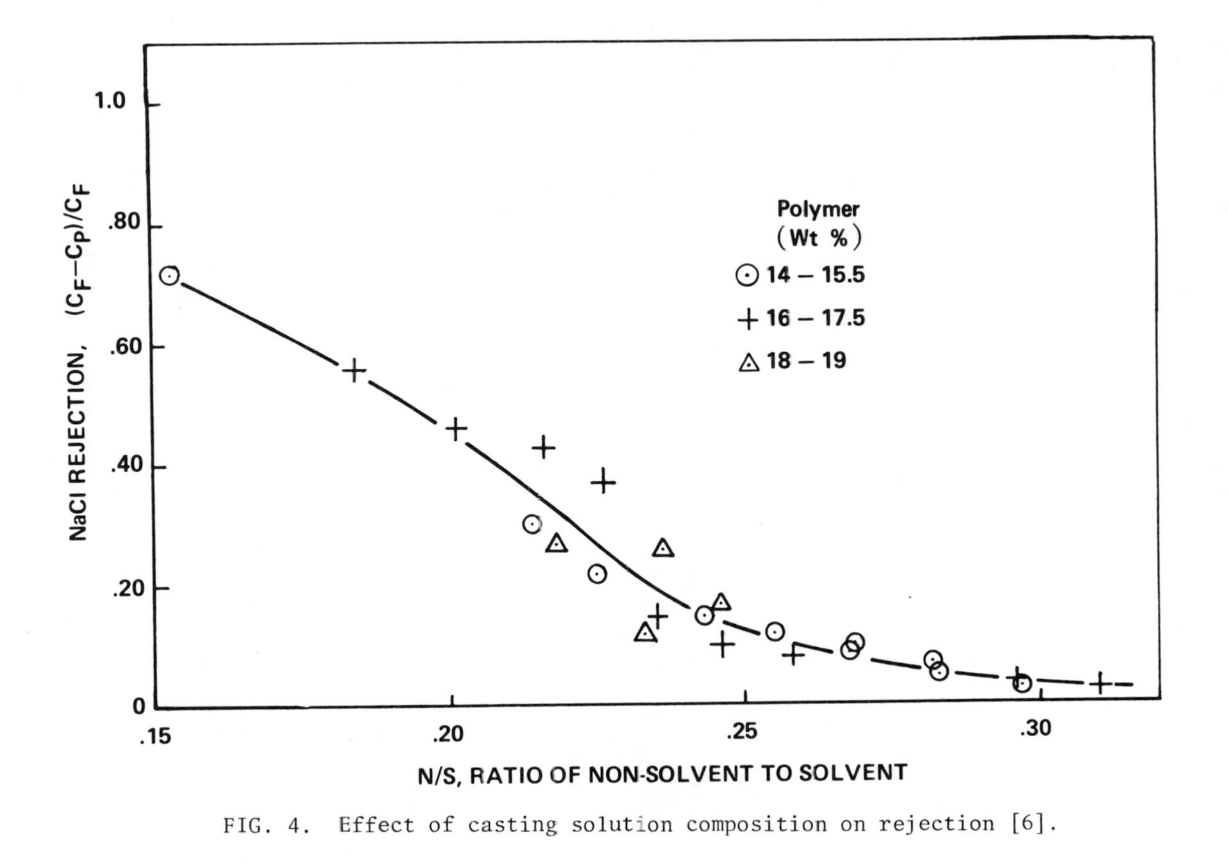

FIG. 4. Effect of casting solution composition on rejection [6].

of solvent volatilization from the surface, apparently producing a thicker and/or more consolidated active layer, and thereby causing decreased flux and increased solute rejection.

IV. MASS TRANSFER

The development of high-flux ultrafiltration membranes has resulted in increased importance of fluid-phase mass-transfer resistances. Indeed, in most practical ultrafiltration applications, transport in the liquid phase is the rate-controlling step, and membrane resistance is relatively unimportant. Thus, a reasonable understanding of factors governing mass transfer is essential for ultrafiltration process design.

A. Flux

Consider a system operating at steady-state under isothermal conditions, with a semipermeable membrane separating two fluid compartments, such as that illustrated in Fig. 5. An aqueous solution containing a single solute concentration c_1 and elevated pressure P_1 flows continuously by the upstream surface of the membrane. The membrane is permeable to the solvent (water) but only partly permeable to the solute, so that the solute concentration c_2 in the permeate is less than c_1. Permeate flowing at a membrane flux J is continuously removed from the downstream compartment at a lower pressure P_2. The flow of solution across the upstream face of the

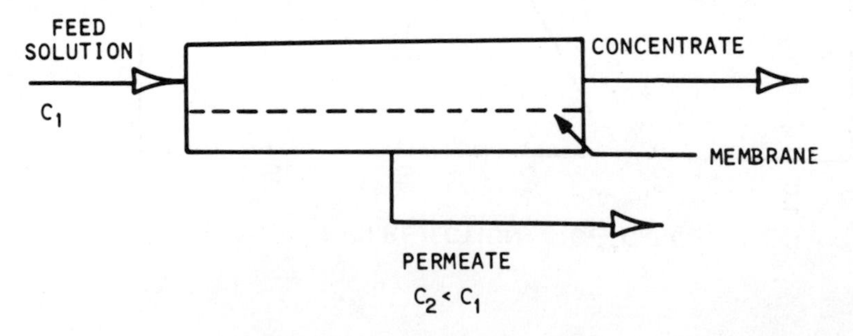

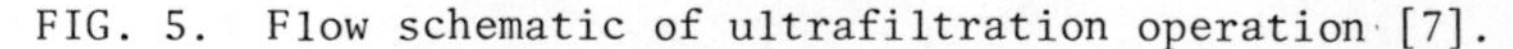
FIG. 5. Flow schematic of ultrafiltration operation [7].

membrane is high relative to the permeation flux J, so that the concentration change across the upstream compartment is negligible. Under these conditions, the flux J can be described by the relationship,

$$J = K_m[(P_1 - P_2) - (\Pi_1 - \Pi_2)] \qquad (1)$$

where Π_1 and Π_2 are the solution osmotic pressures corresponding to the concentrations c_1 and c_2, respectively, and K_m is a constant over a wide range of pressures and concentrations. This relationship is justified on thermodynamic grounds, recognizing that the difference in chemical potential for each solution component is the driving force for membrane transport.

In cases where one is dealing with a membrane capable of rejecting small molecules which exert significant osmotic pressures at normal concentrations, the membrane process described above is known as reverse osmosis. For such solutions, transport would occur from the dilute to the concentrated solution side of the membrane if the applied hydrostatic pressure were less than the osmotic pressure difference, due to simple osmosis. When the applied hydrostatic pressure exactly balances the osmotic pressure difference, no net transport will take place. When the applied pressure difference exceeds $\Delta\Pi$, however, solution will flow from the high-pressure, concentrated solution to the low-pressure, dilute solution, the reverse of normal osmotic transport. If the membrane is made selective to water only with high rejection of solute, then water may be recovered from the aqueous solution. This is the basis of desalination of brackish and seawater by reverse osmosis.

When the solute is a large uncharged molecule, such as a macromolecule with a molecular weight of 1,000 or greater, the osmotic pressure difference across the membrane will usually be negligible relative to the applied pressure. Under this circumstance, flux through the membrane, in the absence of any fluid-phase mass-transfer resistances, is simply proportional to the applied pressure difference. If the membrane is selective to large molecules only, and freely transmits low molecular-weight solutes, the concentration of small molecules on both sides of the membrane will be equivalent, and the transmembrane osmotic pressure difference will still be negligible. Thus, for a two-solute system, one macromolecular and the other of low molecular weight, fractionation of the solutes can be achieved, and flux is still proportional to the applied ΔP only. The term ultrafiltration is customarily applied to these situations where negligible osmotic pressure differences exist.

It was discovered in a number of early studies of ultrafiltration that the simple proportionality between membrane flux and applied pressure difference was valid for pure solvent, solutions of small molecules where no rejection takes place, and dilute solutions of macromolecules [8,9]. However, as macromolecule concentration was increased, an anomalous result was observed, as illustrated in Figs. 6 and 7 for two different systems. The linear flux-ΔP relationship is evident only at very low fluxes; as applied pressure

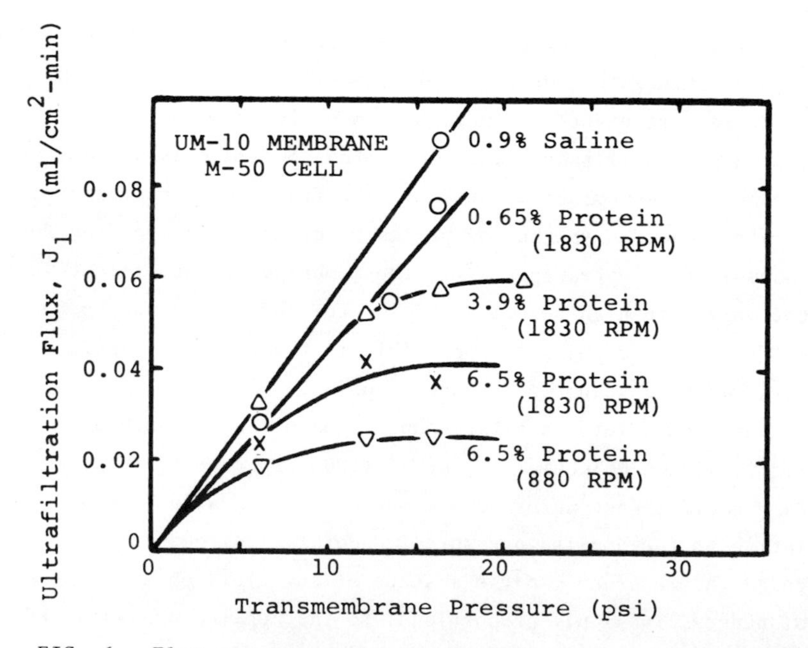

FIG. 6. Flux-pressure relationships for bovine-serum albumin solutions in a stirred batch cell [5].

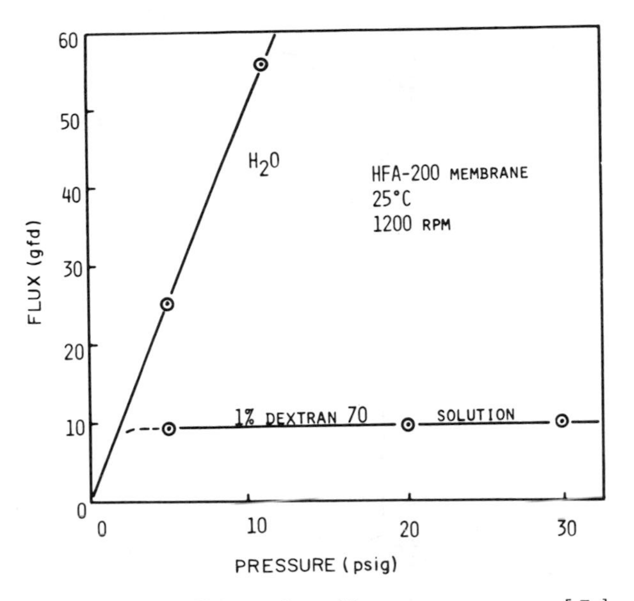

FIG. 7. Ultrafiltration flux vs. pressure [7].

difference is increased, the ultrafiltration flux levels off at a value far below that for the solvent. As the concentration of the rejected solute increases, the asymptotic flux assumes a lower valuc.

Thus, for more concentrated solutions, which represent major areas of interest for industrial applications, two observations are evident which contradict the assumption of a simple linear flux-ΔP relationship: (1) ultrafiltration fluxes are below those observed for pure solvent; and (2) in spite of the fact that the solvent flux is pressure dependent, the solution flux becomes insensitive to changes in applied pressure as pressure increases. Therefore, some additional mechanism must be invoked to explain transport behavior in real systems.

To account fully for these apparently anomalous flux characteristics, the mass transfer phenomena occurring must be examined further. This can be done by considering the concentration profile shown in Fig. 8. In the bulk fluid stream flowing by the upstream side of the membrane, the concentration of partly rejected solute

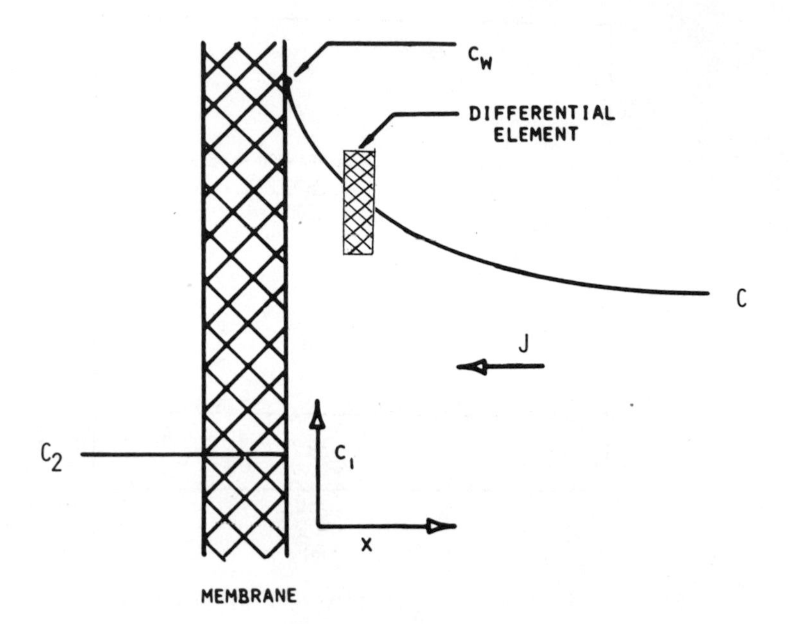

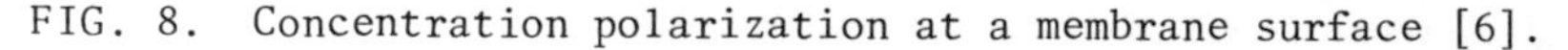
FIG. 8. Concentration polarization at a membrane surface [6].

is c_1. At the interface of the membrane and feed solution, however, it is evident that the rejected solute concentration at steady state must be at some value c_w greater than c_1. Solute is transported to the membrane surface with solution due to permeation through the membrane. At steady state, this must be balanced by an equal and opposite transport of solute by diffusion away from the membrane surface driven by a concentration gradient of solute. Thus, there is a solute concentration profile ranging from c_1 to some elevated value c_w at the upstream membrane surface.

For a single solute system, neglecting concentration gradients in the dimension parallel to the membrane surface, one may write a simple mass balance may be written for solute in the mass-transfer boundary layer (i.e., the region where concentration is changing with position). At steady state, since the convective flux of solute to the membrane surface is exactly counterbalanced by solute diffusion away from the membrane, this relationship is

$$Jc - D\frac{dc}{dx} = 0 \tag{2}$$

The solute diffusivity in solution is represented by D, and x is the direction perpendicular to the membrane surface. J is the solvent flux through the membrane. Assuming a highly selective membrane, i.e., where the permeate concentrate of solute c_2 is negligible, Eq. (2) can be integrated across the boundary layer to give

$$J = \frac{D}{\delta} \ln \frac{c_w}{c_1} \tag{3}$$

By definition, the quantity δ represents the distance over which the concentration changes from c_w to the bulk stream concentration c_1 on the upstream side. The quantity c_w/c_1 is known as the concentration polarization ratio. In turbulent flow, δ is defined as the mass-transfer boundary-layer thickness, and the quantity D/δ is the mass-transfer coefficient, k [10]. Thus,

$$J = k \ln \frac{c_w}{c_1} \tag{4}$$

In laminar flow, terms can be included in Eq. (2) to account for convective transport parallel to the membrane surface and the mass balance solved analytically assuming a Newtonian fluid. This more rigorous solution can be approximated by again replacing D/δ with k, as in the turbulent flow case. By relying on known relationships for mass transfer in laminar flow, Eq. (4) has been found to be adequate [10].

Equation (4) is descriptive of any membrane transport system with solute rejected at the membrane surface. Its applicability to ultrafiltration systems is a special case. Ultrafiltration membranes have relatively high fluxes, which lead to relatively high values of the ratio c_w/c_1. At some value of the flux, J, a value of c_w will be reached which will cause precipitation or gelation of the rejected solute, resulting in the formation of a solute precipitate at the membrane surface. This circumstance is depicted in

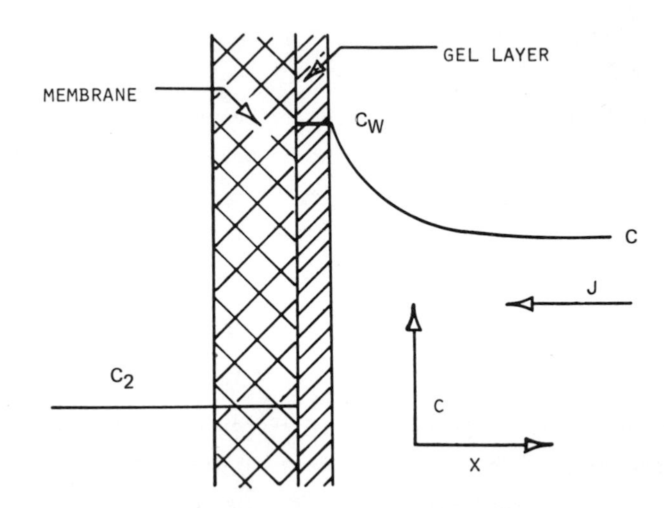

FIG. 9. Gel formation in concentration polarization [6].

Fig. 9. When this occurs, the wall concentration c_w is no longer independently controlled by flux but is now a fixed value. Since Eq. (4) still holds, it must be true that for a given bulk upstream concentration c_1, and for flow conditions leading to a fixed mass-transfer coefficient, k, the flux J must also be fixed, and is <u>independent of the applied pressure.</u> Thus, the observation of pressure-independent flux for ultrafiltration of concentrated solutions, where gel formation at the membrane surface is likely, appears to be borne out by this mass transfer analysis.

It is apparent that the physical occurrence limiting flux is the formation of a gel layer, which in itself has a hydraulic resistance to solution transport. This gel layer does not accumulate as a filter cake would, because of mass transfer of solute back from the gel into the bulk fluid stream. After operating at steady state at a given pressure, if the pressure is increased, a transient increase in flux is observed which ultimately decreases to the same value observed at the prior steady state. During that transient period, solution is brought to the gel layer at a more rapid rate because of the pressure increase. At the same time, solute deposition in the gel increases until the hydraulic resistance of the gel slows solution transport just to the point where the solute mass balance in Eq. (2) is satisfied. Similarly, when pressure is reduced, flux goes through a transient reduction and then rises to the same steady-state value. In this instance, mass transfer of solute away from the gel by diffusion remains the same, while convective transport of solute to the gel surface decreases until removal of solute from the gel causes sufficient reduction in hydraulic resistance to elevate the solution flux to its prior steady-state value.

In the ultrafiltration of some solutions, such as certain proteins, a significant osmotic pressure can prevail at the membrane surface when concentration polarization is high. Under these circumstances, even without gel formation, there is a significant decrease in flux relative to that observed with water at the same pressure [10].

The relationship among flux, mass transfer coefficient, and concentration expressed in Eq. (4) has several useful implications. First, it can be used to describe the relationship between solution flux and feed concentration. Since ultrafiltration is used to concentrate a rejected solute, knowledge of flux variation with concentration is essential for process design. Equation (4) indicates that flux decreases linearly with increasing logarithm of concentration. Data obtained over a wide range of concentration, such as those shown in Fig. 10, verify the relationship. It also implies that flux will go to zero when the concentration reaches c_w, as extrapolated on a plot such as Fig. 10. Evidence at these high concentrations is scanty, but in some cases flux appears to approach zero asymptotically as concentrations in excess of c_w are reached. Indeed, the achievement of concentrations greater than c_w implies that the latter value is not a true saturation concentration.

Second, the presence of a fluid-phase mass-transfer coefficient, k, in the flux equation indicates that control of fluid properties and flow conditions has an important effect in ultrafiltration system design. Known mass-transfer relationships can be used

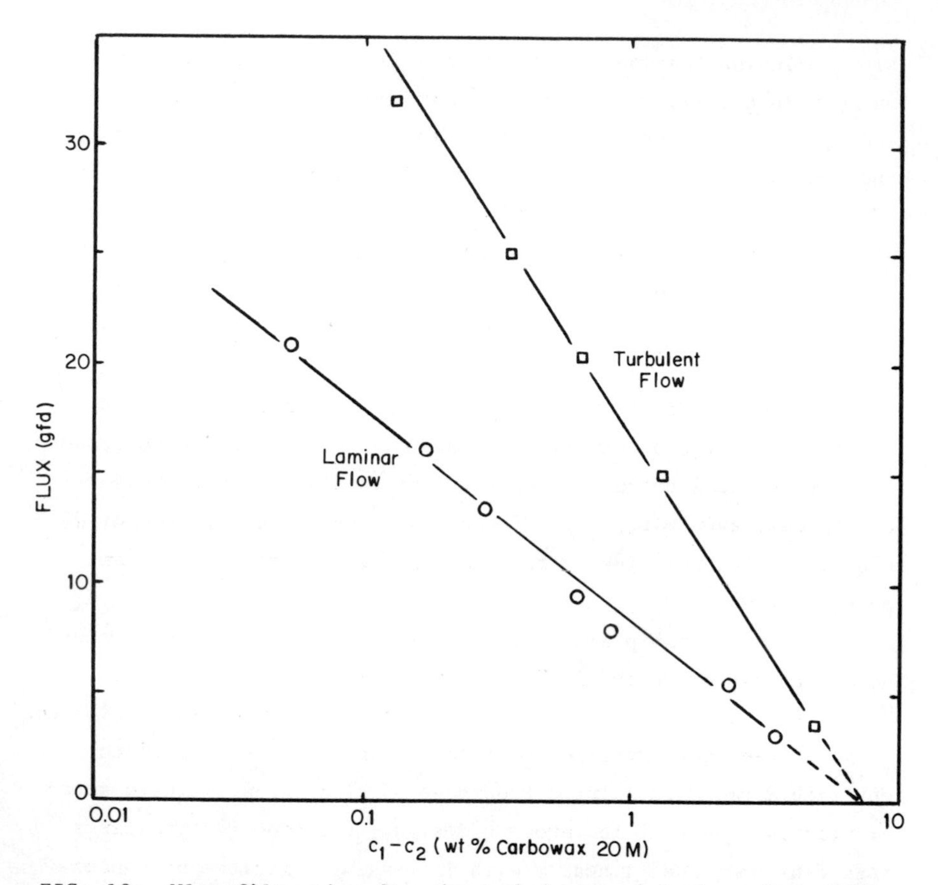

FIG. 10. Ultrafiltration flux in turbulent and laminar flow [11].

at least for purposes of data correlation. For example, in turbulent flow, the mass-transfer coefficient can be related to fluid properties and conditions by equation [7]:

$$\frac{kd}{D} = (\text{constant})_1 \ (\text{Sc})^{0.33} \ (\text{Re})^{0.9} \tag{5}$$

where d is the flow channel height, Sc the Schmidt number (kinematic viscosity divided by solute diffusivity), and Re the Reynolds number. If ultrafiltration flux is directly proportional to the mass transfer coefficient for a fixed polarization ratio (c_w/c_1), then the relationship between flux and flow Reynolds number should also conform to this same functionality. This has been demonstrated in a number of systems, such as that shown in Fig. 11.

In laminar flow, k can be related to flow conditions as follows [12]:

$$\frac{kd}{D} = (\text{constant})_2 \left(\text{Re Sc} \frac{d}{L}\right)^{1/3} \tag{6}$$

where L is the channel length. This has also been borne out experimentally (Fig. 12).

Convection promotion devices placed in the flow channel lead to enhanced fluxes, as would be expected [13]. In Fig. 13, a

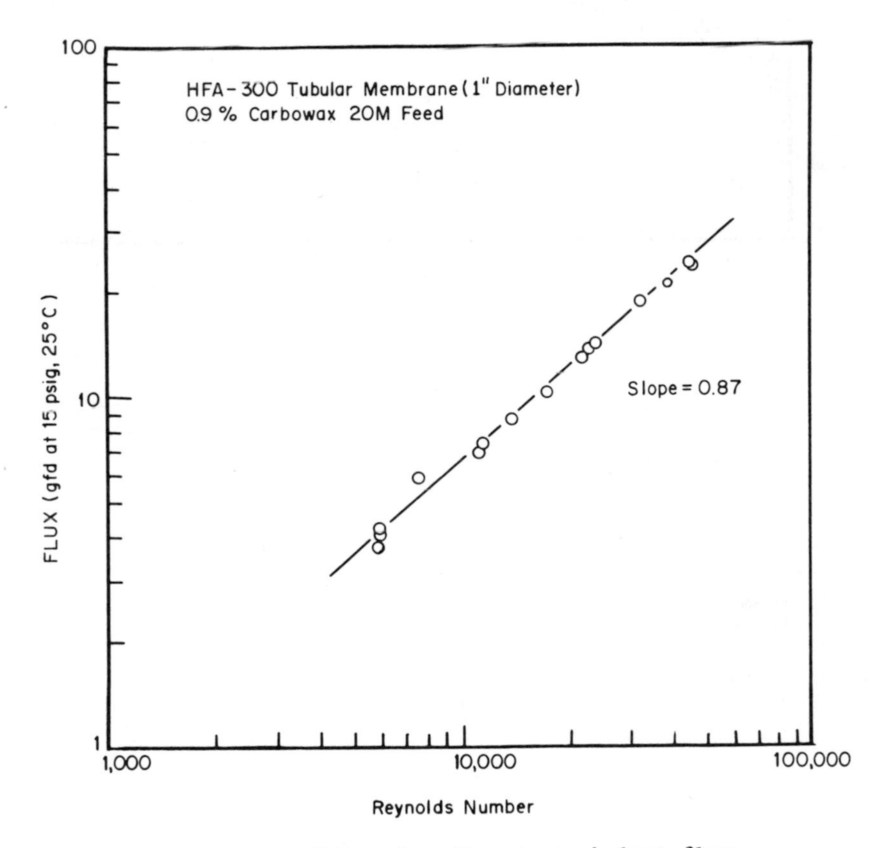

FIG. 11. Ultrafiltration flux in turbulent flow.

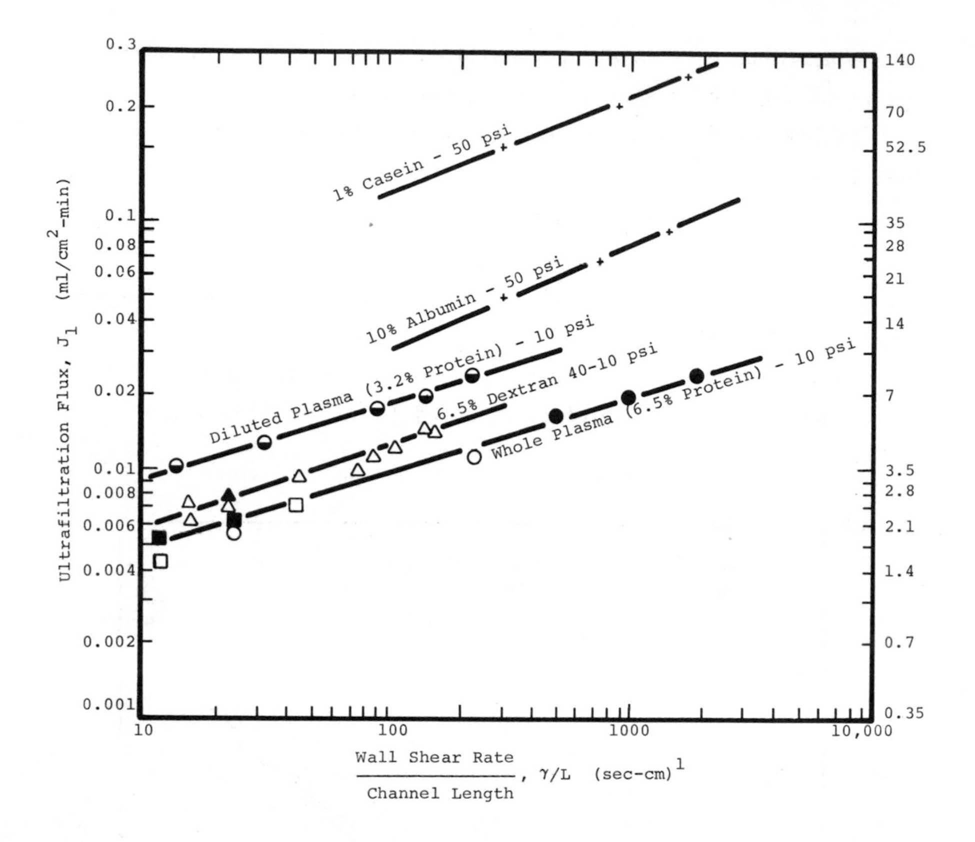

FIG. 12. Ultrafiltration rates in the gel-polarized region—dependence on wall shear rate and channel length. Data are for various macromolecular solutions in thin-channel laminar-flow recirculating cells [5].

Channel type		Channel length (cm)	Channel depth (in.)
+	Rectangular	41	0.010
●	Rectangular	6.5	0.005
▲	Triangular	6.5	0.017
■	Triangular	6.5	0.035
○	Triangular	13	0.005
△	Triangular	13	0.017
□	Triangular	13	0.035
◒	Tubular	40	0.008

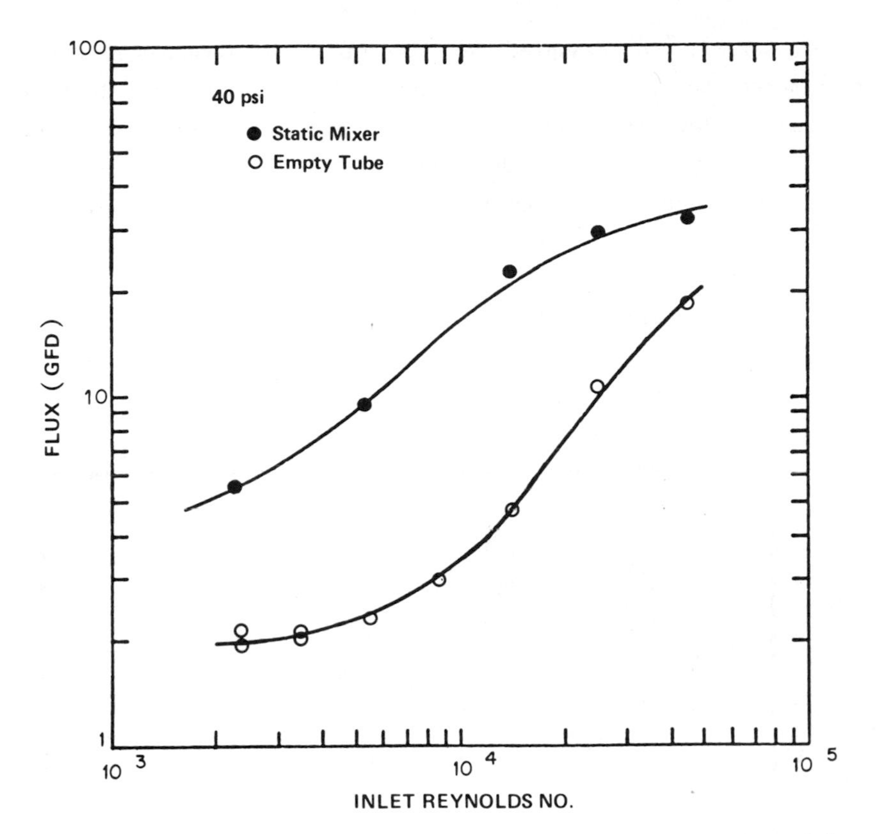

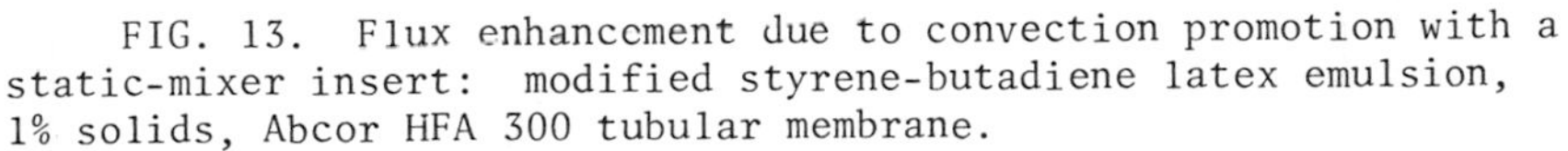
FIG. 13. Flux enhancement due to convection promotion with a static-mixer insert: modified styrene-butadiene latex emulsion, 1% solids, Abcor HFA 300 tubular membrane.

substantial flux improvement is shown both in transitional and turbulent flow using a static-mixer insert in a tubular membrane. The enhancement factor appears to decrease at flux levels that are substantially below the unpolarized membrane flux. The factors leading to this premature leveling of flux in the promoted membrane are yet to be explained. It should be noted that convection-promotion devices lead to increased flow resistance, which may be quite significant at high Reynolds numbers.

In ultrafiltration of suspensions of fine particles, such as colloids and latices, some deviations from predicted behavior have been observed. For example, the exponent on the Reynolds number-flux dependence is higher than expected [14]. For polymer latices in laminar flow, flux is proportional to Reynolds number raised to the 0.8-0.85 power, rather than the predicted 1/3 power. In turbulent flow, exponents as high as 1.3 have been observed, compared to the predicted value of 0.9. Also, the magnitude of diffusivities that are observed in these systems are much higher than those calculated for particles from the Stokes-Einstein relationship. However, the expected relationship between flux and logarithm of concentration is still obtained, as shown in Fig. 14.

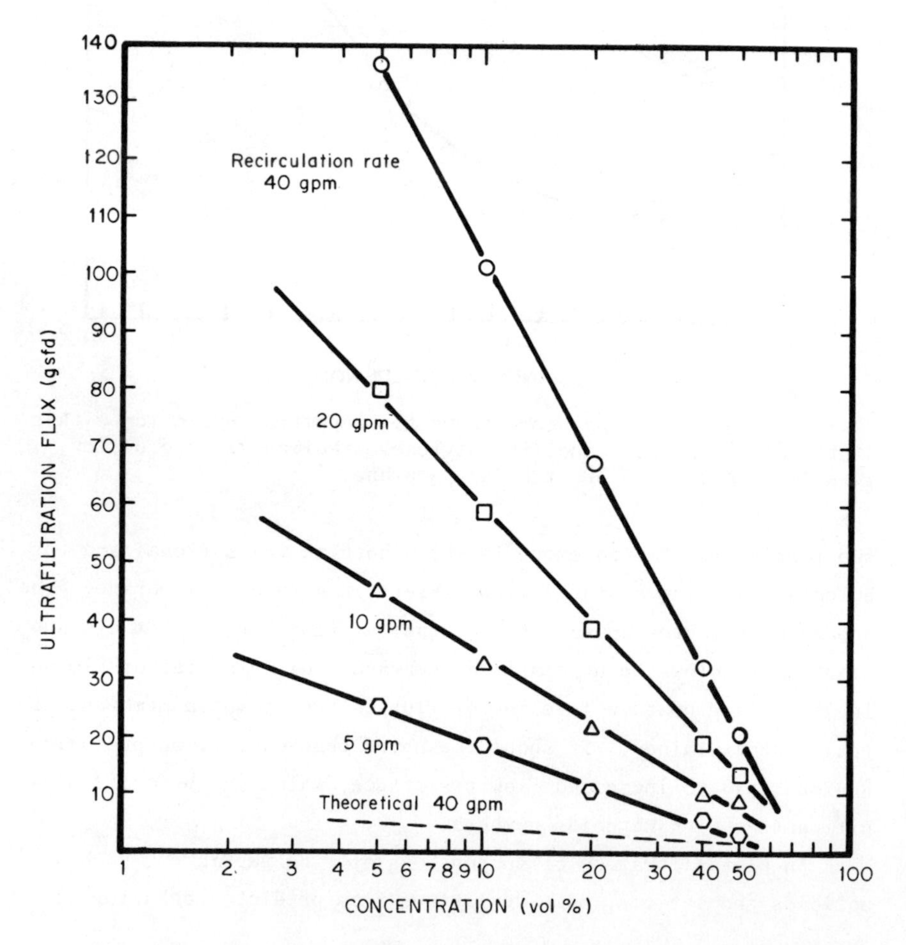

FIG. 14. Ultrafiltration of styrene-butadiene polymer latex vs. concentration.

The evidence seems to indicate that a nonaccumulating layer of solids forms at the membrane surface when ultrafiltering fine-particle suspensions, as is true with solutions of gel-forming macromolecular solids, and this layer is a major resistance to solution flux. However, the back diffusion of particles is significantly enhanced, probably by inertial forces aiding radial migration. This enhancement leads to greatly improved ultrafiltration fluxes, particularly at higher Reynolds numbers.

From a practical viewpoint, it is important to note that some ultrafiltration applications involve dissolved or suspended solids that tend to deposit at the membrane surface irreversibly. This fouling phenomenon is often slow enough to allow economic commercial operation by periodically cleaning the membrane surface and regaining high flux. Intermittent cleaning is carried out, not for solids recovery, but solely for removing accumulated foulants from the membrane.

In some membrane separation processes, the temperature dependence of flux has been observed to be quite strong. This strong dependence, however, is characteristic of solution-diffusion processes and not pore flow processes, such as ultrafiltration. In ultrafiltration, the temperature dependence of flux is relatively low, as shown in Fig. 15 [15]. Cost factors generally favor operation at ambient temperature.

B. Rejection

Membrane rejection, or the ability to reject or retain a given solute, is defined as

$$R = \frac{c_1 - c_2}{c_1} \tag{7}$$

A membrane with a rejection of 1.0 will produce a permeate in which $c_2 = 0$. A membrane with a rejection of 0 will produce a permeate with $c_2 = c_1$. Since it is believed that rejection of solutes in ultrafiltration is based on a sieving mechanism, it is appropriate

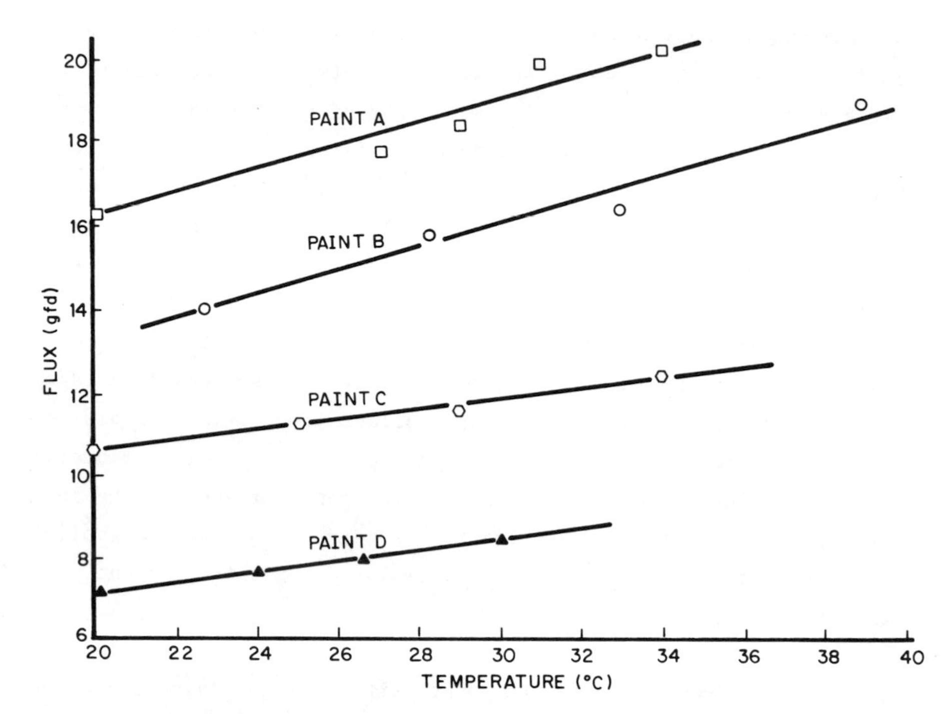

FIG. 15. Membrane flux vs. temperature for colloidal suspensions of electrocoat paint [14].

Paint	% Solids	Type
A	10	Koh
B	12	Amine
C	8	Amine
D	10	Koh

to define a sieving factor as [16]

$$\phi = \frac{c_2}{c_w} \tag{8}$$

where ϕ is dependent on molecular and pore dimensions only. It is evident that concentration polarization, where the upstream concentration at the membrane interface c_w is greater than the bulk stream concentration c_1, has a detrimental effect on rejection.

The constancy of the sieving factor ϕ can be tested by analysis of rejection data for dilute solutions where gel formation does not occur. Under these circumstances, c_w will vary as a function of upstream pressure and fluid dynamic conditions. For any system where c_2 is not negligible (rejection is less than 1.0), Eq. (4) becomes

$$J = k \ln\left(\frac{c_w - c_2}{c_1 - c_2}\right) \tag{9}$$

Combining Eqs. (7), (8), and (9), one obtains

$$\frac{J}{k} = \ln\left[\left(\frac{1 - \phi}{\phi}\right)\left(\frac{1 - R}{R}\right)\right] \tag{10}$$

When operating in turbulent flow, the mass-transfer coefficient k may be assumed to be proportional to the Reynolds number to the 0.8-1.0 power. Therefore, a plot of $J/Re^{0.9}$ vs. $(1 - R)/R$ should lead to a straight line, and this is illustrated in Fig. 16.

The validity of this relationship seems to confirm the concept of a constant sieving factor over a wide range of flow and pressure conditions for dilute solutions.

In concentrated solutions, where gel formation occurs and c_w is fixed, a constant sieving factor would lead to the conclusion that the permeate concentration c_2 should be constant, independent of c_1, since $c_2 = \phi c_w$. However, this is not borne out experimentally, as shown in Fig. 17. Rather, it appears that c_2 increases with c_1, giving a nearly concentration-independent rejection. Indeed, under some circumstances, R actually increases with increasing c_1. It is evident that the simplified sieving model does not hold under conditions of gel formation; rather, the presence of the gel has a role in determining rejection which is not yet understood. Indeed, in some cases, such as certain protein solutions, the gel layer appears to exhibit a solute rejecting capability over and above that of the membrane [17,18].

Observations of rejection as a function of pressure show other anomalies. In some concentrated solutions, for example, rejection has been observed to increase with increasing pressure, rather than remaining constant as the simple sieving model would dictate. In other systems, under different flow conditions, rejection has been

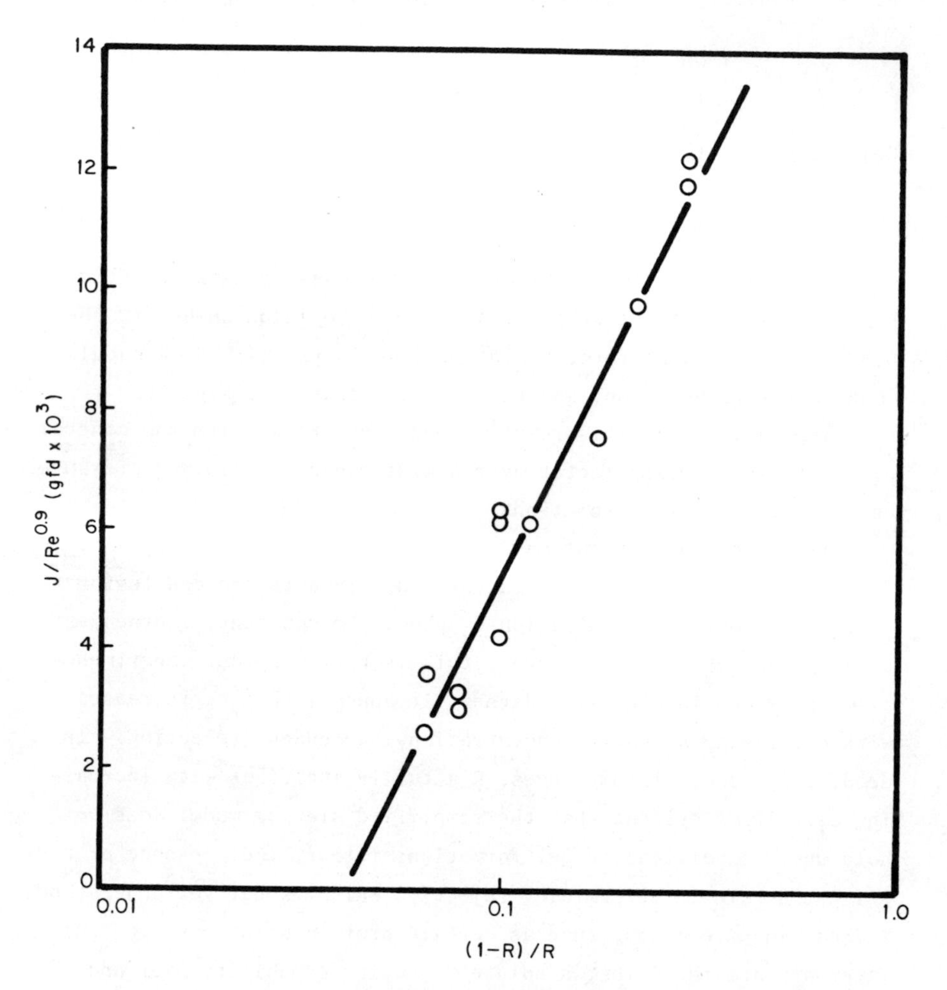

FIG. 16. Flux-rejection relationship for dilute solution. HFA-200 tubular membranes (1" diameter); 0.2% Dextran yellow feed; 25°C.

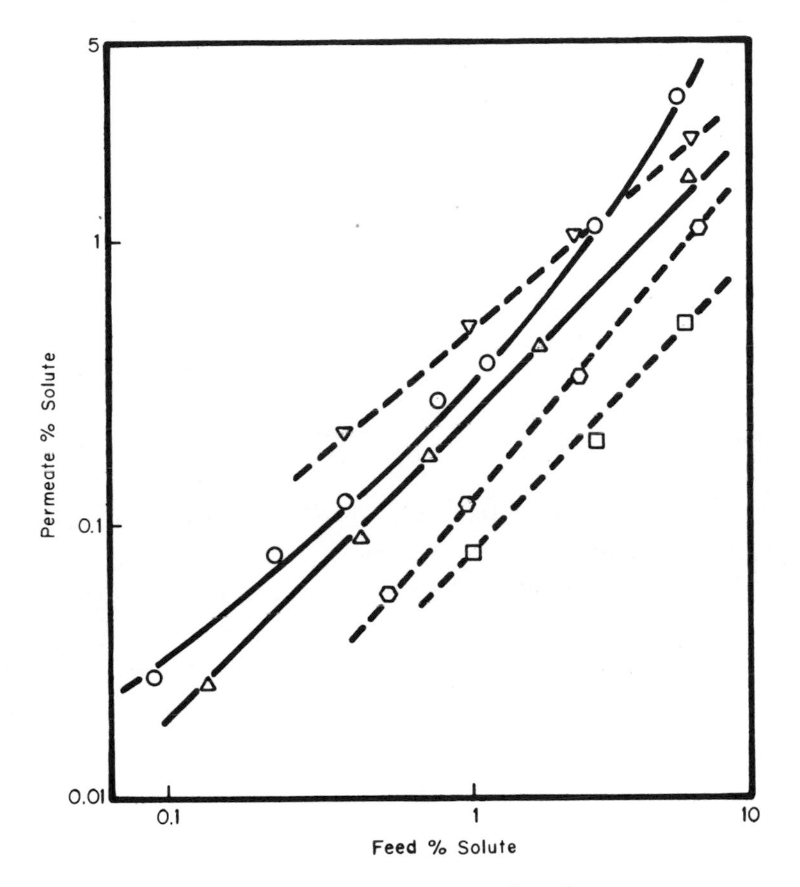

FIG. 17. Solute retention data.

△ Turbulent flow; HFA-200 tubular membrane; constant feed flow; 15 psi; 25°C

○ Laminar flow; HFA-300 membrane in flat cell; 60-mil channel height; constant feed flow; 10 psi; 25°C

▽ Stirred cell; HFA-200 membrane; 1200 rpm; 10 psi; 25°C; Dextran 20

⬡ Stirred cell; same conditions; Dextran 70

□ Stirred cell; same conditions; Dextran 110

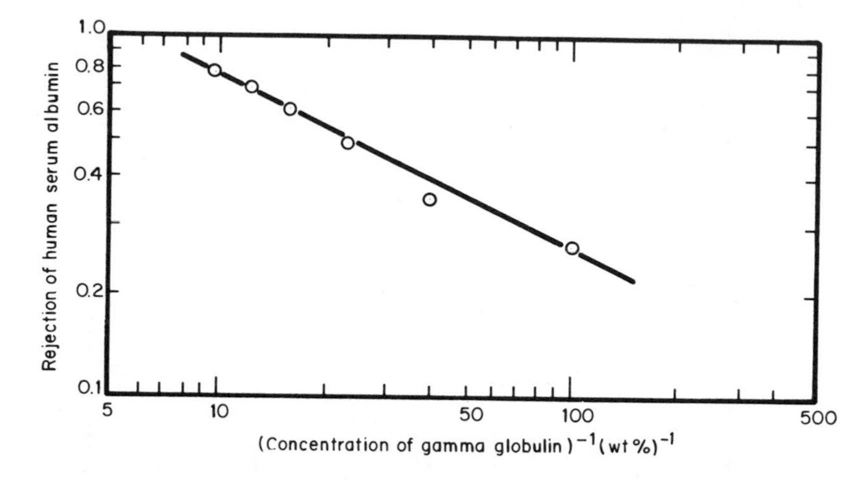

FIG. 18. The interference of gamma globulin with the passage of human-serum albumin through a Diaflo® XM-100 membrane.

observed to decrease with increasing pressure, also in a regime where gel formation would be expected.

The rejection behavior of membranes for solutions of mixed macromolecule solutes has been found to be different from that observed with solutions of each solute singly. For example (Fig. 18), with blood plasma proteins, the rejection of albumin in the absence of gamma globulin was approximately zero. As gamma globulin concentration was increased, albumin rejection increased. Apparently, the formation of a gel of gamma globulin at the membrane surface enhanced the rejection of the lower molecule-weight protein albumin.

In summary, while the influence of gel formation at the ultrafiltration membrane surface on flux is reasonably clearly established, the more complex effects of the gel structure on rejection have yet to be fully resolved.

V. ENGINEERING AND DESIGN OF SYSTEMS

At the present time, ultrafiltration membrane modules based on several different configurations are manufactured commercially. Each of these configurations is a compromise between performance characteristics, such as mass transfer, mechanical integrity, and membrane durability, on the one hand, versus minimum cost on the other. Some represent strongly contrasting approaches: laminar versus turbulent flow; and thin-flow channel versus large-diameter tubes.

It is difficult to readily show the influence of all mass transfer and hydrodynamic variables on ultrafiltration operating performance in different membrane configurations. However, by narrowing the number of independent variables through some arbitrary choices, an analysis of contrasting systems can be performed. Where gel formation is a limitation on flux, which is true for most cases of practical interest, increased ultrafiltration flux can be achieved at the expense of pumping power required to achieve high velocity, and thus high mass-transfer coefficients. Thus, one can examine at least the effects of (1) Reynolds number, and (2) flow-channel dimension. If fluid properties such as viscosity and solute diffusivity are fixed, then ultrafiltration flux and power consumption per unit volume of permeate are fixed once the channel dimensions are established. The results of one such analysis [8] show that both the ultrafiltration flux and power consumption are of the same orders of magnitude for two significantly different types of systems: turbulent-flow large-diameter tubular membranes versus laminar-flow thin-channel devices. Changes in certain parameters, such as fluid viscosity, will change the relative advantages of one versus the other. However, there are many cases for which the choice of membrane configuration will be dictated by other considerations, such as unit cost, hold-up volume, and ability to operate with suspended solids in the feed without plugging flow channels or fouling the membrane surface.

A. Tubular Membrane Configuration

Membranes in the form of open tubes of 0.5-1.0 in. diameter are in commercial use. Typically, the membrane is cast on the inside of a porous support structure which is capable of containing operating pressures of 50-100 psi. Pressures of this magnitude are often required to achieve high flow Reynolds numbers for high mass-transfer rates. In some cases, membranes are precast as cylinders and then inserted into the porous support tubes. In other cases, a "dynamically-formed" membrane is gelled in place by filtering a precursor suspension directly through the porous substrate and then initiating the membrane gelation reaction in situ. An example of a tubular ultrafiltration membrane system is shown in Fig. 19.

FIG. 19. Tubular ultrafiltration membrane unit.

Tubular membranes of large diameter have several advantages. Most notably, they are capable of processing streams with suspended solids without plugging of the flow path. Also, they can be cleaned fairly easily by flushing, an important consideration in food and dairy applications. Since tubes of this size must be operated in turbulent flow to achieve high mass-transfer coefficients, high wall-shear rates may prohibit their use for certain shear-sensitive macromolecules such as some enzymes. Also, large-diameter tubular systems have a large holdup volume, which can represent a problem in some batch operations.

B. Thin-Channel Membrane Modules

In this geometry, flat membrane sheets are fabricated into modules with careful control of the spacing between the membranes. Spacing must be sufficiently small to achieve high mass-transfer coefficients, but large enough so that pressure drop will not be overwhelming. A thin-channel system is shown in Fig. 20.

Thin-channel systems are operated in either laminar or turbulent flow. Laminar flow is used where construction allows careful control of channel height on the order of 0.01-0.02 in. With these lower flow rates, a high concentration per pass through the equipment can be achieved, and this sometimes eliminates the need for recirculation. Other thin-channel systems have spacings of the order of 0.05-0.1 in. and operate in turbulent flow to achieve good mass transfer. Both designs have the advantage of fairly low holdup volume and decreased shear, relative to the high Reynolds number used in tubular systems. Laminar-flow systems are also advantageous for highly viscous solutions. The disadvantage of any thin-channel system is its susceptibility to plugging with suspended solids and the attendant difficulty in cleaning.

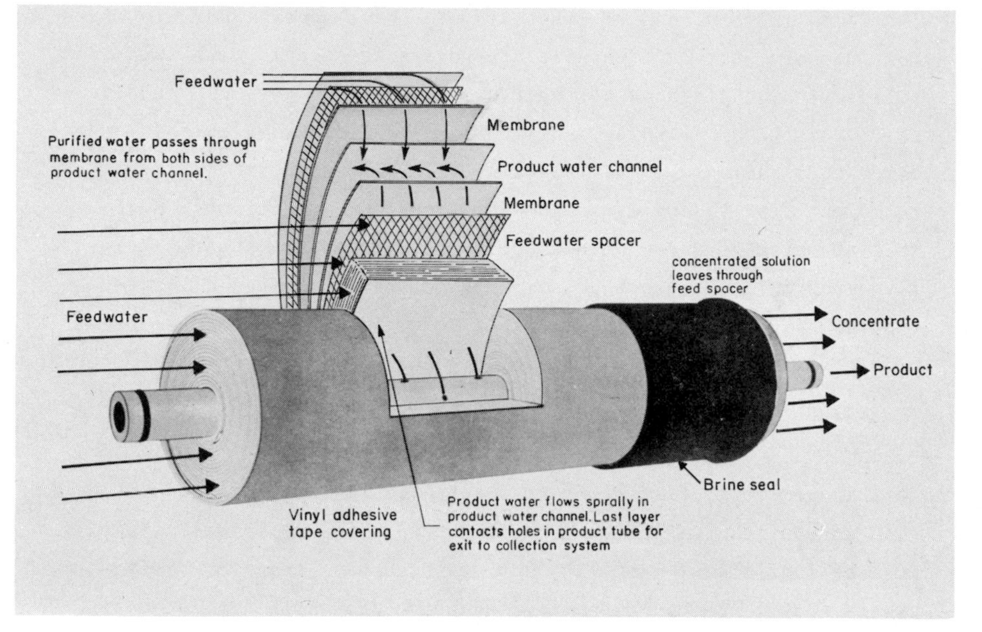

FIG. 20. Schematic cross-section of a spiral element showing water flow patterns.

C. Hollow Fiber (Capillary Membrane) Systems

High-flux hollow fiber membranes for ultrafiltration are a relatively recent development. These fibers have diameters as small as 0.001 in. Operated in laminar flow, hollow fiber systems have the advantages of thin-channel construction and are lower in cost per unit membrane area. As with other thin-channel systems, they suffer the disadvantage of plugging by suspended solids, and they may not conform to food and dairy sanitary standards. For applications where these criteria are unimportant, however, it is likely that hollow fiber ultrafiltration systems will be important in the equipment field in the future.

VI. APPLICATIONS

Present applications of ultrafiltration typically involve conversion of a low-value feed solution to one or more products of moderate to high value. For example, waste streams of zero or negative value are converted to concentrates of valuable by-products plus effluents which are acceptable for direct disposable. The higher cost of ultrafiltration, relative to ordinary particle filtration, is due to the low filtration flux and the special nature of the highly selective membrane. Use of this process for ordinary low-cost products, such as in water purification, is acceptable only where other alternatives simply are not feasible. Further extension of ultrafiltration to low-cost products will require new methods of flux improvement and/or reduced costs of manufacturing of membranes and related hardware.

A. Comparison with Alternative Processes

Industrial applications of ultrafiltration may fall into one of three categories: concentration, where the desired component is rejected by the membrane and taken off as a fluid concentrate; fractionation, for systems where more than one solute is to be recovered and products are taken from both the rejected concentrate and permeate; and purification, where the desired product is purified solvent (water), for applications such as potable water production.

For concentration, ultrafiltration has several advantages over competitive processes such as evaporation or solvent precipitation. Ultrafiltration requires no heat, which is therefore eliminated as a process utility requirement and allows higher yields of heat-sensitive materials. Relative to vacuum evaporation, existing information indicates that costs for ultrafiltration are less [19]. Its principal disadvantage relative to evaporation is the limitation on concentration levels that can be achieved, often due to viscosity constraints.

As a fractionation process, ultrafiltration often competes with adsorption processes. For solutions with solutes having very substantial differences in molecular weight, such as an order of magnitude or more, ultrafiltration may be the preferred alternative. Typically, adsorbents exhibit much higher selectivity than ultrafiltration membranes, and consequently are more effective for close fractionations. Ultrafiltration offers the advantage of being a continuous, noncyclic process, as opposed to adsorbents which require a regenerative cycle.

In contrast to biological oxidation processes for waste-water treatment which destroy organic matter, ultrafiltration allows by-product recovery from wastewaters. However, where recoverably by-products are not present, higher costs are a drawback against ultrafiltration relative to standard treatment processes.

B. Concentration and Fractionation of Cheese Whey

In many industrial processes, wastewater streams contain one or more valuable products that have been considered uneconomical to recover. This is particularly true in the food and dairy industries, where wastes from processing of various grains, soy beans, and potatoes, for example, represent major pollution problems for which no economic alternatives have existed to date. Because it can recover high molecular-weight by-products which can help pay for the pollution control cost, ultrafiltration represents an attractive alternative in this area.

Perhaps one of the most severe of these problems is that of disposal of cheese whey, a by-product from the production of cheese. For every pound of cheese produced, 5-10 lb of raw whey result. One-thousand gal/day of whey represents a sewage-treatment plant load equivalent to that of 1,800 people. About 5 billion pounds of whey are produced from cottage-cheese production operations in the United States. This so-called acid whey represents only a portion of the problem. In addition, cheddar-cheese manufacturing operations produce a similar pollutant known as sweet whey.

Whey contains 6% to 7% solids, with about 0.7% to 0.9% being nitrogenous, mostly protein. These proteins, β-lactoglobulin and lactalbumin, are high in nutrient value. In addition, whey contains 4.5 to 5% lactose, 0.2 to 0.6% acid, 0.5 to 0.6% ash, and between 0.05 and 5% fat. Recovery of all of these solids by evaporation is possible, giving a dried whey product with 12 to 14% protein. The value of this product, however, is low because of its relatively low protein content. If a whey solids product with higher protein content could be obtained, it would represent a valuable by-product from cheese-making operations, and its sale would recover part or all of the cost of the pollution control operation.

A two-stage membrane process, combining ultrafiltration with reverse osmosis, is capable of producing a protein concentrate of high value while reducing the pollutant content of whey by about 95% or more [20]. A schematic flow sheet of the process is shown in Fig. 21. The ultrafiltration step utilizes membranes that freely transmit lactose, nonprotein nitrogen, lactic acid, and ash, while producing a concentrate with a high protein content relative to other solids. The permeate from the ultrafiltration step is then processed by reverse osmosis, which produces a lactose solution

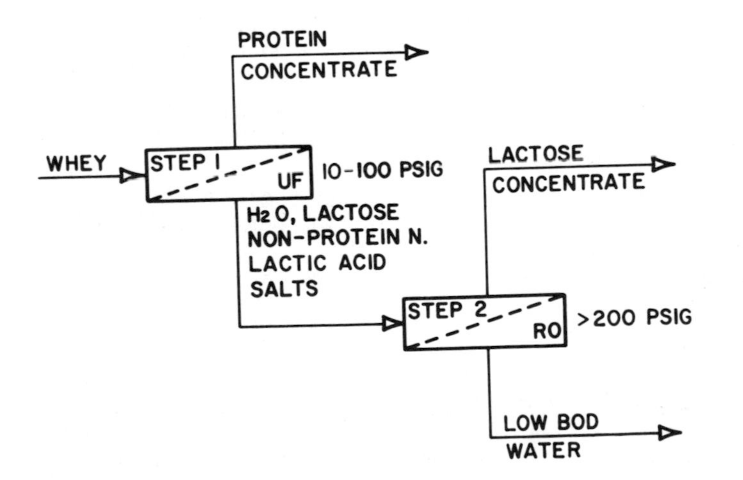

FIG. 21. Membrane process for whey treatment—flow schematic.

as a concentrate and water low in organic pollutants as a permeate.

The selectivity of the ultrafiltration membranes is shown by feed and permeate chromatographic analyses in Fig. 22. The permeate is essentially free of whey proteins but contains most of the low molecular-weight components. The ultrafiltration membrane has a protein rejection of 0.98, a lactic acid rejection of 0.1 to 0.2, and lactose, ash, and nonprotein nitrogen rejections of essentially zero. The protein can be concentrated about 20-fold, and the resulting concentrate contains about 80% protein based on dry solids (Fig. 23). This product, which contrasts to a solid product of 12 to 14% protein content (based on total whey solids) obtained by evaporation, has shown acceptable nutritional and functional properties as a food additive.

C. Recovery of Electrocoat Paint

Several industrial applications of ultrafiltration focused largely on process rather than pollution control have been quite successful. An example is the use of ultrafiltration in conjunction with electrocoat painting [20]. In electrocoating, the metal part to be

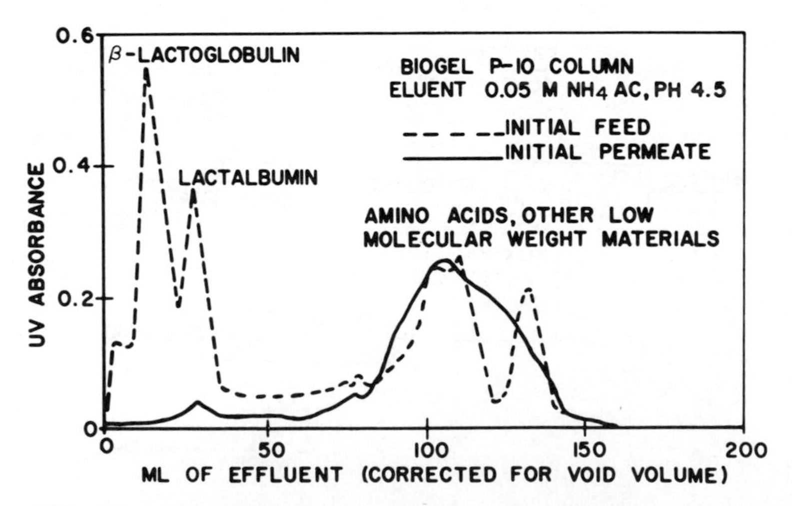

FIG. 22. Ultrafiltration section, gel permeation chromatography data.

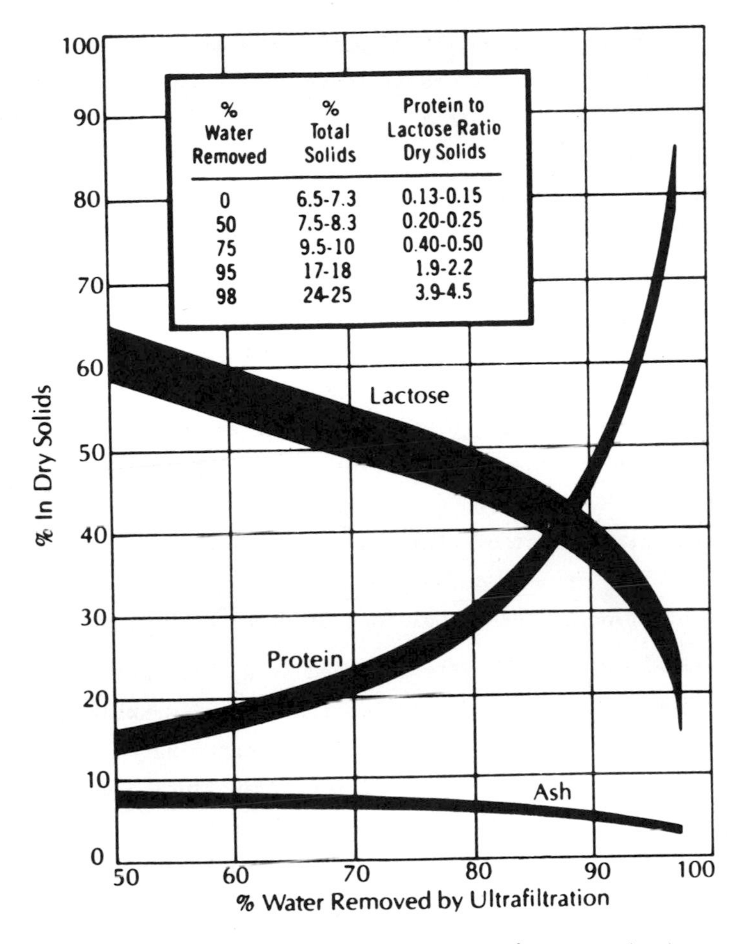

FIG. 23. Composition of whey protein concentrate.

coated is immersed in a bath containing charged paint-resin particles. A potential difference is established between the metal part and the bath tank wall, causing migration of paint particles to the part surface.

Three problems exist with the electrocoating process which are alleviated by the application of ultrafiltration. First, foreign ions are brought into the tank bath by the metal part, causing an upset in the ionic balance which maintains the colloidal suspension of paint particles, and eventually destroying the entire paint batch.

Second, a significant quantity of paint is lost as "drag out" entrained with the metal part as it leaves the bath. This loss is reflected both as a paint cost as well as a pollution problem. Third, large quantities of deionized rinse water are required to remove excess paint from the metal part.

An ultrafiltration unit can be placed in a recirculating line from the main paint tank (Fig. 24) and serves to effectively remove foreign ions. The paint particles are completely rejected by the ultrafiltration membranes, and a solution of low molecular-weight ions is removed as permeate. Fresh deionized water is used to make up for losses in solution volume due to permeation. Paint drag-out can similarly be recovered by treatment with ultrafiltration. Finally, permeate can be used as a prerinse for the metal part, reducing the total amount of rinse water required.

D. Concentration of Spent Emulsions of Metal Machining Oil

Metal cutting and rolling oils are used as dilute emulsions (so-called soluble oils) to cool and lubricate metal-working machine surfaces. The spent lubricant is heavily contaminated with metal fines and dirt and represents a major water-pollution problem.

After pretreatment to remove solids by settling, ultrafiltration is capable of concentrating the emulsion from several percent oil to levels as high as 90% [21,22]. The concentrate can be disposed by combustion or further concentrated by centrifugation or evaporation. The permeate contains less than 50 ppm oil, and it can be suitable either for reuse with fresh oil or direct disposal by drain.

Compared to other emulsion-breaking methods, ultrafiltration appears to have several advantages. No chemical addition is required, and the process appears to be insensitive to the chemical type of lubricant. Both the high concentrations achievable in the oil-rich stream and the relative purity of the water effluent minimize the costs of posttreatment or disposal. Also, the presence of a mechanical barrier, the membrane, reduces the potential problem

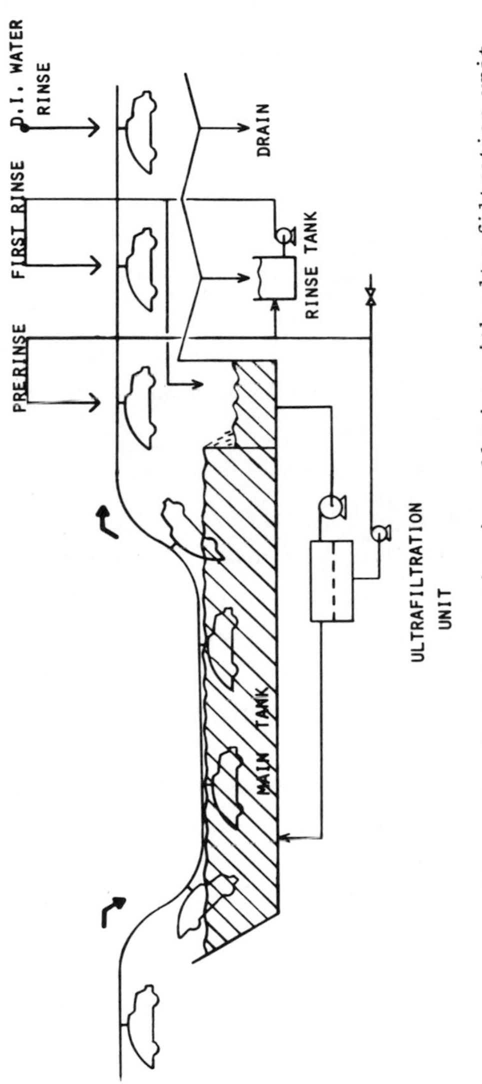

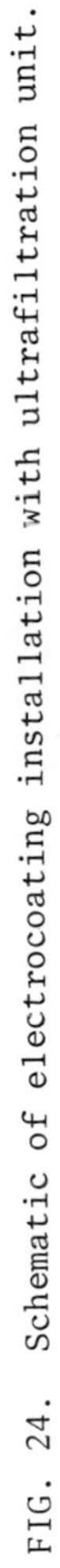

FIG. 24. Schematic of electrocoating installation with ultrafiltration unit.

of stream contamination (release of oil emulsion into the water effluent) by operating error.

E. Recovery of Polyvinyl Alcohol Sizing from Textile Mill Desizing Wastes

The textile industry uses on the order of forty million pounds of polyvinyl alcohol (PVA) as warp sizing. After production of the woven goods, the sizing is washed from the fabric, and the wash water is discarded.

Ultrafiltration has been introduced to concentrate PVA in desizing wastes from 0.5-1.0% to about 10% [23]. The recovered PVA, with 10% makeup added to compensate for losses, has been successfully recycled for reuse as sizing. The removal of PVA from the water effluent also relieves the load on downstream treatment of the textile-mill wastewaters.

F. Other

A number of other ultrafiltration applications have achieved at least pilot-scale operation. In the food processing industry, these include concentration of skim milk proteins, fractionation of soy whey, concentration of gelatin, and recovery of yeast processing wastes. In the production of biologicals, recovery of enzymes from fermentation broths and concentration of virus have been described. The process has also been shown to be effective for concentrating textile dye suspensions, acrylic lattices, and pulp and paper manufacturing wastes. Industrial laundry waste treatment, involving the concentration of surfactant molecules by ultrafiltration, has also been demonstrated.

NOTATION

c Concentration

d Flow-channel height

D Diffusivity

J Flux

k Mass-transfer coefficient (in linear velocity units)

K_m Membrane permeability coefficient

L Flow-channel length

P Pressure

R Rejection (Eq. 7)

Re Reynolds number

Sc Schmidt number (kinematic viscosity divided by diffusivity)

x Distance normal to the membrane surface

Greek Letters

δ Mass-transfer boundary-layer thickness

ϕ Sieving factor (Eq. 8)

Π Osmotic pressure

Subscripts

1 Feed (upstream) side of membrane

2 Permeate (downstream) side of membrane

w Wall (upstream membrane surface)

REFERENCES

1. S. Loeb, in *Desalination by Reverse Osmosis* (U. Merten, ed.), M.I.T., Cambridge, 1966, p. 55.
2. D. I. C. Wang, A. J. Sinsky, and T. A. Butterworth, in *Membrane Science and Technology* (J. E. Flinn, ed.), Plenum, New York, 1970, p. 108.
3. C. J. Van Oss and P. M. Bronson, in *Membrane Science and Technology* (J. E. Flinn, ed.), Plenum, New York, 1970, p. 141.
4. H. K. Lonsdale, in *Desalination by Reverse Osmosis* (U. Merten, ed.), M.I.T., Cambridge, 1966, p. 93.
5. B. Kunst and S. Sourirajan, *J. Appl. Polym. Sci.*, *18*, 3423 (1974).
6. O. Kutowy and S. Sourirajan, *J. Appl. Polym. Sci.*, *19*, 1449 (1975).
7. R. E. Treybal, *Mass Transfer Operations*, McGraw-Hill, New York, 1955.

8. W. F. Blatt, A. Dravid, A. S. Michaels, and L. Nelson, in *Membrane Science and Technology* (J. E. Flinn, ed.), Plenum, New York, 1970, p. 47.

9. R. P. de Filippi and R. L. Goldsmith, in *Membrane Science and Technology* (J. E. Flinn, ed.), Plenum, New York, 1970, p. 33.

10. P. L. T. Brian, in *Desalination by Reverse Osmosis* (U. Merten, ed.), M.I.T., Cambridge, 1966, p. 161.

11. A. A. Kozinsky and E. N. Lightfoot, *A.I.Ch.E. J.*, *18*, 1030 (1972).

12. T. K. Sherwood, P. L. T. Brian, R. E. Fisher, and L. Dresner, *Ind. Eng. Chem., Fundam.*, *4*, 113 (1965).

13. A. L. Copas and S. Middleman, *Ind. Eng. Chem., Process. Des. Develop.*, *13*, 143 (1974).

14. M. C. Porter, *Ind. Eng. Chem., Prod. Res. Develop.*, *11*, 234 (1972).

15. B. J. Weissman, "Optimization of Ultrafiltration in Electrocoating," presented at *Electrocoat '72*, sponsored by the Electrocoating Seminar Committee and the National Paint and Coatings Association, Inc.

16. R. L. Goldsmith, *Ind. Eng. Chem., Fundam.*, *10*, 113 (1971).

17. M. C. Porter and L. Nelson, in *Recent Developments in Separation Science*, vol. II (N. N. Li, ed.), CRC Press, Cleveland, 1975, 239.

18. F. Bellucci and E. Drioli, *J. Appl. Polym. Sci.*, *19*, 1639 (1975).

19. M. C. Porter and A. S. Michaels, *Chem. Tech.*, Jan., 56 (1971).

20. R. L. Goldsmith, R. P. de Filippi, S. Hossain, and R. S. Timmins, in *Membrane Processes in Industry and Biomedicine* (M. Bier, ed.), Plenum, New York, 1971, p. 267.

21. R. L. Goldsmith and R. P. de Filippi, in *Membranes in Separation Processes*, Proceedings of a Workshop Symposium sponsored by the National Science Foundation, Case Western Reserve University, 1973, p. 157.

22. R. L. Goldsmith, D. A. Roberts, and D. L. Burre, *J. Water Pollut. Contr. Fed.*, *46*, 2183 (1974).

23. Anonymous, *American Dye Reporter*, October, 1974, p. 71.

AUTHOR INDEX

Abrikosova, I. I., 116
Adzumi, H., 133, 138
Aiba, S., 6, 79, 82, 264
Albrecht, E., 133
Albrecht, F., 19, 43
Albrecht, J., 123, 124
Alciatore, A. F., 445
Aleinikova, I. N., 116, 117
Algren, A. B., 33
Almlof, J. W., 38, 83
Amelin, A. G., 47
Anderson, D. M., 55, 128
Anderson, W. L., 47, 82, 91, 92
Andrews, B. D., 264, 267
Aoi, T., 22, 23
Aragaki, T., 394, 403, 417
Arkell, G. M., 227
Asset, G., 48
Aynsley, E., 117

Bachmann, W., 129, 130
Backer, S., 262, 274, 275, 276, 280
Backheuer, K., 101, 102
Bairstow, L., 22
Bakanov, S. P., 61, 64
Baker, W. E., 440
Bakker, P. J., 257, 260
Ball, W. H., 47
Bartell, F. E., 129
Baskerville, R. C., 185, 234
Basmanov, P. I., 55, 56
Bauer, J. H., 129, 130
Baumann, E. E., 290
Bays, L. R., 242
Bechhold, H., 129, 130
Bellucci, F., 501
Belyakov, M. I., 47
Ben Aim, R., 270
Benarie, M., 13, 14, 16, 90, 96, 98, 99
Benn, D., 242
Beutelspacher, H., 134, 143
Bevis, D. A., 329
Bigelow, S. L., 129
Bijosowa, W. I., 129
Biles, B., 101, 103
Billings, C. E., 29, 31, 33, 118, 120, 123, 125, 310, 320, 346
Binek, B., 37, 95, 97, 98, 142
Bixler, H. J., 440
Bjerrum, N., 133
Blaschke, R., 143
Blasewitz, A., 33, 82
Blatt, W. F., 488
Blodgett, K. B., 43, 44, 49
Bondarenko, V. S., 127
Booth, R. B., 219, 220, 221, 222
Borisichina, V. J., 129
Botterill, S. M., 117
Boudreau, W. A., 13
Boussinesque, J., 36
Boylan, D. R., 189
Bradley, D., 185
Bradley, R. S., 115
Bredee, H. L., 293
Brenner, H., 22, 23, 26, 27, 30, 268, 269, 284
Brian, P. L. T., 491, 495

Brink, J. A., 95, 337
Brinkman, H. C., 22, 33, 269, 270, 284
Bronson, P. M., 483
Broughton, G., 272
Brown, N., 129
Brubaker, D. W., 137
Brun, E. A., 70
Brun, R., 43, 44
Brun, R. J., 43
Burchsted, C. A., 329
Burggrabe, W. F., 95, 337
Burre, D. L., 514
Burton, E. F., 211
Butler, G. M., 97, 98
Butler, J. A. V., 115
Butterworth, T. A., 481

Camp, T. R., 205
Campbell, W. B., 21
Caplan, K. J., 320
Carman, P. C., 20, 21, 176
Carnell, P. H., 129
Casimir, H. B. G., 115
Cassidy, H. G., 129
Cavanagh, P., 37
Chamberlain, A. C., 41
Chandrasekhar, S., 10
Chapman, D. L., 205
Chapman, S., 61, 94, 95
Chen, C. Y., 6, 19, 31, 33, 50, 76, 79, 82, 83, 84, 85, 88, 92, 93
Cherry, G. B., 82
Chrosciel, S., 4, 107, 122
Church, T., 48
Clarenburg, L. A., 12, 13, 17, 21, 65, 66, 94, 264, 267
Clifford, R., 47
Cline, H. E., 129, 130
Cochet, R., 54
Collins, R. E., 20
Conley, W. R., 235
Conners, E. W., 55, 128
Cooper, H., 187, 189
Cooper, H. R., 399, 405, 417
Copas, A. L., 495
Corn, M., 115, 116, 117
Corte, H., 13
Coudeville, H., 70
Coulson, J. M., 176
Cove, B. M., 22
Cowling, T. G., 61, 94, 95
Cox, H., 138
Creager, M. O., 72
Croll, B. T., 227
Culhane, F. R., 310, 320
Cunningham, C. E., 272
Cussler, E. L., 140

Dahlstrom, D. A., 185, 440
D'Amico, G., 143
D'Ans, A. M., 134
Darcy, H. P. G., 173
Daubner, I., 129
Davies, C. N., 6, 19, 22, 23, 27, 29, 31, 33, 37, 43, 45, 80, 84, 126, 194, 263, 267, 273, 284
Davis, R. E., 141
Dawkins, G. S., 55
Dawson, S. V., 10, 28, 29, 30, 33, 46, 80
Deb, A. K., 290
de Filippi, R. P., 488, 505, 511, 512, 514
de Loureiro, J. A., 129
Denton, M. J., 270
Deryaguin, B. V., 11, 54, 61, 64, 116, 117, 198
Desorbo, W., 129, 130
Detwiller, C. G., 86, 88, 144
Devienne, M., 18, 61, 62, 133, 137, 138
Dianova, E. V., 129, 130
Di Bari, I. L., 143
Dickey, G. D., 253
Dmitrieva, T. F., 295
Dobie, W. B., 253
Dobry, R., 37
Doe, P. W., 242
Dofield, J. D., 329
Doi, H., 427
Dollinger, L. S., 310
Donovan, D. T., 129, 130
Dorman, R. G., 6, 16, 62, 81, 83, 118
Dravid, A., 488
Dresner, L., 495
Drioli, E., 501

Duchin, S. S., 11, 54
Duclaux, J., 138
Dupuit, A. J. E. J., 176

Eggerth, A. H., 129
Elford, W. J., 129, 130
Ellison, J. McK., 101, 103
Elphick, A., 229
Emersleben, O., 26, 268, 284
Emi, H., 43, 44, 45, 46, 49, 55, 81, 104, 118, 121, 126
Engelbrecht, H. L., 337
Engelhard, H., 6, 93
English, J. E., 296
Epstein, S. P., 61, 63
Erbe, F., 132
Errera, J., 138
Ettinger, H. J., 329

Farrow, R. M., 12
Faxen, H., 22
Fairs, G., 127, 129
Ferry, J. D., 129, 140
Finn, R., 37
Finn, R. K., 27
First, M., 33
First, M. V., 97, 98
Fischer, W. H., 134
Fisher, R. E., 495
Fitzgerald, J. J., 86, 88, 144
Fleischer, R. L., 146
Flood, J. E., 369
Fortier, A., 7
Fowler, J. L., 14, 21
Frank, E., 134
Frank, E. R., 148
French, R. C., 253
Freshwater, D. C., 48, 85, 93
Freundlich, H., 86, 88
Friedlander, S. K., 6, 35, 47, 74, 77, 78, 79, 84, 85, 86, 87
Friend, A. G., 82
Fromme, H. G., 134
Fuchs, N. A., 6, 15, 19, 24, 26, 27, 28, 31, 32, 36, 37, 38, 39, 40, 44, 46, 47, 50, 68, 76, 78, 81, 82, 86, 88, 90, 92, 139, 203
Fujikawa, H., 23, 27
Fukaya, S., 427
Fukushima, M., 44, 45, 49
Fuller, A. B., 329

Gaden, E. L., 82, 92
Gale, R. S., 234, 235
Gallily, I., 100
Gardner, J. O., 134
Gärtner, H., 134
Gelman, Ch., 129, 130
Gemberling, A., 129
Gillespie, T., 40, 51, 52, 55, 117, 118
Glauert, M., 43
Glavin, T. P., 82
Glucharev, G. P., 129
Glushkov, Yu. M., 13, 73
Goldschmidt, W. W., 45
Goldsmith, R. L., 488, 500, 505, 511, 512, 514
Golovin, M. N., 41, 44
Gonsalves, H., 289
Goodwin, G., 72
Goren, S. L., 7, 14, 25, 29, 30, 32, 36, 40, 60, 64, 65
Gouy, G., 205
Goyer, G. G., 55
Grabar, P., 129, 132
Grace, H. P., 189, 260, 263, 267, 292, 397, 446
Graham, J. B., 97, 98
Green, D. J., 258, 266, 273, 283, 284, 287
Green, H. L., 6, 82
Green, T. C., 397
Greenburg, L., 119
Greenwell, L. E., 95
Gregory, J., 115, 116, 196
Gregory, P. H., 47
Griffiths, P. V. R., 272, 273, 279, 283, 284, 297, 298
Gruen, R., 55
Guerout, M., 133, 136
Grüner, P., 100
Gupalo, Yu. P., 7, 26, 111
Gutowski, W., 107, 109

Haagen-Crodel, B., 143
Hackel, A., 427

Hall, J. R., 129, 130
Hamaker, H. C., 59
Hameed, M. S., 462
Han, C. D., 440
van der Hans, H., 290, 293
Hampl, V., 12, 133, 134, 135, 137, 138
Hansmann, G., 134
Happel, J., 22, 23, 24, 26, 27, 30, 31, 268, 269, 284
Hardy, W. B., 211
Harris, W. B., 320
Harrop, J. A., 46, 48, 85, 93, 195
Hasenclever, D., 82
Hasimoto, H., 28, 146
Hatch, T., 119
Hatschek, E., 291
Hattori, M., 55
Haupt, C. G., 47, 100
Havlicek, V., 56, 130
Hayashi, F., 343
Healey, R., 197, 219, 235
Heertjes, P. M., 257, 260, 261, 262, 290, 293, 294
Helmcke, J. G., 134
Hemeon, W. C., 119
Henneberg, G., 143
Hermann, K., 47
Hermans, P. H., 293
Hertel, K. L., 14, 21, 33, 266, 273
Hess, G. E., 134
Hibou, J. L., 257, 260
Hidy, G. M., 58
Hixson, A., 300
Hochrainer, D., 58
Hodge, R. P. J., 227
Hofmann, R., 140
Honold, E., 13, 132
Horrak, H., 427
Hossain, S., 511, 512
Hounam, R. F., 101, 103
Householder, M. K., 45
Hrbek, J., 137, 138, 146, 148
Huang, C. J., 403
Hughes, T. P., 129, 130
Huitt, H. A., 133, 138
Humphrey, A. E., 6, 82, 92
Husmann, W., 243
Hutchins, T. G., 48
Hutson, V. C. L., 286
Hutto, F. B., Jr., 189, 417
Hyde, R. R., 138

Iberall, A. S., 30, 178, 284, 286
Ichimura, K., 417
Ignatiev, V., 47
Iinoya, K., 56, 324, 329, 340, 341, 343
Ikamura, S., 189
Illingworth, C. R., 22
Imaizumi, T., 243
Imamura, T., 329
Imai, I., 22, 23
Ingmanson, W. L., 264, 267
Inone, T., 243
Inoue, O., 329
Ionya, G., 264
Ison, C. R., 195
Ives, K. J., 195, 196, 235, 290

Jakob, M., 94
Jacobs, S., 129
Jander, G., 129
Jarman, R., 47
Jech, C., 101, 102, 139
Jelenz, H., 427
Jilk, L. T., 129
Johnson, R. C., 264, 267
Johnstone, H. F., 35, 47, 53, 79, 82, 84, 111
Jones, G. G., 130
Jordan, R. C., 33, 47
Juda, J., 4, 107, 122
Judson, B., 33, 82

Kaplan, A., 296
Kaplun, S., 22, 23
Kammermayer, K., 137
Kangro, C., 6
Kanig, G., 134
Kato, H., 394, 403, 417, 427
Kehat, E., 296
Keller, J. B., 16, 23, 26, 109
Kempke, L. L., 462
Kester, B. E., 310, 320
Kestner, N, R., 115
Khoo, H. E., 266, 273, 283, 284, 287

Kimura, N., 33, 264, 340, 341, 343
Kintner, R. C., 267, 273
Kirsch, A., 24, 28, 32, 37, 38, 40, 46, 47, 50, 68, 81, 82, 86, 88, 90, 92, 146, 148
Kitchener, J. A., 116, 235
Knuth, R. H., 141
Kolganov, V. A., 38, 40, 72, 79, 86, 88, 89, 125, 127
Komagata, S., 133
Kormendy, I., 427
Kottwitz, F. A., 189
Kovacs, J. P., 253
Kozeny, J., 21, 175
Kozinsky, A. A., 493
Kraemer, H. F., 53
Kragh, A. M., 225, 227
Kraybill, R. R., 272
Krüger, A. P., 129
Krupp, H., 115, 116, 117, 118
Kubalski, F. E., 134
Kubie, G., 139
Kuloor, N. R., 43, 44
Kunst, B., 483, 488, 496
Kurvabara, S., 13, 16, 23, 24, 28
Kurz, R., 134
Kutowy, O., 483, 484, 485, 490
Kuzutaka, M., 56
Kyan, C. P., 267, 273

Labrecque, R. P., 13
Lake, L. J., 229
Lamb, H., 22, 27
La Mer, V., 55, 197, 219, 235
Lancaster, B. W., 97
Landahl, H., 47
Landau, L., 198
Landt, E., 84, 86, 87
Lane, W. R., 6
Lang, E. D., 22
Langbein, D., 116
Langmuir, I., 21, 35, 40, 43, 44, 49, 62, 77, 83, 84, 86, 87
Langston, W. B., 225, 227
Langstroth, G. O., 117
Lapple, C. E., 82
Lazarev, K. A., 118
Leers, E., 85
Leers, R., 123
Le Goff, P., 270
Le Lec, P., 270
Lemlich, R., 26
Lennard-Jones, J. E., 115
Leonard, R. A., 26
Lerk, C. F., 29
Leveque, M., 139
Levich, V. G., 139
Levin, L. M., 7, 41, 45, 48, 53
Lewis, J., 47
Lewis, J. B., 242
Lewis, W., 43, 44
Lewis, W. K., 36
Lifshitz, E. M., 116
Lightfoot, E. N., 493
Lin, A., 296
Lindeken, C. L., 93
Linke, W. F., 219, 220, 221, 222
Liu, C. Y., 69
Liu, Y. H., 51, 52, 55
Lloyd, P. J., 440
Lodge, J. P., 133, 134, 138, 143, 148
Loeb, A. L., 210
Loeb, S., 481, 483
Loefler, A. L., 26, 31, 268, 269, 284
Loesche, W. J., 130
Löffler, F., 118, 121
London, F., 115, 199
Lonsdale, H. K., 483
Lord, E., 267, 273
Lössner, V., 101, 102
Lu, Wei-Ming, 397
Lundgren, D. A., 47, 55
Lyon, W. A., 243

Maas, O., 21
Mackrle, V., 290
Machacova, J., 137, 138, 148
Madelaine, G., 101, 103, 143
Madock, M. J., 47
Maier, K. H., 134
Makino, K., 56, 324, 329
Manegold, E., 129, 133, 140
Margenau, H., 115
Martius, Ch. H., 6
Mason, M. G., 320
Matherson, G., 12
Matsumura, H., 45, 46, 81, 104

Maxson, W. D., 82
Maxwell, C., 61, 63
May, K. R., 47
Mazin, I. P., 7, 41, 43, 44
McFarland, A. R., 47
McGregor, R., 266, 269, 271, 283
Megaw, W. J., 37, 139, 142, 145
Meissner, H. P., 335
Mergler, H. W., 43
Merten, V., 134
Meyer, H. C. W., 112
Michaels, A. S., 440, 488, 509
Mickley, H. S., 335
Miczek, G., 120
Middleman, S., 495
Miles, F. D., 130
Millis, N. F., 6
Mitchell, R. N., 329
Miyagi, T., 23, 27
Mohrman, H., 6, 93, 128
Mokruschin, S. G., 129
Moll, G., 134, 143
Montillon, G. H., 185
Montonna, R. E., 129, 185
Morgan, B. B., 115
Morgin, R. L., 93
Moridera, H., 427
Mott, R. A., 284
Müller, H., 203
Munson, J. S., 354
Murase, T., 417, 427
Murdoch, R., 242
Muskat, M., 20

Nagamoto, C., 47
Natanson, G. L., 35, 36, 39, 51, 52, 53, 54, 60, 62, 64, 68, 77, 140
Negawa, M., 427
Nelson, L., 488, 501
Nelson, P. A., 440
Neu, E., 445
Nikitine, S., 132

Odell, J., 300
Ogorodnikov, B. I., 55, 56, 70, 74, 94, 95, 98, 99
Okamura, S., 397, 422
O'Leary, F. M., 134
Olivier, J. P., 208
O'Melia, C. R., 213
Ootsuji, N., 417
Oulman, C. S., 290
Overbeek, J. Th. G., 198, 210

Packham, R. F., 214
Pakshver, A. B., 295
Parnianpour, H., 101, 103
Pasceri, R. E., 6, 47, 78, 79
Passamaneck, R., 69
Pate, J. B., 133, 138
Patterson, G. N., 62
Pazar, C., 310
Peck, R. B., 431, 435
Pedersen, G. C., 276
Peetz, C. V., 43, 45
Pelzbauer, Z., 134
Penner, S. E., 276
Perkins, P., 43, 44
Petras, E., 143
Petrock, K. F., 93
Petryanov, I. V., 59, 70, 74, 94, 95, 98, 99, 127
Pfefferkorn, G., 143
Pich, J., 6, 7, 12, 21, 34, 40, 52, 63, 64, 65, 66, 68, 70, 73, 86, 87, 91, 94, 95, 97, 98, 129, 139, 141, 142, 143, 144, 145
Piekaar, H. W., 13, 17, 21
Pietsch, H., 134
Pirie, J. M., 176
Pisa, M., 132
Polder, D., 115
Poiseuille, J. L. M., 173
Polydorova, M., 130
Ponroy, J., 6
Poppele, E. W., 335, 337
Porter, H. E., 369
Porter, M. C., 498, 500, 501, 509
Preusser, H. J., 134, 135
Price, P. B., 146
Pring, R. T., 97, 120, 310
Prosser, A. P., 116
Przyborowski, S., 142
Purchas, D. B., 254, 256
Putnam, A. A., 41, 44
Pyne, H. W., 111, 112

Quetier, J. P., 96, 98, 99

Radushkevich, L. V., 17, 18, 38, 40, 55, 69, 70, 71, 72, 77, 79, 86, 88, 89, 104, 114, 123, 125, 127
Ramskill, E. A., 47, 82, 91, 92
Ranz, W. E., 35, 41, 44, 47, 50, 52, 75, 82, 84, 111, 142
Rauscher, J. A., 337
Rennie, F. W., 369
Richards, R. T., 129, 130
Richardson, J. F., 176
Riley, R., 134
Rimberg, D., 328
Risbud, H., 418, 443, 466
Ritter, R. C., 129
Rivers, R. D., 56
Roberts, D. A., 514
Roberts, M. H., 35
Robertson, A. F., 262, 263, 274, 275, 276
Robinson, A., 43, 44, 75
Rodebush, W. H., 39
Rogers, E. E., 47
Rosenblum, N. A., 127
Rosinski, J., 47, 48
Rossano, A., 55
Ruehrwein, R. A., 215
Ruggieri, R., 47
Rukina, A. E., 129
Rushton, A., 258, 261, 266, 273, 279, 283, 284, 287, 291, 297, 298, 299, 301, 462
Ruth, B. F., 185, 462

Sabitt, W. L., 94
Sadoff, H. L., 38, 83
Sambuichi, M., 394, 403
Sato, H., 118, 121, 126
Sazanova, V. G., 86
Scarlett, B., 179
Scheidegger, A. E., 20, 27, 137, 138, 177, 179, 259, 266, 268
Schekman, A. I., 31, 65, 67, 92, 94, 95
Schiereck, F. C., 21
Schmid, G., 130
Schnabel, W., 118
Schülze, H., 211
Schwarz, H., 130
Schyma, D., 134
Seal, S. H., 148
Sedunov, I. S., 45
Sedunov, Yu. S., 11
Sell, W., 19, 43
Senda, T., 427
Sennett, P., 208
Serafini, J., 43, 44
Settineri, W. J., 118
Shau, E. L., 13
Sheesly, D. C., 148
Sherwood, T. K., 495
Shibata, M., 417, 427
Shimaoka, H., 427
Shimasaki, S., 82
Shirato, M., 189, 394, 397, 403, 417, 422, 427
Shleien, B., 82
Shoemaker, W., 253
Silberberg, A., 217
Silbereisen, N., 129
Silverblatt, C. E., 443, 466
Silverman, L., 33, 55, 128
Simon, A., 12
Sinsky, A. J., 481
Sisefsky, J., 101
Sittkus, A., 101, 102
Skau, E. L., 132
Slater, R. W., 235
Smilga, V., 117
Smith, E. G., 290, 292
Smith, J. L., 354
Smith, J. M., 36
Smith, S. E., 310
Smith, W. J., 92, 329
von Smoluchowski, M., 201
Smyth, M. J., 242
Snell, H. A., 354
Snyder, C. A., 97, 120, 310
Socransky, S. S., 130
Solbach, W., 121
Sone, H., 43, 45
Soole, B. W., 111, 112
Sourirajan, S., 483, 484, 485, 488, 490, 496
Spaite, P. W., 313, 316
Spandau, H., 134
Sparrow, E. M., 26, 31, 268, 269, 284

Spear, M., 299
Spielman, L., 7, 14, 25, 29, 30, 32, 36, 40, 60, 64, 65
Spurny, K., 21, 64, 129, 130, 133, 134, 135, 137, 138, 139, 142, 143, 144, 145, 146
Spurny, K. R., 146, 148
Squires, B. J., 352
Stadler, J. R., 72
Stafford, E., 329
Stairmand, C. J., 36
Starr, J. R., 47
Stechkina, I. B., 7, 31, 34, 35, 36, 40, 46, 47, 50, 78, 81, 82, 86, 88, 90, 92
Stein, P. C., 203
Stenhouse, J. I. T., 46, 47, 48, 85, 93, 195
Stephan, D. G., 313
Stern, A. C., 310
Stern, O., 206
Stern, S. T., 31, 65, 67, 92, 94, 95
Stevens, D. C., 101, 103
Stöber, W., 134
Stokes, I. M., 242
Strauss, H. J., 324
Strauss, W., 97, 310
Subramanyam, M. V., 43, 44
Sullivan, R. R., 14, 15, 21, 31, 33, 266, 267, 268, 273
Suncov, A. S., 70, 74, 94, 95, 98, 99
Suprenant, N. F., 92
Sutherland, D. N., 219
Suzuki, S., 82
Symes, E. M., 146

Tabor, D., 118
Tamada, K., 23, 27
Tamori, I., 55
Taylor, F. R. S., 43, 48
Terjesen, S. G., 82
Terzaghi, K., 431, 435
Thalhammer, T., 37
Thom, A., 22, 23
Thomas, D. G., 82
Thomas, D. J., 82, 120
Thomas, J. M., 56
Thomas, J. W., 50, 82, 87, 88, 141
Thring, M. W., 97
Tiller, F. M., 187, 189, 260, 397, 399, 403, 405, 407, 411, 412, 417, 218, 443, 466
Timmins, R. S., 511, 512
Toporov, Yu. P., 116, 117
Tomotika, S., 22, 23
Torgenson, W. L., 35, 40, 45, 53, 78, 80
Toth, A., 143, 144
Tovarnickij, V. I., 129
Trepaud, P., 70
Treybal, R. E., 486, 494
Tritton, D. J., 27
Trowbridge, M. E. O'K., 185
Tunitskii, N. N., 59
Turner, N., 299
Twomey, S., 139

Ushakova, E. N., 77
Uzelac, B. M., 140

Van der Wal, J. F., 13, 21
Van Dyke, M., 22
Van Loo, M., 129
Van Oss, C. J., 483
Velichko, M. V., 69, 70, 71
Venturi, J. L., 352
Verwey, E. J., 198
Vieth, W. R., 440
Vincent, B., 214
Visser, J., 118
Vold, M. J., 116
Voloschchuk, V. M., 49
Voorhoeve, R. J. H., 51
Voroshilova, A. A., 129, 130

Wakeman, R. J., 299, 440
Walkenhorst, W., 56, 58, 145
Walling, J. C., 320
Walsh, G. W., 313, 316
Walworth, C. B., 97, 98
Wang, D. I. C., 481
Ward, D. W., 215
Warren, R. P., 97, 98
Wasan, D. T., 267, 273
Wajsfelner, R., 139
Weissman, B. J., 499

Werner, R. M., 12, 21, 65, 66, 94, 264, 267
Wheat, J. A., 65, 66, 94
Whitby, K. T., 33, 47, 51, 52, 55, 86, 88, 103
White, C. M., 27
White, P. A. F., 310
Wiersema, P. H., 210
Wiffen, R. D., 139, 142, 145
Wiggins, E. J., 21
Wilder, J. E., 310, 320, 346
Williams, C. E., 119
Wilson, L. G., 37
Wilson, L. H., 94
Wilson, R. B., 111, 112
Winterton, R. H. S., 118
Witzmann, H., 99
Woodfin, E. J., 56
Wong, J., 47, 50, 52, 75, 82
Wong, J. B., 79, 84, 111
Work, L., 300
Wright, A., 362
Wrotnowski, A. C., 272

Yao, K. M., 196
Yasuda, T., 79
Yasunami, M., 45, 46, 81, 104, 118, 121, 126
Yeomans, A. H., 47
Yoder, R. E., 50, 82, 87, 88
York, O. H., 335, 337
Yoshioka, N., 43, 44, 45, 46, 49, 55, 81, 104, 118, 121, 126
Yuryev, I. M., 43, 44

Zakowski, J., 129
Zebel, G., 51, 56, 57, 58, 85
Zeller, H. W., 31, 65, 67, 92, 94, 95
Zievers, J. F., 259
Zimon, A. D., 115, 118
Zsigmondy, R., 129, 130
Zumach, W., 82

SUBJECT INDEX

Absorption coefficient, 84
Adhesion number, 60
Adhesion, particle, 114
Admix, 370, 447
Aeration, 243
Agent
  adsorption, 217
  diffusion, 217
  dispersion, 217
  flocculation, 215
Agitation, 220
Air mixing, 347
Analogy
  orifice, 274
  packed bed, 283
Approach angle, 48
Arrangement, fiber, 14, 15

Blake number, 193
Bleed, 256, 300, 301
Blinding, 255
Body aid, 236, 241
Bridging, 301, 302
Brownian motion, 74

Cake build-up, 368
Cake
  minimum thickness, 466
  washing, 437
Capacity, 256
  filter, 3
  vacuum pump, 442
Capillary phenomena, 127
Characteristic, selective, 86
Charge
  electrostatic, 50, 311
  filter, 85
  image, 51
  neutralization, 128
Chromatography, gel permeation, 512
Cleaning, reverse jet, 318
Clogging, 118
  value, 292
Coagulant, 211
  assessment, 231
  usage, 227
Coagulation, 197, 198
  orthokinetic, 204
  rate equation, 201
  secondary minimum, 200
Coefficient
  absorption, 84
  capture, 6, 9, 34, 38, 39, 40, 41, 44, 50, 54, 67, 70, 71, 72, 74, 76, 78, 80, 82, 84, 89, 90, 92, 97, 107, 123
  capture, fiber, 7
  capture, total, 75
  clogging, 120
  collision, 108
Colloid, 497
  hydrophilic, 214
  hydrophobic, 211
Concentration
  fiber, 16
  solids, 364
Convection, 495

Cooling
convection, 347
radiation, 347
water spray, 347, 348
Corona, 324
Cost
initial, 352
maintenance, 351
Coulombic attraction, 51

Darcy
defined, 174, 261
equation, 263
law, 20, 29, 33, 173, 179, 183, 272, 394
Davies
correlation, 285
equation, 264, 266, 272, 273, 283, 284
Deflection number, 273
Deliquoring
hydraulic, 423
suction, 440
Dependence, temperature, 96
Deposition
diffusion, 8
electrostatic, 9, 142
gravitational, 9, 140
inertial, 8
internal, 255
London-van der Waals, 9
particle, 6, 138, 288
windward, 53
Deviation, geometric standard, 13
Diameter, Fiber, 12
Diatomite, 370
Dielectric constant, 52, 324
Diffusion, 34, 94, 138, 195
agent, 217
boundary layer, 66
coefficient, 195
Diffusivity, 491, 498
Dioctylphthalate (DOP), 329
Discharge
belt, 388, 389
cake, 255
coefficient, 263, 277
scraper, 387
string, 387
Dispersion agent, 217
Distance
interfiber, 14
stopping, 42
Distribution
Clausius, 17
diameter, 12
log-normal, 12
pore size, 13
Drag, 284
coefficient, 341
fluid, 270
frictional, 395

Effect
leeward, 48
interception, 83
interference, 19, 82
slip, 62
temperature, 95
turbulence, 262
windward, 48
Efficiency, 84, 94, 95
average, 343
changes in, 123
collection, 337, 343
fiber collection, 7
fiber diameter dependence, 100
filter, 3, 6
instantaneous, 343, 344
porosity dependence, 100
target, 42
temperature dependence, high, 98
temperature dependence, low, 99
thickness dependence, 99
Electric double layer, 205
Electric field, external, 56
Electrocoating, 512
Electrolysis, 243
Electrophoresis, 207, 208
Emulsions, 514
Equipment
classification, 377
initial selection, 369
Erosion, blade, 351
Exchanger, heat, 348
Expression, 427

Fabric, antistatic, 311
Factor, hydrodynamic, 25, 37, 40, 63, 68
  depth, 172, 191, 364
  diagram, 91
  general design, 365
  incompressible cake, 179
  liquid-solid, defined, 170
  local rate of, 7
  medium, 171
  nonstationary, 113
  process, 403
  process types, 171
  rate, cake-weight form, 185
  secondary processes, 114
  solidarity, 84
Finishing, 346
Flocculant, 370
  assessment, 231
  usage, 227
Flocculation, 197, 214
  agent, 215
  effect of ions, 223
  effect of molecular weight, 223
  effect of pH, 221
  effect of solids concentration, 224
  reaction, 216
Flow
  axial, 26
  creeping, 43
  fibrous filter, 20
  free molecule, 62, 93
  potential, 43, 46, 60, 104
  pulsating, 317
  rate, 337
  reverse, 313
  slip, 93
  transverse, 23, 26
  velocity, 90
  viscous, 45, 49, 60, 103
Fluidized bed, 335
Flushing, back, 256,257
Flux, gel layer formation, 492
Force
  coulombic, 52
  double layer, 199
  drag, 27, 64
  electrostatic, 116
  image, 53, 54
[Force]
  London-van der Waals, 59, 199
  van der Waals, 115
Form filtration rate, 184
Fouling, 499
Freezing, 242
Fuchs number, 85

Gelation, 491

Hagen-Poiseuille equation, 177
Heat treatment, 242
Holding capacity, 313
Hydraulic diameter, 175
Hydraulic radius, 13, 21
Hydrocycloning, 364
Hydrodynamic action, 195

Impaction, 46, 141
Index, economic, 3
Inertia, 194
Inertial impaction parameter, 42
Interception, 68, 193
  diffusion and direct, 77
  direct, 8, 39, 71, 140
Ionizing radiation, 244
Isoefficiency, 81, 92

Knudsen number, 18, 61
Kozeny-Carman
  constant, 179
  equation, 33, 176, 179, 187, 258, 266, 270, 283

Latex, 497
Load, maximum dust, 346
Loading, 119
  surface, 121
  volume, 122

Magnetization, 243
Mass transfer coefficient, 491
Mechanism, 255
  combined, 75
  deposition, 293

[Mechanism]
diffusion, 83
electrostatic, 55
filtration, 293
gravitational, 49
inertial, 41, 83
Media
bridging, 299
chemical resistance, 253
classification, 254
felt, 320
index, 260
monofilament, 274
multifilament, 262
nonwoven, 272
stability, 253
strength, 253
structure, 296
Membrane, 128, 255
characteristics, 144
elastic, 427
evaluation, 132
hollow-fiber configuration, 508
metallic, 130
pores, 478
preparation, 129
production, 482
properties, 129
rejection, 499
retention size, 478
selectivity, 145, 512
structure, 132, 481
thin-channel, configuration, 507
tubular, configuration, 505
two-stage, process, 511
Mercury intrusion, 259, 292
Methodology, testing, 442
Microorganism, resistance to, 256
Mist collector, 335
Mixing, air, 347
Mobility
electrophoretic, 210
particle, 60
Monte Carlo technique, 10

Navier-Stokes equation, 18, 22, 43, 61, 177, 268, 284
Newton law, 10
Number
adhesion, 60
Blake, 193
deflection, 273
Fuchs, 85
Knudsen, 18, 61
Peclet, 11, 34, 70, 196
Reynolds, 15, 18, 177, 261, 263, 273, 494
Schmidt, 34, 494
Stokes, 10, 48, 194, 295

Operation
pressure, 351
suction, 351
Orifice analogy, 274
Oseen approximation, 22, 23
Osmosis, reverse, 478, 487, 511

Packed bed analogy, 283
Packing density, 16
Particle
characteristic, 368
test, 260
Peclet number, 11, 34, 70, 196
Penetration, 3, 39, 55, 58, 76, 84, 91, 92, 101, 123
Permeability, 21, 29, 65, 173, 174, 238, 256, 259, 260, 262, 266, 270, 273, 274, 283, 284, 286, 288, 289, 340, 364, 374, 394, 480
cell, 188
specific, 263
Plates, perforated, 255
Poiseuille relation, 21, 174
Polyelectrolyte, concentration, 219
natural, 224
synthetic, 226
Pore
bridging, 253
number, 273
penetration, 255
plugging, 295
size, 13, 257
Porosity, 16, 396
cake, 416
defined, 174

[Porosity]
function of distance, 417
function of pressure, 419
Post treatment, 364
Potential
sedimentation, 208
streaming, 208
Powder, standard, 259
Precipitation, 491
Precoat, 370, 447
Pressure
compressive, 188
drop, 29, 119, 337, 339
Pressure drop, filter, 6
Pretreatment, 364, 370
Protein recovery, 511

Quality, 95

Radius, effective fiber, 42
Ratio
air-to-cloth, 311, 313
concentration-polarization, 491
molecular speed, 73
Reentrainment, particle, 114
Reflection
diffuse, 63, 68, 70
specular, 64
Region, continuum, 18, 19, 21, 23, 64, 93
noncontinuum, 19, 61
transition, 69
Rejection, 504
Relaxation time, 42
Repulping, 376
Resistance, 260
coefficient of, 29
local, 188
medium, 181
specific, 340
Resistivity, 29, 32, 64, 66, 110
Retention
particle, 138, 257, 259
surface, 103
Reynolds number, 15, 18, 177, 261, 263, 273, 494

Schmidt number, 34, 494
Schülze-Hardy rule, 211
Sedimentation, 195
velocity, 108
Selectivity, 87, 105
Separation
solids, 364
stages, 363
Service life, 337, 346
Shaking, mechanical, 313
Shape
fiber, 13
particle, 90
Sieve constant, 140
Sieve effect, 8
Sieving, 140
factor, 500
Sizing, 516
Slip, 63
coefficient, 62
flow, 93
Solidity, 16, 33, 66
Sonic cleaning, 320, 323
Specification, product, 366
Stability, colloid, 198
Stagnation point, 53
Sticking coefficient, 108
Stokes number, 10, 48, 194, 295
Structure, filter, 5, 12, 99
Surface, fiber specific, 15
Suspension, coarse, 255

Test
bubble point, 258
permeability, 258
Tester, Dill dust spot, 344
Theory
channel, 21
drag, 21
phenomenological, 104
probability, 107
statistical, 107
Thickening, 364
Thickness, boundary layer, 491
Tortuosity, 177

Ultrafiltration
  applications, 480, 509
  defined, 476
  flux, 486
  principle of, 476
  protein, 493

Vibration, 243
Voidage, defined, 174

Wash time, 439

Zeta potential, 205, 290